The Arkana Dictionary of Astrology

Fred Gettings is the author of over fifty books, including *The Hidden Art*, *The Dictionary of Hermetic, occult and Alchemical Sigils* (Arkana 1988) and *The Secret Zodiac* (Arkana 1989. An art historian, he is also an expert on astrology, with a special interest in the history of symbolism and arcane lore.

The Arkana Dictionary of

ASTROLOGY

Fred Gettings

ARKANA

ARKANA

Published by the Penguin Group
Penguin Books Ltd, 80 Strand, London WC2R ORL, England
Penguin Putnam Inc., 375 Hudson Street, New York, New York 10014, USA
Penguin Books Australia Ltd, 250 Camberwell Road, Camberwell, Victoria 3124, Australia
Penguin Books Canada Ltd, 10 Alcorn Avenue, Toronto, Ontario, Canada M4V 3B2
Penguin Books India (P) Ltd, 11 Community Centre, Panchsheel Park, New Delhi – 110 017, India
Penguin Books (NZ) Ltd, Cnr Rosedale and Airborne Roads, Albany, Auckland, New Zealand
Penguin Books (South Africa) (Pty) Ltd, 24 Sturdee Avenue, Rosebank 2196, South Africa

Penguin Books Ltd, Registered Offices: 80 Strand, London WC2R ORL, England

www.penguin.com

First published by Routledge & Kegan Paul Ltd, 1985
This revised edition first published by Arkana, 1990

4

Copyright © Fred Gettings, 1985
All rights reserved

Filmset in Monophoto 9pt Garamond
Printed in England by Clays Ltd, St Ives plc

To my father Alfred Gettings –
born 19 July 1898,
died 8 December 1984

Contents

Introduction

The entries and text in this dictionary have been designed to form a reference, guide and source-book for those involved in general astrological studies, as much as for those interested in specialist researches into the many realms of this art and science. Towards this end the text presents, under nearly 4,000 headings, terms derived from the major post-medieval tradition, as well as the most important words and concepts evolved as a result of recent fissiparous tendencies within astrology. The very nature of the history of astrology requires that the post-medieval sources must incorporate a few Hellenistic and Arabian terms, and even a few words derived from Babylonian and Egyptian streams: to these must be added a number of Indian (Sanskrit) concepts which have entered the mainstream of European astrology by way of theosophy and related occult sources.

The modern revival of interest in practical and theoretical astrological studies, and the proliferation of its modern forms, are sufficient reason for the publication of this work. The truth is that all available dictionaries dealing specifically with astrological terminology are in some respects wanting: none of them serves any useful academic purpose, and most of them are frequently misleading in regard to the terms they define. It is hoped that this present work will remedy such deficiencies, and provide useful guidance for those involved in practical astrology or in the study of its various historical forms, as well as for those peripherally interested in its obvious relevance to literature and art forms prior to the 18th century.

My selection of terms from what is in effect a lexicographic labyrinth has been determined almost entirely by my own personal view of what is likely to be needed by a student who is alert to the antiquity of the art, and wishes to use astrology in a practical way, or in order to throw light on related humanistic studies. It is of paramount importance nowadays (when the fissiparous tendencies common to our entire age are eroding the traditional basis of the art) that the student be encouraged to develop an awareness of the rich antiquity of astrology. On the whole it is only specialists who are aware of just how many different forms of astrology exist, all uncomfortably lumped together under the innocent-seeming single term 'astrology'. This has always been the case: we find, at the very beginnings of our records, the Dorothean astrology was different from the Ptolemaic, and the fragmentary art known to us from

Introduction

the ancient mystery centres of Egypt, Babylon and Greece was different from these. There were several different forms of Arabic astrology, various forms of medieval astrology, and the modern syncretism has encouraged the growth of many different 'astrologies'.

The nature of the modern astrologies will of course tend to interest us more than the natures of the ancient ones. The fact is that the numerous forms of 'modern astrology' are to be distinguished from the so-called 'traditional' corpus, which tends to make use of generally accepted terminologies and concepts. In a wide generalization, I might say that the modern systems of astrology have been derived in three different ways. The first is by means of what has been called 'inspiration', by techniques resting upon various clair-voyant faculties: the INTUITIONAL ASTROLOGY of Bailey is of this kind. The second is by means of the syncretic approach to the esoteric and traditional forms – a co-mingling of ideas not originally designed to be yoked together: the ASTRO-PSYCHOLOGY of Morrish is of this kind. The third is derived from what I can only term (with as much kindness as I can muster) 'intellectual dreaming'. Many of the modern systems, often erected behind a façade of impressive-sounding terminologies and 'research programmes', are of this kind, and are really astrological systems ungrounded in any reality. It would perhaps be unfair of me to identify such systems by name in the present context: however, one might note a peculiar intensification of this sort of astrology in recent decades by the application of computing techniques to certain branches of astrology which are – in their origin, essence and spirit – inimical to such techniques. What is often called 'traditional astrology' is being practised with excellent results in Europe today as a fourth group of 'astrologies'.

Strictly speaking, however, there is no such thing as 'traditional astrology' – there exist only the various astrologies of certain periods, or of certain schools of thought. It is this simple fact which so often foxes the academic and popular historian alike. Almost every astrologer of import, from Ptolemy and Valens, through Al Biruni, Campanus, Morin and Sibly, into the grey twilight of modern systems, have developed their own 'personal' astrology. Because of this, it is essential when attempting to evaluate an astrological term or idea to establish which of the many astrologies was involved. Dante had his Al-fraganus, Blake his Boehme and Varley, and (not to put too fine a point on things) Mondrian had his Blavatsky, Kandinsky his Steiner, and so on. It is by personality working into personality that the history of astrology has been made, not through the influence of this mythical astrological system which is called 'traditional'. Many individual terms may be called traditional, of course, but any reference to 'traditional astrology' is usually to the Ptolemaic which, due to a series of historical accidents, has left so strong a mark on European astrology.

Introduction

One may not write on the contemporary situation in astrology without feeling the urge to lament the rapid passing of a true feeling for the ancient forms – were this passing due to a wish on behalf of the student to reject that which has been tested and found wanting, then all would be as it should be. However, the fact is that the passing of the ancient forms appears to arise from ignorance rather than from perceptivity. In a perceptive comment on the modern penchant for astrological 'research', the astrologer Davison writes of those who eagerly follow up every new method suggested, in the hope that at last they will find the solution to all their problems, 'not realizing that most of these problems spring from an imperfect acquaintance with the ordinary processes of astrology. There would be far less dissatisfaction with existing methods if students took the trouble to master them thoroughly before turning their attentions further afield.' Davison is quite right – so many astrologers follow the most arcane-seeming methods, and find themselves involved in what they happily call 'esoteric astrology', without knowing anything about the art as a whole.

The Bibliography at the end of this dictionary is designed as supportive reference material, yet will also serve as an index to all the literary references made in the text. If an astrologer or historian is mentioned by name, without further commentary, then that name will refer to a relevant text in the Bibliography. Within the entries themselves I have adopted the convention of using small capital letters to indicate cross-references and – unless otherwise indicated in the text – I have marked with an asterisk those which *need* to be consulted for a full understanding of the term defined. Mindful of the large number of specialist ABBREVIATIONS available today, I have used the following:

A	Ascendant	JU	Jupiter	PL	Pluto
AQ	Aquarius	LB	Libra	RA	Right Ascension
AR	Aries	LE	Leo	SA	Saturn
AS	asteroids	MA	Mars	SC	Scorpio
CN	Cancer	MC	Medium Coeli	SG	Sagittarius
CP	Capricorn	ME	Mercury	SU	Sun
DH	Dragon's Head	MO	Moon	SUr	sunrise
DS	Descendant	NE	Neptune	TA	Taurus
EA	Earth	OR	Orion	UR	Uranus
EC	ecliptic	OS	Old Style	VE	Venus
EQ	equator	PF	Part of Fortune	VG	Virgo
GE	Gemini	PG	Pegasus	VP	Vernal Point
IC	Imum Coeli	PI	Pisces	ZO	zodiac (tropical)

Introduction

The general reader may feel that some of the data presented in tabular form is inconsistent in that it combines systems of measurement, merging the metric with the more ancient traditions. Such inconsistency as there may be found is intentional, the figures (where not directly quoted, with accredited sources) being chosen for practical purposes, designed (for example) to help the astrologer in his calculations, which normally involve the use of astrological rather than astronomical ephemerides.

A

Ab See BABYLONIAN CALENDAR.

Abbreviations From relatively early times astrologers have complemented their wide use of SIGILS with alphabetic abbreviations intended to denote principles or data. The IC, short for the latin IMUM COELI, is an example derived from late medieval astrology: the use of H to stand for the ASCENDANT is perhaps less evident, until one realizes that the Ascendant was originally the HOROSCOPOS. In modern times, no doubt under the impress of the use of the typewriter in professional reporting – and also, no doubt, because of a feeling that the ancient sigils are somehow unscientific – there has been a move to establish a system of alphabetic abbreviations which might be recognized internationally among astrologers. A so-called internationally agreed system of 1975 was reported by Dean and Harvey. However, there are so many different schools of astrology, each with their own personalized approach to the subject, that there is little hope of international agreement as to which system of abbreviation should be adopted. The list in table 1 is culled from modern astrological sources, and sets out the main abbreviations used in post-medieval astrology. For abbreviations in connection with meridian times, see TIME ZONES. For abbreviations for horoscopic calculation and time, see TIME. With the exceptions in table 2 the standard modern abbreviations to denote constellations consist of the first three letters of the name – thus Andromeda is And., whilst Vulpecula is Vul.

─────────────── TABLE 1 ───────────────

A	Amplitude	aq	Aquarius
	Ascendant	AQ	Aquarius
a.d.	*ante diem* (see KALENDS)	ar	Aries
AD	Admetos	AR	Aries
	Anno Domini	AS	Ascendant
	(see FOUNDATION DATES)		asteroid
Ap	solar apex	AU	astronomical unit
AP	Apollon	AUC	*ab urbe condita*
			(see FOUNDATION DATES)

AV	antivertex		horizon
AX	solar apex		Ascendant (Horoscopus)
b	birth (place and/or time)		Mercury (Hermes)
B	birth (place and/or time)	HA	Hades
ba	birthplace approximate	HG	hypergalactic centre
BQ	biquintile	IAU	international astronomical union
Bi	biquintile		
ca	Cancer	IC	Imum Coeli
CA	Capricorn	IMC	Imum Coeli
CE	Ceres	J	Jupiter
CN	Cancer	JO	Juno
cp	Capricorn	JU	Jupiter
CP	Capricorn	K	Saturn (Kronos)
CU	Cupido	KdG	(see COSMOBIOLOGY)
d	day, days	KR	Kronos
	declination	L	Moon (Luna)
	dexter	Lb	Libra
	died	LB	Libra
D	Demeter	le	Leo
	direct	LE	Leo
	Dragon's Head	li	Libra
DH	Dragon's Head	LI	Libra
DR	Dragon's Head	LMT	local mean time
DS	Descendant	LST	local sidereal time
DSC	Descendant	LY	light years
DT	Dragon's Tail	m	minute, minutes
EC	ecliptic	M	Mars
EH	equal house		Midheaven
EP	East Point	MA	Mars
EQ	equator	MC	Medium Coeli
FS	fixed star	ME	Mercury
GC	galactic centre	MH	Midheaven
ge	Gemini	MK	Dragon's Tail (Mond-knoten)
GE	Gemini		
GMT	Greenwich Mean Time	MO	Moon
GST	Greenwich Sidereal Time	N	harmonic number
GOH	(see KOCHIAN SYSTEM)	NE	Neptune
h	hour, hours	NN	Dragon's Head (north node)
H	harmonic	ns	New Style (of calendars)
	heliocentric	NS	New Style (of calendars)
	Hermes	OA	Oblique Ascension

2

OD	One Degree Method	SG	Sagittarius
OR	Orion	sh	Speculative horoscope
OS	Old Style (of calendars)	SN	Dragon's Tail (south node)
p	progressed	SO	Sun (Sol)
P	progression	Sr	stationary, becoming retr.
	Persephone	St	stationary
	phase	ST	Saturn
P	Pluto	SU	Sun
PA	Pallas	SUr	sunrise
PF	Part of Fortune	SVP	Syncretic Vernal point
PG	Pegasus	Sx	sextile
pi	Pisces	SyZ	synetic zodiac
PI	Pisces	SZ	sidereal zodiac
	Pluto	t	transit
PO	Poseidon	ta	Taurus
prid	Pridie (see KALENDS)		time approximate
Q	quintessence	TA	Taurus
	quintile	TE	Earth (Terra)
r	radical	TP	transpluto
	radix	TZ	tropical zodiac
R	retrograde	U	Uranus
RA	Right Ascension	UR	Uranus
s	second, seconds	V	Venus
	sign, signs		Vulcan
	sinister	VA	Vesta
	solar-arc directions	VE	Venus
	stationary	vi	Virgo
S	Sun	VI	Virgo
sa	Sagittarius	VG	Virgo
SA	Saturn	VP	Vernal Point
	semi arc	VU	Vulkanus
sc	Scorpio	VX	vertex
SC	Scorpio	y	year, years
Sd	stationary, becoming direct	ZE	Zeus

Abhijit One of the terms used for a supposed Hindu NAKSHATRA. The nakshatras are obtained by dividing the zodiacal circle into 27 arcs of 13 degrees and 20 minutes. This Abhijit is interpolated between Uttara Ashadha and Sravana, thus completing the traditional number of 28 asterisms associated with systems of lunar mansions. It seems that no explanation is

3

─────────────────── TABLE 2 ───────────────────

Aps	Apus	LMi	Leo Minor
Aqr	Aquarius	Phe	Phoenix
Aql	Aquila	Psc	Pisces
Cnc	Cancer	PsA	Piscis Australis
CVn	Canes Venatici	Sge	Sagitta
CMa	Canis Major	Sgr	Sagittarius
CMi	Canis Minor	Scl	Sculptor
CrA	Corona Australis	Sct	Scutum
CrB	Corona Borealis	TrA	Triangulum Australe
Crv	Corvus	UMa	Ursa Major
Crt	Crater	UMi	Ursa Minor
Hya	Hydra	Vol	Piscis Volans
Hyi	Hydrus		

available for this interpolation, which is used in some systems of horary astrology.

Abscission Term used of effect during a multiple ASPECT, when the aspecting planet is forming contemporaneously one or more aspects to two or more planets. The aspect which culminates first is said to produce an abscission of influence ray, which mitigates the influence of the second aspect. The term FRUSTRATION is sometimes used as synonymous with abscission, though it is applied tó the effect of the later aspect, rather than to the activity of the prior culmination.

Absolute stellar magnitude A scientific formulation denoting the apparent magnitude of a star in relation to distance from the Earth. It is based on an estimate of how bright a star would be if it were located at a distance of 10 parsecs from the Earth.

Acceleration See TIME.

Accidental ascendant A term used in * HORARY ASTROLOGY of an Ascendant constructed for the moment at which the question is posed. A relevant natal figure (for example, that of the querent) is adjusted to this accidental Ascendant, and the resultant chart interpreted.

Accidental debility Each planet is said to have its own * DEBILITY, which may be essential or accidental. Wilson gives a long list of the accidentals, along with a useful indication of specific values, of which the following is merely a digest:

Besieged by SA and MA	6	In partile conjunction with	
In conjunction with Algol	6	SA and MA	5
If retrograde	5	If combust	5
Partile opposition Cauda	4	In the 12th house	5
		Partile opposition SA or MA	4

Accidental dignity   Each planet is said to have its own *DIGNITY, which may be either essential or accidental. An accidental dignity may arise due to the planet being in a particular house, degree of the zodiac, or due to its bearing certain special aspects. The rules for interpreting accidental dignities are complex and even obscure, though Wilson has made some attempt to tabulate them, along with an interesting system of specific values. The following is merely a digest of the more important:

In Ascendant or Midheaven	5	Free from combustion	5
In Cazimi	5	Besieged by JU and VE	6
In partile conjunction of JU	5	In partile conjunction of VE	5
In conjunction with Cor Leonis	8	In conjunction with Spica	5
In 4th, 7th or 11th house	4	In direct motion	4

Achernar   Fixed star of 1st magnitude, the alpha of Eridani. The Arabian words from which the term is derived means 'end of the river'. Being of the nature of Jupiter, it is said to exert a beneficial influence. Ebertin shows that it inclines towards religion.

Acimon   See SPICA.

Acquirius   Name given by Simmonite to a fixed star in the face of Aquarius – it is probably *SADALSUUD. Simmonite says that it 'denotes erudition', but Ptolemy gives it an evil influence.

Acrab   A modern name for the fixed star *GRAFFIAS.

Acronycal   Term applied to a planet in opposition to the Sun, rising after sunset or setting before sunrise. The acronycal place is the degree in direct opposition to the Sun.

Acrux Fixed star of 1st magnitude, the alpha of Crucis – in fact it is a triple, the brightest star in that constellation. Of a beneficial influence, linked with the power of Jupiter, it inclines towards occultism.

Actinobolia Term applied to the rays of the *HYLEG, when it is placed on the diurnal arc of the figure, to the east of the Midheaven. Perhaps in origin it was loosely related to the term ACTINOBOLISM.

Actinobolism The term appears to have been used in astrology in connection with the theory of ASPECTS, when it was believed that planets projected rays of influence down to the Earth (the etymology of the word meaning 'throwing out beams'): Ptolemy uses the word in this sense. The term has been used to denote the idea of spirits (or even materialities) shooting forth from some source, such as the Sun. In more recent times occultists have adapted the word to techniques of hypnosis, and it is used at times even to denote forms of divination involving trance, such as scrying.

Active planets Planets acting as *PROMITTORS.

Active virtues In their specialist astrological sense, these are heat and cold – see VIRTUES.

Acubens Fixed star (double), the alpha of constellation Cancer – the Arabian word from which the term is derived means 'claws'. Powell uses it to mark the termination of his sidereal Gemini (see SIDEREAL ZODIAC). It is said to be of an evil nature, exuding an influence of Saturn conjunct Mercury.

Aculeus A cluster of stars (the 6M in Scorpius), often treated by astrologers as a single FIXED STAR influence. Set in the sting of Scorpius, it is said to have a deleterious effect on eyesight.

Acumen A cluster of stars set in the sting of Scorpius, and treated by astrologers as of the same nature as *ACULEUS.

acvini The Sanskrit name for the constellation Aries – but see MESHAM.

adad A Babylonian term for meteorological phenomena and astrology.

Adar See BABYLONIAN CALENDAR.

adhafera Fixed star (double) of 3rd magnitude, set in the mane of LEO. It is said to be of an evil influence, save when in the rising degree.

adjusted calculation date Since the planetary positions in most EPHEMER-
IDES are given for noon, astrologers find it convenient to convert the data of
the radical horoscope to an adjusted date, so that the progressed annual
figure may be directly related to the noon date. The adjusted calculation
date is therefore also called the noon date, or the perpetual noon date. The
noon date may be calculated, following the normal rules of conversion
involved in the DAY FOR A YEAR method, or may be read from a special table.

Admetos Name given to one of the HYPOTHETICAL PLANETS used in
URANIAN ASTROLOGY. This transplutonian is included in an ephemeris, and
is claimed to relate to raw material, circulation and death.

Adnachiel The ruling Angel of zodiacal Sagittarius.

Adonis Name given by the astrologer Sutcliffe to one of the HYPOTHETICAL
PLANETS – said to be intramercurial.

advantage See LINE OF ADVANTAGE.

aestival signs Sometimes the ESTIVAL SIGNS (from the Latin *Aestas*) are the
zodiacal signs of the summer period: Cancer, Leo and Virgo.

Aesula See HYADES.

aether Name given by early astrologers to the *QUINTESSENCE, the term
probably being coined to contrast with the Latin *aer* (air), which properly
belonged to the Earth, the aether being supralunar. This ancient aether is
not to be confused with the ether of the modern physicists, even though this
is probably directly related to the ETHERIC of the occultists, as to the
ASTRAL LIGHT of the ancients (see Wachsmuth for astrological connections).
The aether must be distinguished from the *ETHERIC SPHERE.

aethereum See MACROCOSM.

Aetherius Name of a constellation, the Balloon, or the Globus Aerostaticus,
formed by La Lande in 1798, between the tail of the Southern Fish and
Capricornus.

affinity The term is used in traditional astrology to apply to the affinity or
attraction which is said to exist between certain pairs of planets. For example,
Mars is in affinity with Venus by virtue of the sexual forces, Venus is in
affinity with Mercury in an artistic sense, and so on. The whole realm of

astrology may be defined in terms of such affinities-and repulsions, for it is deeply rooted in the theory of *SYMPATHY, as Lovejoy makes clear in his study of the CHAIN OF BEING. In modern specialist astrological use, the term is applied to the widely recognized fact that astrologers appear to attract clients of a particular type – through some sort of affinity, which is usually reflected in a COMPARISON CHART.

affliction Generally, a planet is said to be in affliction with another when it is in unfavourable aspect with it: the planet receiving the aspect is said to be afflicted. However, the term may also be used of the conjunction of planets with certain degrees of the zodiac – as, for example, when the degree is marked by a malevolent fixed star, which then afflicts the planet. Planets may also be afflicted by *PARALLEL aspect, or through *MUNDANE ASPECT. A planet, or a nodal point, is said to be afflicted when it is *BESIEGED by two evil planets. In a popular or even poetic sense, a native may be afflicted by an evil star, or even by his entire horoscope.

agathos daimon The Greek term, meaning 'good genius' (see DAIMON), is applied to the 11th locus (see LOCI), which appears to have ranked in importance second only to the Ascendant or Midheaven. The same Greek term was applied to the 11th locus – measured not from the Ascendant, but from the PART OF FORTUNE. See also AGOTHODEMON.

Age along the zodiac Term originated by Frankland as an alternative name for his *POINT OF LIFE arc.

Agena Fixed star of 1st magnitude, the beta of Centaurus. Ptolemy gives it the beneficial influence of Venus conjunct Jupiter, but others relate it to Mercury conjunct Mars.

Age of Aquarius See AGES, EPOCHS, PLATONIC YEAR and PRECESSION.

Age of Jupiter See AGE OF SATURN.

Age of Pisces See AGES, EPOCHS, PLATONIC YEAR and PRECESSION.

Age of Saturn A term derived from the poetic historicism of Virgil. He divides the historical period of mankind into two ages – those of Saturn and Jupiter, the former corresponding to the Golden Age, familiar to us from the description of Hesiod. The Age of Jupiter is merely a later age, one of expansion, yet involved with a reform of those darker elements developed towards the end of the preceding age. Both of these two ages contributed to the development of the Ovidian *FOUR AGES.

ages   There are several different systems of ages in astrology. Most often the term is applied to the so-called zodiacal ages, the EPOCHS, which are said to arise due to the phenomenon of the *PRECESSION of the equinoxes. The entire period of this precession is an age of 25,920 years, which is divided into twelve sub-periods, of 2,160 years' duration, allocated to each of the twelve signs of the zodiac (see PLATONIC YEAR and THREE PROTOTYPES). There is little agreement among astrologers as to when the twelve epochs begin and end: some schools insist that we are now well into the Age of Aquarius, others that we are in the Age of Pisces, and will so remain for another century or so, whilst yet others claim that we are in the Age of Capricorn. It is, of course, a question of which co-ordinates and philosophical outlooks are adopted (see, for example, AYANAMSA). Accurate dating of the epochs are a source of much argument: confusion arises in this connexion because the constellational system of measurement is spatial, and gives periodicities which are of uneven length as well as of indeterminate beginning and end. The zodiacal ages are non-constellational, and there is argument as to which fiducials determine the zodiac proper: see both SIDEREAL CORRE-SPONDENCES and SIDEREAL ZODIAC. An important astrological tradition linked with Ages is that concerned with the periodicities during which the Archangels have rule over the destiny of the world, and shape human history – see SECUNDADEIAN BEINGS. In relation to personal astrology, much attention is placed on the crisis points in the ages of the individual – see CLIMACTERICS. There is also another system of personal ages linked with the sequence of the planets, which are allocated periods of rule over the growing human – see SEVEN AGES. Yet another system is derived from the *PLANETARY PERIODS.

ages of man   See CLIMACTERICS and SEVEN AGES.

Agiel   Name given by Agrippa to the Intelligency of Saturn. He gives the magical number of Agiel as 45, which is 3 times the linear sum (15) of the MAGIC SQUARE of Saturn.

Agothodemon   Term derived from Ptolemy for the *ELEVENTH HOUSE, properly *agathodaemon*, from the Greek toast, '*agathou daimonos*' (to the good daemon), the Greek DAIMON being a higher spirit. See also LOT OF DAIMON.

Aidonius   A name given by the esotericist Thierens to a future development of his planet Pluto (see THIERENS-PLUTO) as a realization of the modern Mars, in his ETHERIC SPHERES. It is said to represent the principle of initiation.

Aiel According to grimoires, this is the name given to the Governor of zodiacal Aries: he is said to rule Sunday. See ZODIACAL SPIRITS.

Aigokeros Greek name, meaning 'horned creature', used to denote the sign and constellation Capricorn. See also KRIOS.

air element The air element finds expression in the zodiac through the three signs Gemini, Libra and Aquarius. It is the element most deeply associated with thought, and with volatility: this is recorded in the esoteric tradition by Philalethes, who says that the air element is 'the envelope of the life of our sensitive spirit', by which he means that the human being swims in the world of thought, as though in a sea. Indeed, Philalethes even calls air the 'sea of things invisible'. It is evident from such a view that the air of the esoteric astrologer, and indeed of astrology in general, bears little other than an analogous relationship to the air of the physicists. This analogy is often pursued by astrologers anxious to express in material terms the nature of the air element: it is the element which carries sound, and thus permits communication between people; the oxygen of air enlivens or feeds the flames of fire – which is to say that the air element has an affinity with the FIRE ELEMENT. See also SANGUINE.

Airesis See SECTA.

air signs The three zodiacal signs Gemini, Libra and Aquarius are expressions of the AIR ELEMENT. These are sometimes called the Air asterisms, rather than the air signs. Gemini is air expressed through MUTABILITY, Libra is air expressed through CARDINALITY, and Aquarius is air expressed through FIXITY. By analogy with the air of the physical world, therefore, Gemini may be pictured as air in constant movement, an untamed wind, blowing in many directions. Libra is air blowing in a purposive single direction, and therefore capable of establishing specific relationships. Aquarius is static air, a condition which is somewhat contrary to nature – air under pressure, which tends under certain conditions to explode outwards, in rapid dispersal of energies. In modern astrological parlance, these three signs are the intellectual or mental signs, yet in the ancient astrological tradition, air was associated with the thought process – perhaps once more by analogy, for it was said that just as man breathes in life-saving air from the world around, so he breathes in the nutrient of thought or ideas from the world outside: by such standards, thinking was a cosmic process. Since air was regarded as the most restless of the four elements, the three air signs were seen as being dominated by various degrees of restlessness and vacillation. See also SANGUINE for a portrayal of air in human personality. The air signs are sometimes called the NERVOUS SIGNS, the SWEET SIGNS or the WHOLE SIGNS. See also AIR TRIPLICITY.

air triplicity This is the generic term for the three signs linked with the AIR ELEMENT, manifesting different aspects of the SANGUINE temperament – Gemini, Libra and Aquarius. Air is all-pervasive, a palpable link between the inner world of man (lungs) and the outer environment. It is this tendency, towards unification of the individual with the whole, which is manifest in each of the signs of the air triplicity. Gemini seeks to unify people, to establish useful and creative relationships. Libra is intent on establishing unity through discovering harmonies, shared experiences and communal responsibilities. Aquarius seeks to establish unity by bringing together large groups of people, especially with a view to establishing large-scale social ideals. Each of the air signs is versatile and dedicated to human relationships, tending however to be lost in superficial and transient matters. Since the keyword of the air triplicity is 'unification', each of the signs is involved with establishing a relationship with the world at large. Gemini works through Platonic relationships (predatory at worst), Libra is concerned with unity through bonds between different sexes (notably through marriage, or business ties), while Aquarius is concerned with friendships, especially 'distant friendships', and unions established towards a common ideal of progress or moral good.

akasha A modern theosophical term, taken from the Sanskrit as an equivalent for the QUINTESSENCE (fifth element), which was seen as binding together in union or pact the other four elements. A complete definition of the term would carry us out of the astrological domain, into the realm of occultism, but it is sufficiently relevant to mention in this context the so-called Akashic Chronicles (sometimes the Akashic Records) which are the historical records (that may be read only by adepts and initiates) of all thoughts and deeds which have taken place on the Earth, and which are indelibly imprinted upon the Akasha.

Akrab A variant name for the star *GRAFFIAS, the beta of Scorpius, incorrectly applied by Ebertin to the delta of Scorpius, properly DSCHUBBA.

Ala Corvi   See ALGORAB and FIFTEEN STARS.

Alacrab   See SCORPIUS.

Alamac   Fixed star (binary) set in the left foot of *ANDROMEDA. The name has not been convincingly explained, and is given in several variants, such as Almach and Alamak: the entire asterism of Andromeda is sometimes called Alamac. It is said to give eminence or creative ability, and Ebertin links it with popularity and cheerfulness.

Alangue   See OPHIUCHUS and RASALHAGUE.

Alatus   See PEGASUS.

Alayodi   See CAPELLA.

Albategnius system   A method of *HOUSE DIVISION based on the division of the Prime Vertical into diurnal and nocturnal arcs, each in turn divided (respectively) into three equal parts by two declination circles. The system, ascribed to the 10th-century Arabian astrologer-prince Muhammed ben Gebir al Batani (called Albategnius in Europe), appears to have given rise to the *PLACIDEAN SYSTEM, held in such apparent esteem today: however, there is much argument among scholars as to the origin of this latter system, even though it is generally agreed that it was not originated by the monk Placidus.

Albireo   The name given to the binary star, the beta of Cygnus, the word not, as one might imagine, from the Arabic, but from the Latin *ab ireo*. It is set in the beak of the asterism, and is named by the Arabic astrologers as Al Minhar al Dajajah (Hen's Beak).

Albohazen system   See HALYIAN SYSTEM.

Alcabitius system   A method of *HOUSE DIVISION, the principle of which is the trisection of the semi-arcs by a time division. The sidereal time is taken for the degree of the Ascendant to revolve to the cusp of the Midheaven, and this is divided into three equal parts. These divisions are then added to the sidereal time at birth, to reveal the positions of the diurnal arcs. A reverse procedure is used to determine the nocturnal arcs. It is claimed that the house grid offered by this system, unlike that of the related trisectional method of the *PLACIDEAN SYSTEM, is valid for polar regions. This system is ascribed by some historians to one Alcabitius, who is supposed to have lived in the

1st century AD. However, it appears that the method is really linked with an Arabic system derived no earlier than the 10th century (see ALBATEGNIUS SYSTEM): its name probably refers not to Alcabitius, but to the Arabian astrologer, Alchabitus, of the 12th century.

Alcalst   See VINDEMIATRIX.

Alchameth   See ARCTURUS and FIFTEEN STARS.

alchemical Mercury   The term Mercury is often used in esoteric and occult circles coterminous with astrology in a sense which does not refer to the purely astrological conception of *MERCURY. This alchemical Mercury is really the Hermetic Mercury, which figures in alchemical treatises under a variety of spagyric names. In the hermetic tradition, which informs exoteric astrology, Mercury is cold and moist, and therefore aqueous: it is therefore the Permanent Water, the vitalizing spirit of the body, and is linked with what the medieval occultists would call the ENS VENENI, the approximate equivalent of the modern ETHERIC. In this capacity, Mercury has been given many names suggestive of potent liquidity – Blessed Water, Virtuous Water, Philosopher's Vinegar, Dew of Heaven, Virgin's Milk, and so on. It is sometimes said that the whole alchemical art depends upon a true understanding of the nature of this Mercury, which is directly linked with the QUINTESSENCE.

Alcohoden   Arabian astrological term for the HYLEG.

Alcor   A fixed star, set in the constellation of *URSA MAJOR. It is not a conspicuous star (perhaps more so now than in the 8th century), and was used by the desert Arabs as a test of good vision. Allen records that they were used to regard *CANOPUS as the highest, and Alcor as the lowest in visual scale. It seems that the adage '*Vidit Alcor et non lunam plenam*' (He sees Alcor, but not the full Moon) was at one time the equivalent of our proverb, 'He cannot see the wood for the trees'.

Alcyone   Fixed star of 3rd magnitude, the brightest of the *PLEIADES, set in the shoulder of constellation Taurus. Early astronomers regarded it as a sort of central sun to the universe, and accorded it considerable importance: the Arabian astrologers called it Al Wasat (the central one), while the Babylonian astrologers called it Temennu (foundation stone). Powell commences his Taurus with this star in his *SIDEREAL ZODIAC, locating it in the 6th degree. It is said by modern astrologers to be of the nature equivalent to Moon conjunct Mars.

Aldebaran   Fixed star of 1st magnitude, the alpha of constellation Taurus, set in the left eye of the Bull, its name (perhaps) derived from the Arabian Al Dabaranu (the forecaster). It was the Star of the Tablet for the Babylonians (see MUL-APIN TABLES) and because of its almost perfect opposition to *ANTARES was used as a fiducial in the BABYLONIAN ZODIAC: see figure 28, under FIFTEEN STARS. Aldebaran has many names, reflecting its important role in astrological observation – see, however, STELLA DOMINATRIX. Aldebaran was one of the four ROYAL STARS of Persian astrology. It appears to bring good fortune and honours, when emphasized by benefic conjunction: in Mundane astrology, however, it is a cause of catastrophies when in aspect with a malefic. Cornell provides a formidable list of disasters associated with the star, and claims that when afflicting the planets it tends to cause extreme suffering to humanity, sickness or death, and even directions to the Angles produce periods of terrible stress.

Alexandrian calendar   See EGYPTIAN CALENDAR.

Algebar   See ORIAN.

Algedi   A multiple star, the alpha of Capricornus, set in the south horn of the Goat. The name is from the Arabian Al Jady (the goat), and it is often called Giedi. Ptolemy accords it the nature of Venus conjunct Mars, but when conjuncted with planets in a chart its influence is towards the peculiar and unexpected (similar to that of Uranus, as Alvidas notes).

Algenib   Fixed star of 3rd magnitude, the gamma of Pegasus, set in the wing-tip of the horse, its name probably from the Arabic Al Janah (the wing). It is said to be an unfortunate influence, projecting the nature of Mercury conjunct Mars.

Algenubi   Fixed star of 3rd magnitude, the epsilon of constellation Leo, set in the mouth of the Lion. It is said to exude a difficult influence, being of the nature of Mars conjunct Saturn – though (by analogy?) it gives great power of expression.

Algibbar   See ORION and RIGEL.

Algol   Fixed star (binary), the beta of Perseus, set in the head of the gorgon Medusa. It is said to be the most evil of all stars – the 'most unfortunate, violent, and dangerous star in the heavens', as Allen puts it, a view supported by most of its traditional names. The word Algol is from the Arabic Ra's al

Ghul (demon head), for which reason it is sometimes called Caput Algol, almost a modern version of the Caput Medusae. It is the Demon Star, the Satan's Head (*Rosh ha Satan*) of the Hebrews, whose astrologers also call it Lilith after their own female demon. Ebertin gives several interesting charts of murderers involved with dominant Algol, and, on a more prosaic level, notes that the star appears to bring the need for false teeth – a pale modern equivalent of the ancient tradition which has the star afflicting the face and head to the point of decapitation or strangulation. It is one of the *FIFTEEN STARS.

Algomeyla See PROCYON.

Algomeysa See PROCYON.

Algorab Fixed star (double) of 3rd magnitude, the delta of Corvus, set in the Crow's right wing, the term being from the Arabic Al Ghirab (the crow). It is sometimes called Ala Corvii (crow's wing). Astrologers give it a harmful nature, akin to Mars with Saturn. It is one of the *FIFTEEN STARS.

Alhayhoch Late-medieval name for the star *ALPHECCA – see *BIRTH STONES.

Alhecka Fixed star of 3rd magnitude, set in the south horn of constellation Taurus. Powell adopted this star as the termination of his Taurus asterism, locating it in the 30th degree of Taurus – see SIDEREAL ZODIAC. It is said to give a quarrelsome and difficult nature.

Alhena Fixed star of 2nd magnitude, the gamma of constellation Gemini, set in the left foot of the southern twin. The term is said to be from the Arabic Al Han'ah (a burn brand-mark). Ptolemy ascribes it a nature equivalent to Mercury conjunct Venus.

Alidade The rotable metal bar on the ASTROLABE, which, being centred on the Mater, is used as the sighting line for taking readings. In Latin, it is the Ostensor.

Alien signs A term used in Ptolemaic astrology, approximately the equivalent of the modern *DISSOCIATE SIGNS. A synonym was DISJUNCT SIGNS.

Alioth One of the seven fixed stars forming the constellation *URSA MAJOR, said by Ptolemy to be of the influence equivalent to Mars.

Allore

Allore   See WEGA.

Almagest   Sometimes called the *Syntaxis Mathematica*, this is a treatise in 13 books, dealing with astronomy and astrology, attributed to the 2nd-century astrologer Ptolemy, about whom virtually nothing is known. Much of the astrological star-lore in this present text is derived from the Almagest. See also CENTILOQUIUM and TETRABIBLOS.

Almanac   Name given to a booklet or set of tables, sometimes accompanied by predictions of a public nature, or relating to weather and matters affecting agriculture (see ASTRO-METEOROLOGY), setting out the phenomenal aspect of future astronomical and astrological conditions. Some almanacs offer material on a monthly basis, others on an annual. The earliest almanacs were prophetic in nature, many of them written around simple progression techniques, and (in connection with weather forecasts) around ingresses into cardinal signs. A comparison of almanac weather predictions for specific periods in the 17th century has revealed remarkably little accuracy, and no particular agreement for the same days among contemporaneous almanac makers. See also EPHEMERIS.

Almochoden   An Arabian astrological term for the *HYLEG.

Almucantar   Name of the circle of constant altitude above the horizon.

Almuten   Term derived from Arabian astrology, signifying the most influential planet in a chart, the strength being determined from essential and accidental dignities. See DIGNITY and LORD OF THE HOROSCOPE.

Alnasl   Fixed star, the gamma of constellation Sagittarius, its Arabian name meaning 'point', probably in reference to the head of the arrow. Powell chooses this star to mark the beginning of his sidereal Sagittarius (see SIDEREAL ZODIAC), locating it in the 7th degree.

Alnilam   Fixed star of 2nd magnitude, the epsilon of Orion, set in the belt of the giant, the name being derived from the Arabic Al Nitham (the string of pearls). It is said to be a difficult influence in a chart, and Ptolemy gives it a nature equivalent to Jupiter conjunct Saturn.

Alpharatz   The name Schedar, with many variant spellings, is also given to this multiple star in *CASSIOPEIA: the name is said to come from the Arabic Al Sadr (the breast), where it is located. Distinguish from ALPHERATZ.

16

Alphard Fixed star of 2nd magnitude, the alpha of Hydra, set in the neck of the constellation, and sometimes called Cor Hydra (hydra's heart). It is said to be of the nature of Venus conjunct Saturn, but of an evil influence, even though the afflicted native may be artistic or musical.

Alphecca Fixed star of 2nd magnitude, the alpha of Corona Borealis, the term being from the Arabic Al Na'ir al Fakkah (the bright one in the dish). It is sometimes called Elpheia, and, by Ebertin, Gemma, due to a misunderstanding. It is said to be a beneficial influence in a chart, and Ptolemy says it is the equivalent of Mercury with Mars. See FIFTEEN STARS.

Alpheichius See OPHIUCHUS.

Alpheras See SCHEAT.

Alpheratz Fixed star (double) of 2nd magnitude, the alpha of Andromeda, set in the hair of the figure. The Arabian astrologers placed this star in Pegasus, and called it Al Surrat al Faras (the horse's navel), from which our own term was derived. It is sometimes called Caput Andromedae (Andromeda's head). Of a beneficial nature, especially giving a good intellect, it is linked with Venus conjunct Jupiter.

Alrescha Fixed star (double), the alpha of constellation Pisces, marking the knot in the cord, and hence called Nodus Piscium, Nodus Coelestis, and the like. Powell adopted this star to mark the completion of sidereal Pisces, locating it in the 5th degree of Aries. See SIDEREAL ZODIAC.

Altair Fixed star of 1st magnitude, alpha of Aquila, set in the neck of the Eagle, the name being derived from the Arabic Al Tair (the eagle). A powerful influence, concerning which nature few authorities agree, though Ptolemy gives it the equivalent of Mars conjunct Jupiter.

Altarf Fixed star, the 4th magnitude beta of constellation Cancer, the name in Arabic meaning 'the end', in reference to its position at the southern foot of the asterism. Powell adopted this star as the commencement of his sidereal Cancer, locating it in the 10th degree. See SIDEREAL ZODIAC.

Alterf Fixed star, the lamda of constellation Leo, located by Ptolemy in the mouth of the Lion: the name derives from the corresponding MANZIL. Powell locates it as the beginning of his *SIDEREAL ZODIAC Leo, in the 24th degree of Cancer.

Altitude In astrology, altitude is usually expressed in degrees, minutes a..d seconds, and signifies the point of elevation above or below the horizon, along the arc of the zodiac. A planet is on MERIDAN ALTITUDE when it reaches the cusp of the Midheaven. Altitude above the horizon is usually noted positively (thus + 90 for zenith) and negatively (thus − 90 for nadir).

Ambient Little-used term in modern astrology, originally applied to the TENTH SPHERE, the Ambient Sphere, which in the Ptolemaic cosmoconception was visualized as moving all the other SPHERES with it. The term is now sometimes applied to the Heavens in general.

Ambriel The name given in certain grimoires and methods of ceremonial magic to the Governor of the zodiacal sign Gemini. The name appears in a confusing variety of contexts, however, not all of them relating to this rulership. In certain qabbalistic texts, for example, Ambriel is associated with Mars.

Ambrosia See HYADES.

Amplitude A term derived from the *HARMONICS of the modern astrologer Addey, relating to the strength of each particular harmonic, shown as a percentage of mean distribution.

Anabibayon Term given by Simmonite as a synonym for *DRAGON'S HEAD. Perhaps a mistake for *ANAHIBAZON.

Anael The name of the Archangel associated with the Sphere of Venus.

Anahibazon Arabian astrological term for the *DRAGON'S HEAD.

Anareta See ANARETIC and ANARETIC PLACES.

Anaretic Term derived from the Greek, meaning 'destroyer', and applied to the planet or degree which is for one reason or another regarded as the destroyer of life in a particular chart. The rules for determining which planet is the anareta, or in the anaretic place (sometimes called the 'killing place' by modern astrologers), are complex, and most astrologers warn of the difficulty of making predictions based on such rules. In some contexts the term is used of a planet which forms a difficult aspect to the *HYLEG, though this does not necessarily mean that it is therefore the 'killer'. In other contexts, the term is also used of certain fixed stars which, by virtue of conjunctions in a natal chart, or by progression, pour their baneful and killing influence into that chart, and bring about the death of the native. The anaretic point, or place, is the degree occupied by such a planet or fixed star in a chart but see also ANARETIC PLACES.

Anaretic places The anaretic places are often said to be the degrees in a horoscope figure occupied by the radical Mars or Saturn – or by the radical Sun, Moon and Mercury, should these be badly aspected by Mars or Saturn. However, many different rules exist for determining the 'killing place', as the anaretic place is sometimes called. Cornell simplifies a complex issue by speaking of the anaretic place as that occupied by an ANARETIC planet in the 4th, 6th, 8th or 12th houses. The issue remains complex, however, for besides establishing the identity of the anaretic from such places, there is the problem of determining also not only how, but when, it will bring death.

Anatomy See BIOLOGICAL CORRESPONDENCES and MELOTHESIC MAN.

Ancient moon Another term for *OLD MOON.

Ancient saturn Another term for *OLD SATURN.

Ancient sun Another term for *OLD SUN.

Androgyne The term means 'having characteristics of both sexes', but in an astrological context this sense is often limited in application to the planet Mercury (see ANDROGYNOUS PLANET), which has the facility for receiving the colour and sexuality of both male and female planets (such as Mars and Venus, respectively), through aspect to these. In an astrological context, the Androgyne is Mercury.

Androgynous planet A name for Mercury, a planet said to participate in both the feminine and masculine natures – but see, for example, FEMININE PLANETS. In the ancient tradition Mercury was said to be both dry and moist, and thus linked with both the male and female natures (see SECTA). In fact, even in accord with traditional astrology, it is clear that all planets manifest a sexual duality (save for the Sun and Moon) so that when Mercury is termed androgynous, or is called the Androgyne, it is in reference to the effects observed through aspects cast to this planet. Thus, an aspect formed between the feminine Moon and a planet will tend to feminize that other planet: however, an aspect received or projected by Mercury will result in a polarity which depends entirely upon the sexuality or polarity of the other planet.

Andromeda Constellation, approximately from 12 degrees Aries to 15 degrees Taurus, 20 to 55 degrees north of equator (but see CONSTELLATION MAPS). In Greek mythology, Andromeda was the daughter of Cepheus, rescued from a monster by Perseus, who rode the winged horse, Pegasus. As an asterism, it is said to imbue purity of thought, honour and virtue – though Manilius tells us that when Andromeda is setting, it causes death from the bite of an animal, or by crucifixion. Among the many names for this constellation are Chained Woman (Mulier Catenata), Persea (perhaps meaning 'bride of Perseus'), Cepheis and Alamac (after the binary of this constellation). The important stars set in the figure are ALPHERATZ, MIRACH and ALAMAC.

Andromeda's head See ALPHERATZ.

Angels A name given to the NINTH HIERARCHY of incorporeal spiritual beings above the realm of man – see SPIRITUAL HIERARCHIES. Whilst the word applies to a specific rank of the Hierarchies, it is also used in a general sense of all nine levels, with resultant confusion – for example, the so-called Angels of the zodiac are actually ARCHANGELS. According to the esoteric tradition, the Ninth Hierarchy is charged with the guidance of individuals through incarnation after incarnation. In the ecclesiastical tradition (which has deeply influenced astrological lore) St Gregory associates the Angels with the emerald, whilst biblical exegesis associates them with the colour green: in esoteric lore, they are linked with the Sphere of the Moon. The order of the remaining eight levels of Hierarchies, from the list provided by Dionysius the Areopagite, is set out in table 3. See also PLANETARY ANGELS, SECUN-DADEIAN BEINGS and ZODIACAL ANGELS.

TABLE 3

Rank	*Sphere*	*Name (Greek)*	*Other names*
Ninth	Moon	Angeloi	
Eighth	Mercury	Archangeloi	
Seventh	Venus	Archai	Principates or Principalities
Sixth	Sun	Exsusiai	Potestates or Powers
Fifth	Mars	Dynamis	Virtues or Virtutes
Fourth	Jupiter	Kyriotetes	Dominions or Dominations
Third	Saturn	Trones	Thrones

Angels of the zodiac See ZODIACAL ANGELS.

Angelus symbols Name given to a series of short readings ascribed to each of the 360 degrees of the zodiac (in accordance with a system of *DEGREE SYMBOLS) attributed to the 16th-century astrologer Johannes Angelus.

Angle In relation to the horoscope figure, any of the four cardinal points are called Angles. The Eastern Angle is the Ascendant of the horizon line (AS in figure 1): in early astrology this was called Oriens, and even Horoscope (Greek) or Horoscopus (Latin). The Western Angle (DS in figure 1) is the Descendant of the horizon line, in early astrology called Occidens (the setting of the Sun). The South Angle (MC in figure 1) is the cusp of the 10th house,

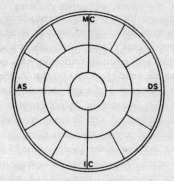

Figure 1: The Angles of the horoscope.

sometimes called the Medium Coeli (Latin for middle of the skies), the Midheaven, zenith or South Vertical, though technically these points do not always correspond. The North Angle (IC in figure 1) is the cusp of the 4th house, sometimes called the Imum Coeli (Latin for lowest part of the skies), the nadir or North Vertical. See also ANGULARITY.

Anguis See DRACO.

angular See ANGULARITY.

Angular houses These are the houses which fall on the *ANGLE of the figure – namely, the 1st, 4th, 7th and 10th houses, the strongest of all houses in a horoscope. In modern astrology (see ANGULARITY) the Angular houses are often said to be those which immediately follow the Angles (in clockwise direction), the cusps regarded as marking the outside limits of the houses, rather than the most powerful part (the centres).

Angularity A planet is said to be Angular when located on one of the ANGULAR HOUSES. In many systems of interpreting houses, the cusps were themselves regarded as the centres of houses, and thus marked the most powerful part of the house under question. However, in modern times, the distinction between the segment of the house and the cusp marking it has almost been lost, with the result that in modern astrology an Angular planet is defined in terms of Angles (the two main Ascendant axis and Midheaven axis), rather than in terms of houses. However, in accordance with most astrological terms, the actual sense of the definition to be applied by the term rather depends on the system of astrology being used. There is some agreement among astrologers that planets on the Angles (however defined) are usually powerful. Ptolemy is often quoted as supporting this idea, but in fact it is difficult to be sure what Ptolemy really had in mind in regard to this matter: the few readings which have survived from Greek horoscopes, as from early Arabic figures, support the general importance of the Angles. There is also much argument as to what arc of *ORB should be applied to Angularity. If it were a question of mundane position, as the original use of the term indicates, then the question of orb need hardly arise. However, if the modern interpretation is adhered to (there is no general consensus of opinion) an orb equivalent to the 1st magnitude fixed star recognized in the *MORINEAN ORBS (that is, 8 degrees) might well be regarded as applicable. The most important and influential research on the significance of angularity in modern times is undoubtedly that done by Gauquelin, who has tabulated and published useful material dealing with planets sited within certain angular orbs. The interesting result is that Gauquelin shows that such orbs are not located centrally upon the Angles, but exert a pronounced clockwise-direction tilt of approximately 5 degrees. An excellent summary of Gauquelin's findings in regard to what is questionably called 'planetary heredity', and to the professions, along with a survey of the useful extension of Gauquelin's work into harmonics by Addey, is given by Dean, 1977.

Angular signs These are the *CARDINAL SIGNS.

Aniadus A term derived from Paracelsus, which appears to be the equivalent of the stellar *VIRTUE from which we receive celestial influences by the medium of fantasy and imagination.

anida Term used by certain alchemists to signify the VIRTUE in things, sometimes called (as, for example, by Sendivogius) the Astral Virtue or the Celestial Virtue. Though etymologically linked with ANIADUS, it appears to be the equivalent of the *ILECH.

animal signs An alternative term for FOUR-FOOTED SIGNS. It is interesting to observe that if the supposed etymology of the word 'zodiac' (as being from the Greek *zoon*) is correct, then all twelve signs would be animal signs. For the correct etymology, see Smith.

Anima Mundi The Latin term for 'the femine Soul of the World', relating to the enveloping divine essence which vitalizes all beings. As Blavatsky writes, it is 'the essence of seven planes of sentience, consciousness and differentiation, moral and physical'. It is said to be of an igneous, ethereal nature in the world of form, but entirely spiritual in higher realms. The higher element of the human being is said to be derived from the Anima Mundi, which is itself an emanation of a higher spiritual realm. In astrological imagery, the Anima Mundi is often given in the form of a naked woman, surrounded by the Spheres, in which are embedded the planets – one of the most famous images is that given by Fludd.

animonder A term apparently derived from Egyptian astrology by way of Ptolemy, and often misunderstood by modern astrologers. The term is used to denote a sort of rectification principle, designed to determine the exact degree of Ascendant by means of the previous syzygy to the birth. It is often wrongly confused with the TRUTINE OF HERMES.

Annael A variant for ANAEL, ruler of Venus.

annual chart A term used by Bradley for the *SOLAR RETURN.

Annulus Platonicus See CHAIN OF BEING. Distinguish from ANNUS PLATONICUS.

annus Platonicus The Latin equivalent for *PLATONIC YEAR.

anomalistic period The period of time elapsing in the movement of a planet from one perihelion to the next. When applied to the Moon, the period marks the time elapsing between one perigee to the next.

anomalistic year One of the three lengths of the year, defined as the time interval between the perihelions of the orbit of the Earth: it is 365.23964 days. See also TIME, TROPICAL YEAR and SIDEREAL YEAR.

anomaly In relation to planetary movement, the Angular distance of a planet from its perihelion or aphelion.

Anser Americanus See TOUCAN.

Antar See ANTARES.

Antares Fixed star (binary) of 1st magnitude, the alpha of Scorpius, set in the body of the Scorpion. This was one of the ancient *ROYAL STARS, and is sometimes called Cor Scorpionis (Scorpion's heart) and Shiloh. Sometimes called Antar, in confusion with a literary hero (see Allen), the modern name is said to be derived from its red colour, in that it was rival even of the planet Mars – the Greek, *anti-Ares*. It is of an extremely powerful nature, both martial and malefic, with the equivalent influence of Mars conjunct Jupiter: Cornell (clearly unmindful of Algol) writes of it as 'the most mentioned star in the textbook of Astrology, for his evils'.

antecanis See PROCYON.

anti-aspects A term derived from research into the relationship between aspects and harmonics, which has suggested that there are two major anti-aspects, and two minors. The majors are those of about 79 degrees and 101 degrees (allowing an orb of about 3 degrees). Such Angular relationships play no part in traditional aspectal theory, though they are actually accommodated in the Ptolemaic elemental aspectal theory (see ELEMENTAL ASPECTS). The same is true of the so-called minor anti-aspects, which are of about 66 degrees and 114 degrees (and allow an orb of 1 degree).

antichthon Term of Greek origin, used to denote the 'anti-Earth', the 'second Earth' which was supposed to lie on the opposite side of the Sun to our own Earth, revolving invisibly to man. Historically, the idea of the antichthon is linked with the school of Pythagoras, but it was especially widely circulated in medieval astrological lore, even though it contradicted the assumptions of the geocentric PTOLEMAIC SYSTEM, then held as a valid model of the cosmos.

anti-Earth See ANTICHTHON.

antigenetic chart   Term used for a chart (usually one of a related series) cast for anniversaries of a radical natal chart, or for related periods (months, weeks, etc.) derived from that radical in a calendrical sense.

anti-Midheaven   See IMUM COELI.

Antinous   Constellation, approximately from 13 degrees Capricorn to 10 degrees Aquarius, 8 to 13 degrees south of the equator – but see CONSTELLA-TION MAPS. Named after the favourite of Hadrian, its proximity to the constellation Aquila has led Bayer to call it Ganymede, after the youth carried off by an eagle, to become Jupiter's cupbearer: it is also Pocillator (cup-bearer).

antipathy   Planets may be said to be in antipathy to each other, under certain circumstances – most usually when they are in difficult or inharmonous aspect. However, planets do have a natural antipathy by virtue of their potency: for example, the Moon is by nature antipathetic to Saturn, since the former is a dissolving influence, while the latter is a fixer (rigidifier) of form. In SYNASTRY, planets (and even the same planet in two charts) may be antipathetic when involved in difficult aspects between the two charts. In a specialist sense, the word is sometimes applied to two planets ruling opposing signs – thus the Mars of Aries is in antipathy with the Venus of Libra, and the ruler of Capricorn, Saturn, is said to bear an antipathy to the ruler of Cancer, the Moon. See also SYMPATHY.

antiscion   In modern astrology, the antiscion of a planet is the degree equidistant to that planet on the opposite side of the Cancer–Capricorn axis of a birth chart (the line falling through the first degrees of these signs). This antiscion degree is a sort of mirror-image of a planetary position, or of a nodal point, and is regarded as being effective when progressed or transitted by another planet. In traditional astrology, the same term was used of two planets with the same degree of declination on the same side of the ecliptic: the same declination degree on the opposite side of the ecliptic was called a contra antiscion. But for the modern version of this term, see CONTRASCION. The astrologer Antonio Bonattis was at pains to distinguish the modern sense of the antiscion, which he called the *antiscium* (plural *antiscia*) from the declination of *antiscium*: interestingly, he writes of the planetary antiscion 'vibrating' to that planetary influence which mirrors it.

antiscium   See ANTISCION.

antisedent   An antisedent planet is one in retrograde motion.

antivertex See VERTEX.

Antlia Properly speaking, Antlia Pneumatica (Air Pump), a constellation named by La Caille in the 18th century. This inconspicuous asterism is just south of Crater and Hydra.

Anuradha The 15th of the Hindu NAKSHATRAS.

Aorion See ORION.

apex See SOLAR APEX.

apex ephemeris See GALACTIC CENTRE.

Aphelion The point in the orbit of a planet (or comet) furthest from the Sun. The plural is Aphelia.

apheta Term applied to a planet or to a nodal point in a horoscope which sustains the life of the native. See also APHETIC PLACES. Alternative terms from different astrological traditions are *HYLEG and *PROROGATOR.

aphetic places In Greco-Roman astrology, the aphetic places were those areas in a chart which were said to maintain or enhance life. These are the central arc of 20 degrees in the 1st, 7th, 9th and 11th houses. The rules for determining which planets are aphetic or *HYLEG are directly related to these aphetic places, though complex in the extreme. The Sun in an aphetic place is always given preference, followed by the Moon, and then by the planet which carries most dignity. If no planet qualifies, then the horoscope degree itself is the aphetic (incidentally, this rule confirms that the relative house system places the Ascendant in the middle of the 1st house, otherwise it could not be in the aphetic place – see HOUSE SYSTEMS).

aphorisms A term used in astrology in relation to the collection of pithy sayings, often in the form of concentrated astrological lore, and contained in * CENTILOQUIUM texts. An example from the *Centiloquium* of Gadbury is the aphorism, 'The Art of Astrology is certain and most indubitably true; but there are few that Practice it, who rightly understand it.' Many aphorisms are merely simplistic generalizations of astrological traditions.

Apis See MUSCA AUSTRALIS and MUSCA BOREALIS.

apocalyptic star   An esoteric term used by Paracelsus for the one 'personal' star which 'exists higher than all the rest' – see ESOTERIC ASTROLOGY.

apogee   The point in the orbit of a planet which is farthest from the Earth. The term is often used in a wide and even figurative sense, but it was originally restricted to the motion of the Moon.

Apollo   The name of this Roman god is frequently used as a synonym for the Sun in late medieval texts. The fixed star CASTOR was also called Apollo by Ptolemy, a name adopted by many later astrologers. The Greek equivalent, Helios, was also used in a similar dual way.

Apollon   Name given to one of the *HYPOTHETICAL PLANETS used in the URANIAN ASTROLOGY. This transneptunian planet is accorded an ephemeris, and is claimed to relate to expansion, science, commerce and peace.

aporhoea   Little-used term, used of the Moon as it separates from aspect to one planet, and applies to another, when it is said to be in aporhoea.

apotelesmatic   A term used as almost synonymous with *GENETHLIACAL. It appears to have been derived from the Greek *apotelesma* (the sum, or completion), perhaps with reference to the completing or casting of a horoscope. Bosc suggests that the original application was intended to convey the idea of the science of influences. In a non-specialist sense, the word is applied to the casting of horoscopes of all kinds, well beyond the limits of NATAL ASTROLOGY.

apparent motion   In astrology, the apparent motion of celestial bodies is regarded as being due to the rotary motion of the Earth, as distinct from their own PROPER MOTION.

apparent solar time   See TIME.

appearance   Much credence is given in popular astrology to the notion that the appearance of the native is a reflection of the zodiacal or planetary dominants in his chart – that a horoscope is a significator of the physical as well as the psychic life of the native. The earliest printed textbooks set out fairly simplistic rules for determining appearances, and what has more recently been called facial types, in the aphorismic astrology which bypasses the more profound laws of horoscopic delineation. From such notions, a series of word-portraits, and even graphic images of facial types, claimed to correspond to the planets or the signs, have been published – almost all of them of dubious validity. The word-portraits derive from principles set out in medieval astrology, the simple rule (given, for example, by Lilly) being that one may assess appearance from the sign ascending and from the two signs containing the Moon and the 'Lord of the Ascendant' – an axiom which is repeated even into modern textbooks. The finest of the short word-portraits (though descriptive more of personality than actual appearance), linked with the Ascendant and/or Sun-sign astrology, is given by Waite, the typical longer portraits are given by Pagan. However, the literature relating to types is of much the same order, and is self-proliferating. The graphic portraitures (limited to planetary types, signs and Sun-signs) appears to have been derived from medieval physiognomy: Delaporta did not hesitate to derive facial appearances from the images of the corresponding zodiacal signs. The most sophisticated early examples may be seen in Sibly, who gave copperplate portraits set in relevant horoscopes – a considerable advance on the crude woodcuts. A planetary physiognomy developed around the AL-MUTEN, conveniently dividing mankind into seven physical types – though the astrologer-artist Varley prepared some interesting drawings to support his (unfinished) studies which seem to have been based on direct observations. The widely published planetary and zodiacal types associated with the popularization of astrology by the theosophists, or with such writers as Muchèry, are little more than crude continuations of the medieval tradition: the modern practice (in popular textbooks, at least) of using photographs of famous personalities to demonstrate Sun-sign or Ascendant facial types (for example, Lynch) is little more than a technical development of this ancient practice. The root of the problem is that appearance is just about as unique as the horoscope itself, and therefore does not lend itself easily to simplistic methods of classification into seven, ten or twelve types (or for that matter into the 36 decanate types, as has been proposed). In view of the uniqueness of the chart, it is hardly surprising that Gauquelin should come to the conclusion that it was not possible 'to establish a link between the human physique and planetary type' – thereby brushing aside a thousand-year-old tradition. There is certainly much need for a comparative study of facial appearance in relation to well-documented horoscopes, perhaps along the lines indicated by Toonder, who observes the similar appearance of the Nobel Prize winners Einstein and Hahn, the tenors Gigli and Laurintz, and the writers Stephens and Joyce. The German astrologer Wangemann has (on the account of Dean) spent many years researching facial appearance in relation to house cusps, which marks a departure from the traditional approach – working mainly from the KOCHIAN SYSTEM which seems to lend itself to this form of research.

application In a strictly specialist sense, a planet is said to be applying to another body, or to a nodal point, when it is on the point of forming an aspect to it. Usually, it is the faster body which is said to be applying to the slower, since it is this which is building up or forming the aspect. The importance of the term lies in the fact that traditional astrology insists that aspects are more powerful when in application, so that the orb of application is more powerful than the orb of separation: but see ASYMMETRIC ORB. The term is directly from the Latin *applicatio*, itself a translation of Ptolemy's *synaptein*. See DIRECT APPLICATION and RETROGRADE APPLICATION.

appulse Strictly an astronomical term for a partial occultation, but sometimes used to denote a conjunction of one orbital body with another, or even the crossing of the meridian by such a body. In a specialist sense, it is applied to the entry of the Moon into the Earth's shadow.

April The fourth month of the *GREGORIAN CALENDAR. The etymology of the word is probably linked with the Latin verb *aperire* (or modern 'aperture') and exoterically explained as the time when the 'flowers open', and so on. In esoteric astrology, however, the month is linked with the opening of the Earth plane to the influences of the stellar world, from which it has to some extent been isolated in the winter time (see MARCH). It has been argued that, in spite of the early etymologies, this term is from the Greek 'Aphrodite' (as Aphrilis), whose festivals were celebrated in this month. Certainly, the pagan Anglo-Saxon goddess of the spring, Eastre, gave us the Easter-month.

apsis In early astrology, apsis was sometimes used as synonymous with 'orbit'. In modern astronomy, however, it is used as descriptive of one of the two points in the elliptic orbit of such a body, at which it is at its greatest and least distance from the centre of its motion. These points (plural: absides) are the *APHELION and *PERIHELION.

Apus Constellation, some 75 to 80 degrees south of the equator, called the Bird of Paradise, and said to exude a beneficent influence.

aquarids See METEORS.

Aquarius The eleventh sign of the zodiac. It corresponds as a zodiacal sign neither in location nor extent with the constellation of the same name (see AQUARIUS CONSTELLATION). The modern sigil for Aquarius ♒ is said by some to represent the flow of water, especially that of the Nile, and this sign

Aquarius

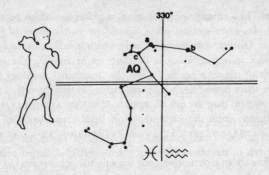

Figure 2: Aquarius – 13th-century image from SAN MINIATO ZODIAC, *and constellation map after the* MODERN ZODIAC *of Delporte.*

(for all it is an air sign) has from the earliest days of astrology been associated with water: on a symbolic level, however, the stream of water which is contained in the urn of the Aquarian image (figure 2) is said to represent spiritual knowledge, the Celestial Waters of the ancients. Some astrologers prefer to interpret the sigil as representative of the invisible waves of electricity (in occult terms, the Fohat) – see, however, SIGILS – ZODIAC. Aquarius is of the air triplicity, and of the Fixed Quality, the influence being erratic, refined, artistic, tenacious, perverse, intuitive, independent and original. The nature of Aquarius as it manifests in humans is expressed in the many keywords which have been attached to it by modern astrologers: friendly, humanitarian, progressive, persistent, inventive, perverse, creative, tolerant, fond of science and literature, discreet and optimistic – in a word, all the qualities which may be associated with an air type working with a view to establish freedom for self and others. In excess, the Aquarian tends to be something of a crank, an opportunist, a rebel, with both an unsympathetic and irresponsible attitude to life. Aquarius speaks of 'freedom' more insistently than others, usually without being able to explain fully what is meant by the word – often he or she will attempt to destroy the status quo, rather than work at establishing something beneficial for the future. According to the majority of modern astrologers, Aquarius is ruled by the planet *URANUS, though before this planet was discovered in the 18th century, the sign was ruled by Saturn (see SECTA). Mercury is Exalted in Aquarius, whilst the Sun is in his Detriment in this sign. In addition to the special terms already introduced above, the various traditional classifications present Aquarius as Diurnal, Fixed, Fortunate, Warm, Fruitful, Human, Masculine, Mental,

Moist, Nervous, Obeying, Rational, Strong, Sweet, Vital and Whole. See also SANGUINE.

Aquarius constellation Linked in name, if not in influence, with zodiacal Aquarius, located as indicated in figure 2, straddling the equator from 5 degrees north to 30 degrees south. The image associated with this constellation (and hence with the sign) was even in Babylonian times that of a man pouring water from an urn, and thus understandably in Greek astrology was called Hydroxous (water-pourer – the image in figure 2). The numerous names given to the constellation by astrologers of different eras and places (listed by Allen) generally allude to the idea of water, or to the image of a water pot. The fixed stars within the asterism which are of importance to astrologers are *SADALMELIK and *SADALSUUD, the alpha and beta of figure 2.

Aquila Constellation, approximately from 12 degrees Capricorn to 15 degrees Aquarius, from 22 degrees north to 13 degrees south of the equator. This asterism was even in early astrological texts confused with another bird, the Vultur Volans, and translated into English (inexactly) as the Flying Grype. The Roman astrologers knew it as Aquila, however, or as the bird of Jupiter (which was an eagle), whilst most other names from different times and countries link the asterism with the image of a flying bird. The important fixed star in the constellation is *ALTAIR.

Aqrab Al Aqrab, the Arabic term for Scorpio.

Ara Name given to a constellation, the Altar, located to the south of Sagittarius and, as Aratos says, 'beneath the glowing sting of that huge sign the Scorpion'. The classical mythology is merely putting a seal on a much more ancient idea, for it was the Tul-Ku '(Holy Altar) in Euphratean astrology, but the story has it that an altar made by the Centaurs was taken to the heavens after being used by the Gods for their oath to withstand the Titans – hence the name Ara Centauri. Medieval astrologers would sometimes call it Puteus Sacrarius or Templum, all in reference to a sacred place, perhaps following Manilius, for whom it was Mundi Templum (Temple of the World). Ptolemy links its influence with that of Venus and Mercury combined, and it is said to give scientific ability and concern for ecclesiastical matters.

Arabian Directions Another term for *SECONDARY DIRECTIONS.

Arabian Parts   See ARABIAN POINTS.

Arabian Points   A general name for a number of 'parts', some derived from early Greco-Byzantine astrology, some from Arabic astrology, but transmitted to Europe through Arabian astrological literature. Few of these numinous points or parts are used in modern astrology, but for a complete list, see the entry under PARS.

Arabic astrology   The images and associations for the twelve signs of the zodiac, used in Arabic astrology, as well as many of the names for fixed stars, were derived from Greco-Roman astrological sources. However, with the growth in sophistication of Arabic astrology, particularly during the Baghdad Caliphate, many of the names and associations were completely Arabicized. Some indication of the extent of the Arabic influence may be gleaned from a study of the names still used in connection with the lore of *FIXED STARS. The following table gives the Arabic names for the signs and constellations, along with the Arabic itself:

Aries	Al Hamal	الحَمَل
Taurus	Al Thawr	الثَّور
Gemini	Al Jawza	الجوزاء
Cancer	As Saratan	السَّرَطَان
Leo	Al Asad	الأسَـد
Virgo	As Sunbula	السُّنْبُلة
Libra	Al Mizan	المِيزَان
Scorpio	Al Aqrab	العَقْرَب
Sagittarius	Al Qaws	القَوْس
Capricorn	Al Jadiyy	الجَدْي
Aquarius	Ad Dalw	الدَّلْو
Pisces	Al Hut	الحُوت

No short entry may do justice to the importance of Arabic astrology to the Western form of the art – it is almost true to say that everything which was used in medieval European astrology (the basis of most modern forms) was actually derived from the Arabs, mainly by way of Moorish Spain. The sophistication of Arab astrology, especially in its horoscopic applications, was of the highest order, and incorporated such notions as various theories of precession, the idea of conception-charts, a specialized series of house systems, based on both spatial and temporal divisions, through to the superstitious levels of amuletic astrology, the use of zodiacal images, and so on. A glance at the section on *HOUSE DIVISION will reveal something of the contribution of Arabic lore to the Western tradition. The entry on the important *ASTROLABE indicates something of the role played by Arabic astrology in the transmission of classical ideas to the West. This contribution to the West should not obscure the fact that Arabic astrology contained elements and emphases which were special to Islam – a few astrological symbols, such as the so-called *ZAWZAHR, were Arabic in both symbolic content and in astrological implication. The images and sigils used for the planets are surprisingly different from those used in the West, though the seven classical planets are still employed. The names of the seven traditional planets in Arabic are: Shams (Sun), Utarid (Mercury), Zuhra (Venus), Mirrih (Mars), Mushtari (Jupiter) and Zuhal (Saturn). Usually the Moon is Qamar, but the full Moon is Badr, the new Moon is Hilal, while the change of Moon is Hulul al Qamar.

Araboth The Seventh Heaven of the Cabbalistic system, ruled by CASSIEL.

Arachne A name applied by Vogh to what is claimed to be a 'lost' 13th sign of the zodiac. The new model of the zodiac which results from this addition has been called the Psychic Zodiac: needless to say, neither it nor the 13th sign finds a correspondence or mention in ancient astrological sources. It is simply not possible to have a sign of less than 30 degrees. The lunar zodiac with which Vogh makes such play was a constellational lunar zodiac, but there was no Arachne in any of the lunar asterism, nor in any of the *SIDEREAL ZODIAC constellations.

Arahsamma See BABYLONIAN CALENDAR.

Aratron First of the seven supreme Angels of the qabbalistic system, one of the powers of the Arbatel corresponding to the Sphere of Saturn. He was said to be able to petrify living organisms, and to possess the power to change higher forms of life into lower forms – the Saturnine powers.

arc A segment of a curved line – in a specialist use, an arc of the zodiac, usually expressed and measured in degrees, minutes and seconds. The measurement of arc may also be translated into temporal co-ordinates, as for example in primary directions, where degree of arc is often expressed in time. See also ARC OF DIRECTION, DIURNAL ARC, NOCTURNAL ARC and SEMI-ARC.

Arcanum In general the term means anything hidden, but Paracelsus adopts the word for specialist astrological and alchemical use, using it to denote the secret incorporeal virtue behind or within natural forms. It is therefore a synonym in this sense for both *ILECH and *VIRTUE. The word is also used to denote any of the series of so-called Major Arcana of the Tarot Pack – see therefore ASTROLOGICAL TAROT.

archaeo-astronomy Term used in a special sense by Krupp to designate the study of the astronomies of ancient and prehistoric times. Many of the subjects included under Krupp's editorships, such as the British stone circles and menhirs, the Mesoamerican remains, the medicine wheels of North America and the Egyptian pyramids and temples, are certainly linked with important and intriguing calendrical and astrological factors.

Archai Name given to the incorporeal beings forming the Seventh Hierarchy of the spiritual realm, according to Dionysius the Areopagite (see ANGELS), the beings of the Sphere of Venus. The Archai are said to guide the motion of time itself, and in some esoteric systems they are called the Spirits of the Revolution of Time. The Golden Legends of Voragine link them with the mystery of the Eucharist, which explains why they are often pictured as holding Chrismatories. Sometimes the Archai are called Principalities, an attempt to Latinize the Greek word. St Gregory calls them the Virtutes (sometimes translated Virtues), and associates them with the sapphire and the colour blue. In other systems, however, the Virtutes are linked with the *DYNAMIS of the Martian sphere – though the connection between the Archai and Venus is widely accepted in esoteric circles.

Archangels Name given to the incorporeal beings of the Eighth Hierarchy, according to Dionysius the Areopagite (see ANGELS), the beings of the Sphere of Mercury. The Archangels are said to guide the spiritual destiny of groups of people, of nations, rather than individuals – this probably explains why the Archangels are often pictured as carrying formalized models of cities in their arms. St Gregory associates the Archangels with the carbuncle stone, and biblical exegesis links them with the colour red. Since the Archangels are often called Angels, much confusion has arisen in astrological lore: see ARCHANGELS OF THE PLANETS, ARCHANGELS OF THE ZODIAC, CELESTIAL HIERARCHIES and TWELVE CREATIVE HIERARCHIES. For the individual Archangelic rule over elements, see GABRIEL (water), MICHAEL (fire),

Archangels of the Sephiroth

RAPHAEL (air) and URIEL (earth) – for Archangelic periodicities, see SECUNDADEIAN BEINGS.

Archangels of the planets Sometimes called by the generic term *ANGELS, these spiritual beings were originally charged with the regulation of the *SPHERES (rather than the planets), though this distinction is now virtually lost. The most frequently used names and planetary spheres are given in table 4. See also ARCHANGELS OF THE SEPHIROTH and PLANETARY SPIRITS.

--- TABLE 4 ---

Gabriel – Moon	Samael – Mars
Raphael – Mercury	Zadkiel – Jupiter
Anael – Venus	Cassiel – Saturn
Michael – Sun	

Archangels of the Sephiroth The occult and qabbalistic tradition has preserved several lists of the names of the Archangels which have rule over the ten Sephirah in the Tree of Life, which is linked with various astrological correspondences. Few of the traditional lists agree in detail, but those shown in table 5, based on Grey, may be taken as representative: see QABBALISTIC ASTROLOGY. The Archangels are opposed by their Adversaries.

--- TABLE 5 ---

Sephirah	Archangel	Adversary	Sphere
Kether (Crown)	Metatron	Thaumiel	Primum Mobile
Chockmah (Wisdom)	Raziel	Chaigidiel	Zodiac
Binah (Understanding)	Zaphkiel	Sathariel	Saturn
Chesed (Compassion)	Zadkiel	Gamchicoth	Jupiter
Geburah (Severity)	Camael	Golleb	Mars
Tiphereth (Beauty)	Michael	Togarini	Sun
Netzach (Victory)	Hamiel	Harab Serap	Venus
Hod (Splendour)	Raphael	Sammael	Mercury
Yesod (Foundation)	Gabriel	Gamaliel	Moon
Malkuth (Kingdom)	Metatron	Lilith	Man and Elements

Archangels of the zodiac The astrological tradition has preserved many lists of ANGELS, Archangels and Spirits, which are said to be specifically in control of the twelve signs of the zodiac. Unfortunately, few of these lists agree, but those in table 6, derived from Haywood, are representative. Those marked with an asterisk are said by some authorities not to be Archangels. The overall ruler of the zodiac is sometimes named as Maslem, but in the esoteric view, the zodiac falls under the Cherubim or Seraphim.

--- TABLE 6 ---

AR	Malshidael, Aiei*	LB	Zuriel, Jael,* Joel*
TA	Asmodei, Tual*	SC	Barcheil, Sozol
GE	Ambriel	SG	Adnacheil, Ayil
CN	Muriel, Manuel, Cael*	CP	Hamael, Semakiel,* Casujoiah*
LE	Vercheil, Ol,* Voel*	AQ	Cambiel, Ausiel*
VG	Hamaeliel	PI	Barchiel, Varchiel*, Pasiel*

Archer Popular name for SAGITTARIUS, whose image is sometimes wrongly given as that of an archer, or a mounted archer, both forms being derived from the Babylonian image of a kind of man-beast centaur, shooting from a bow.

Arcitenens See SAGITTARIUS CONSTELLATION.

arc of direction An arc measuring distance between a planet (significator) and the point where it will form an aspect with a promittor – see DIRECTIONS. The arc may be measured either in time, or in degrees of Right Ascension.

arc of dwarfed stature An arc marked from the middle of the 7th house to the middle of the 2nd house (in clockwise direction), with its most powerful influence at the cusp of the 6th house. The arc is interpreted only when the DRAGON'S HEAD is placed within its compass: such a position is supposed to stunt growth of the native.

arc of increased stature An arc marked from the middle of the 1st house to the middle of the 8th house (in clockwise direction), with its most powerful influence at the cusp of the 12th house. The arc is interpreted only when the DRAGON'S HEAD is placed within its compass, and such a position is supposed to increase growth of the native.

arc symbols   See DEGREE SYMBOLS.

Arctophilax   See ARCTURUS.

arc transform chart   A figure introduced by Williamsen, as a variation of a HARMONIC CHART, derived from the theory of harmonics, in which the determinant harmonic is the angular separation between any two planets in the radical chart. The system has been much criticized on the grounds of unwieldiness, as the simple radical gives rise to a very large number of interpretative charts.

Arcturus   Fixed star of 1st magnitude, the alpha of Bootis, set in the left knee of the figure. The term comes from the Greek Arktouros (Bear-guard), but many names have been recorded for this important star, including Arctophilax (Bear-watcher), Alchameth, Azimech, Al Simak, and (in Egyptian astrology) Bau. It is a highly beneficial star, especially when rising or culminating in a chart, and Ebertin says that its Jupiter-Mars influence has a reputation for achieving justice through power. See FIFTEEN STARS.

Arctus   See both URSA MAJOR and URSA MINOR.

Ardra   The 4th of the Hindu NAKSHATRAS.

Ares   One of the ancient Greek terms for Mars. Paracelsus used the same term to denote the formative power which creates differences (*differentia*) among species. The choice of word is probably guided by the esoteric astrological notion that Aries, ruled by POSITIVE MARS, is the realm of incarnation, where differentiation of spirit into form begins.

Argion   See ORION.

Argo Navis   Constellation, approximately from 10 degrees Cancer to 20 degrees Libra, from 15 to 65 degrees south of the ecliptic. Not surprisingly of an asterism linked with the ship of the Golden Fleece, the constellation is said to give prosperity in voyages, though it is linked with drowning. Robson suggests that this explains why the first decanate of Virgo is connected with drowning, but this area of the zodiac is riddled with powerful stars which could easily bring disasters of this kind, even though they are not all set in Argo. The only Argoan fixed star which is openly associated with shipwreck is FORAMEN.

Argos See PYXIS NAUTICA.

Argus A frequent error for ARGO NAVIS.

Ariadne's Hair See COMA BERENICES.

Ariel Name given to one of the satellites of Uranus.

Aries The first sign of the zodiac. It corresponds as a sign neither in location nor extent with the constellation of the same name – see ARIES CONSTELLATION. The modern sigil for Aries ♈ is said by some to be a vestigial drawing of the horns of the Ram (see figure 3) with which the sign is associated, but

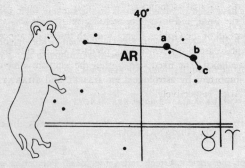

Figure 3: The image of Aries, from the 13th-century SAN MINIATO ZODIAC, *and the constellation from the* MODERN ZODIAC *of Delporte.*

some astrologers claim that it is a drawing of the human eyebrows and nose, for Aries has rule over the human head and face in the MELOTHESIC MAN. Some esoteric astrologers explain the sigil as a primitive diagram of the implosion and explosion of spirit in and out of the material body – but see SIGILS – ZODIAC. Aries is one of the fire elements, of the Cardinal Quality, and its influence is outgoing, pioneering, self-reliant, idealistic, enthusiastic and exaggerative. There is a strong element of selfishness, and the Arietan type tends to be insensitive to the needs of others. This outgoing nature of Aries is expressed in many keywords attached to the sign by modern astrologers: initiatory, freedom-loving, active, intense, resourceful, aggressive, impulsive, inspirational, courageous, spontaneous, audacious – in a word, all those qualities which may be associated with a fire nature expressing itself with untrammelled confidence. In excess, the Arietan nature may be described in terms which express its underlying destructive nature, the keywords being: overbearing, dictatorial, resentful, sarcastic, jealous, coarse, argumentative, violent, uncontrolled, impatient and egotistical. Aries is ruled by the planet Mars – indeed, in the ancient tradition by POSITIVE MARS (see also SECTA), as opposed to the NEGATIVE MARS which has rule over Scorpio. The sign marks the Exaltation of the Sun, the Fall of Saturn, and the Detriment of Venus. In addition to the special terms already introduced, the traditional classifications present Aries as Barren, Bitter, Commanding, Diurnal, Dry, Equinoctial, Fortunate, Four-booted, Hoarse, Hot, Hurtful, Inflammatory, Luxurious, Masculine, Mental, Northern and Violent. See also CHOLERIC.

Aries constellation   Constellation linked in name, if not in influence, with zodiacal ARIES, located as indicated in figure 3. In popular language this is often called the Ram, and in ancient astrological texts it was associated with the Greek legend of the Golden Fleece, though the image appears in the very earliest Egyptian astrology which long predates Greek culture: the image in figure 3 is medieval. The fixed stars within the asterism which are generally considered important by astrologers are HAMAL and SHARATAN, the alpha and beta of Aries, respectively.

Arietan   Pertaining to the sign or constellation of *ARIES: the construction is from the Latin genitive *Arietis*.

Arion   See ORION.

Arista   See SPICA and VIRGO CONSTELLATION.

Aristarcheian system   Name given to an imperfectly recorded model of the solar system postulated by Aristarchus of Samos in the 3rd century BC. See HELIOCENTRIC SYSTEM.

Aristotelian cycle   The Aristotelian cycle, linked with the cycles set out in the Great Year of the PLATONIC CYCLE is a later gloss on Aristotle, and insists that two great catastrophies mark the direct and retrograde motions of the spheres. When the spheres are in conjunction in Cancer, then the world is destroyed in fire. When the spheres are conjunct in Capricorn, then the world is destroyed by water or flood. Here there is a clear deviation from the original idea of sphere cycles to planetary cycles. The Aristotelian cycle has been associated (quite needlessly) with the ordinary seasonal solar transit of the solstices, when an esoteric origin is more apposite.

Aristotelian principles See PRINCIPLES.

Aristotelian system A model of the universe proposed (chiefly) by Aristotle, which postulates a geocentric solar system, with all the planets making revolutions in perfect motion, within a series of concentric spheres. The Earth itself was visualized as a sphere, bathed in three sublunar spheres relating to the elements – first that of water, then of air, and then of fire. Set in the CRYSTALLINE HEAVEN were the Moon, Mercury, Venus, Sun, Mars, Jupiter and Saturn. These were bounded by the Sphere of Fixed Stars (STELLATUM) which marked the end of the universe: it was this sphere which transmitted movement to the other spheres, for which reason it was called the PRIMUM MOVENS or Primum Mobile. Everything beneath the Sphere of the Moon (which of course embraced the Earth) was corruptible, but all above that sphere was bathed in incorruptible purity of the QUINTESSENCE. This model was essentially a physical one, permeated to some extent by esoteric lore derived from Greek Mystery schools: it formed the basis for the Christian and Neoplatonic cosmoconception of the CELESTIAL HIERAR-CHIES, the Sphere of the Earth being divided into a reflective series of concentrics to accommodate the infernal regions and adversaries. The importance of the Aristotelian system was not so much its practical value to astrologers, as its spiritual heritage.

Arkab Fixed star (double), the beta of Sagittarius; its name and variants (such as Urkab) from the Arabic Al Urkub, of disputed meaning.

armillax See ASTROLABE.

Armus Fixed star, the eta of Capricornus, set in the heart of the Goat. Said to be of the nature of Mercury conjunct Mars and to give a disagreeable, contentious nature.

aromal planets Name given by 19th-century astrologers, mainly working within the framework of theosophy, to a large group of planets said to orbit within our solar system, though invisible to normal sight. The astrologer Harris claims that there are more than 1,000 such aromals in our system. It is possible that certain of the * HYPOTHETICAL PLANETS perceived clairvoyantly belong to this supposed class – but see also ETHEREAL PLANETS.

Arrow Star A name used to denote the fixed star SIRIUS, when regarded as a fiducial, derived by Fagan from the Assyrian term *gag-i-sa*.

artificial gamalei See GAMALEI.

Aryan   In an astrological context, this term is usually an unfortunate mistake for ARIETAN. The word Aryan is, of course, properly applied to a member of a supposed race which made use of a particular form of language, sometimes called the Indo-European language – but see FIFTH ROOT RACE.

Asad   Al Asad, the Arabic term for Leo.

Asaru   One of the Chaldean names for the planet Jupiter.

Ascella   A fixed star (binary) set in the armpit of constellation Sagittarius. It has a beneficial influence, said to be of a nature equivalent to Mercury conjunct Jupiter.

Ascelli   Pair of stars, treated by astrologers as a single (though in adjacent degrees), both set in the constellation Cancer. Robson properly gives separate influences – the northern Ascellus tends towards defiance and heroism, while the southern is unfavourable. Cornell, who gives both a thoroughly evil influence, appears to have confused this pair with the contiguous PRAESAEPE cloud.

Ascendant   The term is properly applied to the degree of zodiac arising on the eastern horizon of a figure, or indeed to the degree of zodiac arising over the eastern horizon. This degree was originally called the Horoscopos in Greek astrology, from which the modern word HOROSCOPE was derived. The term Ascendant is often used of the whole sign arising, rather than of the degree. To avoid confusion in ordinary parlance, modern astrologers often distinguish between the rising sign and the rising degree, both of which may be referred to as the Ascendant. Sometimes a planet is misleadingly termed the 'Ascendant planet', when it is in the last few degrees of the nocturnal semi-arc. In earlier forms of astrology, it was common for all the planets near the Easter Angle to be described as 'Ascendant', but at that time the special meaning of 'horoscope' in reference to a particular degree was not yet obsolete, so that the distinction was clear. See also ASCENDANT SIGNS.

Ascendant axis   Term used of the horizon axis, which in the horoscope figure usually corresponds to the axis of the Ascendant and Descendant.

ascendentia signa   A curious term listed by Sendivogius and said by him to represent either the 'Stars of the Firmament' or the 'Sydereal Spirits': but see ASCENDANT SIGNS.

Ascendant signs*

Ascendant signs A technical term indicating those signs which contain the Sun when its declination is on the increase – that is, Aries, Taurus, Gemini, Libra, Scorpio and Sagittarius. However, the term is used in a general sense in reference to a particular Ascendant of a horoscope, in which case any of the twelve signs may be described as being Ascendant signs – see ASCENDANT.

ascending A fairly general, if imprecise, application of this term is to any planet or nodal point on the eastward arc of the figure. Properly speaking, however, this pair of semi-arcs should be termed 'the ascending part of the heavens', and a planet placed within the arcs described in such a lengthy way. More precision is implied when reference is made to a planet within orb of the *ASCENDANT itself. See also ASCENDANT SIGNS.

ascending arc This term, which is sometimes applied to the degree of zodiac on the *ASCENDANT, must be distinguished from the theosophical term which is the equivalent of the so-called 'luminous arc', relating to the upward passage of life-waves from the physical realm.

Ascension See RIGHT ASCENSION and OBLIQUE ASCENSION.

ascensional difference The difference between the RIGHT ASCENSION and OBLIQUE ASCENSION of a body.

Aschemie See PROCYON.

Aschere See PROCYON and SIRIUS.

Asclepius See OPHIUCHUS.

Ashadha The Sanskrit name for Capricornus: see also MAKARAM.

Asimec See SPICA.

Aslesha The 7th of the Hindu NAKSHATRAS (the embracer).

Asmodel The ruling Angel of zodiacal Taurus. For sigils, see SIGILS, ANGELS OF THE ZODIAC and ARCHANGELS OF THE ZODIAC.

aspectal pattern See CONFIGURATION and MAJOR CONFIGURATION.

aspectarian A list, usually in chronological order, of all significant aspects formed during a particular period, sometimes in relation to a particular chart. In modern times such aspectarians are often found in EPHEMERIDES. Special symbols (apart from the ABBREVIATIONS and SIGILS recorded elsewhere) include:

Declination	☌	Minute of arc	′
Eclipse of Moon	♪	New Moon	●
Eclipse of Sun	◖	North declination or latitude	+
First quarter	☽	Second of arc	″
Full Moon	○	South declination or latitude	−
Latitude	℘	Third quarter	☾
Longitude	λ		

aspects Term applied to a large number of specific Angular relationships between planets and other celestial bodies or nodal points. Most aspects are considered ecliptically, and are generally regarded as symbolic, rather than as having any geophysical validity. In the majority of cases, aspects are also regarded as being geocentric – which is to say in terms of Angular relationships relating to the Earth (the actual place of birth) as the receptive matrix. However, a number of astrologers who favour HELIOCENTRIC ASTROLOGY do make use of heliocentric aspects, and Thorburn, an astrologer who has investigated both systems, is clearly in favour of heliocentrics. The traditional forms of astrology are concerned with nine Angular relationships only, these being divided into the MAJOR ASPECTS and the MINOR ASPECTS, to which the mathematician-astrologer Kepler introduced a series of QUINTILE aspects (see KEPLERIAN ASPECTS), as listed in table 8. The astrologer Placidus introduced an important *PARALLEL aspect. There are two basic groups of traditional aspects – the MUTUAL ASPECTS which are those formed between two or more bodies at a specific time, and those formed between a single moving body and given points within a chart: this latter group is the DIRECTIONAL ASPECT, the PROGRESSED ASPECT and the TRANSITORY ASPECT, all used mainly in predictive astrology. For an aspect to be fully effective, it is required that it be precise (at least, as an ecliptically symbolic consideration), but it is taught by those who have researched the traditional doctrines (for example, Carter) that aspects which are FORMING or SEPARATING will still exert influences. The degree of arc wherein the aspect may be said to be operative as an influence depends upon the nature of the planets concerned, and is called an ORB. Unfortunately, there appears to be

aspects

TABLE 7

Aspect name	Aspect (degrees)	Sigil		General meaning
Opposition	180	☍	☊	Tension
Trine	120	△		Expansion
Square	90	□		Difficult, but energizing
Conjunction	0	☌		Intensifies planets involved
Quincunx	150	⊼		Strain
Sesquiquadrate	135	⟎	⟎	Difficult (less than square)
Semi-square	45	∠		Difficult
Sextile	60	✳		Expansive (weaker than trine)
Semi-sextile	30	⊻		Slight strain
Bi-quintile	144	±		Strengthening?
Quintile	72	Q	★	Strengthening?

no agreement as to what orbs may be applied to the various aspects: for major aspects, some use orbs with as little as a 5- to 10-degree orb; see, however, ASYMMETRIC ORB and MORINEAN ORBS. The body with the faster mean motion is said to aspect the slower, which usually means that the inner planets aspect the outer planets. Forming aspects are generally regarded as being more powerful than separating aspects. Aspects are said to generate beneficial or harmful influences, depending upon the natures of the planets involved – the former are the BENEFIC ASPECTS, the latter the MALEFIC ASPECTS. The benefics contribute towards harmony, but may lead to a dispersion of energies. The malefics create tensions, which may be resolved and balanced, sometimes leading to a concentration of energies. The original theory of aspects, announced somewhat inconclusively by Ptolemy, appears to have been based on the interactions of the zodiacal elements, rather than being based on geometric or symbolic considerations. Thus, it was said that the harmony of the trine was a manifestation of two planets working through the same triplicity, whilst the malefic aspect of the square was a manifestation of two planets working through triplicities opposed in essential natures. In Hindu astrology there are two terms which distinguish the aspects of an elemental nature from those of a purely geometric (or symbolic) kind (see JAIMINI and TAIJAK), and a useful word has been coined from the former to denote elemental aspects – jaiminic. Modern researchers into astrological principles have tended to ignore the symbolic nature of aspects and the jaiminic importance of the theory of aspects, and have concentrated on the study of Angular relationships as geometric propositions. The problem is that if aspects are examined in terms of three-dimensional geometry, involving the astronomical conception of an absolute plane, then the approximate conjunction (which is the basis of certain cycles) would occur only rarely. Aspects have been submitted to a certain number of statistical tests, difficult as these have proved to be. Nelson has done some useful work on the relationship between certain aspects and their influences on radio reception, whilst Dieschbourg has spent many years investigating aspects from an unconventional statistical standpoint. Almost all statistical and computerized studies have yielded disappointing results, which perhaps reflects more upon the methodology itself, than upon aspects. It is significant, for example, that most of those who have sought to establish connections between aspects and statistical interpretation, reduce aspects to Angular relationships, rather than in a jaiminic sense. In contrast, Addey's researches into a field of astrology related to aspects shows signs of revolutionizing modern astrology – see HARMONICS. In addition to the terms already mentioned, see also AB-SCISSION, CAZIMI, COMBUST, DECLINATION, DEXTER ASPECT, OCCULTA-TION. PARALLEL IN MUNDO, PARTILE ASPECT, PLATIC ASPECT and SINISTER ASPECT.

association houses   An alternative name for the RELATIVE HOUSES.

Asterion   See CANES VENATICI.

asterism   Literally a collection or group of stars, but usually a configuration of such stars, a constellation. Since the LUNAR MANSIONS are determined by star-groups, they are sometimes called asterisms. Occasionally, the term is wrongly applied to one or other of the signs of the zodiac, which are not groups of stars.

asterism image   A term coined to designate the abstract graphic image traced in imaginary lines between the stars in a particular constellation. The asterism image in figure 3 under ARIES, for example, is derived from the system of Delporte incorporated into the MODERN ZODIAC. The term is used partly to distinguish such abstract figures from the realistic *IMAGE linked with the majority of the constellations, and which are in some cases associated with the signs.

asterogram   A neologism (sometimes 'asterogramme') probably originated

by Sucher, and applied to horoscopes used in connection with *ASTROSOPHY.
The asterogram may be geocentric or heliocentric (as in the sample chart in
figure 4), but generally employs the *SIDEREAL ZODIAC. The term is also

Figure 4: Asterogram for Goethe's birth (after Sucher).

extended to 'pre-natal asterograms', as a specialist application of the EPOCHAL
CHART, and to the 'death asterogram', cast for the moment of death.

asteroid periodicity See ASTRO-ECONOMICS.

asteroids Minor planets, the name meaning 'star-like' – the alternative (and
now rare) 'planetoid' being etymologically better. Shortly before the discovery
of the so-called NEW PLANETS, towards the end of the 18th century, it was
suggested that the series of numerical progressions called Bode's Series
seemed to postulate the existence of another planet between Mars and Jupiter
(corresponding in the series to the progression 28). The first of these
planetoids was not located by Piazzi until 1801, however: he named it CERES.
In the following year another, PALLAS, was seen, and then JUNO and VESTA.
Almost four decades passed before the fifth, ASTRAEA was located, but since
then 2,000 have been named or (mainly) numbered, and it is estimated that
there are about 50,000 in orbit. The asteroid Hidalgo (with an orbit of 13.7
years) has the longest period. Most asteroids are small by astronomical
standards, though Ceres has a diameter of nearly 500 miles. The importance
of the asteroids in astrology is that they were eventually vested with a
supposed influence on human affairs, and in some systems of astrology have
been accorded rule over certain signs; for example, Wemyss gives Pisces to
the asteroids, while Dobyns links only Ceres and Vesta with this sign.

Additionally, many of the HYPOTHETICAL PLANETS have been variously identified and confused with asteroids: see LUCIFER.

asteroid scheme   This term and the variant 'asteroid chain', are derived from the theosophical SCHEME OF EVOLUTION, provisionally linked with the eighth of the so-called schemes, on the grounds that the present asteroidal belt will eventually form a globe – an idea foreign to occult lore.

asteron   The ancient Greek term for stars, and often used also for planets. Ptolemy uses the term in this general sense.

Astraea   Name of one of the *ASTEROIDS. The Romans used the same word (meaning 'Starry Goddess') for the constellation and sign Virgo.

astral   The term appears to be derived from the Latin for 'star', and is sometimes applied to the stellar world as descriptive of the fabric of the heavens. Blavatsky suggests an origin from the Scythian '*aist-aer*' (star), but such an etymology should not confuse the astral with the fixed stars. In occult and astrological terminology the astral plane is that contiguous in space (if not in time) with the material realm: it is one which the spiritual part of man enters during periods of sleep and after death. The astral realm is one normally invisible to ordinary sight, yet it is the proper dwelling of the higher spiritual bodies of man (the 'astral body' and 'ego' of modern occultism). It is significant that Paracelsus should coin his term ENS ASTRALE to denote the Desire Body of man. The realm is said to be permeated by the ASTRAL LIGHT, which is the equivalent of the ancient AKASHA, and so the Paracelsian '*ens*' does link with the stars, and clairvoyants describe the astral body as consisting of light, like the stars themselves, though in continuous motion. The model of man projected in astrology is not far removed from such occultist cosmoconceptions, and inevitably the link between the world of fixed stars and the astral has found expression in several terms. Cornell, for example, calls the astral body the 'sidereal body', naming it the seat of the sensations through which the planetary influences work upon man (he appears to confuse it with the occult Etheric Body, however, though he is on more sure grounds when he says that the Astral Body is the realm in which Jupiter has direct relation to the breathing and blood, and where Mars and Venus relate to the sensations and passions). The ASTRUM of the alchemists relates to the old conception of hidden VIRTUES, which are in turn said to proceed from the stars. In occultism, the 'astral realm' is not linked with the stars, but with the sphere of the Moon, which explains why this is the 'realm of Purgatory' (where astral deficiencies are said to be cleansed, during the post-mortem state) – see, for example, EIGHTH SPHERE.

astral globe See GLOBES.

astral light See ASTRAL and AETHER.

astral sphere Leadbeater, himself no astrologer, tells us that the astral sphere of the Earth extends nearly to the mean distance of the Moon's orbit. It is not therefore possible (as Leadbeater claims) that the astral planes of the two worlds (Moon and Earth) touch one another when the Moon is in perigee, but do not so touch when the Moon is in apogee. The definition implies that the two spheres interpenetrate.

astral spheres A term coined by the esoteric astrologer Thierens relating to the future evolution of the SPHERES, and the consequent development of new planetary bodies – see FUTURE PLANETS. These astral spheres consist of the traditional spheres of Aquarius and Pisces, ruled by the planets NEREUS and COELUS respectively. Together with the ETHERIC SPHERES, these make up the LUNAR BODY of Thierens's cosmoconception. See also ASTRAL SPHERE.

astral year See SIDEREAL YEAR.

astro-alchemist A person who combines the practice of astrology with alchemy. Many of the operations of alchemy involve the casting of charts for propitious moments, and the careful consideration of planetary hours.

astro-alchemy A term derived to denote the astrology linked with the late medieval form of alchemy: this consisted mainly of a prolix terminology that made use of astrological words and concepts in a way which recognized the occult philosophical basis of the art. The literature of such occultists as Boehme, Dee, Trithemius and Welling is representative of this approach to both alchemy and astrology.

astro-archaeology A term coined in modern times to designate the quasi-archaeological theory which relates the design (and hence location and orientation) of megalithic monuments to the celestial phenomena observed during the long period of construction. The treatment by Michell limits the term to megalithic circles and related structures, but it might usefully be extended to the study of orientations which are astrologically conceived. For example, the orientation of the SAN MINIATO ZODIAC (towards the daily sunrise over Florence), and its demonstrable link with a FOUNDATION CHART of 1207, might be covered by the term astro-archaeology, or by some other neologism, such as ASTRO-ORIENTATION.

astro-biochemistry   A modern term used to cover the newly developing study of astrology in relation to the *TWELVE CELL-SALTS.

astro-cartography   The name given to a modern system of chart extrapolation involving the imposition of the four Angular positions of the ten planets of the natal chart on a world map, in order to determine geographic 'power zones' which will correspond to potentials within the native. A location, such as a city, town or rural area, falling upon the longitudinal equivalent of Jupiter's Midheaven position (for example) is said to evoke for the native all the qualities associated with his or her particular natal Jupiter in regard to his or her Midheaven propensities. The system, devised and promulgated in America by Lewis, is also subject to the study of the unfolding of radical potentials in terms of transits and progressions relating to the natal chart.

astro-chronology   Relating to the chronology of heavenly bodies – for example, to the timing of CYCLES.

astro-diagnosis   Medical diagnosis, either from a natal chart, or from a chart set up for the time of illness. The latter method involves determining the significator of the illness: this, along with the sign occupied by it, may be then used to determine the part of the body afflicted. Reference was usually made to what is called the Coley Table, devised by the astrologer of that name. The significator of illness is said to be determined from examination of the progressions and transits of malefics or malefic aspects. Several methods for calculating the courses of chronic diseases have been preserved from medieval astrology, of which the most influential was the Galenic or Galen's method, a horary method, now long out of use. The planetary and aspectal views of astro-diagnosis may be seen contrasted in Wemyss and Cornell. See also MEDICAL ASTROLOGY.

astrodyne   A term originally derived from the work done by groups of astrologers working within the Church of Light, centred in Los Angeles, relating to the estimation of PLANETARY STRENGTHS in aspects (including the parallels) and angularity, which are accorded a numerical value. In modern astrological literature the astrodyne is often confused with the COSMODYNE, and such usage would suggest that for all intents and purposes the two should be regarded as being synonymous. However, the original difference was that the cosmodyne was a graphic representation, or tabulation, of a particular group of planetary factors which included astrodynes. Doane's work on the subject (published in 1974) relates to the computerization of astrodynes, which demand considerable insight into the significance of orbs, angularity and planetary weighting. The theory of the astrodyne, while in modern times well served by computerization, is in fact rooted in traditional astrological theory which may be traced back to Ptolemy, who rightly insisted on the importance of angles, orbs and planetary identity as factors in chart interpretation.

astro-economics A term related to the study of supposed business cycles, which are not all in fact astrologically derived. A science or art of astro-economics would presumably seek to establish a connection between recognized cycles and astrologically recognized data. For example, sunspot cycles, which are supposed to correlate with a 17-week cycle of stock-prices, as well as other natural phenomena, are often dealt with in astrological literature, yet are not really astrologically attested. A certain degree of accuracy in establishing valid links is evident in the work of Williams, however. Among the cycles observed by Collin are the 9-year stock market cycle (linked with the multiple of the so-called asteroid periodicity), an 18-year cycle affecting house prices (again linked with the asteroids), and a complex 41-month 'prices and industrial product cycle'.

astro-genetics Term probably originated by Deusen, ostensibly to designate a field of research involved with the study of the relationship between genetics and astrology. The statistical findings are based on Sun signs and known careers of natives. As Davison rightly points out, many of Deusen's findings duplicate the work of Krafft.

astro-geomancy See GEOMANCY.

astrognosy Knowledge of the stars and, by extension, of the planets.

astrogony The doctrine pertaining to the generation of the stars: most of the modern esoteric systems of astrology have their own astrogonic approach.

astrography The science (once an art) of describing or mapping the heavens: see CONSTELLATION MAPS.

astro-kinetics A term adopted by Whitman in his extensive study of planetary influences and aspectal relationships. Whitman's work (thorough and informed) has resulted in his postulating that the Ptolemaic planetary rulerships over the signs should be maintained, with the NEW PLANETS (Uranus, Neptune and Pluto) adopted as independent planets, relegated to sub-rulerships over Aquarius, Pisces and Aries respectively. It is clear that Whitman sees the Plutonian rule over Aries as displacing POSITIVE MARS, whereas in fact the displacement (by Pluto) is usually of the NEGATIVE MARS of Scorpio – even so Whitman's general concept would respect the Ptolemaic, and insist that the three extrasaturnians should be regarded as 'sub-rulers' of the related signs – to which Pluto over Scorpio should be added as such a sub-ruler.

astrolabe   This instrument is a small model of the geocentric cosmos, visualized as a stereographic projection. It is usually made in cast brass, and is used as a kind of computer for determining stellar, zodiacal and planetary positions, as well as for calculating heights and directions – in the medieval period it was also used for telling the time. The portable astrolabe consists of a number of articulated plates, secured by a central pin: it is said to have been invented by Hipparchus, in the 2nd century BC, but it was not perfected as an instrument and work of art until the 8th and 12th century by Islamic scientists, during a period when it was of great importance in astrology, as well as for determining the position of Mecca. As the following list indicates, many of the terminologies still used as descriptive of the astrolabe reflect the Arabian origin – the Latin terms are often translations of the earlier Arabic, used by medieval Western astrologers. For example, the basal plate, called in Latin the Mater (mother) is in Arabic Al Umm, which still means 'mother', while the Spider plate, which carries the projection of zodiacal arcs and a certain number of star positions, is called the Rete in Latin, but the Al Ankabout, which still means 'spider' in Arabic. Not all astrolabes are

———— TABLE 8 ————

Latin	Arabic	English
Ostensor	Al Idada	Alidade
Facies	Wajh	Face
Clavus	Qutb (watad)	Rib
Armilla fixa	Kursi	Throne
Equus	Faras	Horse

portable, and several gigantic examples are still preserved in the ancient Islamic observatories, the most impressive being that in Jaipur. The portable stereographic astrolabe is called the Musattah, while the rare spherical model is called the Kurri. The equally rare linear form is called the Khatti.

astrolatry   Worship of the stars as divine beings.

astrologer   In modern times, one who practises a form or forms of ASTRO-LOGY. In earlier times, an observer of the stars.

astrologic   Pertaining to the realm of astrological theory or practice.

astrological age   A name wrongly given to a period usually supposed to be of a 2,160-year duration – but see PRECESSION.

astrological clocks   See ZODIACAL CLOCKS.

astrological frescoes   See ZODIACAL CYCLES.

astrological Tarot   The many different systems of divinatory Tarot usually proclaim a relationship with astrology, especially in regard to the zodiac and constellations. There appears to be little or no agreement as to the astrological connections which may be drawn with the 22 Major Arcana, but the schema proposed by Wirth as set out in table 9 has been widely accepted. The dubious quality of this influential correspondence may be gleaned from the fact that Wirth associates the Devil arcanum with the fixed star CAPELLA, the alpha of Auriga, an association which is seemingly derived only from the image of GOAT – the fact, is however, that in the astrological tradition, this Capella, or Alayodi (of medieval astrology), was regarded as a beneficial influence, in no way demonic. It is no accident that this alpha was linked with Amalthea, after the nurse who raised Jupiter on goat's milk, for this namesake was the mediator between man and god. Such attempts to link the Major Arcana with the signs and constellations is based on little esoteric knowledge.

astrological year   See PRECESSION.

astrological zodiac   One of Blavatsky's terms for the TROPICAL ZODIAC, which she distinguishes from the NATURAL ZODIAC: see, however, INTELLECTUAL ZODIAC, which is a term synonymous with the astrological zodiac.

astrologize   In fact, to examine by means of astrology, though in modern times the word is sometimes used scurrilously.

astrology   The study of the relationship between the macrocosm and the

--------------------------------- TABLE 9 ---------------------------------

Card	Card name	Zodiacal sign	Constellation
1	Juggler	Taurus	Orion
2	Lady Pope		Cassiopeia
3	Empress	Virgo	
4	Emperor		Hercules, Corona Borealis
5	Pope	Aries	
6	Lovers	Sagittarius	Aquila, Antinous
7	Chariot		Great Bear
8	Justice	Libra	
9	Hermit		Bootis
10	Wheel of Fortune	Capricorn	
11	Force	Leo	
12	Hanging Man		Perseus
13	Death		Draco
14	Temperance	Aquarius	
15	The Devil		Auriga
16	House of God	Scorpio	Ophiuchus
17	The Star	Pisces	Andromeda
18	The Moon	Cancer	Canis Major and Minor
19	The Sun	Gemini	
20	Judgement		Cygnus
21	The World	Entire planisphere and Polar Star	
0	The Fool		Cepheus, Little Bear

microcosm, which (in materialistic terms) is often defined as the study of the influence of the celestial bodies on the Earth and its inhabitants. Astrology appears to be one of the most ancient of the surviving occult sciences, and a vast evidence of a highly sophisticated system in Babylonian and Egyptian cultures has survived – though the lore appears to have been well guarded by the ancient Mystery schools. The oral traditions, as well as the development of astrological lore in ancient times, suggest an even older origin than even Babylon and Egypt, however, as recent findings in the realm of ARCHAEO-ASTRONOMY have indicated. There are many different forms of astrology, though in popular terms the most important of these is GENETHLIACAL ASTROLOGY, which deals with the casting and interpreting of horoscopes for individual humans – this does not appear to have been practised (save for kings, as representatives of the state) much before the 4th century BC. Two important forms (which to some extent overlap) are those called ESOTERIC ASTROLOGY and EXOTERIC ASTROLOGY, the former of which tends to deal with the individual incarnation in spiritual terms, to see the world populated by spiritual entities, and which tends to treat astrological doctrine as a sort of philosophical machine. Exoteric astrology (the most widely practised, and usually the only one known to the average practitioner) tends to be involved with personal readings and horoscopic data, and with the predicting of events and experiences in the life of individuals. Attached to the esoteric astrological traditions in the present time are a large number of (often syncretic) teachings concerning the spiritual nature of man, as well as a corpus of traditions dealing with the hidden meanings of the signs, sigils, glyphs and symbols by which the horoscope may be understood as an esoteric diagram. Such esoteric forms are not always confined to the study of the events of a lifetime, but may be directed towards the study of pre-natal, pre-conceptional and post-mortem experiences, as well as to the study of the karmic consequences operative between lifetimes. In the exoteric astrology, the most important forms are ASTRO-METEOROLOGY, HORARY ASTROLOGY, MEDICAL ASTROLOGY, POLITICAL ASTROLOGY and SYNASTRY.

astro-magical   Relating to divination from the stars (and planets).

astro-meteorology   A term used of that branch of astrology concerned with the predicting of weather and telluric conditions, such as earthquakes and volcanic eruptions. An alternative name for this study is meteorological astrology. There are few modern practitioners of this once widely spread art, due to the encroachment of science into a realm once considered proper only to astrological investigation. Astrological weather prediction appears to have been founded upon the study of the solar ingresses into the four Cardinal signs, particularly in connection with conjunctions and eclipses (and in some cases, lunations). Robson records that in some cases daily horoscopes were cast for such predictive purposes, and points to one or two of the rules, especially those touching upon fixed-star lore. Whilst Pearce gives interesting material on the subject, it is significant that the last full treatment which was not fanciful was made by Goad in the 17th century. The interest in earthquakes and volcanic activities (and indeed in the appearance of COMETS) still persists in popular astrological circles. Pearce published some horoscopic data relating to earthquakes, and seems to have predicted some of the more important ones (such as the disaster in San Francisco, in 1906). The modern tendency is to cast INCEPTIONAL CHARTS as a means of studying telluric disasters and freak weather.

astronomical unit A measure based on the Earth's mean distance from the Sun (see SOLAR DISTANCE), which is given as 93 million miles.

astronomical zodiac In his useful account of the early zodiacs, Powell names that zodiac revealed in Ptolemy's star catalogue (which consists of twelve unequal constellations) the 'Astronomical Zodiac'. It seems to have been the prototype of the MODERN ZODIAC, which has been mathematically formulated. See, however, CONSTELLATION MAPS in connection with the IAU ZODIAC. Ptolemy's catalogue actually gives 48 constellations, and whilst only twelve of these are listed as zodiacal, one or two of the other asterisms do actually cross the determinative line of the ecliptic, as with the modern astronomical zodiac. Ptolemy almost certainly inherited the twelve from the BABYLONIAN ZODIAC, or perhaps he intended to relate them to the equal-arc (contemporaneous) GREEK ZODIAC.

astronomien An early term for astrologer (astronomer) – also 'astronomian'.

astronomicon A term usually used in reference to one of the chief early Latin works on astrology – a poem in five books (unfinished) by the poet-astrologer Manilius, written during the period spanning the reigns of Augustus and Tiberius. The text is coloured by Greek mythology, and has indeed become a standard source for popular star-lore legends: the astrological material is said (though on dubious grounds) to be based on the work of Asclepiades of Myrlea. Many of the important doctrines differ from those set out by Ptolemy in his TETRABIBLOS. An interesting deviation is that Manilius establishes a system of DECANS with sign rulerships (rather than the traditional planetary rule), with Aries for the 1st decan of Aries, Taurus for the 2nd, and so on. The house system is an OCTOPODOS (though perhaps based on a misunderstanding – see, for example, EIGHT PLACES).

astro-orientation A term coined to designate the scarcely explored realm of art history and architecture, relating to FOUNDATION CHARTS, and the orientation and design of buildings, as touched upon by Gettings, Nissen and Taylor. But see ASTROSOPHIC GEOGRAPHY.

astropalmistry A clumsy neologism used to designate the modern study of the relationships between astrology and palmistry. It has been taken for granted from very early times that the human hand is a physical model of the zodiac, though most of the early texts on the subject tend to establish links between the planets and parts of the hand, as in the diagram in figure 5. For a modern survey, see PALMISTIC ZODIAC.

ME:	Little finger
SU:	Ring finger
SA:	Middle finger
JU:	Index finger
MA:	Centre of palm
VE:	Root of thumb
MO:	Lower hand mount

Figure 5: Planetary rulership over the hand (after Agrippa).

astro-psychology A term coined (seemingly) by Morrish in connection with his attempt to create a syncretic astrology uniting various streams of psychology, which are essentially Jungian, Gestalt and theosophically derived esoteric psychology, with a surprisingly traditional astrology. An important element in his structure is the (ancient) idea of polarities, which he terms 'conflicts': he notes such conflicts in the SECTA of the planets, and in the opposites, as well as in the conventional aspects. In this syncretism, the 'opposition' between Aries/Libra is the familiar theme of 'War-Peace', which he links with the Freudian 'Love-Destruction' motif. The reconciliation of the conflict is said to come about through the germinating force of Aries becoming 'organic structure' in Libra. Aries represents the 'Individual Potentiality', Libra represents the 'seed-process of Collective Possibilities'. The various systems of astrological correspondences suggested by Morrish often simplify beyond recognition some of the more profound esoteric teachings concerning the relationship between the microcosm and the macrocosm (as for example in the diagram – figure 6 – of the correspondences between the spiritual and material planes, derived from superficial appraisal of theosophic teachings, and graphically presented by other esotericists in a different way). See also ASTRO-SYMBOLISM.

astrosonics A name given to the branch of science (rather than astrology) concerned with the investigation of the supposed relationship between planets and frequencies. The term should not be applied to the investigation of the ratios involved in the harmonies and intervals between planets, for which there is no satisfactory term. See CELESTIAL HARMONICS, MUSES, MUSIC OF THE SPHERES and PLANETARY TONAL INTERVALS.

astrosophia see ASTROSOPHY.

astrosophy A name given to a modern form of astrology, based on the

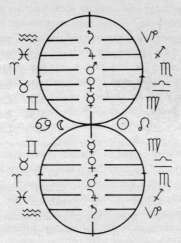

Figure 6: Astro-psychological correspondences with the theosophical model of man and cosmos (after Morrish).

indications given by Steiner, formulated and expounded by such researchers as Vreede, Thun, Sucher and Powell. Whilst essentially syncretic in form, and rooted in traditional methods, it makes frequent use of the SIDEREAL ZODIAC and of HELIOTROPIC CHARTS, as well as sophisticated application of EPOCH CHARTS and DEATH CHARTS. It is rooted in a spiritual cosmoconception, and in a view of history which has given rise to such interesting generalizations as the THREE PROTOTYPES. The connected work of Powell (in relation to the study of ancient zodiacs) has been of great importance: see, for example, SIDEREAL CORRESPONDENCES. The practical application of the sidereal zodiac to bio-dynamic farming is also associated with astrosophy, as for example in the work of Thun: see SIDEREAL MOON RHYTHMS. For all the system of astrosophy is modern in formulation and outlook (though fortunately not beset by the modern penchant for statistics), the term has long been in use in regard to ordinary traditional astrology, and indeed figured in the title of an important (and influential) astrological text, in the Latinized *Astrosophia*, published in 1687: the *Universa Astrosophia Naturalis* of Bonattis. The same term 'astrosophy' has been used by Volguine (*c.* 1946) in a more general sense, almost as a synonym for astrological, as for example in his term ASTROSOPHIC GEOGRAPHY.

astrosophic geography   Term used by Volguine to denote the now defunct practice of siting buildings and cities in accordance with astrological principles, aimed at reflecting the heavenly patterns in the earth's geography. Such practices, relating to geographic patterns and structures, must not be confused with those relating to the FOUNDATION CHART. Volguine quotes the federation of twelve cities of the Etruscans, whose representatives gathered annually at the temple of the god Voltumna. Perhaps more apposite is the claim made by modern astrologers that the positions of churches dedicated to the Virgin Mary on the map of France are located in the same spatial relationship as that figured in the constellational Virgo, the earth thereby reflecting the heavens.

astro-symbolism A term coined by Morrish in connection with his ASTRO-PSYCHOLOGY. The symbolism described is of a highly syncretic type, mingling half-digested symbols from a variety of cultures (such as the ancient Egyptian and the theosophically derived Buddhist). These are not very convincingly married to traditional astrological structures, such as the zodiac, the planets, and so on, within a framework of merged Jungian, Adlerian and Theosophical psychologies of a highly personal level of interpretation. Morrish suggests, for example, that the symbol (presumably he means 'sigil') for Taurus may be taken as a coiled serpent, indicating Fohat, or Kundalini lying 'coiled' or inert and unawakened in the root of the spine. Now, whilst this interpretation is entirely personal, and bears no relation to the astrological tradition (esoteric or exoteric), it is quite erroneous to so confuse Fohat with Kundalini.

astro-theosophy A term derived from the attempt to link certain aspects of European astrology with the Oriental tenets implicit in theosophy, towards the end of the last century. The astrologer Leo may have been the first to popularize the term (1912), but it was given earlier by Blavatsky.

astro-vocational guidance A term coined by Morrish to denote the aspect of chart interpretation relating to vocation and occupation.

astrum This term is usually restricted to alchemy, but is also associated with the ancient astrological doctrine of VIRTUE. The astrum is a secret regenerative principle, almost the ancient equivalent of the Etheric of modern occultism (see AETHER). The astrum contains within itself all created things, and may be abstracted by alchemical processes from these things – it is the quintessence of things, the binding principle of life.

Asvini The 27th of the Hindu NAKSHATRAS (the horseman).

asymmetric orb Within the astrological doctrine of ASPECTS it is maintained that the applying aspect is stronger than the separating aspect, which would indicate that the orb allowed an aspect need not be distributed evenly around the precise aspect itself. The work of Wangemann appears to support the contention that such orbs should be asymmetrically disposed, to allow a wider orb for the applying aspect.

Atalia One of several names given to the dragon who was believed by the ancients to swallow the luminaries, and thus cause what we now call eclipses. Beck sees an image of a dragon in the Ponza zodiac (see MITHRAIC ASTROLOGY) as this Atalia, and observes that the Assyrian word for eclipse was *atalu*.

athazer The Moon is said to be athazer when it is in precise conjunction with the Sun, or when it bears one of a number of precise Angular relationships to the Sun (some of which are actually *ASPECTS). These Angles, for which no orb is permitted, are 12, 45, 90, 150, 168, 180, 192, 215, 270 and 348 degrees. It is evident from these figures that the athazer is to be measured in arc along the zodiac (the last four being something of a puzzle, in that they represent 168, 135, 90 and 12 degrees, respectively – could there be an error in the dissident?).

Atlantic Sisters See PLEIADES.

Atlas See PLEIADES.

Atlas Céleste The name sometimes given to the MODERN ZODIAC of Delporte – but see also CONSTELLATION MAPS.

audacity Name given to one of the SEVEN LOTS of Greek astrology, the Lot of Toma (Daring). The degree is the same distance from the Ascendant as Mars is from the PART OF FORTUNE.

August The eighth month of the GREGORIAN CALENDAR named after minor calendrical reforms by Augustus Caesar, who is said to have refused a month shorter than that named in honour of Julius (see JULY), and therefore extended this ancient Sextilis month by one day, and gave it his honorific name. The emperor and the month have, however, a pertinent communal etymology, as the Latin *augustus* originally had a religious meaning – sacred and worthy of honour, linked with the idea of augury.

Auphanim One of the variant names for the OPHANIM.

Aurea Catena Homeri   See CHAIN OF BEING.

Auriga   Constellation approximately 10 degrees Gemini to 3 degrees Cancer, from 30 to 60 degrees north of the equator. The image for this asterism is a charioteer without his chariot, but according to Greek legend it represents Erichtonius, who first mounted a four-horse chariot. Almost certainly it was derived from Babylonian astrology, and a large number of names have been attached to the image, linking with the idea of a wagoner. Within the constellation is found an image of a goat and a kid, variously explained, but of interest to astrologers because of the emphasis placed on the relevant fixed star CAPELLA (little goat).

Ausiel   According to certain grimoires, the name given to the Governor of zodiacal Aquarius. The name is not to be confused with Auriel, which is an alternative for URIEL. See ZODIACAL SPIRITS.

austral signs   The so-called southern signs, which are those from Libra to Pisces, inclusive.

autozoon   A name derived from gnostic and Neoplatonic sources for the image of the *TETRAMORPH. The etymology of this Greek term is linked with that of the zodiac – but see Smith.

autumnal equinox   The time of year, around 22 September, when the Sun reaches the 1st degree of zodiacal Libra, and thus crosses the COLURE – see EQUINOX and EQUINOCTIAL POINT.

autumnal quadrant   See QUADRANTS.

autumnal signs   These are Libra, Scorpio and Sagittarius.

Avis   See CYGNUS.

Awwa   The 11th of the Arabic MANZILS, from Al Awwa (the barker).

ayanamsa   A term derived from the Sanskrit (itself meaning something like 'precession') to denote the difference in degrees for any given point in time between the fiducial of the tropical zodiac and those of the various sidereal zodiacs. Since there is no basic agreement as to a fiducial, or even to when the zodiacs were actually contiguous, there is no standard arc of ayanamsa. In modern times, the ayanamsa of Markandyar gives a difference at the beginning of our century of 22 degrees, 26 minutes and 14 seconds, while the Fagan-Bradley ayanamsa (used by many Western siderealists) gives one degree in excess of this. The astrologer Sucher appears to suggest an ayanamsa of 8 degrees for the middle of this century. The result of these (and many other figures) is that the study of the ayanamsa is one of the most perplexing realms for any astrologer who finds himself occupied with the problem of converting the sidereal to the tropical for any given time. It would seem that the vexing question of ayanamsa was no more simple in ancient times than it is now, for at least five different systems have been recorded. The Babylonian astrologers knew of two systems, one of which corresponded to a solar longitude of Aries 10 degrees, and one of which corresponded to Aries 8 degrees: this latter system is found in astrology well into the medieval period. In his study of Egyptian astrology, Fagan gives the ayanamsa of 786 BC in Aries 13.8 degrees: see, however, PRECESSION. Dean, following the general consensus of historical opinions, notes it significant that the zero ayanamsa coincided with the period during which the earliest astrological records were made, and suggests that the sidereal zodiac may be 'an artefact of civilization rather than a genuine entity'. It would seem likely that the two systems – tropical and sidereal – were used contemporaneously for two separate purposes. The fact is that the GREEK ZODIAC and the BABYLONIAN ZODIAC, whilst relating to different co-ordinates, required only minute adjustment even in the days of Ptolemy to bring them into correspondence: an addition of the ayanamsa of 2 degrees and 20 minutes permits the stellar longitude of Ptolemy's star catalogue to be almost precisely related to the earlier Babylonian catalogue. However, the historian Neugebauer points out that due to an error in determination of the Vernal Point, the epoch for Ptolemy's catalogue should be AD 48, and not AD 138, as is often believed. This would imply that in terms of a precessional rate of 1 degree in 72 years, the Babylonian stellar zodiac and the Greek tropical zodiac coincided in AD 216. The German Kugler, working from Babylonian tablets of lunar computations (based on the tables of the Babylonian Kidinnu) showed that the longitude of the Vernal Point was in Aries 8 degrees. The historian Weidner located the point from later astronomical computations tables (according to the Babylonian Naburiannu) as 10 degrees. By a systematic comparison of the two systems, Schnabel computed the positions for the equivalent equinoctial points, and obtained for the Naburiannu tables the epoch 508 BC and for Kidinnu 379 BC. These figures have since been adjusted to 500 BC and 373 BC. See HYPSOMATIC AYANAMSA, SIDEREAL CORRESPONDENCES and SYSTEM A.

Ayar See BABYLONIAN CALENDAR.

Ayil   According to certain grimoires, this is the name given to the Governor of zodiacal Sagittarius – but see also SIZAJASEL.

Azimech   See ARCTURUS.

azimene degrees   A term derived from Arabian astrology, now rarely used, and applied to certain degrees or arcs in the zodiac which are regarded as being debilitating to the native when on the Ascendant, or otherwise brought into prominence in a chart. The azimene degrees are sometimes called the deficient degrees, the weak degrees or lame degrees, and may have been derived from consideration of the influence of powerful (malefic) fixed stars, now precessed. Figure 7 sets out the individual degrees and arcs accorded each sign.

azimuth   A term used to denote the Angle (with apex at zenith) formed by a star's vertical circle with the meridian.

Figure 7: The azimene degrees and arcs.

Azure Dragon

Azoth   Name applied to Mercury by the alchemists, but used rarely by astrologers – for an exception, see Cornell.

Azure Dragon   Name of the so-called 'constellation' of the Chinese Heavens, which actually consists of an arc of seven SIEU, from the asterisms 'Kio' to 'Ki', as indicated in table 48. Properly speaking the Azure Dragon is a quadrant or arc. See also BLACK WARRIOR, RED BIRD and WHITE TIGER.

B

Babylonian calendar   The data for converting the ancient Babylonian calendar of 786–5 BC (the so-called HYPSOMATIC YEAR) to the modern Gregorian system, is provided by Fagan in table 10 below. The Babylonian equivalents are given for the first day of the ancient NEOMENIA.

TABLE 10

Babylonian term		Gregorian date
Nisannu	1st Nisan	27 March
Ariu	1st Ayar	26 April
Simanu	1st Sivan	25 May
Du'uzu	1st Tammuz	24 June
Abu	1st Ab	23 July
Ululu	1st Elul	22 August
Teshritu	1st Teshrit	21 September
Arakhasamma	1st Arahsamma	20 October
Kislimu	1st Kislev	19 November
Tebitu	1st Tebit	18 December
Shabatu	1st Shebat	16 January
Addaru	1st Adar	15 February

Babylonian zodiac   The earliest formulated Babylonian zodiac appears to be an 18-asterism lunar zodiac, which is peripherally linked with the MUL-APIN TABLES. The first recorded use in Babylonian astrology of a system of twelve fixed-star zodiacal asterisms has been dated to the first half of the 5th century. In this system, the zodiacal belt is divided into twelve equal sectors, the signs themselves measured in terms of the fixed stars of the zodiacal belt – essentially in terms of the NORMAL STARS. The historian Powell has suggested that the transition from the system of normal stars to the Babylonian equal-sign zodiac was by means of the axial fiducial of the fixed stars ALDEBARAN and ANTARES, which divide the zodiac in half, and therefore provide an excellent framework for a system of arc measurements. It would seem that the 30-degree sign was derived from analogy with the twelve month schematic year of the Babylonians (see MUL-APIN TABLES). The later GREEK ZODIAC, formally influenced by this fixed-star Babylonian zodiac, was based on a division of the ecliptic 'line', whilst the Babylonian zodiacal belt was about 16-degree wide – literally a pathway through the asterisms.

Bacchus   Term applied by the astrologer Thierens in his esoteric cosmo-conception to a *HYPOTHETICAL PLANET, with the (inapposite) alternative name of Horus. The planet is related to the future evolution of our Cosmos, and is said to be invisibly present in the *ETHERIC SPHERE.

background   A specialist term used of planets near to the cusps of the cadent houses, which are by virtue of this placing regarded as weak or ineffectual, and are said to be in the 'background' of the chart. See also FOREGROUND and MIDDLEGROUND.

Bailey-Pluto   See ESOTERIC PLUTO.

Balance   Popular name for the zodiacal sign and constellation LIBRA.

Baldah   Al Baldah (the city or district), the 19th of the Arabic MANZILS.

Baraquel   One of the Grigori (angels of Jewish legend – the term being derived from the word Egregori, meaning 'the watchers'), who instructed men in the art of astrology, according to their own higher wisdom.

barren node   An alternative term for the DRAGON'S TAIL.

barren planets   According to Cornell, the barren planets are the Sun, Mars, Saturn and Uranus, which tend to 'deny or kill' children when located in the 5th or 11th houses of parents – an idea which finds little support in traditional astrology. The Dragon's Tail is classed also as a 'barren' influence, and hence may be described as a 'barren node', being regarded as an influence equivalent to that of Saturn. Cornell's view of barren planets is highly specialized, though obviously based on a profound study of available literature.

barren signs   These are Gemini, Leo and Virgo, but it is maintained that the native may be said to be barren only if one of these is upon the Ascendant, or (as some astrologers insist) on the cusp of a house of the same elemental nature. In fact, the traditional term is something of a misnomer, for these positions do not confer barrenness, so much as limit the number of children.

Baru   Babylonian name for an omen-reader – see OMNIA.

Barzabel   In medieval occultist literature, Barzabel is the name of the ruling Daemon of the planet Mars. Agrippa accorded him the number 325, the same as was linked with Graphiel.

Basilica Stella   See REGULUS.

Basiliscus   See REGULUS.

Basilisk   See SCORPIUS.

Baten Kaitos   Fixed star set in the body of the WHALE – the term being derived from the Arabic Al Batn al Kaitos (the belly of the whale). It is regarded as an evil nature, associated with shipwrecks.

Batn al Hut   Term from the Arabic Al Batn al Hut (the belly of the fish), the name given to the 26th of the MANZILS, now located by the beta of Andromeda.

Bau   See ARCTURUS.

beastly   An alternative term for BESTIAL or FERAL. In his highly personal approach to medical astrology, Cornell comes to the conclusion that a 'beastly form' may develop if at conception (presumably he refers to a pre-natal chart), the Sun and Moon are in the 6th or 12th houses, in one or other of the FOUR-FOOTED SIGNS.

Beehive   See PRAESAEPE.

Beelzebub   See MUSCA BOREALIS.

Befor   One of the occult names given to the spirit of Jupiter.

beholding   Those pairs of signs, which are of equal distance from the tropics, were said to 'behold' each other, and are sometimes called the beholding signs. These pairs are Aries-Libra, Taurus-Virgo, Gemini-Leo, Aquarius-Sagittarius and Pisces-Scorpio. Naturally, the Sun in any of these pairs gives equal ratios of day to night. Ptolemy explains the term as being connected with the fact that the pairs rise from the same part of the horizon, and set in the same parts (in a geographic sense, of course). The beholding signs are sometimes called the 'signs of equal power', though the terminology is now largely archaic.

Bellatrix The fixed star, gamma of Orion, set in the left shoulder of the constellation. The name, which means 'female warrior', expresses its war-like nature, for it is said by Ptolemy to be of the influence of Mars with Mercury – not altogether beneficial in a chart.

benefic aspect A number of ASPECTS which promote a general well-being between planets are termed 'benefic' – the trine and sextile being the most notable examples. However, the entire nature of a particular aspect must be taken into account if the term is to be used with any precision, since some planets, whilst fulfilling the requirements of angular geometry (such as a precise trine, for example) may themselves be of an intrinsic nature which does not promote a beneficent influence. Benefic aspects create harmonies which may, under certain circumstances, lead to a dispersion of energies (towards a general laziness, for example), and this throws doubt upon the applicability of this ancient term.

benefics The term benefic, whilst sometimes used of aspects (see BENEFIC ASPECT) is also used of planets and nodal points. Whilst Venus and Jupiter are the lesser benefic and the greater benefic respectively, it is often said that the Sun, Moon and the Dragon's Head are also benefics, when well aspected.

Bentenash One of the seven major fixed stars forming the constellation URSA MAJOR (the eta of the asterism), said by Ptolemy to be of the influence equivalent to that of Mars. Sometimes called Benetnash.

besieged When a planet is hemmed in by two malefic planets, it is said to be besieged by them, and is regarded as being in an unfortunate position. In the medieval terminology, a planet so besieged by two benefics was usually regarded as being fortunate in influence. In fact, the theory underlying the tradition is merely an extension of that theory of aspects relating to the CONJUNCTION, since the planet besieged is induced into a state of double conjunction by forming a link between two planets which would otherwise not necessarily be in orb, and which then mingle their influences.

Bestia See LUPUS.

bestial   Originally, the bestial signs were those which had as zodiacal image a whole animal. Thus Aries as Ram, Taurus as Bull, Leo as Lion and Scorpio as Scorpion, were bestial signs. Due to linguistic change, however, the term now carries a somewhat unfortunate connotation. There have also been significant changes in the imagery of the signs: the original Capricorn (once the goat-fish) has changed, so that it is now often regarded as being entirely bestial, in its guise of Goat. Sagittarius, which was properly half-human, half-horse, was usually termed a bestial sign (though see BICORPOREAL SIGNS). The fishes of Pisces were also regarded as beasts, indicating that it is likely the classification was adopted to distinguish the signs into the pairs of bestial and HUMAN SIGNS.

Betelgeuze   Fixed star, the alpha of Orion, set in the right shoulder of the giant. The name is derived from the Arabic Ibt al Jauzah (armpit of the Giant). Ptolemy gives it the nature equivalent to Mars conjunct Mercury: it is said to confer martial honour and wealth.

Bhadzapada   The Sanskrit name for constellation Pisces – see MEENAM.

Bharani   The 28th of the Hindu NAKSHATRAS (the bearer).

biblical astrology   There have been very many attempts to relate astrology to certain passages within the Bible. Examples may be found in Beckh (in connection specifically with the Gospel of St Mark) and in many passing references in Butler, Heidenreich, Libra, Steiner, Wemyss, and so on. There is scarcely a system of astrology which has not attempted to establish its own list of biblical (or even theological) correspondences. The available literature may be observed in the Rosicrucian tradition (for example, in Boehme) or in the specialist Christological assumptions developed by Dee in connection with his magical sigil of the MONAS. On a less exalted level, the present author has contributed somewhat to this extensive astro-biblical literature in his examination of esoteric traditions in *The Hidden Art*. In some astrological systems, associations have been drawn between the twelve signs of the zodiac and the twelve Apostles, the twelve Tribes of Israel and the twelve stones on the breastplate of the high priest, while the Angels of the Seven Churches are often interpreted in planetary terms, but there are too many variations in such lists to draw valid conclusions from them. However, certain works of art of the medieval period do evince physiognomies in the portraits of disciples which are rooted in such astrological lore. We find that the association of Judas with Scorpio is the most consistent. The Hierarchies of the pseudo-Dionysius are obviously astro-theological at base, as are the

Seven Days (more properly, the seven epochs) of Creation, which are linked with the esoteric cosmogenesis adequately set out in such modern occult sources as Blavatsky and (specifically in astrological contexts) Heindel. See therefore BIRTH STONES, CELESTIAL HIERARCHIES, TRIBES OF ISRAEL and SECUNDADEIAN BEINGS. The specific symbolism and correspondences established in theological and astrological literature between signs, planets and biblical references is far too complex to be examined here, but it is sufficient to note the ancient connection between Christ and Pisces (and more especially in popular astrological literature with the Age of Pisces), due to the communality of the image of the fish, which so intrigued Augustine, and which was even a part of the pre-Christian ethos relating to the coming Messiah. In this category falls the association established between the Virgin Mary and the celestial Virgo, that gave rise to a crop of symbolism which survived (though largely misunderstood) well into the 16th century. The speculations regarding the STAR OF BETHLEHEM and the Magi, those Zarathustrian initiates who are so often wrongly called 'astrologers' in popular lore, have also played a prominent part in astrological literature. There appears to be no adequate and comprehensive study of this subject of biblical astrology, though passing references in Blavatsky, Testa, Steiner and the remarkable Massey, for all their rejection by the academic world, are worth serious attention. In spite of the many exegetical essays relating to the astrological view of the Bible, it is sobering to note the short (though probably comprehensive) index to the biblical astronomical references given by Allen (R.H.) – 18 only are culled from the Old Testament, and three from the New. None of these is specifically astrological, and one or two of the 'traditional' exegetical quotations merely arise from a misunderstanding in the English translation: for example, the Genesis 1:14 'let them be for signs, and for seasons, and for days and years', is by no means a reference to the zodiacal signs, as is often fondly imagined by some astrologers. The link drawn in esotericism between the Secret Wisdom of the Bible and the esoteric astrological lore is however reliable and affords a rich field for speculative research. See also COSMIC PANTHEONS.

bicorporeal signs   The bicorporeal signs were those with images consisting of two separate forms – Gemini (as two humans), Sagittarius (originally a human and horse), and Pisces (two fishes). Ptolemy included in his list of bicorporeals the sign Virgo, an inclusion which may be explained either from an esoteric point of view (since what was of prime importance was the corn or child held in the hands of the celestial virgin) or from a point of view of exoteric theory, since the bicorporeals of Ptolemy appear to have been the equivalent of the modern MUTABLE SIGNS. In fact, Ptolemy called these the

Disoma, and explained the term as relating to the two kinds of weather which the group shared: the medieval terminology, from which the modern term is derived, was certainly limited to the consideration of the images. Ptolemy says the bicorporeals make souls complex, changeable, hard to apprehend, fond of music, and both amorous and versatile.

Big Dipper   One of the several names for the constellation URSA MAJOR.

biological correspondences   In recent astrological development, the traditional image of melothesic man formulated in Greek astrology, has been changed to adapt to the modern (popularist) image of man. A good example of such a modern system, which incorporates not only a new series of connections, but also the new planets, is that given by Ebertin in connection with cosmobiology set out in table 11. To the planets, Ebertin adds the

---

TABLE 11

SU	Cell, body, heart and circulatory system.
MO	Bodily fluids, blood, lymph, the soul.
ME	The nervous system, youth.
VE	Glands and hormonal system, growth and maturity.
MA	Muscles, middle age, prime of life.
JU	Organic system (lungs, liver, etc.), the climacteric.
SA	Bones and old age.
UR	Rhythmic system (pulse, breathing, peristaltic).
NE	Weakness and slackness, paralysis and (possibly) excretory.
PL	Change, transformation and regeneration.

---

following nodal influences: Dragon's Head relates to hereditary factors, the astral body and the autonomic system; the Midheaven relates to ego-consciousness, reason as cerebral function, whilst the Ascendant relates to environment. In terms of glandular systems, the esotericist Collin, working mainly within the tradition established by Ouspensky, gives the melothesia in table 12. See also MELOTHESIC MAN.

bi-quintile   An aspect of 144 degrees, arising from a division of the zodiac into a two-fifth arc. See QUINTILE.

birth stones

TABLE 12

SU	Thymus glands	JU	Posterior pituitary
MO	Pancreatic system	SA	Anterior pituitary
ME	Thyroid	UR	Sex glands
VE	Parathyroid	NE	Pineal gland
MA	Solar plexus		

birth   The human horoscope, around which the whole edifice of genethliacal astrological theory has been constructed, is usually a symbolic representation of the moment of birth. Inevitably, there is much argument as to precisely what is meant by 'birth', as several stages and processes are involved, each giving significantly different times. In fact, few serious astrologers will use a chart which has not been subjected to careful RECTIFICATION, so that argument as to precise time of birth is often irrelevant. Rectification aside, however, most modern astrologers would time a birth (and cast an appropriate chart) from either the severing of the umbilical cord, or from the first cry.

birthplace houses   English name applied to the GOH SYSTEM devised by Walter Koch: see KOCHIAN SYSTEM.

birthplace system   English name applied to the method of house division proposed by Walter Koch – see KOCHIAN SYSTEM.

birth star   Sometimes defined as the most powerful fixed star conjunct a planet or angle in a natal chart, or (failing such a conjunction) a star in close orb with such a planet or Angle. Sometimes also defined as the fixed star (of astrological significance) nearest to the Ascendant: this must be contrasted with the insistence by many astrologers that no orb should be allowed for a fixed-star conjunction – in this connection, see MORINEAN ORBS. In popular journalistic astrology, the birth star is sometimes merely the Sun-sign, sometimes the planetary ruler of that sign. See also PERSONAL STAR.

birth stones   Traditional astrological doctrines associate a number of precious stones with each of the twelve signs of the zodiac: most of these associations are derived from TALISMANIC MAGIC, however. The stones were said to transmit a specific VIRTUE (in which connection, see GAMALAI), when used

birth stones

to make Seals, or even when worn unsealed as magnetic centres, to attract the corresponding virtues of the stars. An extensive literature, derived mainly from the Arabian-inspired medieval lore, presents numerous correspondences, and interesting related documents have been published by Evans. The most reliable lists are those derived from the SCALA and CALENDARIA MAGICA of the late medieval tradition, but in modern times (as the representative collections in table 13 indicate) many stones are attributed to several different signs, and there

---
TABLE 13

AR	Amethyst, diamond, garnet, bloodstone, jasper, malachite.
TA	Alabaster, emerald, coral, jade, lapis lazuli, sapphire.
GE	Aquamarine, agate, beryl, chalcedony, chrysolite, crystal, marble, topaz.
CN	Emerald, black onyx, selenite, pearl, crystal.
LE	Carbuncle, diamond, hyacinth, ruby, sardonyx.
VG	Agate, aquamarine, cornelian, hyacinth, marble, pink jasper.
LB	Alabaster, beryl, crysolite, coral, cornelian, diamond, jade, lapis lazuli, opal.
SC	Beryl, bloodstone, flint, jasper, magnet, malachite, topaz.
SG	Amethyst, carbuncle, hyacinth, moonstone, turquoise.
CP	Chrysoprase, jet, moonstone, sapphire, white onyx.
AQ	Chalcedony, hyacinth, opal, sapphire, slate.
PI	Amethyst, chrysolite, moonstone, topaz, hyacinth.

---

appears to be little order or sense in some of the correspondences. A medieval tradition which is a little more certain is that recorded by Agrippa in his Scala Duodenarii, which gives only one jewel (*lapis*) to each of the signs, as in table 14. Kircher gives the interesting correspondences

---
TABLE 14

AR	Sardonyx	TA	Sard	GE	Topaz
CN	Chalcedony	LE	Jasper	VG	Smaragdine
LB	Beryl	SC	Amethyst	SG	Hyacinth
CP	Chrysoprase	AQ	Crystal	PI	Sapphire

---------------------------- TABLE 15 ----------------------------

Sign	Gem	Creature	Colour
AR	Amethyst	Sheep	Red
TA	Jacinth	Cattle	Dark
GE	Chrysoprase	Apes	Yellow
CN	Topaz	Water creatures	Blue
LE	Beryl	Forest creatures	Golden
VG	Chrysolite	Dogs	Green
LB	Sard	Birds	Purple
SC	Sardonyx	Birds of prey	Flame
SG	Emerald	Military beasts	Deep blue
AQ	Sapphire	Marine creatures	Ash
PI	Jasper	River beasts	

shown in table 15. The modern esoteric astrologer Thierens provides quite a different list, which he appears to have from Pavitt: (see table 16). Similar

---------------------------- TABLE 16 ----------------------------

AR	Jasper	TA	Sapphire	GE	Chalcedony
CN	Emerald	LE	Sardonyx	VG	Cornelian
LB	Chrysolite	SC	Beryl	SG	Topaz
CP	Chrysoprase	AQ	Hyacinth	PI	Amethyst

lists, with many varieties of correspondences, could be provided, but the above is sufficient to show that there is really no traditional system of birth-stone associations upon which one might rely. Two other lists of magical jewels, once pertaining to a separate tradition, have been incorporated into the general lists relating to talismanic magic. The first is that tradition derived from the attempt to link a magical jewel with the twelve stones set in the breast-plate of the Jewish high priest. The three lists in table 17 below represent the better type of such correspondences: the two tables derived from Wemyss are a result of an attempt to equate the priest stones with the Foundation Stones of the New Jerusalem.

--- TABLE 17 ---

Sign	Thierens	Wemyss	Wemyss
AR	Jasper	Red Jasper	Sard
TA	Sapphire	Topaz	Topaz
GE	Chalcedony	Carbuncle	Chrysolite
CN	Emerald	Emerald	Emerald
LE	Sardonyx	Lapis Lazuli	Sapphire
VG	Cornelian	Amethyst	Amethyst
LB	Chrysolite	Carbuncle	Sardonyx
SC	Beryl	Agate	Jacinth
SG	Topaz	Yellow Agate	Chrysoprasos
CP	Chrysoprase	Beryl	Beryl
AQ	Hyacinth	Chalcedony	Chalcedony
PI	Amethyst	Jasper	Jasper

Unfortunately, many other variant lists might be given from related sources. Another tradition, related to zodiacal associations, is that linked with FIXED STARS (a subject with which Evans has dealt in a most scholarly manner), though it must be remarked that even in medieval talismanic magic, these

--- TABLE 18 ---

Ala Corvi (Algorab) – Onichus (onyx)
Alayhoch (Alphecca) – Saphirus (sapphire)
Aldebaran – Carbuncle or ruby
Pleiades – Cristellus (crystal)
Algol – Diamas (diamond)
Arcturus – Iaspis (jasper)
Cauda Capricorni (Deneb Algedi) – Calcedonius (chalcedony)
Cauda Urse Majoris (Polaris) – Magnes (see below)
Cor Leonis (Regulus) – Gergonsa or Granatus (garnet)
Cor Scorpionis (Antares) – Sardinus and Amatisto (sardonyx/amethyst)
Lucida Corone Scorpionis (Capella) – Topasius (topaz)
Spica – Smaragdus
Vulture Cadens (Wega) – Crisolitus (chrysolite)

fixed stars were often associated with different jewels, a representative list given by Evans is from a 15th-century manuscript, which provides also corresponding sigils (*carectus*) in Latin. The *magnes* associated with Deneb Algedi is usually translated as 'lodestone' or 'magnet stone', but the term has several different meanings in the esoteric tradition. A similar series of (varied) associations to those above, though between the planets and the range of magical stones, is too varied to be recorded here.

bi-septile An aspect of 102.8 degrees, arising from the multiple of the SEPTILE by two, and used especially in the theory of *HARMONICS.

bitter sign An archaic term, applied to each of the fire triplicities – Aries, Leo and Sagittarius – which were hot, fiery and bitter.

Black Warrior One of the so-called 'constellations' of the Chinese Heavens, which actually consisted of an arc of seven Sieu, from 'Tow' to 'Peih', as indicated in table 48 under LUNAR MANSIONS. There are three other such arcs, the AZURE DRAGON, the RED BIRD and the WHITE TIGER.

blind trial Within an astrological context, a term used to denote a test designed to establish whether an astrologer may correctly interpret a chart cast for a person unknown to him. In order to ensure that there is no unconscious bias, the blind trial is often conducted by means of a third party (in the so-called 'double blind trial'), who prepares the astrological test material on behalf of the investigator.

blossom-light trigon See SIDEREAL MOON RHYTHMS.

Blunsdon chart A term sometimes used for the stereographic projection of the *THREE-DIMENSIONAL CHART proposed by the astrologer Blunsdon – see figure 17 under CHART SYSTEMS.

Bonatis directions Name given to a system of secondary directions associated with the 13th-century Italian astrologer Bonatus, and concerned with the use of a daily solar chart, cast for the return of the Sun to its own radical position, each daily chart being related to the sequence of years. See also HIEROZ-BONATIS DIRECTIONS.

Bootis Constellation approximately from 27 degrees Virgo to 7 degrees Scorpio, from 10 to 55 degrees north of the equator. Several different mythical accounts of the origin of the constellation have survived. Bootis is identified with Icarius, the Athenian who taught the use of grape to make alcohol, and with Arcas, who almost killed his mother, transformed as she was into the shape of a bear. However, the name was used in the *Odyssey*, and the constellation is associated with the idea of a Herdsman. Very many alternative names exist, including Bear-Watcher, Wagoner, Arctophylax, Atlas and (because of its nearness to the north) Septentrio. It is said to give a fondness for rural work, as well as earthly desire. The important stars in the astrological lists are ARCTURUS, PRINCEPS and SEGINUS.

boreal signs A term derived from the Greek personification (Boreus) of the North Wind, and used of the six 'northern signs', Aries to Virgo.

bowl One of the *JONES PATTERNS, in which the planets are gathered into half of the chart figure (within an arc of 180 degrees): it is said to denote a high degree of equanimity and self-containment.

Brachium Meaning 'forearm', the Latin name given to the 7th of the LUNAR MANSIONS in some medieval lists. Probably located by the fixed stars CASTOR and POLLUX.

Brhajjataka The title of a Sanskrit work on popular astrology, written in the mid-6th century by Varahamihira. This remains the basic textbook for the popular interpretation of horoscopes in India. The historian Pingree says that such horoscopes are cast for perhaps 90 per cent of the Indian population, and depend largely upon this text for interpretation. See NADIGRANTHAMS.

broken When a planet comes to the Descendant (by transit or progression), it is said to be broken, and its power wanes for the entire semi-arc.

broken signs See IMPERFECT SIGNS.

Bronze Age A term derived from Hindu chronology and modern esoteric lore, and relating to an epoch, see DWAPARA YUGA and FOUR AGES.

Brothers and Sisters The Point of Brothers and Sisters, not to be confused with 3rd house (sometimes called the House of Brothers and Sisters) is one of the so-called Arabian PARS in a chart. It is determined in a natal chart by taking the degree of Saturn to the Ascendant point, and then adjusting the position of Jupiter to give the Point. The connection between Saturn and fraternity appears to have stemmed from Ptolemy, who associated the sign on the Midheaven (linked with Capricorn, ruled by Saturn) with Brethren.

brutish The term brutish signs is traditionally (though inexplicably) applied to Leo, and to the last DECAN (sometimes the 'last half') of Sagittarius. Like so many similar archaic terms, the word appears to be derived from consideration of the image of the corresponding sign: those born under these two being said to be brutish, savage and inhuman. Cornell refines the tradition somewhat, however, by claiming that the brutish signs on the Ascendant, or the luminaries in either of them, and with malefics in the Angles, 'render the native fierce, cruel and brutish'. See BESTIAL.

bucket Name given to one of the *JONES PATTERNS, in which a single planet is found approximately opposite all the other planets, which are themselves gathered within an arc of 180 degrees or less. This isolated planet (often called the 'singleton') is regarded as a sort of handle, and is said to bestow the capacity for special activity, in terms of its own nature.

Budham The Sanskrit word for Mercury, as used in Hindu astrology.

Bull A name given to both zodiacal and constellational Taurus.

bundle One of the *JONES PATTERNS, in which the planets are grouped into a third of the horoscope circle (that is, into an arc of 120 degrees). Such a figuration is said to denote a limited response to external stimuli, and a greater dependence on self.

Bungula A name often wrongly applied to the fixed star, alpha of Centaurus. See RIGIL KENTAURUS.

Butain Al Butain (the belly), the 28th of the Arabian MANZILS.

C

cabbalistic astrology See QABBALISTIC ASTROLOGY.

Cacadaemon See PERSEUS.

cacodemon A term once applied to the TWELFTH HOUSE, derived from a Greek compound meaning 'evil influence'. See also DAIMON.

cadent The houses of the horoscope which 'fall away' from the ANGLES are called cadent, from the Latin verb *cadere* (to fall): these are the 3rd, 6th, 9th and 12th houses. Planets in such houses are called cadent planets, and it is believed that the position marks a diminution in their influence.

caduceus The symbolic staff or wand of the god Mercury (Hermes), and sometimes the SIGIL for the planet Mercury: ☿ In the ancient Mysteries this appears to have been made from a three-headed snake, but in relatively early times it was changed to a single rod intertwined by two serpents, linked esoterically with the human spinal column. In a specialist sense, the word is used of an Arabian PARS, the part of Mercury: this point is determined by revolving the solar degree to the Ascendant – the position of Mercury then marks the caduceus within the houses, a point which relates to commerce and business dealings. In his study of the MONAS, Dee formulated a specialist theory relating a caduceus derivation to sigillic forms.

Cael In certain methods of ceremonial magic, this name is given to the Governor of zodiacal Cancer.

Caelum Constellation, approximately from 12 degrees Taurus to 15 degrees Gemini, and from 35 to 50 degrees north of the equator. Formed only in 1752, the name means 'graving tool' – it is said to give artistic taste.

Cagaster A term (probably) originated by Paracelsus, from the Greek *kako* (evil) and *astron* (star). It is applied to that spiritual force in matter – derived from the stellar realm – which strives towards the destruction of form. It works against the ILIASTER.

calendaria magica

calendar A system of co-ordinating and recording Earth time with the apparent and real cycles of cosmic bodies. See BABYLONIAN CALENDAR, CHINESE CALENDAR, EGYPTIAN CALENDAR, EUCTEMONIAN CALENDAR, GREGORIAN CALENDAR, JULIAN CALENDAR and ROMAN CALENDAR. See also ZODIACAL CALENDARS.

calendaria magica The name used in medieval occultism and astrology for a series of tables setting out the correspondences, numerological connections and sigillic forms between zodiacal signs, planets, elements, celestial hierarchies and so on. The best of the surviving calendaria is that attributed to Trithemius, which has influenced the associations passed into astrology by the electicism of Agrippa. The example in table 19 is from a 15th-century *Calendarium Naturale Magicum* reproduced by Nowotny. The calendaria are usually replete with magical seals, sigils, symbols, and the like, and constitute one of the most fascinating realms of research into astrological and occult symbolism. See also SCALA.

TABLE 19

One The One Principle of all things: God. The 'Archetype' itself.

Two The polarities of Masculine (*agens*) – Feminine (*patiens*)
Sun – Moon
Feeling (*cor*) – Thinking (*cerebrum*)

Three The Trinity of the Godhead, and the trinity in all things. The three ranks of Hierarchies (*suprema*, *media* and *infirma*).

Hierarchies	Triune man	Principles	Elements
Suprema	Thinking (*intellectus*)	Salt	Fire
Media	Feeling (*sensitiva*)	Mercury	Air
Infirma	Body (*vegetabilia*)	Sulphur	Water

Four

ELEMENTS	Fire	Air	Water	Earth
ANGELS	Michael	Raphael	Gabriel	Uriel
CARDINALS	Oriens	Occidens	Septentrio	Meridies
SOUL POWERS	Intellectus	Ratio	Phantasi	Sensus
VIRTUES	Prudence	Justice	Temperance	Fortitude
HUMOURS	Cholera	Sanguis	Pituita*	Melancholia
QUALITIES	Hot	Humid	Cold	Dry

* The medieval term for Phlegma.

Five

Water	Air	Fire	Earth	'Mixtum' (QUINTESSENCE)
Saturn	Jupiter	Mars	Venus	Mercury

Six	Seraphim	Cherubim	Thrones	Dominions	Powers	Virtutes
	Saturn	Jupiter	Mars	Venus	Mercury	Moon

(This is an entirely fanciful set of associations – see table 3.)

Seven	Saturn	Jupiter	Mars	Sun	Venus	Mercury	Moon
	Aratron	Befor	Phaleg	Och	Hagith	Ophiel	Phul
	Oriphiel	Zachariel	Samuel	Michael	Anael	Raphael	Gabriel
	Zazel*	Iophiel	Graphiel	Nachiel	Hagiel	Tiriel	Malcha
	Agiel*	Hismael	Barzabel	Sorath	Kedemel	Taphitar-tarat	Hasmodai

* These groups are planetary spirits.

Eight   The Stellatum (*coelum stella*) and the seven planets.

Nine	*Spheres*	*Hierarchies*	*Rulers*	*Precious stones*
	Primum mobile	Seraphim	Metatron	Saphyrus
	Stellatum	Cherubim	Ophaniel	Smaragdus
	Saturn	Throni	Zophkiel	Carbunculus
	Jupiter	Dominations	Zadkiel	Berillus
	Mars	Potestates	Camael	Onijx
	Sun	Virtues	Raphael*	Crysolithus
	Venus	Principalities	Haniel	Iaspis
	Mercury	Archangeli	Michael*	Topasius
	Moon	Angeli	Gabriel	Sardius

* Michael and Raphael appear to have been confused.

Ten   Primum Mobile, the zodiac (*sphaera zodiacus*), the seven planetary spheres, and the Earth spheres (*sphaerae elementaris*).

Eleven   is regarded as having no value in the system, and has no listing.

Twelve   gives the associations at their richest with the twelve signs of the zodiac, and their SIGILS, the names of the zodiacal Angels (Malchidiel, Asmodel, Ambriel, Muriel, Verchiel, Hamliel, Zuriel, Barbiel, Adnachiel, Hanael, Gabriel and Babchiel), the twelve months and the Roman gods (actually called 'numina') listed under COSMIC PANTHEONS, the twelve IMAGES of the signs, the corresponding MAGICAL SQUARES and BIRTH STONES.

---

calendrical orientation   A term used to denote architectural orientations in which the axes of buildings (or some significant feature within a building) are directed towards a calendrically determined point, such as the arc of sunrise, or heliacal risings for a particular epoch. Such orientations are usually derived from special astrological considerations. An example is the SAN MINIATO ZODIAC, which has a calendrical orientation of an annual and diurnal kind. See, however, ASTRO-ARCHAEOLOGY and FOUNDATION CHART.

Callesta   Name given to one of the satellites of Jupiter.

Camael   A variant for Samael, the ruler of Mars.

Camelopardalis   Constellation, approximately from 2 degrees Gemini to 15 Cancer, from 53 to 90 degrees north of the equator, the Giraffe. Formed only in 1614, it is said to confer patience and wisdom.

Campanean system   A method of HOUSE DIVISION, adapted from an earlier Arabian system by the 13th-century astrologer-mathematician Giovanni Campano. The principle underlying the method is the projection onto the ecliptic, from the pole of the Prime Vertical, an equal division of the Prime Vertical; thus, the planes of the six intersecting great circles are always at intervals of 30 degrees. The system is one of the several quadrant systems of division regarded as satisfactory by many astrologers – especially since Campanus insisted that the cusps of the houses mark the centre of the house, the strongest area of influence. One of the main objections to it is the distortion of the houses in charts cast for births near to the poles. Unfortunately, the Campanean tables are not so generally available as those drawn from the PLACIDEAN SYSTEM.

Campanus houses   See CAMPANEAN SYSTEM.

Cancer   The fourth sign of the zodiac. It corresponds as a zodiacal sign neither in location nor extent with the constellation of the same name – see CANCER CONSTELLATION – though the fixed stars in the asterism, along with those in part of constellation Gemini, almost entirely account for the erratic genius which is associated with natives of this sign. The modern sigil for Cancer ♋ has been variously explained. For example, Hall sees it as 'two spermatozoa twisted together', intended to signify the male and female seed: however, the sigil is relatively modern, and had a different form in medieval astrology. The image for Cancer is nowadays a crab, as in the medieval form in figure 8, however, in early astrology the sign and constellation was presented more frequently in the image of a crayfish. Cancer is of the WATER ELEMENT, of the Cardinal Quality, and the influence is emotional, sensitive, imaginative, gregarious and cautious. This nature of Cancer is manifest in human beings in the many keywords attached to it by modern astrologers: romantic, social, shrewd, domesticated, passive, strong in feelings, and so

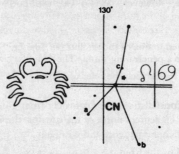

Figure 8: The image for Cancer, from the SAN MINIATO ZODIAC, *and the constellation from the* MODERN ZODIAC *of Delporte.*

on – in a word, all those qualities which may be associated with a sensitive water type seeking emotional unfoldment through experiencing the material realm. In excess, or when under pressure, the Cancerian nature may be described in terms which express its underlying fear of coming out of its crab-like shell: it is timid, self-absorbed, opinionated, acquisitive, pedestrian, driven by a fear of emotional vulnerability, moody, superficial, tending to live in the past, and so on. Like the other water signs, it may at times wish to withdraw completely from the material world. Cancer is ruled by the Moon, marks the Exaltation of Jupiter, the Detriment of Saturn and the Fall of Mars. In addition to the special terms already mentioned, the sign Cancer is said to be Cardinal, Cold, Commanding, Crooked, Destructive, Feminine, Fruitful, Moist, Mute, Negative, Phlegmatic, Psychic, Sensitive, Unfortunate, Tropical and Weak. The powerful fixed stars associated with Cancer, which make its influence especially remarkable in the zodiacal series, are ALHENA, CANOPUS, CASTOR, DIRAH, POLLUX, PROCYON, PROPUS, SIRIUS, TEJAT and WASAT.

cancer constellation Linked in name, if not in specific influence with the zodiacal CANCER, located as indicated in figure 8. The image of the crab, which is pre-Grecian, is said to be that of the crab elevated to the firmament by Juno for attacking the heel of her enemy Hercules – however, see CANCER. Of particular interest to astrologers are the large number of powerful fixed stars in this constellation, and in the neighbouring Gemini, which appear to account for the erratic and eccentric genius which seems to flourish under the sign Cancer: see the Cancerian asterism stars in the list furnished for CANCER and FIXED STARS.

Canes Venatici Constellation, once linked with the neighbouring URSA MAJOR, is presented in the modern image (evolving 1690) of two greyhounds, Asterion and Chara (themselves asterisms), held in the rein of the constellation Bootis. Inevitably, it is linked by astrologers with a love for hunting, and it is said also to give a penetrating mind. The (double) star COR CAROLI is set in this asterism.

Canicula See SIRIUS.

Canis The 13th of the LUNAR MANSIONS in the medieval lists, the Dog – probably identified by the fixed star VINDE MATRIX.

Canis Major Constellation, approximately from 1 degree of Cancer to 1 degree of Leo, from 15 to 40 degrees south of the equator. This is the 'greater dog' distinguished from the 'lesser' of CANIS MINOR. The fact that these two constellations have for centuries sandwiched the zodiacal Cancer has often been used to explain the love which the native of this sign exhibits for dogs (though in this connection, see SIRIUS). The asterism was once simply called Canis, and was said to be the hound of Actaeon, though others gave it to Orion: there are, however, many names and legends attached to the asterism. Ptolemy likens the stars of the constellation (with the necessary exception of Sirius) to the nature of Venus, which makes it generally beneficial.

Canis Minor Constellation approximately 18 to 28 degrees of Cancer, from 1 to 10 degrees north of the equator, the 'lesser dog', in early astrological texts called Procyon, along with derivatives of this Greek form meaning 'before the dog' – Prochion, Antecanis, Procanis, and so on. See, however, PROCYON in its modern use, for it is the influence of this fixed star which dominates the asterism.

Canopus Fixed star of 1st magnitude, the alpha of Argus, set originally in one of the rudders, but in later charts in one of the oars of the ship. The name is after the chief pilot of the fleet of Menelaos (at least in Greek sources), but in fact a variant name of almost similar sound (Kahi Nub) is from the Egyptian, and it is certain that some Egyptian temples were orientated towards the heliacal Canopus of the autumnal equinox, and later (such as the Karnak temple) to its setting. To judge from the Arabian tradition, the conjunction of Canopus seems to identify the nature of any planet, and when culminating it brings glory, fame and wealth.

Capella Fixed star of 1st magnitude, the alpha of constellation Auriga, set in the body of the Goat, its name meaning 'little female goat'. Sometimes it is called Amalthea, after the nurse who reared Jupiter on goat's milk (see ASTROLOGICAL TAROT). In medieval charts and manuscripts it is called Alayodi (with variant spellings), derived from the Arabic. Ptolemy gives it the beneficial nature of Mercury conjunct Mars. See PORRIMA.

capillary dynamolysis A term introduced by Kolisko to denote the study of capillary action of plant saps in relation to the effects of planetary aspects. The litmus paper images made as an aid to this study are called MORPHOCHROMATOGRAMS. See KOLISKO EFFECT.

Capillus Leonis Name meaning 'hair on the head of the Lion', given to the 11th of the LUNAR MANSIONS in medieval lists. Probably identified by the fixed stars Zosma and Coxa to the rear of the Lion.

Capricorn The tenth sign of the zodiac. It corresponds neither in location nor extent with the constellation CAPRICORNUS. The modern sigil for Capricorn ♑ is said by some to be a drawing of the horns of the Goat, with which the modern image is associated. However, Capricorn was never properly a goat, but a goat-fish, as the Greek term AIGOKEROS itself implied: the Babylonian prototype for the image was that of a goat with a curled fish tail, to which the modern sigil clearly still refers – see the medieval image in figure 9. Capricorn is of the earth element, and of the Cardinal Quality, the influence being practical, industrious, prudent, persevering, diplomatic, cautious, methodical and ambitious. The nature of Capricorn as it manifests in human beings is expressed in many keywords attached to the sign by modern astrologers: dependable, concentrative, trustworthy, efficient, just, industrious, honest, undemonstrative, conservative, responsive, patient, systematic – in a word, all those qualities which may be associated with an earth type working with integrity to achieve some particular aim. In excess, or under pressure, the Capricornian nature may be described in terms which express its underlying need for security: secretive, fearful, miserly, unsympathetic, rigid, suspicious, selfish, materialistic, brooding and egotistical. Like the other earth signs, it is subject at times to deep melancholia. Capricorn is ruled by the planet Saturn, marks the Exaltation of Mars, and the Fall of Jupiter. In addition to the specialist terms already mentioned, Capricorn is traditionally called Cardinal, Changeable, Cold, Crooked, Domestic, Earthy, Egotistical, Feminine, Four-footed, Hoarse, Hurtful, Melancholic, Negative, Nocturnal, Obeying, Southern, Tropical, Unfortunate, Violent and Wintery.

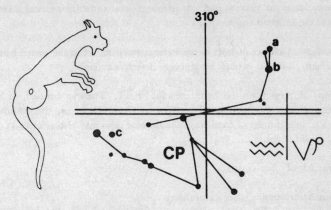

Figure 9. Image for Capricorn from the SAN MINIATO ZODIAC, *and the constellation from the* MODERN ZODIAC *of Delporte.*

Capricornus The constellation linked in name, if not in influence, with the zodiacal CAPRICORN, located as indicated in figure 9. In relation to its image of goat-fish, see CAPRICORN. The two images of Capricornus in figure 10 tell their own story – the first is Babylonian, the second, 12th century European – the fish in the mouth of the monster denoting Pisces is a delightful zodiacal curiosity. The Greek myth connected with the naming of the asterism tells that Pan, pursued by Typhon, escaped by leaping into the Nile: his upper body turned into a goat, the lower part into a fish, and in this guise he was taken into heaven by Jupiter – however, as the CONSTELLATION NAMES entry indicates, the name is pre-Grecian. The most powerful fixed stars in

Figure 10. Images of Capricorn from a 12th-century BC *Babylonian stele and a 12th-century* AD *astrological manuscript (after Saxl).*

Capricornus to play a part in astrology are ALGEDI, ARMUS, CASTRA, DABITH, DORSUM and NASHIRA.

Capulus A cluster of stars in the Perseus asterism, set in the sword hand of the hero. It is said to cause blindness or defective eyesight.

Caput A term from the Latin meaning 'head'. When so isolated, in most astrological contexts it usually refers to Caput Draconis, the DRAGON'S HEAD. In a context of fixed stars, the isolated term may also refer to ALGOL, the Caput Algol.

Caput Algol See ALGOL.

Caput Andromedae See ALPHERATZ.

Caput Canis Validi Name given to the 5th of the LUNAR MANSIONS in some medieval lists, meaning (probably) 'the head of the mighty dog'. The mansion was probably identified by the stars in the belt of Orion – for example MINTAKA or HEKA.

Caput Draconis See DRAGON'S HEAD.

Caput Medusa See ALGOL.

Caput Serpens See SERPENS.

Caput Tauri Name given to the 3rd of the LUNAR MANSIONS in some medieval lists, 'the head of Taurus' – probably located by the PLEIADES.

Cardinal The Cardinal signs are Aries, Cancer, Libra and Capricorn, and the Cardinal houses are their associates – the 1st, 4th, 7th and 10th, all so named because they fall on the Cardinal ANGLES, Aries to the east, Cancer to the north, Libra to the west, and Capricorn to the south.

Cardinal Cross In traditional astrology, this is the Cross formed in the zodiac by Aries, Cancer, Libra and Capricorn (see CARDINAL). However, in connection with the esoteric system proposed by Bailey in her INTUITIONAL ASTROLOGY, the Cross is associated with the crisis of initiation, and with the life of spirit and the monad. Bailey calls it the Cross of Transcendence, and traces in its sequence an evolutionary development. Aries is the place where the initial idea to institute activity takes form: it is the birthplace of ideas.

What appears in Aries as spiritual energy meets the soul stage of Cancer, and finds incarnation for the first time in form. It reaches a point of equilibrium in Libra, in which soul and personality achieve a balance of co-operation, so that in Capricorn the will nature arrives at fulfilment, and a visioned goal is reached. See also FIXED CROSS and MUTABLE CROSS.

Cardinality   A term applied to one of the three QUALITIES (see FIXITY and MUTABILITY) which acts as the mainspring of action, as the fount of energy. The four Cardinal signs (Aries, Cancer, Libra and Capricorn) are each in their own way involved with initiating action, with making a movement into the world: indeed, the alternative names for the Cardinal quadruplicity all suggest this idea – they are the initiating signs, the leading signs, the moving (or movable) signs. Sometimes, because they fall upon the Angles of the macrocosmic chart, they are called the Angular signs.

Cardinal quadruplicity   See CARDINALITY.

Carina   See PUPPIS.

Cassiel   The name given to the Archangel of Saturn. In certain occult systems, Cassiel was the Angel of Solitude and Tears. In the superficial grimoire tradition he is as a demon with bat-wings, seated on a dragon.

Cassiopeia   Constellation, stretching from about 25 degrees Aries to the beginning of Gemini, some 50 to 70 degrees north of the equator. The asterism was called from very early times by this name (and variants), as well as She of the Throne, sometimes Mulier Sedis (the woman of the chair), or even as Sedis. Cassiopeia is supposed to have boasted that she was more beautiful than the Nereids, and was bound in punishment to a chair, condemned to circle the pole-star with head downwards, an eternal lesson in humility. The astrological tradition insists that the constellation promotes pride and boasting, and it was said by Ptolemy to exude a nature similar to Venus with Saturn. The multiple fixed star ALPHERATZ (sometimes called Schedar), ALMACH, MIRACH, and the nebula VERTEX (all set in the asterism) are regarded by astrologers as powerful influences when prominent in a chart.

Castillo   One of the moons of Jupiter.

casting   The concept of casting a horoscope is probably derived from the Middle English verb 'to cast', relating to calculations made by means of counters. Strictly speaking, the term should be limited to the procedure preliminary to astrological interpretation, by which a horoscope figure is drawn by means of calculations, usually derived from consultation of an ephemeris. However, in modern use the term is employed to cover the entire process of both drawing up a figure, and interpreting it. See COMPUTATION.

cast no rays   See RAY.

Castor   Fixed star (binary), the alpha of constellation Gemini, set in the head of the mortal twin, sometimes called Eques or Apollo, along with related variants. There is little agreement as to its intrinsic nature, though Ptolemy likens its influence to that of Mercury, and it seems to threaten blindness and wounds when rising.

Castra   Fixed star, the epsilon of Capricorn, set in the belly of the Goat: an unfortunate influence.

Casujoiah   In certain forms of ceremonial magic, this is the name given to the Governor of zodiacal Capricorn.

Catabhishaj   The 28th of the Hindu NAKSHATRAS (hundred physicians).

cataclysmic planet   A name given to the planet URANUS.

Catahibazon   See KATABABAZON.

catholic astrology   A term derived from the Greek word used by Ptolemy (*katholike*) to distinguish a form of astrology of the genethliathic kind, which deals with impersonal matters, such as the astrological influence which affects all mankind, whole countries and races. See MUNDANE ASTROLOGY.

cauda   A term from the Latin for tail, which in most astrological contexts refers to Cauda Draconis, the DRAGON'S TAIL. In a context of fixed stars, the term sometimes refers to Cauda Ursae – see POLARIS.

Cauda Capricorni   See DENEB ALGEDI and FIFTEEN STARS.

Cauda Dragonis   See DRAGON'S TAIL.

Cauda Leonis   Name (tail of the Lion) given to the 12th of the LUNAR

MANSIONS in some medieval lists. Probably located by the fixed star DENEBOLA.

Cauda Scorpionis   Name (tail of the Scorpion), given to the 19th of the LUNAR MANSIONS in some medieval lists. Probably identified by the stars around SHAULA.

Cauda Ursae   See FIFTEEN STARS and POLARIS.

cazimi   An Arabian name given to an extreme case of COMBUSTION, when a planet is within 17 minutes of the Sun's longitude, and thus in conjunction with the Sun. Authorities differ on how the cazimi is to be interpreted, but in the tradition the placing was seen as strengthening for the planet.

Celaeno   See PLEIADES.

celestial ecliptic   A term coined to distinguish the ecliptic proper, which is the apparent path of the sun (see ECLIPTIC) from the path of the Moon's orbit within the ecliptic belt, which is really an orbit around the Earth, and hence called the terrestrial ecliptic.

celestial equator   The great circle of the Celestial Sphere, whose plane is perpendicular to the axis of the Earth, is called the celestial equator.

celestial figure   A synonym for CHART. See also PLANISPHERE and FIGURE.

celestial harmonics   It has been maintained from the very earliest times that there is a relationship between the planetary intervals, and even between the planetary movements, and earthly music – an idea which may be traced back to the Pythagorean concept of MUSIC OF THE SPHERES. The developed form of the Pythagorean conception deeply influenced the esoteric astrology of the Renaissance (mainly through the writings of Alberti and Gafurus), especially in regard to the new awareness of the conversion of ratios and harmonics based on planetary harmonics into spatial equivalents. In an attempt to unify the esotericism of theology with the new materialism of science, Kepler pointed to musical analogies in the ratios of aphelia and periphelia, which he linked with a notion of the major scale, but his ideas appear to have received scant understanding. In modern astrology, similar analogies have been explored by the esotericist Collin, who claims to have established an analogy between the tonic scale, and what he calls 'planetary octaves', directly linked with certain multiples of planetary cycles and groups of planetary cycles, whilst more recently Schmidt has linked musical intervals with synodic periodicities. Some fields of modern music – for example, certain compositions of the German Stockhausen – appear to be a result of the search for basic cosmic harmonies linked with astrological and esoteric notions. The earlier 'planetary music' of such composers as Holst appear to have been more concerned with establishing soul-moods analogous to the supposed natures of the planets, and therefore are not directly connected with celestial harmonics as such. Musical synthesizers are now being used to explore the musical potential in the extrapolation and analogy-making between mathematics and planetary motions.

Celestial Hierarchies The term is from the Greek compound of *ieros* (sacred) and *archen* (rule), and refers as a whole to the chain of spiritual beings which are visualized as ranged in descending order from the realm of God to the world of man. They were given an official nomenclature by Dionysius the Areopagite, who is said to have derived them from much earlier sources. While this list of names and their symbolic attributes are still used by occultists, esoteric astrologers and ecclesiastical writers, there has been considerable elaboration on their functions since the early days of Christianity. Each group has its distinctive function within the spiritual realm, and since (according to the laws of sympathy) each spiritual activity finds expression on the material realm, each group may be seen manifest on the Earth. The schema, in accordance with the theory of number-magic, is presented as three groups of three. In descending order of power and responsibility, the 1st Hierarchy consists of the SERAPHIM, who are directly involved with maintaining the relationship of the solar system to the entire stellar universe, and who are often accorded rule over the STELLATUM. Then follows the CHERUBIM, said by some authorities to be responsible for ordering the planetary movements within the solar system, but often accorded rule over the realm of the ZODIAC, as distinct from the fixed stars. The third of the First Hierarchy are the THRONES, said by some to have rule over the realm of Time itself: esoterically, Time is regarded as being operative only below the Sphere of Saturn, with which the Thrones are associated – this probably explains the connection in Greek mythology between Chronos (Time) and Saturn. The Second Hierarchy consists of the DOMINIONS, in Greek the Kyriotetes, who are said to be involved with the metamorphosis of matter, and who are linked with the Sphere of Jupiter. Then follows the DYNAMIS, sometimes called the Powers or Mights, who are directly charged with rule over the Sphere of Mars. The third of the Second Hierarchy are the POWERS, in Greek the Exsusiai, sometimes also called the Elohim, who are the creative beings who lend form and substance to the Earth itself, and who have rule over the Sphere of the Sun. The Third Hierarchy consists of the ARCHAI, sometimes called the Principalities, who have charge over the rise and fall of the epochs of civilization, and the history of the world: these beings have rule over the Sphere of Venus. Then follows the ARCHANGELS, who have rule over individual nations of the Earth, and who are in charge of the Sphere of Mercury. The last of the Third Hierarchy is the class of ANGELS, who have charge over the lives of individual humans (according to the esoteric lore, over one individual from lifetime to lifetime): these are linked with the Sphere of the Moon. At the bottom of the celestial chain is the realm of man, which has been called the Tenth Hierarchy. Below man are the various infernal spheres of the demonic and purgatorial realms, which in some systems are a mirror-image distortion of the Celestial Hierarchies: fortunately, however, the demonic sphere plays little part in astrology. The Celestial Hierarchies are sometimes called the Dionysian Hierarchies, even in those cases where the lists do not perfectly tally with the sequence set out by Dionysius. They are also called the Nine Orders. In the present century, following the indications of Steiner, Vreede (whilst adhering to the traditional nomenclature and cosmic functions) has proposed a series of group rulerships which departs from the standard, giving rule of pairs of hierarchies over cosmic phenomena:

COMETS	Seraphim and Cherubim
FIXED STARS	Thrones and Spirits of Wisdom (Kyriotetes)
PLANETS	Spirits of Movement (Dynamis) and Spirits of Form (Exsusiai)
MOON	Archai and Archangels
EARTH	Angels and Man

celestial horizon   The notional horizon derived from the spherical model of our Earth in relation to the CELESTIAL SPHERE, which is used in all TABLES OF HOUSES as a basis for computation. The celestial horizon passes through the centre of the Earth. The sensible horizon is parallel to the celestial horizon, but passes through the eyes of the observer at any point on Earth.

celestial latitude   A term used in contradistinction to GEOGRAPHICAL LATITUDE, in reference to the measurement of distance north or south of the zodiacal circle or ecliptic. See also CELESTIAL LONGITUDE.

celestial longitude   A term used in contradistinction to GEOGRAPHICAL LONGITUDE, in reference to the measurement of distance along the zodiacal circle, or ecliptic. By means of celestial longitude and celestial latitude, it is possible to define the position of a stellar body in relation to the ecliptic and hence in relation to both space and time.

celestial meridian   The great circle of the Celestial Sphere, which passes through the celestial poles and the zenith and nadir of a given place.

celestial poles   The point where the celestial axis and the Celestial Sphere intersect.

Celestial River   See VIA LACTEA.

Celestial Sphere   An imaginary projection of the Earth, to include the visible heavens with a view to establishing a system of co-ordinates. The system of co-ordinates is designed to locate any point within this projected sphere in terms of the circumference of the periphery. Several such systems have been established, and are either directly or indirectly used in astrological computation: these include the ECLIPTIC SYSTEM, the EQUATORIAL SYSTEM, the HORIZONTAL SYSTEM and the VERTICAL SPHERE.

Celeub   See PERSEUS.

cell-salts   See TWELVE CELL-SALTS.

Centaur   Sometimes used as a popular name for zodiacal SAGITTARIUS, but also of CENTAURUS.

Centaurus   Constellation, approximately from 2 degrees Libra to 28 Scorpio, and from 28 to 68 degrees south of the equator (see also CENTAUR). The mythopoetic origin of the asterism is usually linked with the story of the centaur Chiron (after whom it is sometimes named), but it has many different designations, including Minotaur and Semi Vir (half-man). Ptolemy divides the nature of the asterism in regard to influences, and sees the human half as of the nature of Mercury conjunct Venus, and the animal part as Venus conjunct Jupiter: in general, the asterism inclines to vengeance and energy. A fixed star in the foot of the image is Rigil Kentaurus, which is sometimes shortened to Rigil, and hence confused with the RIGEL of Orion. For the fixed stars of importance in this asterism, see AGENA and RIGIL KENTAURUS.

centiloquium   A series of 100 astrological aphorisms (plural *centiloquia*), usually from the pen of one writer, but sometimes (as, for example, in the series attributed to Gadbury) collated from a variety of sources. The most important and influential collection, to which the term usually refers, is that wrongly attributed to Ptolemy, which in fact evinces a late medieval astrology quite foreign to him. The most widely known of the centiloquia in use up to the 19th century were those attributed to Hermes Trismegistus and Bethem, each of which was incorporated into the Pruckner edition of Firmicus.

Cepheis See ANDROMEDA.

Cepheus Constellation, approximately 17 degrees of Pisces, to the beginning of Cancer, and from 55 to 85 degrees north of the equator. This asterism, named after a King of Ethiopia who figured among the Argonauts, is linked with the mythology of the adjacent constellations, Cassiopeia and Andromeda, who were his wife and daughter respectively. The asterism is regarded as exuding an influence akin to that of Jupiter with Saturn, inducing sobriety and a sense of authority, save when afflicted.

Cerebus A name given at one time to an asterism in HERCULES.

Ceres Name given to the first of the *ASTEROIDS to be discovered by modern astronomers, in 1801. It lies in orbit between Mars and Jupiter, and has a diameter of about 490 miles. The name is derived from that of the Roman goddess of vegetation – but see also COSMIC PANTHEONS. The astrologer Dobyns associates the asteroid with the sign Virgo, along with a co-ruler VESTA. Sometimes the constellation Virgo (and by extension, the sign Virgo) is called Ceres (more particularly Ceres Spicifera Dea – see SPICA), as well as Demeter, which is the Greek equivalent goddess, whose Mysteries were linked with the pre-Christian image of Virgo.

Cetus Constellation, from about 17 degrees Pisces to 13 degrees Taurus, and 10 to 30 degrees south of the equator, the Whale or Sea Monster for which early pictures sometimes portray a sort of sea-dog, hence the Canis Tritonis, and a number of related names. In myth this asterism is the monster sent to devour Andromeda: it has an evil reputation among astrologers, and Ptolemy regards it as of an equivalent nature to Saturn. The three fixed stars within the asterism are also ill-disposed – see BATEN KAITOS, DIFDA and MENKAR.

chain In the theosophical cosmoconception, a chain is a series of seven GLOBES. Each of the seven globes of a particular chain may be regarded as having a definite and distinct existence in space and time, though not all globes in a particular chain are visible to ordinary sight. The chain is properly called a PLANETARY CHAIN, while seven successive chains are sometimes called the Incarnation of a Chain: see, however, GLOBE PERIOD.

Chain of Being A name given to an ancient doctrine which postulates an unfaltering order of created things, ranging sequentially from the highest spiritual levels to the meanest inanimate objects on earth. The Chain was visualized as a perfectly graduated hierarchy of beings, reflected in medieval astrology in the sophisticated theory of SPHERES, peopled by spiritual entities. This 'chain', stretching as it were from the Throne of God to the very centre of the Earth, admitted no breaks, and no chaos. Developed as a philosophical idea by Plato, added to by Aristotle, elaborated by the Neoplatonists, this has become a stock image underlying most philosophies and cosmoconceptions, as Lovejoy has shown. Hell alone (because it had rebelled from the order of things) was disconnected from this chain, yet the vision of Dante, resting as it did upon the redemptive thesis of theology, embraced even Hell in this chain. A symbolic model of this order, sometimes called the Annulus Platonicus, sometimes the Aurea Catena Homeri (Golden Chain of Homer), was given various sigillic forms in astrological and alchemical symbolism.

chain period See GLOBE PERIOD.

Chaldean From late Classical times, the word 'Chaldean' has been synonymous with 'astrologer', mainly because of the fame of the Mesopotamians as astronomers and astrologers: however, distinguish CHALDEAN TERMS. Much of what is now called Chaldean astrology is of the omnia kind, though it is clear from Diodorus of Sicily that the Chaldeans would cast personal horoscopes. Sachs dates the earliest known cuneiform horoscope to 410 BC, at which time a twelve-sign equal-arc zodiac was in existence. It would seem

---

TABLE 20

Planet	Name	Ruling Spirit (God)
MO		Sin
ME	Mustabarru	Nebu
VE	Kilbat	Ishtar
SU		Shamash
MA	Bibu	Nergal
JU	Dapinu	Marduk
SA	Kaimanu	Ninib

---

that the Mystery centres of the Babylonians originated the zodiac which was later amended by the Greeks to provide the names and designations with which we are now familiar – see ZODIAC and CONSTELLATION NAMES for ancient nomenclature. The names and rulerships of the planets are given in table 20.

Chaldean Order   A name given in esoteric astrology for the septenary star, which is (wrongly) believed to have determined the days of the week. The astrologer Libra sets out the order, ascribing a planet to each of the seven points of the star, with the Sun at the top, as follows:

```
    SU
  VE  MA
ME        JU     The sequence to be read: SU MO MA ME JU VE SA
  MO  SA
```

The sequence of planets which emerges from following the inner structure of the septenary star is, of course, different from the traditional order (which follows the sequence of the sphere), yet it accounts for many curious-seeming relationships established in the occult tradition – for example, for the order of planetary rulerships given for the SECUNDADEIAN BEINGS. The sequence does of course give rise to the order of the days of the week, but this was not in fact determined by the Chaldean Order.

Chaldean terms   See PTOLEMAIC TERMS.

Chamael   Like Camael, this is a variant for SAMAEL, the ruler of Mars.

Chamaeleon   Constellation located close to the South Pole, formed in 1604.

Chandra   One of the Sanskrit names for the Moon, as used in Hindu astrology. In this system, the nodes of the Moon are accorded a greater significance in chart interpretation than in Western astrology: the ascending node is called RAHU, the descending node KETU.

changeable signs   Six of the zodiacal signs have been traditionally classified as changeable because (it was claimed) they would change their natures depending upon their positions in relation to the diurnal movement of the earth. The six signs are Taurus, Gemini, Leo, Virgo, Sagittarius and Capricorn – three sets of pairs. When eastern, Taurus was hot, Gemini and Leo were hot and dry, Virgo was hot, Sagittarius was cold and moist, while

Capricorn was cold and dry. When western, Taurus was cold, Gemini was cold and moist, Leo was hot and moist, Virgo cold and moist, Sagittarius hot and dry, while Capricorn was cold and moist. The concept has little or no application in modern astrology, and indeed the inner meanings of the pairs of qualities has in any case been lost. See, however, QUALITY and PRIMORDIAL QUALITIES. The same term, changeable signs, is sometimes applied to the signs of the air triplicity, which is said to promote the changeable qualities associated with the SANGUINE temperament.

Chara   See CANES VENATICI.

characters   A term used of certain symbols (sigils) used in astrological and occult texts, especially so in those of late-medieval derivation. Many such texts provide lists of characters which in form are usually different from the traditional lists of sigils associated with planets and signs: in practice, it is difficult to distinguish the two, save in so far as the former are often vestigial drawings, while the latter are more usually quite abstract in quality. The three sigils in figure 11 relate to the characters for Saturn: the first sigil is derived

Saturni.   Signacula siue charaɕeres, Intelligentiæ Saturni.   Demonij Saturni.

Figure 11: The characters of Saturn (after Agrippa).

from the corresponding MAGIC SQUARE for the planetary spirit of Saturn. Paracelsus defines the characters as being drawn from the highest stars and artificially assumed by the lower realm. Such are therefore SEALS: they are often used in the manufacture of artificial gamalei. It would appear that the characters were originally intended as receptacles of special VIRTUE or prophylactic value. In another context, Paracelsus writes of characters as being words, either spoken or written, which 'produce effects which Nature itself is not able to bring about, but only magical science'. Some of the planetary characters are derived from MAGIC SQUARES involved with occult number symbolism.

Chariots   A synonym for the THRONES of the early Greek astrology.

Chart   A term synonymous with HOROSCOPE in the modern sense. It probably was once cognate with 'chart' in the sense of 'map', of Greek etymology. Certainly the late medieval astrologer often referred to the horoscope figure as a map, by which he meant 'map of the Heavens'. Many synonyms exist, such as FIGURE, HOROSCOPE, GENITURE, MIRROR OF HEAVEN, NATIVITY, SCHEMA, and so on. See also CHART SYSTEMS.

chart systems   All horoscope figures (horoscope charts), which combine the symbolic tabulation of planetary positions, ecliptic and house division into a single figure, are actually stereographic projections on to a flat surface. The choice of the matrix plane determines the kind of projection, and hence the symbolic nature of the chart – in some cases, the ecliptic is fixed, and in other cases the house system is fixed. The earliest (Greek) horoscope charts were usually circular, most often with the Ascendant (Horoscope) to the left, with house cusps (in the dual-house system) symbolically presented in such a way as to mark 'equal' divisions against the ecliptic. This system, albeit in a more sophisticated guise, is still used in connection with the EQUAL-HOUSE SYSTEM, as in figure 12A. By the time the Arabian-influenced Hellenistic astrological

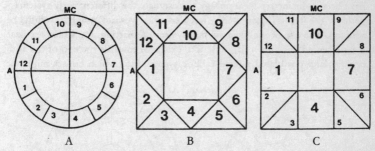

Figure 12: A: Standard circular chart. B: Decussata chart. C: Greek Cross chart.

methods reached Europe (along with sophisticated methods of HOUSE DIVIS-IONS unknown to the Greeks) two different forms of charting were in general use – the circular method, and a form which has been called the QUADRATED CIRCLE – figures 12 B and 12C – in one or two variants. The equal-division house system of both circular and quadrated charts presupposed that the ecliptic may be symbolically accommodated to a fixed system of houses, in the sequence set out in figure 12. The two quadrated forms have been called the 'decussata' and the Greek Cross Chart, for reasons which are clear from their construction. In later methods, especially so in the 18th century, the compartment symbolism was adapted to form a 'cross' from

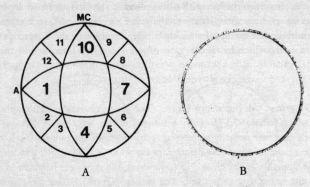

Figure 13: A: Vesican chart. B: Fixed ecliptic system.

simple *vesicae piscis*, as in figure 13A, and in copperplate engravings, such as those given by Sibly, the central areas were often elaborately engraved with portraits (frequently quite fanciful) of the native. There does not appear to have been a distinctive terminology to designate the different early systems – perhaps the terms 'quadrated', 'circular' and 'vesican' might usefully be adopted. The use of the fixed ecliptic system of charting, for which the base is represented in figure 13B, came into existence towards the end of the 19th century, and postulated a fixed ecliptic system, into which houses might be

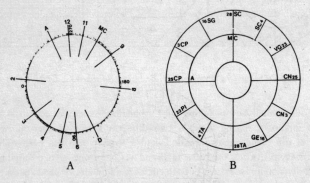

Figure 14: A: Continental chart system. B: Data from Continental chart translated into standard circular chart for ease of reading.

adapted in various degrees of arc. This has floriated into many different systems, of which figure 14A is an example from a Continental method used in France, in which the Vernal Point is placed to the left of the circle, and the houses inserted accordingly: the analysis in figure 14B presents the same data in a more conventional circular chart, for ease of reading. Figure 15A is a variant of this Continental chart, the house cusps being marked to the fixed

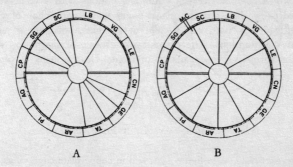

A	B

Figure 15: A: Continental variation, house cusps marked into fixed ecliptic circle. B: Continental method translated into standard 'equal-house' charting, for comparison.

ecliptic, and thus appearing on the chart in unequal divisions, but with the Ascendant to the left. The same data is presented in figure 15B, but with equal-arc houses: one observes that the 10th house cusp (at 90 degrees to the Ascendant) does not correspond to the Medium Coeli (MC), which is inserted separately. These are merely examples of the more frequently encountered methods of conventional charting. Many complex (and indeed unconventional) methods of charting have been developed in modern times, as the grasp on the traditional forms of astrology has been weakened. Under the

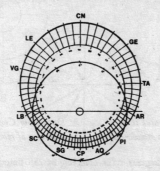

Figure 16: Three-dimensional conversion chart (after Hieroz).

chart systems

impress to establish novel or original methods of astrological symbolizing, such modern forms as the three-dimensional conversion chart (figure 16) have been published, this one influencing the so-called THREE-DIMENSIONAL CHART of Blunsdon. The COSMOGRAM of Ebertin (figure 24) is a complex figure which combines certain traditional notions with a 90-degree circle designed to facilitate transfer of mid-points to the conversion of a Cosmic Picture, which is itself not a chart at all, but a diagram of supposed planetary relationships. The CHOISNARD CHART of figure 17, which is similar in form to several variant Continental chart systems, is convention

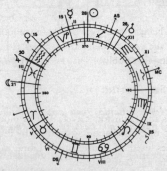

Figure 17: Choisnard chart for Gustav Adolphus, born in Stockholm on 18 December 1594 (OS).

itself in comparison with the cosmopsychogram of figure 25, which for all that it accommodates in its rays astrologically based data, may scarcely be considered a chart in the ordinary sense at all. Many other systems of charting have been developed in modern astrological circles, including those relating to heliocentric astrology, of which the ASTEROGRAM of Sucher is an

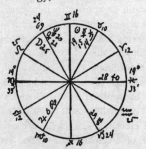

Figure 18: Inceptional chart for the launching of the Titanic, *31 May 1911 (after Libra).*

example, in figure 4. Usually, the systems are designed to present in a clear form the interrelation of the three tabulations of ecliptic, house division, and planetary positions, so that there is generally little difficulty in converting one chart system to another. An example of a simple standard form of charting is given in figure 18, which is the astrologer Libra's INCEPTIONAL CHART for the ill-fated *Titanic* – this, however, relates more to a type of astrology than to a method of charting. See also EPOCHAL CHART, EQUATOR-IAL CHART, HARMONIC CHART, HELIOCENTRIC CHART, HORIZON CHART, JOHNDRO LOCALITY CHART, MUNDOSCOPE, POLAR CHART, PRIME VER-TICAL CHART and ZENITH SYSTEM.

Charubel symbols   Name given to a series of symbolic readings ascribed to the 360 degrees of the zodiac, derived by the astrologer 'Charubel' using clairvoyant means, and first published in 1898. See DEGREE SYMBOLS.

chasing the Moon   See HUNTING THE MOON.

Chelae   The word is the Latin for 'claws', and in modern astrology usually refers to the CLAWS of constellation SCORPIUS. In Ptolemaic astrology, the word was sometimes used of constellation Libra, which was not always distinguished from the adjacent Scorpius.

Chelub   See PERSEUS.

Cherub   In astrology, always a reference to the CHERUBIM.

Cherubim   A term of uncertain Hebraic etymology (but see CHERUB) used to denote the second rank of the Celestial Hierarchies, immediately below the rank of Seraphim. The singular cherub is frequently misused in modern times, for the word should be applied in an astrological sense only to this exalted rank of spiritual beings. Cherubim are often depicted in Christian and occult iconography as having four faces, but traditions vary enormously as to their natures. The occult tradition links them with the Fixed signs of the zodiac, and they are accordingly portrayed as the TETRAMORPHS or as variously combined elements of Lion, Eagle, Bull and Human. In her analysis of the esoteric background to this Hierarchy, Blavatsky points out that the word 'cherub' originally meant 'serpent' and was linked with the idea of the 'serpent of eternity'. The Hebraic characters for the word are derived from an image of a serpent in a circle. Such ideas are in perfect accord with the traditional association of these beings with the circle of the zodiac. Steiner, who refers to them as Lords of Harmony, says that their

Chinese calendar

prime function is that of controlling the relationships between the planets without our solar system. The medieval 'cherubin' sometimes survives in astrological contexts, but this term belongs to a period when the true nature of the Hebraic plural was misunderstood.

Chinese calendar The Chinese system of time measurement is lunisolar, involving intercalation. The year begins with the New Moon following the entry of the Sun into Aquarius. This solar period is divided into twelve months (with an adjustment of an intercalary month every 30 months). Since the calendar began in a year equivalent to our own 2697 BC, the Chinese 4682 is the equivalent of our Gregorian 1985. Little of the Chinese system has penetrated into European astrology: the ordinary and literary names for the months are complex, as are the names and affinities for the ten celestial stems, with the corresponding five CHINESE ELEMENTS. To some extent, the names of the 28 lunar mansions (called SIEU), and their corresponding elements and animals, along with the twelve branches and the associated symbolic animals (see CHINESE SYMBOLIC ANIMALS) have penetrated into European astrological lore in a misunderstood and superficial form. The so-called 'twelve branches' of the Chinese system, linked with the zodiac, combine with the 'ten stems' to establish a 60-year cycle, which is the basis of the Chinese chronology.

Chinese elements The Chinese distinguish five elements which, through a rich system of associations, permeate their astrological and calendrical systems. The exoteric names translate as meaning wood, fire, earth, metal and water, but the associations and correspondences established into the microcosm and macrocosm indicate that (as indeed with the European four elements) these names denote secret principles, rather than materialities. Important to the system are the links between the elements, the planets and the binaries (Yang and Yin), as set out in table 21.

TABLE 21

Element	Bingay: Yang–Yin		Planet	Meaning
Wood	Fir	Bamboo	Jupiter	Wood Star
Fire	Burning Wood	Lamp Flame	Mars	Fire Star
Earth	Hill	Plain	Saturn	Earth Star
Metal	Weapon	Kettle	Venus	Metal Star
Water	Waves	Brook	Mercury	Water Planet

Choisnard chart

Chinese planets See table 21 above.

Chinese symbolic animals The rich system of associations of the Chinese 'twelve branches' (linked with the zodiac) has penetrated popular astrological lore in Europe mainly through the names of the associated symbolic animals in their so-called 'years' which have been misunderstood by popular Western astrologers, who have confused the branches with the months and with the zodiacal equivalents. Mathews gives for the twelve branches the equivalent poetic names, the related 'hours' and directions, as well as the correspondences of table 22. The animals in brackets with asterisk are modern European

	TABLE 22		
Symbolic animal	*Zodiacal sign*	*Symbolical animal*	*Zodiacal sign*
Rat	Aries	Horse	Libra
Ox	Taurus	Sheep	Scorpio
Tiger	Gemini	Monkey	Sagittarius
Hare (Cat)*	Cancer	Cock	Capricorn
Dragon	Leo	Dog	Aquarius
Snake	Virgo	Boar (Pig)*	Pisces

names. It is quite erroneous to confuse the 'annual readings' – which are actually designations of the 60-year cycle of the CHINESE CALENDAR – with the personality readings associated with the symbolic animals. The annual designations begin with Rat (the 'year of the Rat') allocated to a particular year, combining the so-called 'ten stems' (in the cycle 1924–84 – 'Chia') with the so-called 'twelve branches' (in the cycle 1924–84 – 'Tzu') to engineer a cycle of 60 years. Popular western astrology has simplified the cycle to draw attention to the twelve branches (and thus establish an accord with the zodiacal twelve) – however, there is a vast difference between the Chinese 'wooden Rat' of 1984 and the 'metal Rat' of 1922 which is missed in popular systems of interpretation, as attempted by Brau.

Choisnard chart A method of charting horoscopic data, attributed to the modern French astrologer Choisnard (though perhaps used earlier by others). Figure 17 is an example of such a chart: the data is not inserted into a chart system which visualizes the houses to be of equal extent in the symbolic projection, but adjusts the houses to fit the ecliptic, which is itself set out as a framework of 60 × 5-degree arcs (the old FACES), into which the planets and house cusps are inserted. The first point of Aries is always presented on the left of the figure, so that the Ascendant may be anywhere in the circle. This method of charting has gained some popularity on the Continent, but is used only rarely in England and the United States. Unfortunately, the same term (Choisnard chart) has been applied to a system of projecting the heavens (attributed also to Choisnard) which permits a horoscope to be erected when the time of birth is not known, in which the Vernal Point is on the symbolic Ascendant (the actual Ascendant being unknown). It has been called the Vernal Horoscope. See also CHART SYSTEMS.

choleric   Term applied to the TEMPERAMENT linked with the element of fire.

choleric quadrant   See QUADRANTS.

choloric   See QUADRANT.

Chori Angeli   A Latin equivalent for CELESTIAL HIERARCHIES, from which the modern Choir of Angels is derived.

chorography   A term used by Bouche-Leclercq in connection with the various attempts to relate geographic areas to planetary and sign influences. From the days of Ptolemy (and presumably earlier, since Ptolemy was a reporter of astrological lore) lists of planetary and zodiacal rulerships over places – first as geographic areas, later as cities and towns – were established within the astrological tradition. The 'Chorographics' of Ptolemy actually derived from the systematic application of the 'trigons' and their rulerships (in what is now an archaic concept) to the quadrants, corresponding to the longitudinal limits of the ancient CLIMATA. This sytem has all too frequently been repeated in other textbooks in a much simplified form, divorced from its original implications: many of the localities now have different names, and many of the geographic borders differ from those drawn up in Roman times, yet even so the lists of rulerships are still issued in all seriousness. Other systems, derived from sources which must have been contemporaneous with Ptolemy, present quite different chorographic relationships. It is simply not possible to present a digest of all these systems, but the comparative lists in table 23 present a synoptical view of the subject (which is, in fact, exceedingly complex): such a comparative study, resting upon the material published by Bouche-Leclercq, reveals the early conflicts and contrasts which are so often ignored in modern lists (Dorotheus of Sidon, Alexandrian Paul) – the representative modern is Hall. Table 23 is designed to show additions rather than the full lists (most of the authorities correspond in several cases), while

———————————————— TABLE 23 ————————————————

Sign	Ptolemy	Manilius	Dorotheus	Alex. Paul	Hall
AR	Britain Gaul Germany Judaea	Egypt Hellespont Persia	Arabia Babylon	Persia	England Burgundy Japan
TA	Asia Minor Cyclades (Cyprus, Persia)	Arabia Scythia	Egypt Media	Babylon	Greek archipelago
GE	Armenia Lower Egypt	Pontus	Cappadocia Phoenicia		Africa north-eastern coast Wales
CN	Africa Colchis Numidia	Ethiopia India	Thrace	Armenia	Mauritius Paraguay
LE	Apulia Chaldea Gaul Italy (Phoenicia, Sicily)	Armenia Bythnia Cappadocia Macedonia		Asia	Romania
VG	Achaia Assyria Babylon (Crete, Hellas, Mesopotamia)	Arcardy Ionia Rhodes	Cyclades Peloponnese		Brazil Crete West Indies
LB	Bactria	Italy	Cyrene	Libya	Argentine Burma North China
SC	Cappadocia Mauritania	Libya (Islands of the Mediterranean)	Sicily	Italy	Transvaal
SG	Arabia Felix Spain	Crete Sicily	Gaul	Cilicia	Australia Madagascar
CP	Illyria India (Macedonia, Thrace)	Germany Spain		Syria	Afghanistan Albania Greece
AQ	Arabia Mid-Ethiopia	Lower Egypt Phoenicia		Egypt	
PI	Celicia Lydia	Chaldea Mesopotamia Red Sea		India	

those areas which are largely meaningless to the modern mind, or which are merely historical references, have been excluded. The entry for Gaul under Ptolemy is perhaps 'Cisalpine Gaul', for the Gaul of Aries is certainly well ascribed. The 17th-century Lucca manuscript gives Switzerland to Taurus, France to Cancer, Bohemia to Leo, Greece to Virgo, Austria to Libra, Scotland to Scorpio and Portugal to Pisces: it is more usual to ascribe Scotland to Cancer, and Ireland to Taurus. The majority of medieval lists of chorographies were derived from Ptolemy, but under the Arabian impulse introduced cities and towns derived from somewhat complex and unsystematic considerations – such as local dedications to gods, and FOUNDATION CHARTS. For example, the city of Florence appears to have taken its chorographic designation from the earlier rulership of the god Mars, derived ultimately from the days when the city was little more than a Roman barracks. It is likely that the designation for Rome (Leo) is related to the famous foundation chart for the city. In addition to incorporating Ptolemaic material, modern lists tend to incorporate cities and towns derived from foundation charts and ancient traditions. Most modern lists are usually indiscriminate copies of earlier ones: it is significant that as late as 1959 Hall is still ascribing White Russia to Taurus, following the more reasonable ascription of the astrologer Leo, made in the second decade of our century. The rulership of the United States of America, and certain of the cities, have been hotly contested for some years: at least nine different versions for the USA are known to exist (several related to foundation charts), though the most frequently quoted is that given by Sibly, cast for the Declaration of Independence. It must be emphasized that there are many variant chorographies available, and little agreement exists among authorities either as to extant chorographies, or regarding the rules for establishing these. The general rule appears to be linked with catching what can only be called the 'feeling' of a place, which is almost certainly linked either with the zodiacal or planetary force, or even with the Archangelic ruler. In view of the actual origin of the Ptolemaic chorography, the present author has been surprised to discover much of it to be reasonably accurate: from personal experience he would confirm most of the early Ptolemaic system, with the exceptions of Gaul (hence France) which he would place under Virgo; he would place the Cyclades under Leo, and would lump together the two Egypts of the ancients under Capricorn (while admitting that the Egypt familiar to Ptolemy was no doubt very different from the modern Arabized country). Table 24 sets out a few of the city rulerships – an unpublished list from Lucca of the 17th century, a list from the 19th-century astrologer Alan Leo, and the modern Devore, who relies to some extent on Leo yet includes a useful summary of associations drawn up for American cities, recognizing the while that the rulership is 'not always unanimous'.

Sign	Lucca	Leo	Devore
AR	Faenza Florence Naples	Capua, Cracow, Leicester, Marseilles, Padua, Saragossa, Utrecht	
TA	Bologna Capua (Pistoia, Sienna, Verona)	Dublin, Leipzig, Mantua, Palermo, Parma	New York (SU) St Louis (AS)
GE	Paris Trent Turin	Cordoba, London, Louvain, Mentz, Nuremberg, Versailles	
CN	Lucca Venice Vicenza	Amsterdam, Berne, Cadiz, Constantinople, Genoa, Lubeck, Milan, New York, Tunis, York	New York (AS) Philadelphia (SU)
LE	Cremona Prague Ravenna Rome	Bath, Bristol, Damascus, Philadelphia, Portsmouth, Taunton	Chicago (SU) Detroit Miami (SU)
VG	Arezzo Benvenuto Como Ferrara (Navaro, Tarento, Toledo)	Basle, Baghdad, Cheltenham, Corinth, Heidelberg, Jerusalem, Lyons, Padua, Paris, Reading	Los Angeles (SU) Philadelphia (AS) Washington DC
LB	Parma Pesaro Piacenza Sienna	Antwerp, Charlestown, Frankfurt, Fribourg, Lisbon, Vienna	Chicago (AS) Los Angeles (AS) Miami (AS)
SC	Aquila Civitavecchia Cremona (Genoa, Padua, Rimini)	Fez, Frankfurt (Oder), Ghent, Liverpool, Messina	Baltimore (AS) Cleveland (AS) San Francisco Washington (AS)
SG	Buda Asti	Avignon, Cologne, Narbonne, Naples	Toledo (USA) (AS)
CP	(Modena, Prato, Romandola)		Boston (USA) (AS)
AQ		Bremen, Hamburg, Ingoldstadt	St Louis (SU)
PI	Alessandria Ratisbon	Alexandria, Compostela, Ratisbon, Seville, Worms	Cleveland (SU)

Christian zodiac See BIBLICAL ASTROLOGY, COSMIC PANTHEONS and SAN
MINIATO ZODIAC.

chromatic chart See ZODIACAL COLOURS.

chronocrators In general, the term means 'rulers of time', but in specific
application it is used of Jupiter and Saturn. Through the interactions of their
cycles, these planets create successive conjunctions of a 20-year periodicity
in signs of the same element. The Grand Climactic of the chronocrators
marks an epoch, a complete cycle through all the triplicities of these cycles.
In connection with this important Grand Climactic, the two planets are
sometimes called the Great or Grand Chronocrators. Each of the 20-year
conjunctions are called 'minims' or 'specialis', the 200-year 'elemental' cycle
is called 'media' or 'trigonalis', and the complete Grand Climactic of 800
years is called the 'maxima' or 'climacteria'. The same term (chronocrators) is
unfortunately confusingly applied to all the traditional planets in their
capacity of 'markers of time' in relation to individual human life. Systems
vary (see SEVEN AGES for example), but in general tradition has it that the
planets govern the life periodicities as set out in table 25. One of the so-called

---

TABLE 25

MO	First 4 years of life	(in 4-year periodicity)
ME	From 4 to 14 years	(in 10-year periodicity)
VE	From 14 to 22 years	(in 8-year periodicity)
SU	From 22 to 41 years	(in 19-year periodicity)
MA	From 41 to 56 years	(in 15-year periodicity)
JU	From 56 to 68 years	(in 12-year periodicity)
SA	From 68 to 98 years	(in 30-year periodicity)

'modern planets', Uranus, was accorded rule over the next years of life, to a
cycle of 90 years. Quite unaccountable, however, is the establishing of an
extrauranian cycle of 184 years, allotted to Neptune, which would bring the
terminus vitae to about 372 years! See CYCLES.

Chronos In astrological contexts an archaic reference to Saturn.

chronos ageraos A name applied to the primal TETRAMORPH which
united the four Fixed signs of the zodiac. See ZERVAN AKARNA.

Cingula Orionis   Literally, ORION'S BELT in Latin, and usually a reference to the two powerful fixed stars in this belt, MINTAKA and ALNILAM.

Cingulum   See MIRACH.

Circinus   Constellation originated in 1752, the Compasses. Astrologers say that it gives a revengeful and violent nature, but there are no powerful fixed-star influences on record.

Circle of Petosiris   Name applied to an ancient (certainly Hellenistic in origin) system of prognostication by simple numerology, often linked with the name of the querent – see ONOMANTIC ASTROLOGY.

Circle of Position   An obsolete term originally used of a system for determining the position of a fixed star by means of a circle visualized as intersecting the horizon, and passing through the star as one co-ordinate, with another circle passing through the meridian and that star as a second co-ordinate.

circular chart   See CHART SYSTEMS and CIRCULAR FIGURE.

circular figure   A term proposed to distinguish the various circular horoscope CHARTS (based on either equal-arc twelvefold divisions, or on unequal-arc twelvefold divisions, to represent the twelve houses in relation to the signs) from the QUADRATED FIGURE. The circular figure may also be called the circular chart.

circumpolar   Those stars which never set are called circumpolar.

circumpolar constellations   The asterisms which, from a given point on the Earth, do not go below the visible horizon at any time of the year. Some constellations which are circumpolar for part of the year will 'dip' part of their asterism below the horizon, and are not strictly circumpolar.

cities   For a survey of the zodiacal rule accorded cities, see CHOROGRAPHY.

Citra   The 12th of the Hindu NAKSHATRAS (bright). The same Sanskrit term is used to designate the constellation Libra – but see also TULA.

civil day   This day-length is computed to begin at midnight, 180 degrees east or west of Greenwich, continuing its periodicities of one hour with every 15 degrees of arc passage of the Sun. See TIME.

Claws

Claws See CHELAE.

clima An obsolete term (Greek plural *climata*), derived probably from Hipparchus, and applied to a zone of latitude, fixed according to the ancient principle by the maximum length of day, translated into distance. The first of the climata does not begin at the equator, where the day is said to be always of 12-hour duration: it begins approximately 12 hours and 45 minutes from the equator, and extends northwards to a boundary of 13 hours and 15 minutes (approximating to our 13 and 21 degrees of latitude). There were SEVEN CLIMATA, each of different mean time periods, each of different distances. The relevance of the term is mainly historical, since the system was based on geographic and geodesic assumptions which are now known to have been inaccurate. They were, however, widely used in ancient astrological systems. The climata might be projected into a Celestial Sphere, as the images of Dante make clear. However, the use of climata in connection with the longest daylight for a given area, as a base for astrological computation, resulted in the climata being associated with rising times of the ecliptic. The Roman astrologer Vettius Valens defined the climata in terms of rising times, and the various sets of tables for the climata used by earlier astrologers for computations linked the clima with rising times sequential to the longest daylight. A system of SEVEN CLIMATA was derived from both the Babylonian SYSTEM A and SYSTEM B, though it was the latter, with its ayanamsa of 8 degrees, which was most widely used by Hellenistic astrologers. Ptolemy, in the Almagest, gives rising times for eleven zones, between the longest daylights of 12 and 17 hours, and in relation to the seven climata derived a series of astrological tabulations to which was added an interpolation for Byzantium. Neugebauer records useful computations for the main systems. In regard to medieval astrology, perhaps the most influential tables of climata were those issued by the Arabian astrologer Alfraganus, which formed the basis of measurement in Dante's *Divina Commedia*: these were derived from the Hellenistic seven climata, converted into widths in miles of the supposed inhabited area of the Earth, on the basis of $56\frac{2}{3}$ miles being equal to one degree, which naturally leads to a serious underestimation of dimensions in the table.

climacteria A term used for the 'Grand Climactic' of the 800-year cycle of the CHRONOCRATORS. Distinguish CLIMACTERICS.

climacteric conjunction A term used of the conjunction of the two planets Jupiter and Saturn, the CHRONOCRATORS.

climacterics From ancient times it has been believed that certain years in the course of life are more liable to danger and/or change than others. The most important of the climacterics are the septenary years – 7, 14, 21, 28, etc. – which are associated with lunar periodicities in the progressed chart, and which have been linked by such esotericists as Blavatsky and Steiner with the soul's growth. Esoterically, the full spiritual incarnation of the human does not take place until the fourth septenary (28), and some esoteric astrologers insist that properly speaking this is the beginning of life: certainly, this is one of the most important of the climacterics. The important climacteric associated with Saturn usually falls in the 30th and 60th year, when the planet transits its natal point in a chart (just over 29 years) – but see also CHRONOCRATORS. The climacteric period associated with the Moon's node falls on (approximately) 18 years and 2 months, and 36 years and 4 months: see LUNAR NODE CYCLE. Individual climacterics may only be accurately determined from the personal chart, of course. The so-called 'historical climacterics', which have been linked with both PRECESSION and the Arch-angelic periods (see SECUNDADEIAN BEINGS), are really concerned with the sequences of periodicities, rather than with the key-year at the beginning of such a period – for example, in terms of the Archangelic periods, the years 1881 and 1882 have been given as the climacterics for the beginning of the Michaelic Age in which we now live. However, such a key-year is not strictly speaking a year of peril or of change: it merely dates the beginning of a new era. So also with the periods of transition which mark the progression from one Age to another, in terms of precession.

climata   See CLIMA and SEVEN CLIMATA.

clock time   See TIME.

co-ascendant   A term proposed as applying to the Ascendant of the so-called CO-LATITUDE of the place of birth. As Dean says, somewhat ominously, of this and the related CO-VERTEX, 'very little is known about them'.

Coeli Cingulum   See VIA LACTEA.

Coelus   A name given by the esoteric astrologer Thierens to a future development of the planet URANUS in his ASTRAL SPHERES. It is said to represent the principle of emanation.

cognitive correspondence   While a number of variant lists of planetary associations with the cognitive faculties of man have been preserved in the astrological tradition, the generally agreed (if simplistic) series are:

ME: Sight   VE: Touch   MA: Taste   JU: Smell   SA: Hearing

Steiner introduced an entirely new list of cognitive correspondence which are linked not with the traditional seven senses, but with twelve senses, which are associated with the corresponding signs, set out in table 26.

TABLE 26

Sign	*Sense*	*Sign*	*Sense*	*Sign*	*Sense*
AR	Speech	LE	Life	SG	Taste
TA	Thought	VG	Movement	CP	Sight
GE	Ego	LB	Balance	AQ	Warmth
CN	Touch	SC	Smell	PI	Hearing

---

co-latitude   The co-latitude of a place is the complement of its latitude, obtained by subtracting this latter from 90 degrees.

cold   See FOUR QUALITIES and PRINCIPLES.

cold planet   Saturn is sometimes called the cold planet: the Moon is frequently wrongly called 'cold', whereas it is in fact made 'hot' or 'cold' only in terms of the aspects received. See FOUR QUALITIES and PRINCIPLES.

Coley table   See ASTRO-DIAGNOSIS.

Columba   Constellation, from about 5 degrees Gemini to 17 degrees Cancer, and from 27 to 43 degrees south of equator. First noted by Bayer in 1603, it was originally Columba Noae (Noah's Dove) and probably because of its proximity to ARGO NOVIS sometimes called the Ark. Astrologers grant the asterism a gentle nature, and its alpha PHACT is also well-disposed.

colures   Name given to two great circles projected into the Celestial Sphere, cutting the equator at right angles, passing through the poles, one of which cuts the two equinoctial points (called the 'equinoctial colure'), the other of which cuts the solstitial points (the 'solstitial colure').

Coma Berenices   Constellation, from approximately 17 degrees Virgo to 12 degrees Libra, and 25 to 31 degrees north of the equator, with a somewhat chequered history as an asterism. Recognized by Eratosthenes as 'Ariadne's Hair', and for centuries merged with the constellations Leo or Virgo, it was finally catalogued by Tycho in 1602 under the modern name. The hair itself was said to have been that of the wife of Ptolemy Euergetes, a stolen sacrificial gift. The asterism has many different names. It is accorded a beneficial influence, which tends towards dissipation.

combust When a planet is conjunct the Sun, within an orb of 5 degrees, it is said to be 'burnt' by the Sun, and is termed 'combust' or said to be 'in combustion'. Some authorities claim that the planet in question is weakened by this combustion, but a study of Mercury (frequently combust, as one of the inner planets) would suggest otherwise. See also CAZIMI.

comets Luminous bodies, with long tails, in orbit around the Sun. Comets are highly eccentric (at perihelion perhaps only half a million miles away from the Sun, but at aphelion hundreds of thousands of miles beyond even Pluto), and inclined to the ecliptic in almost any angle up to 90 degrees. The shortest known revolution for a comet is just over 3 years, the longest being measured in terms of centuries. Halley's Comet returns in 75-year periods. The term is derived from the characteristic 'tail' (from the Greek kome, hair, relating to the nebulous dust of the tail) the entire phenomenon of head and tail being the *kometes*, or 'hairy star'. The tails may extend over vast distances: Halley's of 1910, for example, spanned an arc of over 150 degrees, its length being 28 million miles. In astrological literature comets have generally acquired a bad reputation, and the earliest classifications of Ptolemy were on the level of omina texts. In medieval astrology in particular the comets took on considerable importance, though generally as presage of evil and change: the appearance of the comet, its position in the zodiac (first appearance), its relation to the ecliptic belt, and other factors, were the determinants of interpretation, and few major comets have failed to make their own special reputations in literature. The comet of 1066, enshrined as detail in the Bayeux Tapestry, may have meant the death of Harold, and the momentous theft of England, but seen from the point of view of the Normans it was clearly a 'good thing' – 'a comet like this is only seen when a kingdom wants a king', William is supposed to have remarked to his courtiers. Usually, however, the presage is evil: the doom of the entire Christian world (at the hands of the Turks) was expected of the 1456 Halley. The comet of 1528 was 'so horrible and dreadful, and engendered such terrors in the minds of men, that they died from fear alone . . .' Typical of the more sane astrological literature, is that of Pearce: the great comet of 1881 was 'first seen in RA 74 75, which is the line of the 16 degrees of Gemini which had rule over Egypt' – the war in Egypt (1885) and in the Sudan (1883, etc.) 'quickly followed'. Pearce records that the astrologer Zadkiel foretold the Italian war of 1859 from the appearance of Donati's Comet of the preceding year. Most of the predictions are of the Mundane astrological type, and the truth is that comets appear to play little part in natal astrology – however, it has been claimed that the transit of a comet over a natal point is significant. Even modern astrological references to comets are not confined to esoteric literature: the Comet Kohoutek of the seventies produced a spawn of pseudo-astrological speculations little advanced on medieval sortilege.

commanding A specialist term much misunderstood in modern astrology. The terms 'commanding' and 'obeying' were applied by Ptolemy and his followers to pairs of signs at equal distance from the equinoctial point. Thus, Taurus has command over Pisces (which in turn 'obeys' Taurus). The pairs, with the commanders first, are Taurus-Pisces, Gemini-Aquarius, Cancer-Capricorn, Leo-Sagittarius and Virgo-Scorpio. Aries and Libra, being on the equinoctial points, are not included in the original schema. As Bouche-Leclercq points out, the original notion appears to have been that these pairs of signs 'heard' each other, the concept of one 'obeying' the other being a mere elaboration. Unfortunately, Ptolemy's ideas were misunderstood, with the result that later astrologers insisted that all the signs from Aries to Virgo were 'commanding' and those from Libra to Pisces inclusive were 'obeying': the astrologer Bonattis, in his attempt to define *antiscia* gives the use of *imperantes* (commanding) and *obedientes* (obeying) as though directly from Ptolemy. As a result, the pairs were totally ignored, and a strange notion arose that the first six signs were somehow more fitted to command or lead, while the last six were fitted to obey or follow. The term is now clearly obsolete, and when used it is almost always wrongly applied.

commence A planet coming to the Ascendant is said to 'commence', and afterwards it is said to 'increase'.

common fire See FIRE ELEMENT.

common planet Mercury is sometimes called the common planet, presumably because of its ANDROGYNE nature.

common signs Another term for the MUTABLE SIGNS – Gemini, Virgo, Sagittarius and Pisces. See MUTABILITY.

comparison chart Term applied to the charts used to combine all the different elements from the two (or more) figures in connection with the study of SYNASTRY. It is sometimes wrongly called the COMPOSITE CHART.

composite chart A single chart detailing astrological material relating to two people, obtained by incorporating into the figure the mid-points between the two sets of planetary and nodal details within two horoscopes. This chart must be distinguished from the RELATIONSHIP CHART.

computation The computing or CASTING of a horoscope usually requires the conversion of a given birth-time and birth-date to Greenwich Mean Time, as a prelude to calculating the Ascendant of the houses (by whatever system) of the chart, and inserting the planetary and nodal positions. For all specialist terms used in the following account of the stages of computation, see TIME. The first stage involves the recording of birth-date and birth-place, from which latter may be obtained the latitude and longitude from a gazetteer. The second stage involves the conversion of birth-time to Greenwich Mean Time: this requires the amendment for Zone standard time (see TIME ZONES), and for Summer Time, Double Summer Time and/or War Time and Daylight-Saving Time. The third stage requires the consultation of an ephemeris, to determine the sidereal time at Noon, according to Greenwich Mean Time, and the adjusting for the interval between this time (given for either noon or midnight, depending upon ephemeris) and the acceleration on the interval (explained in TIME), at the rate of 10 secs per hour, and pro rata. The longitudinal equivalent is then calculated, at the rate of 4 times each degree of longitude, which number of minutes is added if the birth is to the east of Greenwich, and subtracted if to the west. This computation gives the local sidereal time at birth, a figure which may be used to determine (from a table of houses) the Ascendant, Midheaven, and intermediate house cusps, in terms of the latitude of the birth. For latitudes south of the equator, the procedure is to add 12 hours to the local sidereal time at birth, and reverse the signs so that the sign opposite in the zodiac to the one given in the table of houses is used for the Ascendant. It is not possible to construct accurate charts without reliable ephemerides and tables of houses, and in some cases other specialist documentation (such as a gazetteer). For modern horoscopes the series *Die Deutsche Ephemeride* of Otto-Wilhelm-Barth-Verlag is of excellent quality. For historical charts the 5-day and 10-day interval lists of Tuckerman, *Planetary, Lunar, and Solar Positions* is recommended. The problem of time zones and time changes is dealt with by Doane (esp. *Time Changes in the World*). Tables of houses are published by several astrological publishers – those of Raphael, *Tables of Houses for Northern Latitudes*, list from the equator to 50 degrees north, with the houses for Leningrad (nearly 60 degrees north):

the system is Placidean. The Chacornac *Tables de maisons* gives a wider range of latitudes, but is also calculated according to Placidus. Wemyss gives a set of tables (for a limited range of latitudes) according to Campanus.

computerized chart More often than not, the so-called 'computerized chart' is not a chart at all in the sense of being a stereographic projection: it is usually a schematic figuration of the heavens for a particular time – most often a birth-time. In the more sophisticated systems a stereographic projection may be implied in the calculations, but really a computerized chart is a reading rather than a chart. Most of the charts available through computerized systems are little more than a complex extension of popular commercial astrology (of what specialist astrologers are wont to call 'cook-book astrology'). The print-out of computerized material for the purpose of personal interpretation belongs to an altogether different category, however, and may be regarded as relieving an astrologer of much of the tedium of calculation. In regard to the complexity to which computerized charts proper may descend, Dean has a most apposite phrase, for 'computer charts whose capacity for overkill is exceeded only by their unreadability'.

conception chart Usually an alternative name for the *PRE-NATAL CHART cast for the (putative) moment of conception by mathematical computation from the natal chart. The same name has been applied to Collin's schema of logarithmic scales related to the 280-day period of gestation, linked with the 'creation of a body', which might be measured in terms of the Moon and Mercury cycles – though this schema was not a horoscope chart.

conceptive signs Another term for the four *FIXED SIGNS, Taurus, Leo, Scorpio and Aquarius.

conditio See SECTA.

configuration A term sometimes used of a pattern of ASPECTS within a chart, and sometimes of the whole chart itself. In recent years a specific application has been made of the term to cover a number of particular aspects: see MAJOR CONFIGURATION. In traditional astrology, all aspects were sometimes referred to as configurations.

conjunction An aspect maintained by two or more planets posited within the same degree of longitude. It is sometimes called the aspect of prominence, in that it brings the workings of the planets involved into a prominent position relative to the chart: but in this connection, see also COMBUST. The

ORB allowed a conjunction is not agreed among authorities, but some allow as much as 10 degrees for a platic aspect to be operative.

constellation A pattern of stars and star-groups: though the term is very frequently used wrongly to denote the SIGNS of the zodiac, which are not star-groups at all. The majority of the constellations were named and demarcated in ancient times, and all the major ones were ascribed specific influences which were regarded as being projected to the Earth (in addition to the influences cast by the fixed stars of which the constellations were themselves composed – see FIXED STARS and LUNAR MANSIONS). Each of the major constellations, including the important ones which bear the same

Andromeda	Antinous	Antlia
Aquarius Constellation	Aquila	Aries Constellation
Auriga	Bootis	Caelum
Cancer Constellation	Canes Venatici	Canis Major
Canis Minor	Capricornus	Caput Serpens
Centaurus	Cetus	Columba
Coma Berenices	Corona Australis	Corona Borealis
Corvus	Crater	Cygnus
Delphinus	Equuleus	Eridanus
Fornax	Gemini Constellation	Hercules
Hydra	Leo Constellation	Leo Minor
Lepus	Libra Constellation	Lupus
Lyra	Microscopium	Monoceros
Ophiuchus	Orion	Pegasus
Pisces Constellation	Piscis Australis	Puppis
Pyxis Nautica	Sagitta	Sagittarius Const.
Scorpius	Sculptor	Scutum Sobiekii
Serpens	Sextans	Taurus Constellation
Triangulum	Vela	Virgo Constellation
Vulpeca		

names as the twelve signs of the zodiac, is treated separately under the headings. The abbreviations for the above names are set out under ABBREVIATIONS. For a list of the ancient Hebraic, Akkadian, Babylonian and Egyptian (demotic) names for the zodiacal constellations, see CONSTELLATION NAMES. See also CONSTELLATION MAPS.

constellational zodiac A term used of a SIDEREAL ZODIAC divided into twelve unequal asterisms (not signs), the boundaries or cusps of which are to some extent arbitrary, and hence subject to some argument in specialist astrological circles. The definitions in degrees of arc are given as:

| AR 24 | TA 36 | GE 28 | CN 21 | LE 35 | VG 46 |
| LB 18 | SC 31 | SG 30 | CP 28 | AQ 25 | PI 38 |

However, the portrayal of the ecliptical asterisms of the MODERN ZODIAC actually involves 13 asterisms. There are several, differently defined, constellational zodiacs in use today, the main differences arising from the definition of arcs and extent. See CONSTELLATION NAMES.

constellation chart A term sometimes used as synonymous with the CONSTELLATION MAPS, and sometimes in reference to a horoscope chart cast according to one of the several constellational systems, rather than according to the more frequently used zodiacal-sign systems.

constellation maps Sometimes (confusingly) called CONSTELLATION CHARTS or 'star charts', these are maps or delineations of the heavens in terms of individual stars or asterisms, often overlaid by schematic or pictorial IMAGES. There is much confusion in the early terminology: the Greeks called the asterisms by the modern equivalent of 'signs' (*sigmata* and *teirea*), whilst their *zodia*, often in modern times wrongly translated as 'animals', was used of the famous ecliptical twelve (both for sign and constellation). Their *meteora*, 'things in the heavens', was more general, and did not always apply to our modern conception of stars, even when used in a sense to embrace planets. The Romans called the individual bodies *astra* and *sidera*, and took the Greek *sigma* for the modern signs, to be distinguished from the constellations, which referred (as now) to all asterisms, including the so-called 'zodiacal constellations'. That the images should be of 'living things' is more than an etymological nuance, but seems to have arisen from the notion that the stars themselves were living beings (an idea mooted by Aristotle, though often wrongly attributed to Origen, some 7 centuries later). In the 3rd century BC Aratos mentions 45 constellations, and most of these were continued into the Aratos-derived maps and texts, though Hyginus (1st century AD?) was equally influential with his 42 figures, from which the popular woodcuts used in a multitude of 'histories' are derived. Most of the maps which have survived from early astronomy/astrology are mere schemas, conveniently mythopoetic, and even esoteric, picture books, as in the interesting example given by Aldus in figure 19 as a supposed (and not very accurate) summary of the ancient lore. The relationship of the twelve constellations to this planisphere is entirely fictitious, though in some ways little different from the projections of the sophisticated constellation map of the so-called DENDERAH ZODIAC. By comparison, the Arabian constellation

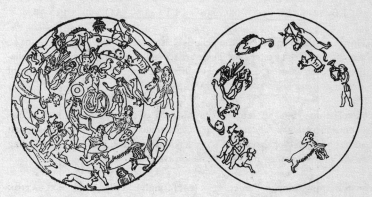

Figure 19: Antique constellation map (after Aldus), with zodiacal constellations separately charted.

maps are usually models of rectitude, and reflect the important contribution made by the Arabs to the lore of fixed-star interpretation in astrological circles. However, the accurate and scientific projection of the Celestial Sphere is a relatively modern thing: the medieval astrologer tended to rely upon tables, and even upon direct observation, rather than upon constellational planispherical models. In his *Syntaxis*, Ptolemy listed 48 constellations, described in terms of 1,028 stars, which has given rise to the absurd notion that the ancients knew only that number of stars. The subsequent history of the naming and listing of the stars in constellation lists and maps is beyond the scope of this present work, falling as it does into the realm of the history of astronomy, yet the following developments may be noted as being of importance to the development of astrology. The finest sky maps which are intended to serve other purposes than entertainment were originated after the 16th century, the earliest of the classics being the *Uranometria* of Bayer, published in 1603, the symbolic images of which influence almost all subsequent atlases. The finest of the 18th-century maps was probably that produced by Flamsteed in his posthumously published *Atlas Coelestis* of 1729, though Bode's *Uranographia* of 1801 was probably the last of the magnificently artistic (as opposed to strictly scientific) productions, before aesthetics fell foul to scientific intent. The Bayer figures (and indeed those of Flamsteed) relied heavily upon the woodcuts made by Dürer for the two hemispherical charts set out schematically by Heinfogel and Stabius, though ever since called the 'Dürer Map' (figure 20). The fluidity of boundaries (seemingly

Figure 20: The so-called 'Dürer Map', cut by the artist Dürer after Heinfogel and Stabius in AD 1515.

favoured even by Ptolemy, perhaps with greater wisdom than the modern mind admits) along with the general lack of consensus as to the nature and form of the IMAGES, and even the feasibility of introducing new asterisms (usually merely severed from the older ones) made each of the early atlases unique as works of art. There were in these classical artistic productions some attempt at objective mensuration, but in the scientific climate their fluidity and lack of consensus was anathema, and Herschel was among the first to attempt to put an end to imprecision by proposing to divide the skies into spherical quadrangles, thus producing somewhat arbitrary (though clearly defined) boundaries. His idea found little favour, however, and it was not until 1927 (after many interesting essays) that the International Astronomical Union accepted the divisions (of surprisingly arbitrary boundaries) proposed by the Belgian Delporte, which are incorporated into the constellations of figure 44. Delporte's *Atlas Céleste* is much used by astronomers, but has been (rightly) the subject of much learned criticism in astrological circles, not least because its complex definition actually incorporates OPHIUCHUS as an ecliptical constellation, and thus makes of it a zodiacal figure. A highly significant change in form from the ancient to the modern is reflected in the abandonment of the ancient images for the rectilinear star-groups. This is particularly interesting, since the early maps (and even the star lists) were linked with astrology more through mythology than any precision of definition in regard to celestial latitude and longitude (though contemporaneous lists did provide more-or-less accurate information relating to the conversion of star lore to astrological practice). It is perhaps significant that the modern constellation maps should attempt to dispense with living images, and have proposed in their place a system of rectilinear abstractions, for it is suggestive of the dehumanization process which has unfortunately accompanied the development of the scientific outlook in astronomy.

constellation names The astrologer Fagan insists that the constellation names (among which he presumably includes the zodiacal names) bore the

TABLE 27

Const.	Babylonian	Meaning	Demotic	Meaning
AR	Lu.hun.ga	hired labourer	Pa.Yesu	fleece
TA	Gud.an.nu	heavenly bull	Pa.ka	bull
GE	Mash.tab.bal.gal.gal	great twins	Na-hetru	two children
CN	Al.lu	crab (?)	Pa.gerhedj	scarabaeus
LE	Ur.gu.la	great dog, lion	Pa-may-hes	fierce lion
VG	Ab.shim	spike of corn	Ta.reply	female
LB	Zibanetum	scales	Ta-Akhet	sunrise place
SC	Gir.tab	scorpion	Ta.Djel	snake
SG	PA.BIL.SAG	overseer*	Pa.nety-ateh	arrow
CP	Sukhuyrmashu	goat-fish	Pa-her-ankh	goat face
AQ	GU.LA	giant	Pa-mu	water
PI	Shim.mah	great swallow	Na-Thebeteyu	fish

same meanings in the 8th century BC as they do today – a point not without dispute in academic circles, as a reading of the scholarly Sachs will confirm. The translation offered by Fagan for the Babylonian terms, and for the

constellation names

Egyptian (demotic) terms of the Greco-Roman period, while not beyond
question, have in recent years been popularized in several histories of the
zodiac. These are set out in table 27, in which the translations under
'Demotic' refer to the hieroglyphic demotic form, rather than to a direct
translation. In any case, it is clear from the translations provided by Fagan
that his basic premise is questionable even in the list he publishes. The
translation for the Babylonian Sagittarius, which Fagan gives as 'overseer
of the Fire Ordeal' is contentious (see Akkadian in table 28). Probably more

TABLE 28

Const.	Akkadian/Persian/Assyrian	Meaning	Hebrew	Chaldean
AR	I-ku-u	prince	Taleh	Kusarikkut
TA	Te te	bull of light	Shor	Te te
	Gut-an-na	heavenly bull		
GE	Do Patkar	two figures	Teomin	Tua-mu
	(Chaldean?) – Tammech	two figures		Masmasu
CN	Nan-garu (and) Puluk-ku	solar north gate(?)	Sarton	Al-lul
LE	Pap-pil-sak	great fire	Ari	A-ru
	(Babylonian) – Aru	lion		
VG	Khusak	ear of wheat	Betulah	Shi-ru
	Bealtis	wife of Bel		Ki
LB	(Euphratean) – Sugi	chariot yoke	Moznayim	Zibanitu
SC	Girtab	stinger	Akrab	Akrabu
SG	Ban	bow star (?)	Kasshat	Pabilsag
	Kakkab kastu	bow constellation		
	(Cuneiform) Utucagaba	smiting sun face		
CP	Shahu	ibex	Gedi	Sakhu
	Munaxa	goat-fish		Enzu
	Su-tul	yoke		
AQ	Ku-ur-ku	flowing-waters seat	Deli	Mulgula
	Rammanu	god of storm		Gu
PI	Nunu (and) Zib	fish*	Dagim	Nunu or Zib

* The 'Zib' of Akkadian Pisces may also mean 'boundary'.

reliable are the early constellation names derived from various (unnamed)
sources by Allen, which correspond fairly closely to those recorded by
Thierens for the Hebrew cosmogony, in table 28. The list given under
'Chaldean' in this table is from Brau: it resembles the Babylonian and

Akkadian list, reminding us that 'Chaldean' was once a generic name for 'astrologer'. But see also ZODIAC.

continuous astrology   A term used by the historian Pingree in a brief but excellent history of astrology, as reference to that form of astrology 'designed to guarantee the astrologer constant patronage'. He is writing specifically of Hellenistic astrology, in so far as it may be gleaned from Dorotheus of Sidon, which proposed that new horoscopic diagrams (antigenesis) must be cast for every anniversary (or even at the beginning of every month, week, day or hour) for comparison with the native's birth chart. The idea is a primal conception underlying the various systems of TRANSITS and PROGRESSIONS, all of which relate to the radical chart. Continuous astrology is really the equivalent of a branch of PREDICTIVE ASTROLOGY which also requires a related series of progressed charts.

contra antiscion   See ANTISCION.

contra-degree   See PROPORTIONAL ARCS.

contraparallels   See PARALLEL.

converse directions   These are DIRECTIONS applied in an order opposite to the natural order of the signs. Many astrologers regard converse directions as an invalid extension of the directing technique. See PROPORTIONAL ARCS.

cooperta   Name, perhaps meaning 'covered', given to the 15th of the LUNAR MANSIONS in some medieval lists. Probably located by one of the less powerful asterisms bordering the constellations Virgo and Libra.

co-ordinate   In a specialist astrological sense, any of a number of magnitudes or geometric constructions – usually projected into the CELESTIAL SPHERE – which permit the astrologer to determine a position precisely.

Copernican system   The cosmoconception which laid the basis for the most widespread modern theory of the nature of the solar system. The name is derived from the Latinized form of the Polish name Nikolaj Kopernik, who set out his proposed model in his *De Revolutionibus Orbium Coelestium* (figure 21), which was not published until the end of his life, in 1543. The original Copernican model (now much modified) proposed that the apparent motions of the planets might be explained within a heliocentric system, with the earth rotating diurnally on its own axis, and with each heavenly body exerting its

Figure 21: Manuscript drawing of the Copernican system from 'De Revolutionibus Orbium Coelestium' manuscript, published in 1543.

own gravitational force. He also placed the fixed stars at a far greater distance from the solar system than had been visualized in the geocentric PTOLEMAIC SYSTEM, the cosmoconception most widely held to be true in his day. In the same extensive text, Copernicus also proposed a new theory to account for the PRECESSION of the equinoxes.

Copula   Name given to the Spiral Nebula, now linked with CANES VENATICI.

Cor Caroli   Fixed star (double), originally included in the image of CANES VENATICI in the collar of the dog Chara, but set apart by Halley in 1725, as a mark of respect for King Charles II.

Cor Hydrae   See ALPHARD.

Cor Leonis   See FIFTEEN STARS and REGULUS.

Cornua Arietis   Name, 'the horns of Aries', given to the 1st of the LUNAR MANSIONS in some medieval lists. Probably located by the star MESARTHIM.

Cornua Scorpionis   Name, curiously 'the horns of Scorpio', given to the 16th of the LUNAR MANSIONS in some medieval lists. Probably identified by the fixed stars ZUBENELSCHEMALI and ZUBENELGENUBI.

correspondencies

Corona Australis Constellation, approximately 2 to 12 degrees Capricorn, and 36 to 45 degrees south of the ecliptic, the southern crown. Once pictured as the crown of Sagittarius, the Corona Sagittarii, the circular shape of the asterism has suggested very many different names, from Tortoise (Al Kubbah in Arabic, Pi in Chinese), to garlands and Rota Ixionis, the wheel of Ixion. Ptolemy likens its influence to Jupiter with Saturn.

Corona Borealis Constellation, approximately 2 to 17 degrees Scorpio, and from 27 to 33 degrees north of ecliptic, the northern crown, said to be the garland given to Ariadne at her marriage. It is actually a wreath in Greek mythology and star lore, and now with many names, sometimes indeed called after its most powerful fixed star ALPHECCA which, like the asterism itself, is largely of a benevolent influence in a chart.

Corona Super Caput Scorpionis Descriptive name, the crown on the head of the Scorpion, of the 17th of the LUNAR MANSIONS in some medieval lists. Identified by the fixed star AKRAB.

correction When limited to specialist application relating to the computation of a horoscope, the adjustment of local mean time to sidereal time in order to obtain relevant birth data. See TIME.

correlation In general astrological application, this term is used to denote the comparison of charts, or selective data taken from charts, and especially to the comparative method involved in SYNASTRY. The term is also used more specifically in the idea of correlation analysis, which is a statistical method of extrapolating astrological data.

correspondence In its occult sense, a term derived from the very earliest astrological lore, and used to denote the 'occult' or 'hidden' relationships between natural forms and spiritual causes, or between the microcosm and the macrocosm, which is observed acting through the SYMPATHY expressed in the CHAIN OF BEING. Swedenborg, who used the term extensively, maintained that every natural object (that is, every sublunar object) is a result of the working of a spiritual cause, and that if the nature of the correspondence between the higher and the lower is understood, then the spiritual may be read through its physical symbol. This doctrine lies at the basis of Rosicrucian meditative endeavour.

correspondencies A translation of the traditional idea of CORRESPONDENCE, intended (presumably) to point specifically to the idea of occult correspondences between things. In his brilliant study of Egyptian symbolism, Schwaller de Lubicz traces the attributes, or correspondencies, of modern astrology to the ancient Egyptian principle of related specific proportions and cubic measurements to definite 'Neters', or principles, which correspond (approximately) to the Platonic ideals.

Cor Scorpii See ANTARES and FIFTEEN STARS.

Cor Scorpionis Name given to the 18th of the LUNAR MANSIONS in some medieval lists. Identified by the fixed star ANTARES.

Cor Serpentis Fixed star, the alpha of SERPENS, set in the neck of the snake, 'the serpent's heart', and sometimes called Unukalhai, from the Arabic 'neck of the snake'. Ptolemy says that it has an influence like that of Saturn and Mars – it is said to bring accidents and danger from poisons.

Cor Tauri Name, 'the heart of the Bull', given to the 4th of the LUNAR MANSIONS in some medieval lists. Probably located by the star ALDEBARAN.

Corvus Constellation, approximately 5 to 15 degrees Libra, and from 7 to 25 degrees south of the ecliptic, a raven to the Greeks, crow to the Romans, and earlier with a variety of bird-like names: it is part of the asterism-arc of the Chinese RED BIRD, and Orebh (raven) to the Hebrews. The various legends of the dishonesty of the bird have been projected into astrological lore, for Ptolemy likens its asterism to Mars conjunct Saturn, and later astrologers link it with craftiness, ingenuity and lying. Only the delta, ALGORAB, plays a part in the astrological lore: the alpha, Al Chiba to the Arabs, appears to have decreased in brilliance, and to have changed colour, during the centuries.

cosignificator A much overworked term. In horary astrology, a planet in conjunction with the SIGNIFICATOR of anything is its cosignificator, simply because it must be taken into consideration when the matter is being adjudged. In another sense, the Moon is regarded as cosignificator with the LORD OF THE ASCENDANT, whatever the placing of the Moon itself. In horary charts relative to marriage (but seemingly not in charts cast for SYNASTRY), the female planets, Moon and Venus, are the cosignificators if the querent is a man, while the male planets, Sun and Mars are cosignificators if the querent is a woman. Further meanings of the term are derived from the ancient tradition, in which a parallel is drawn between the basic chart (here called the MACROCOSMIC CHART) and any particular manifestation of it (that is, any chart at all). Thus, Aries is said to be a cosignificating sign for the Ascendant of any horoscope, simply because this sign stands on the Ascendant of the MACROCOSMIC ·CHART. This theory of cosignification is taken to levels of theoretical refinements which, as Wilson says, 'render the science ridiculous'.

cosmecology   A neologism for a study which attempts to correlate changes and effects of a cosmic nature with observed cycles of behaviour and periodicities on the earth and human beings.

cosmical rising   In a specialist sense, a term applied to the heliacal rising or setting of a planet. See ACRONYCAL.

cosmical setting   A term used of the setting of a planet or a star with the Sun in the evening.

cosmic cross   When two planets in opposition are in square aspect to a third planet, they are all three said to be united in a cosmic cross, sometimes called a T-square. While the formation is undeniably of a dynamic and powerful nature, its force depends upon the qualities of the three planets involved in the aspects. See also GRAND CROSS.

cosmic cybernetics   A term used by the astrologer Landscheidt for his highly personal quasi-mathematical, quasi-mystical, interpretation of certain astrological ideas, connected with the development of *COSMOBIOLOGY.

cosmic night   A modern term derived from the theosophical use of the Sanskrit term *pralaya*, which relates to a period of obscuration or sleep between the 'days' of creative evolution. See YUGA, where the very terms relating to 'twilight' imply a separating *pralaya* between the ages.

cosmic pantheons   Throughout the history of astrology there has been a tendency for astrologers (and for that matter, non-astrologers) to create CORRESPONDENCIES between pantheons and the zodiac. One of the earliest attempts, which has left some mark on the development of astrology, was that of Manilius, which (for all its astrological unlikeliness) at least had the advantage of being related to only one pantheon, the Roman:

AR	Minerva	CN	Mercury	LB	Vulcan	CP	Vesta
TA	Venus	LE	Jupiter	SC	Mars	AQ	Juno
GE	Apollo	VG	Ceres	SG	Diana	PI	Neptune

In this connection, see GABII ALTAR. In fact, a vast number of similar correspondences, more or less poetic and imaginative, have been recorded, and the important ones have been listed by Thierens and Stone, whilst a great number are more or less taken for granted by such esotericists as Blavatsky and Bailey. Surprisingly, many of these cosmic analogies may be traced to that arch-syncretist, Kircher, whose well-known zodiacal dial (figure 22) actually hints at several important esoteric connections, in summary of his vast (if sometimes imaginative) erudition. Thierens neatly summarizes the modern syncretism along the following lines: Aries is personified

Figure 22: The syncretic Kircher zodiac depicting a number of cosmic relationships (after Kircher).

by Osiris, Taurus by Isis, Cancer by the dual Horus and Anubis; Leo is seen as one of the 'many kings in Myths'; Virgo by the Virgin Mary, the Mother of Krishna, and by Christ 'at the back of which the image of Isis reappears'; Libra is the 'ideal ruler', such as we find in the legends of Rama and Solomon; Scorpio is represented in dragon-forms, in the oriental Nagas, and in the Chinese Lung dragons, as well as in the Phoenix (and for that matter, in the Ansa swan); Sagittarius is personified in Zeus (Jupiter); Capricorn in the Nature-god Pan, himself half 'bestial', as well as in Chronos, the ruler of Time; Aquarius is Ouranos, whilst Pisces is inevitably Poseidon (Neptune), and the ruler of the Cosmic Sea. This syncretic cosmic pantheon has been adjusted by numerous modern accretions, many of which go back to the early conflicts between the Romanizing tendency of Christianity and the early gnosticism. The imagery from these conflicts deeply influenced art and literature: thus Christ has been figured as Taurus, as the descending Logos (with Mithraic undertones) as well as Pisces, the sacrificial sufferer, the

cosmic picture

inaugurator of the New Age. Some of this symbolism may be seen in the 12th-century SAN MINIATO ZODIAC. The Virgin Mary is represented in Virgo, and indeed the fixed star Spica of the constellation figured on her dress (as the Stella Maris) well into the 16th century. Each of the four EVANGELISTS are also linked with the Fixed signs of the zodiac – Taurus as St Luke, Leo as St Mark, Scorpio as St John and Aquarius as St Matthew. But see also BIBLICAL ASTROLOGY.

cosmic picture A cosmic picture, or planetary picture, is a term used by Ebertin in regard to diagrammatized images of planetary relationships derived from a COSMOGRAM by means of his NINETY-DEGREE DIAL. Such terms are derived from translations from the *Planetenbilder* proposed by the German Witte, and used by Ebertin in his COSMOBIOLOGY, based on the interpretation of a series of pictures of chart groupings derived from aspects and MID-POINTS. Witte distinguishes three planetary pictures, which may be combined in a variety of ways. The mid-point between planets which is occupied by an actual body is such a picture. Several 'half-sums' (mid-points) with a communal axis, though not necessarily with a planetary body posited on this communal axis, is another picture. A combination of a mid-point between the two planets which is itself aspected from another planet (or planets) to the aspectal order of 45, 90 and 135 degrees represents another such picture. Such planetary pictures give rise to what Witte calls 'equations', which are really keyword equivalents. A representative cosmic picture (which is also called a 'cosmic structure pattern') shown in figure 23, derived from the

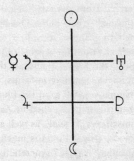

Figure 23: Cosmic picture (after Ebertin).

COSMOGRAM of figure 24, is designed to show the cosmic picture of the Sun in this dial. Uranus and Mercury (conjunct Saturn) are approximately equidistant from the Sun along the dial, and are accordingly recorded on the higher horizontal bar. Jupiter and Pluto are equidistant from the Sun, and are recorded on the lower horizontal bar. Since the Moon is in opposition to the Sun on this dial, it is placed at the bottom of the vertical line of the cosmic picture, completing this graphic assessment of the relationship of the planets to the Sun.

cosmic structure patterns See COSMIC PICTURES.

cosmobiology A term coined by the Austrian doctor Feerhow, and first used in connection with astrology by the Swiss astrologer Krafft. It is especially used by the astrologer Ebertin in relation to his own modernized and highly personalized astrology, relating to a description of the possible correlations between cosmos and organic life, and the effects of cosmic rhythms and stellar notions on man. Ebertin, whilst profoundly aware of the traditional astrology, sought to contrast this system with his own in order to interpret destiny in terms of somatopsychic dispositions, and to 'free astrology from occultism'. He used the term COSMOGRAM for his variations of the horoscope figure, and made much use of PLANETARY PICTURES involving MID-POINTS (which play an important part in the interpretative system). The YEARLY RHYTHM tabulation, used as a basis for chart interpretation, reflects the traditional astrological view, refined by much research, and by the application of a disciplined statistical approach to the subject. The Germanic abbreviation KdG (*Kombination der Gestirneinflüsse*, from the title of a book published by Ebertin in 1940) is often used to denote cosmobiology.

cosmodynes A term derived from the astrological research connected with the Church of Light (published largely under Doane) in regard to a system of numerical values assigned to planets, according to strength of aspect and angularities. These numerical values for aspects and parallels range from 0 to + 10, the closer the aspect, the higher the number – thus the conjunction/ opposition merit the full 10, the square/trine 8, the sextile 6, the semi-square 4, and so on. Similar systems have been suggested by other astrologers (notably Carter), though not on such a firmly based research. These values are weighted (see also INTENSITIES) for orb, by substraction of 1 degree for each degree of orb. There is also a numerical value attributed to planets in regard to proximity to the four Angles (on a + 8 to + 14 range – the closer, the higher). The strongest planet in a chart is that with the highest total. See ASTRODYNE.

cosmogram A method of chart construction (and highly specialized interpretation or adjustment) proposed by Ebertin, far removed from standard

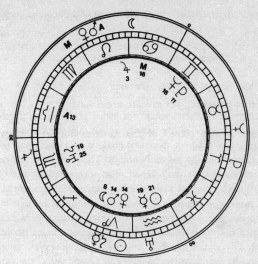

Figure 24: Cosmogram (after Ebertin).

CHART SYSTEMS. The method is based on the conversion of a normal ecliptic chart to a threefold schema, called a NINETY-DEGREE DIAL, which is usually portrayed on the cosmogram as an outer concentric dial, divided into 90 arcs (not degrees, which suggests that the term is something of a misnomer). The data in the inner ecliptic projection is converted to the outer dial in terms of the following rationale, which is here related exclusively to figure 24. The Moon, which is in 8 CA (Ebertin wrongly reports it in 7 CA) is translated to the 8th degree of the dial. This conversion is regarded as valid, since Cancer is one of the Cardinal signs, and the first third of the dial is said to represent the Cardinal nature. On a similar basis, the MC is translated to the dial in this first third, to 16 degrees. Mercury, being in a Fixed sign at 19 AQ is translated to the 49th degree of the dial, since the Fixed signs are related to this second third of the sequence – thus 30 + 18 gives 48. The consequence of this progressive conversion from the ecliptic to the dial may be studied with similar principles in mind. It is argued that from this dial conversion it is possible to note certain emphatic areas, and Angular relationships (not aspectal relationships, of course) based on a simple system of mid-points. The methodology of the interpretation of the cosmogram involves the construction of several analytic diagrams from the 90-degree dials – see COSMIC PICTURES. See also COSMOPSYCHOGRAM: figure 25 under DISTANCE. The French astrologer Volguine has used the term 'cosmogram' in a general sense to apply to any representation in graphic form of the cosmos, or of celestial relationships.

cosmology   An alternative name for COSMOBIOLOGY.

cosmo-psychic   See THREE PROTOTYPES.

cosmopsychogram   A term used of a visual projection (scarcely a chart in the conventional sense of the word) of various intensities of astrological relationships, obtained by a structural analysis of a chart, and incorporating the distance values of the heavenly bodies. The projection was developed by Ebertin (see COSMOBIOLOGY), based on a form used by the German graphologist Wittlich. Each of 13 birth-chart factors are represented by a ray on this visual projection (figure 25): the white inner portion is said to relate to the positive value of the personality, the black portion to its negative value. The length of the ray is determined by evaluating the factors of mid-points and aspects. The total white section is called the positive 'kernel'; the black section is called the negative 'mantle'. In recent years, the internal circle, carrying the distance values, has not always been incorporated into the cosmopsychogram: see DISTANCE.

cosmopsychology   See COSMOBIOLOGY.

counselling   A term used in modern astrological circles as the equivalent for the old 'astrological interpretation' on a one-to-one basis between astrologer and native. The connotation of the term is coloured by the pseudo-psychological approach to astrology which is currently fashionable, the idea being that the astrologer (rightly or wrongly) considers himself to be in a position to both interpret the chart and offer advice.

counterparts   For the specifically astrological sense, see GLOBES.

co-vertex   A term proposed as applying to the VERTEX of the so-called CO-LATITUDE of the place of birth. See CO-ASCENDANT.

Crab   Popular name for both the constellation and sign CANCER.

Crater   Constellation contiguous with CORVUS, extending into LIBRA, and thus linked with the mythology of these constellations: the bird and cup (crater) were transferred to heaven by Jupiter, but the asterism was linked with the image of a cup long before Greek mythology came into existence. While the *kantharos* of the earliest Greek systems, it was also the 'soma-cup' of the prehistoric Indians. It is hardly surprising that one of the fixed stars of this asterism (LABRUM) has been linked with the Grail legends. Astrologers see the asterism as a generally beneficial influence, and Ptolemy links it with Venus.

Creative Hierarchies See TWELVE CREATIVE HIERARCHIES.

creative principles See SEVEN VITAL PRINCIPLES.

crepuscule Term from the Latin *crepusculum* (twilight), and seemingly a variation of crepuscle.

crescent In general use the term is now applied to any planet (more particularly to the Moon) when half the disc of its body is illumined by the Sun. However, the derivation of the term, from the Latin *crescere* (to grow) indicates that the word should more properly be restricted to the partly illumined waxing body. In graphic symbolism, the crescent first quarter is symbolized by the sigil ☾. See also MOON.

Crib See PRAESAEPE.

critical days In the specialist context of horary astrology, and in early forms of MEDICAL ASTROLOGY, the critical days were those marked by the series of directions and transits on which the Moon fell SEMI-SQUARE to its original position. But see also CRITICAL DEGREES.

critical degrees In some forms of early MEDICAL ASTROLOGY, certain cusps of the LUNAR MANSIONS were designated critical or crucial degrees. It seems that some systems counted all these cuspal degrees as critical, whilst others maintained that only the cusps of the 1st, 8th, 15th and 22nd mansions were critical degrees. Since prognosis was usually concerned with the transit of the Moon through these critical degrees, the days on which such transits took place were sometimes termed 'critical-degree days' or even 'critical days' (but see CRITICAL DAYS). A different system of critical degrees of the zodiac is also recorded in some astrological texts: it is said that a planet within orb (usually 2 or 3 degrees) of such degrees is found to exercise a strong influence on the life of the native, somewhat similar to the HYPSOMATA. There are one or two variations in the lists available in the popular textbooks, but the following are representative.

AR	CN	LB	CP	1, 13, 26
TA	LE	SC	AQ	9, 21
GE	VG	SG	PI	4, 17

Crocodile Although exoterically the Hindu astrological equivalent of the sign CAPRICORN (named Makara), the crocodile is linked by Blavatsky with esoteric astrology through the fifth group of the Hierarchies of spirits who had their abode in this constellation. The crocodile appears to be a debased symbol of the dragon, the 'dragon of wisdom', the intelligent principle in the human being, which passes from incarnation to incarnation.

Cross See NORTHERN CROSS and CRUX.

Crow See CORVUS.

crucial degrees See CRITICAL DEGREES.

Crux Constellation, approximately 2 to 12 degrees Scorpio, and from 56 to 65 degrees south of the equator, now our Southern Cross, though the ancients incorporated its stars into CENTAUR. It was perhaps introduced as a separate asterism in the 15th century, though Al Biruni mentions a Hindu name Sula (beam of crucifixion), and Allen records that it was last seen on the horizon of Jerusalem at about the time of the Crucifixion. It is more than a literary puzzle as to why Dante was familiar with its form and symbolism. Robson notes that Cabral, who discovered Brazil in 1500, called the country 'Land of the Holy Cross', and that the asterism has been reproduced on Brazilian postage stamps: Cabral might have had the asterism from his reading of earlier navigators, to whom it was a source of wonder. Astrologers link its influence with burdens and trials, and its star ACRUX plays an important role in stellar lore.

Crystalline Heaven See NINTH HEAVEN.

Crystalline Sphere See NINTH HEAVEN.

Culmen Coeli Little used dog-Latin for the TENTH HOUSE.

culmination The term is used in two different senses in astrological contexts. First, it denotes the completion of a forming aspect to the PLATIC. Secondly, it denotes the arrival of a planet, fixed star, or some nodal point to the Midheaven, or the cusp of the 10th House, by transit, direction and progression. Thus, when any celestial body crosses the meridian it is said to culminate. The upper culmination is the maximum altitude, the lower culmination is the minimum altitude. After culmination, the planet or nodal point is said to decrease.

Cupido Name given to one of the *HYPOTHETICAL PLANETS in the URANIAN ASTROLOGY. This transneptunian is included in an ephemeris, and is claimed to relate to community, marriage and art.

Cura The Sanskrit word for JUPITER, as used in Hindu astrology.

cusp In specialist astrological contexts, the term cusp has a dual reference, which is often confusingly merged into one. First, it refers to the imaginary line which separates one sign of the zodiac from another – thus the cusp between Aries and Taurus is the imaginary line which demarcates the end of the last degree of the former and the beginning of the first degree of the latter. The existence of this term cusp in reference to such a separation between signs has led to a popular notion of the CUSPAL TYPE, which really has no existence in serious astrology. See, however, CUSPAL ARC. The second meaning of the term refers to the line which demarcates the grids of the houses in a chart (see HOUSE SYSTEMS). Thus, the Ascendant line is regarded by many astrologers as the cusp of the 1st house: however, in some systems, such as the CAMPANEAN SYSTEM, the cusp of the house is regarded as marking the centre of that house, and forms its most influential arc of operation. In other systems, such as the PLACIDEAN SYSTEM as it is used by many modern astrologers, the cusp of the house is regarded as marking the beginning of the name house, and the end of the preceding one: in this case, therefore, the 1st house is considered as beginning with the Ascendant line, and extending down to the cusp of the 2nd house. There is much confusion about cusps in the practice of ordinary astrology.

cuspal arc Sometimes the CUSP between the signs of the zodiac, or between the houses, is regarded as consisting of a small arc, rather than of a strictly precise division: such is called a cuspal arc. The concept permits a degree of ambiguity as to the influence of a planet (especially so in reference to the zodiacal cusps), which is hardly necessary in view of the precision permitted by modern ephemerides.

cuspal distance A term defined by the astrologer Leo as denoting the arc of Right Ascension between a planet and the house circle (cusp) of any house to which it is applying. However, the term is also used in a much more general sense in regard to distances between cusps.

cuspal type A term which is scarcely used in serious astrology, but which in popular versions of the art is sometimes used to describe a native born on a day near to the Sun's ingress into a sign. In most cases, the term may only be used of those natives for whom an accurate horoscope has not been cast, for such a figure would show precisely the position of the Sun, and leave no need for the term at all. In effect, the popular concept that there is such a person who may be described as a cuspal type (as though this were some sort of astrological notion) belongs to the realm of superstition.

Custos Messium   Formerly a constellation, the Harvest-keeper, located near Cassiopeia and Cepheus, but no longer recognized as a separate asterism.

cycles   In a specialist astrological sense, the term is generally used in reference to the planetary cycles, or to EPOCHS, though there are actually many other cycles within the microcosm and macrocosm which are of interest to the astrologer – in this connection, see also RHYTHMS. The theory of planetary cycles is based on the idea that each of the planetary bodies mark off periods which are reflected in human life and history. Synodic cycles involve two planets (interval between one conjunction and the next), sidereal cycles involve only one (a complete cycle around the Sun): within the horoscope a third cycle is determined according to PROGRESSIONS – for example, the progression of the Moon to its natal point is equivalent to an important 28-year cycle. Other important cycles are the *LUNAR NODE CYCLE of 18 years and 7 months, and the *LUNAR RETURN. In many respects, the theory of cycles visualizes celestial influences working to some extent outside the spheres of the individual horoscopes, even though the specific cycles within the chart have to be taken into account (and indeed form the very basis of much astrological lore). Among long-term cycles are those marked out by the *CHRONOCRATORS (Jupiter and Saturn), those attributed to the Archangels (*SECUNDADEIAN PERIODS), as well as the *FIRST ORDER CYCLE of the superior planets. Many of the important cycles are involved with periodicities much longer than any human life, and are quite rightly seen as influencing epochs of human history, changes in climate, social revolutions, wars, religious beliefs, political crises, and so on. The long-term cycle most frequently referred to in popular astrological lore is that subsequent to the PRECESSION of the equinoctial point. Many of the ancient cycles, which have often survived in astrological lore in a mutilated form, such as the SAROS CYCLE and the NAROS CYCLE were close approximates to the synodics (though in the latter case this is probably accidental): see MAYAN ASTROLOGY. Some useful and ingenius attempts have been made in modern times to explain or even extend the theory of cycles, among the most notable being those of Collin, who established periodicities in various human and biological phenomena related to multiples of synodic periods, and Wachsmuth, who made complex use of the Precessional periodicities: see THREE PROTOTYPES.

Cygnus

Among the cycles observed by Collin are the 3-year 6-month cycle (prices, industrial production and sales); 4-year cycle; the 7-year 6-month cycle (tree-rings, lake-deposits and barometric pressures); the 8-year cycle (cotton prices, etc.); the 9-year cycle (linked with the asteroids, affecting stock-market prices and suicides); the 9-year 9-month cycle, observable in a wide variety of animals; the 11-year 2-month cycle (weather and sunspots); the important 'war peak cycle' of 15 years (involving Mars and Saturn), and an 18-year cycle (real estate and building). Collin describes also a 36-year cycle affecting European weather, an 84-year cycle relating to sexual mores, and a cycle of 165 years, involving what he terms 'regeneration', linked with the work of esoteric schools, reminding one of the 250-year cultural cycle of Pluto, noted by Rudhyar. The figures in table 29 are derived from Collin's attempts to link the geocentric conjunctions of each of the planets (in terms of the first order cycle), and the longer periodicity of the second order cycle. He interrelates to these the ASTEROIDS, projecting a 'theoretical conjunction' based on averages. The tables constructed from these cycles are designed to suggest a harmonic relationship of numbers and tones which reflect the ancient harmonies called the MUSIC OF THE SPHERES. For further information, see Bradley (solar and lunar returns), Nelson (sunspot cycles) and Ruperti (lunar, solar and planetary cycles).

--- TABLE 29 ---

Planet	First order cycle (days)	Second order cycle (years)
ME	117	8
VE	585	8
MA	780	15
JU	398	12
SA	378	30
UR	369	84
NE	367	163
(Asteroids)	465	9

Cygnus Constellation, from approximately 28 degrees Capricorn to 28 degrees Pisces, and from 28 to 55 degrees north of equator, known as the Swan from Roman times, when it was linked with the myth of Leda and Jupiter. Before that time, however, it was simply a 'Bird' to the Greeks, and still survives in such names as Avis, Volucris and Al Dajaja (hen, to the Arabian astrologers), a term used in degenerate form of its brightest star, the

DENEB ADIGE. Under the impress of those who sought to Christianize the zodiac, it became the Cross of Calvary, and to this day is sometimes called the Northern Cross. Astrologers say that it gives a dreamy but cultured nature, with unsteady affections. See also ALBIREO.

Cynosura   See URSA MINOR.

D

Dabaran Al Dabaran (the forecaster), the 2nd of the Arabian MANZILS.

Dabith Fixed state (multiple) of 3rd magnitude, the beta of Capricornus, set in the left eye of the Goat, the name linked with the Arabian MANZIL by which the latter was identified. Ptolemy accords it the nature of Venus conjunct Saturn: when emphasized in a chart it seems to bring out suspicion and mistrust.

daemon See DAIMON.

daemonum stabula Literally 'dwelling place of the demons', a term used by Kircher for the LUNAR MANSIONS.

Dagon See PISCES CONSTELLATION and PISCIS AUSTRALIS.

daily progressions See DAILY SERIES.

daily series A term relating to a precise formulation of the symbolic time relationship of the 'Day for a Year' method (see MEASURE OF TIME). As used in SECONDARY DIRECTIONS, it takes the true solar day to be the basic unit of time involved (see TIME). The period of a day is – by symbolic projection – to be extended into a WEEKLY SERIES, a MONTHLY SERIES and the YEARLY SERIES. The full term is 'daily series of secondary directions', sometimes reduced to the misleading term 'daily progressions', which might quite easily refer to other conceptions of a daily time span.

daimon The term equivalent to 'daimon' or 'daemon' was used by the Greek astrologers to refer to an ancient PARS no longer used – see LOT OF DAIMON. In late medieval astrology, the same term (as well as 'demon') was also used in connection with spiritual beings (not the 'demon' of popular infernal lore), connected with planetary and zodiacal agencies: the word in this sense was almost certainly derived from the same Sanskrit root which gave 'Deva', now unfortunately wrongly linked with the elemental beings, but once the equivalent of the Angels. For a list of the 'Daemonia' of the planetary Spheres, see INTELLIGENCY.

Dalw Al Dalw   The Arabic name for Aquarius.

Dark Age   See KALI YUGA.

Dark Moon   See LILITH.

day   In ordinary parlance, the day is that period of light (literally 'daylight') between sunrise and sunset, but the term is also used of the 24-hour period marked by the earth's rotation on its axis – see SOLAR DAY. For other definitions, see CIVIL DAY, JULIAN DAY, SIDEREAL DAY and TIME. The Babylonians and Hindus measured the day from sunrise to sunrise: the ancient Greeks (though not always) reckoned the day from sunset to sunset, as do the Jews. Most other nations now measure the day from midnight to midnight – but see ASTRONOMICAL DAY.

day for a year   See MEASURE OF TIME.

day home   See DAY HOUSE.

day horoscope   A general term used of a horoscope for which the Sun is above the horizon line: it is contrasted with the NIGHT HOROSCOPE.

day house   A confusing term arising from an early mistranslation of Greek and Latin astrological texts. In traditional astrology, a planet was considered to have rule over two signs, one being called the 'day house' of that planet,

TABLE 30

Day house	*Planet*	*Night house*
(Cancer	Moon) (Sun	Leo)
Gemini	Mercury	Virgo
Taurus	Venus	Libra
Aries	Mars	Scorpio
Pisces	Jupiter	Sagittarius
Aquarius	Saturn	Capricorn

and the other sign being called the 'night house' ('diurnal' and 'nocturnal' respectively). The Sun and Moon were exceptions, for these were said to concentrate their attentions on Leo and Cancer respectively, and accordingly had only one sign each. Table 30 indicates the order of the ancient rulership. Since a sign is not a house, the term might have been translated in a way which would have avoided confusion. In modern times, 'day home' and 'night home' have been suggested as alternatives, but it is clear from such early sources as Firmicus that the original idea was linked with the equivalent of positivity and negativity – which, of course, explains the misunderstood term POSITIVE MARS when used of the ruler of Aries, in contrast with the NEGATIVE MARS of Scorpio (prior to the introduction of Pluto as ruler of the latter sign). See PLANETARY RULERSHIP.

daylight-saving time Called Summer Time when it originated in 1916, and introduced to provide an hour extra of daylight during working hours. At 2 am on the Sunday following the third Saturday in April, the clock was set ahead by one hour. This was adjusted back one hour on the Sunday following the first Saturday in October (though in practice these days varied, resulting in the need for tables recording the history of daylight-saving regulations). During the Second World War, a similar alteration by two hours was made, called Double Summer Time. A related (though not synchronized) system of War Time was operative in the United States of America. Some countries did not accept these regulations, and in practical terms this means that an astrologer casting horoscopes for the years after 1916 must ensure from reliable sources what sort of time was operative at the place and time of birth. Many of the tables given in standard astrological textbooks are sparse and inaccurate: *World Daylight-Saving Time* by Curran and Taylor should be consulted, but see also TIME and TIME ZONES.

Day of Brahma See KALPA.

days of the week As Roger Bacon intimated in the 13th century, the names of the seven days of the week appear to have been derived from the ancient astrological practice of assigning planetary rulerships to the hours of the day. The first hour of the first day was given to the rule of Saturn – hence the first day was called Saturn-day (Saturday). Each of the planets, in the ancient descending order of the SPHERES (which is not the order of the modern astronomers, since Venus was transposed with Mercury) was then assigned to the sequence of hours, Jupiter ruling the second hour, Mars the third, and so on. By this reckoning, the 22nd hour would again be ruled by Saturn, the 23rd by Jupiter, and the 24th by Mars. The first hour of the second day would thus be ruled by the Sun – hence it was 'Sun-day' (dies Solis). The sequence of hourly rulerships gives in this way the sequence of the seven days, though the survival of the planetary names is more openly reflected in the Romance languages – for example, the French *lundi* is the 'Moon-day' of Monday; *mardi* is the 'Mars-day' (our Tuesday); *mercredi* is the 'Mercury-day' (our Wednesday); *jeudi* is 'Jove-day', the 'Jupiter-day' of Thursday, and *vendredi* is 'Venus-day' of Friday. The notion of a seven-day week developed only late in the Roman Empire, and was legalized by Constantine in the 4th century AD.

day triplicities Almost an archaic term, referring to the ancient tradition that some planets are stronger when they are in certain triplicities in a DAY HOROSCOPE. Such are Venus, when in an earth sign; the Sun when in a fire sign; Mars when in a water sign, and Saturn in an air sign. In fact, this theory runs contrary to some of the traditional teachings regarding DEBILITY and FALL.

death asterogram A term probably originated by the modern astrologer Sucher as synonymous with DEATH CHART, though Sucher's own published death asterograms appear to be based on one of the SIDEREAL ZODIACS with an unconventional ayanamsa. The chart (sometimes the 'death asterogramme') is visualized as a sort of summary of the human life to which it relates, as well as being conceived as having some connection with the various post-mortem experiences of the deceased soul. As with the more conventional death chart, it is apparent that some astrologers connect the configurations of the chart with conditions relating to the next incarnation. The whole field is contentious, and may not be subjected to ordinary investigatory techniques.

death chart A name given to a horoscope figure erected for the date or time of death, and interpreted by curious analogy with the natal chart which is concerned with the relationships a newly incarnated soul holds to the earth and stellar realms. Some astrologers insist that a death chart may reflect man's post-mortem relationship with the spiritual realms (Sucher makes such a claim in connection with the DEATH ASTEROGRAM). Death charts may be cast according to either the tropical or sidereal stereographic projections. However, since the whole structure of astrological lore and symbolism is based upon the idea of the descent of spirit and soul into incarnate form, the application of its rules to an 'ex-carnation' is more than questionable practice.

Death Point The Point of Death, which is not to be confused with the ANARETIC, is one of the so-called Arabian PARS. If the degree of the Moon in a natal chart is revolved to the Ascendant, the cusp of the adjusted 8th house marks the Death Point.

Debility Each planet is said to have its own debility, as a result of which it is generally weakened. A planet in the opposite sign to that which it rules is debilitated by DETRIMENT, for example, but there are many conditions under which a planet may be regarded as being otehwise debilitated. See DIGNITY. As with his list of dignities, Wilson gives a comprehensive list of debilities, and ascribes them a specific value: the three main essential debilities are when the planet is in its Detriment (5), when it is in its Fall (4), and when it is Peregrine (5). The list of accidental debilities is much longer, and the following is merely a digest of the more important:

Besieged by Saturn and Mars	6	In partile conjunction with Saturn	5
Conjunction with Algol	6	In partile conjunction with Mars	5
Combust	5	In 12th house	5
Retrograde	5	In partile conjunction with Cauda	4
In 6th or 8th house	4	In term of Saturn or Mars	1

decan   In general reference, any 10-degree arc, but more specifically a term applied to a 10-degree arc of the zodiac resulting in the division of the twelve signs into 36 equal arcs. In each of the signs, these divisions are called the first, second and third decans (or decanates). Great importance was attached to the supposed influences of these zodiacal arcs in ancient astrology – indeed, the decanate system of the Egyptians appears to have merited as much attention as the 12 signs themselves. In modern astrology, which rests firmly on the ancient tradition, there are two frequently use¹ systems for determining the supposed influences of the decan sequences. In the first, the ruler of the first sign Aries (Mars) is given dominion over the first decan of that sign: then the followers are each assigned to the sequence of decan arcs throughout the zodiac, in a specific order: AR1 – MA; AR2 – SU; AR3 – VE; TA1 – ME; TA2 – MO; TA3 – SA; GE1 – JU, etc. The so-called 'modern planets' play no part in this sequence. The second system does incorporate the moderns, however, since it is based on the adoption of the ruler of the relevant sign being accorded rule over the first arc, with the rulers of the two following signs of the same triplicity being placed over the next two decans: AR1 – MA; AR2 – SU (SU being ruler of Leo); AR3 – JU (JU being ruler of Sagittarius); TA1 – VE, etc. In Arabian and medieval European astrology the decans were highly regarded, and figured frequently in the rich imagery of talismans. Several early systems of decans involved a complex nomenclature, recorded in Latin and Greek in a modern text by Nowotny, but these are long obsolete. Unfortunately, the term FACE (which has quite a different application) has been used by some astrologers as synonymous with decan – this is no modern error: see for example the Ratdolt edition of Angelus.

decanata See DECAN.

December The twelfth month (31 days), its name being derived from the Roman system prior to the introduction of the JULIAN CALENDAR reforms, when it was the tenth month (the Latin *decem* means 'ten'). Esoterically, the month is still linked with the magical significance of the number 10, which Blavatsky records as an image of the standing man (the figure 1) confronting the entire cosmos (the figure 0), and aware of his separation from it, by means of which he defines his own individuality (the space between the two figures). This application to the Arabian numeration was valid within the Roman system, for the letter X (ten) had the same symbolic meaning, though with a different rationale. The month was specifically linked with the festival of the Saturnalia; the esoteric significance of the X-form was associated with the symbolism of the ANGLES of the horoscope, which were in turn an image of the DESCENT OF THE SOUL into matter.

decile An aspect of 36 degrees, arising from a division of the zodiac into tenths. It is sometimes called the semi-quintile. See also QUINTILE.

declination The measurement of Angular distances from the CELESTIAL EQUATOR, north or south. The widest possible declination of the Sun is 23 degrees 28 minutes, when it reaches the solstitial point: since the ecliptic intersects the equator at the 1st degrees of Aries and Libra, the Sun at these points has no declination. The approximate maximum declination of the Moon, Mercury, Venus and Mars is 27 degrees, whilst Jupiter, Saturn, Uranus and Neptune have almost the same declination maximum as the Sun. In modern notation, the positive sign (+) marks the Angle north of the equator, the negative (−) that south of the equator.

decreasing light A term used mainly of the Moon, after direct opposition to the Sun, as it begins to lose light, or to WANE. Sometimes, the term is used of other planets. The opposite term is 'increasing light'.

decrepit quadrant See QUADRANTS.

decumbiture The name given in the quasi-medical branch of HORARY ASTROLOGY to the chart cast for the time when the subject took to bed at the beginning of an illness. The decumbiture chart is usually concerned with medical prognosis, and with determining the so-called 'critical days'.

Decussata chart See CHART SYSTEMS.

degree symbols

Dee zodiac   See GLASTONBURY ZODIAC.

deferent   Term used specifically in connection with the theory of EPICYCLES set out in the PTOLEMAIC SYSTEM. The complex apparent movements of planets in the geocentric model were partly explained in terms of their having two interrelated motions, in a conception elaborated to maintain the Aristotelian doctrine that planets move in perfect circles. The planet was visualized as striving to move in a perfect circle, its centre moving on an imaginary circumference of an eccentrically centred circle (centred on an imaginary line joining the Earth with the Sun, but not on the centre of the Earth itself). This imaginary orbit was called the deferent, around which the moving centre of the planet produced the effect of epicyclic movement. The influential Arabic astrologer Alfraganus appears to have termed the deferent the 'eccentric'.

deficient degrees   See AZIMENE DEGREES.

deflux   A term used as an equivalent of SEPARATION, from the Latin *defluxio*, itself a translation of the Ptolemaic Greek *aporroia*.

degree   An arc of precisely $\frac{1}{360}$th of the circumference of a circle. The visual equivalent of a degree in practical astronomy is about twice the apparent diameter of the Sun or Moon. See also DEGREE SYMBOLS.

degree arcs   See DEGREE AREAS.

degree areas   A term sometimes used as the equivalent of DEGREE SYMBOLS, which in some cases (for example, in the system of WEMYSS DEGREES) relate to areas (really arcs), rather than to specific degrees. Properly, the term should be 'degree arcs'.

degree dial   A term derived from the specialist conversion of the COSMOGRAM, and properly the 90-degree dial – see COSMOBIOLOGY. The term is not to be confused with VOLVELLE.

degree exaltations   See EXALTATIONS.

degree rulerships   See MONOMOIRIA.

degree symbols   A vast astrology literature has tabulated the specific influences said to be exerted by individual degrees of the zodiac, regarded as exerting effects quite independent of occupant bodies, nodes, fixed stars and zodiacal arcs. Whilst the interpretation of individual degrees is probably of Arabic origin, the methods by which the degree symbols (that is, the readings attached to the individual degrees) were determined has never been adequately established. A few of the systems are undoubtedly constructed from clairvoyant readings (as, for example, the modern SABIAN SYMBOLS) but others, undoubtedly ancient, have originated in ways unknown, as, for example, the VOLASFERA SYMBOLS. No satisfactory compendium or comparative study has been made of degree-symbol readings, and many astrologers doubt the validity of all the various systems, maintaining that the individual qualities which are associated with degrees of the zodiac may be explained by consideration of the nature of the fixed stars, constellations and mansions, since there is not a single degree of the zodiac free of such radiations or influences. In some systems, the degree symbols are actually arcs, rather than degrees, or even diametric axes across the zodiac. An example of an arc symbolism would be the association of 25 to 29 degrees of Leo with the art of astrology, an arc which is startlingly recurrent in the charts of astrologers. Well over a dozen sets of degree-symbol systems are currently available: Carruthers lists no fewer than 17 from post-medieval sources alone, for example. In addition to those mentioned above, see also ANGELUS SYMBOLS, CHARUBEL SYMBOLS, THEBAN CALENDAR and WEMYSS SYMBOLS. Each of the preceding systems derived purely from the zodiacal degrees, but there is a system derived from the constellations, which must be distinguished from these, called STAR POINTS, and one composite system, derived from various sources (degree symbols, fixed-star influences, arcs, constellations, etc.), and linked with the constellation zodiac by Devore under the name INDIVIDUAL DEGREES. For additional relevant material, see Carelli, Hasbrouck, Keane, Leinbach and Matthews. See also MONOMOIRIA.

Deimos   The name of one of the two moons of Mars.

dekan   See DECAN.

delation   A term defined by Simmonite as meaning 'restoring of light', and applied to the aspecting of an inferior planet of a superior which is either combust or retrograde. It seems that the superior planet restores to the inferior its VIRTUE.

delineation   Term derived for specialist use from the old idea of drawing, the word being applied to the drawing of a picture, but usually of the portrait revealed in the horoscope figure. Often, the delineation of a horoscope is the interpretation of a horoscope.

Delphinus   Constellation, approximately 8 to 19 degrees Aquarius, from 3 to 19 degrees north of the equator, the Dolphin. This was the Hieros Ichthus (sacred fish) of the ancient Greeks, who accorded it many related names. The Dolphin was said to have been placed in the heavens by Neptune for having persuaded Amphitrite to be his wife – hence the asterism is sometimes called by her name. For all its reputation in mythology, the asterism is held unfavourable by Ptolemy, who links it with the combined natures of Mars and Saturn.

Delporte division   See MODERN ZODIAC.

Demeter   Name given to one of the HYPOTHETICAL PLANETS in the system proposed by the Dutch astrologer Ram, and accorded rule over zodiacal Cancer. This transplutonian is included in an ephemeris, and claimed to relate to 'spiritual aptitudes, art and music'.

demi-decan   See FACE.

Demon Star   See ALGOL.

Denderah Zodiac   Name given to an astrological bas-relief once in Denderah, Egypt, in the Temple of Hathor. This famous circular ceiling relief was carried by Napoleon to the Louvre, and wrongly called the Denderah Zodiac. It is a constellation map, containing the twelve images of the zodiacal asterisms, along with such others as Draco (see POLE STAR), symbolized as a hippopotamus, the ancient ideogram for Sirius, and each of the planets in their EXALTATIONS. The age of this constellation map is in dispute, but many scholars insist that whilst Ptolemaic (Fagan convincingly dates it to April, AD 17), it certainly contains elements from a much earlier Egyptian astrological tradition. Breasted even suggests that the entire Temple of Osiris, and its planet-studded constellation map (now *in situ* as a plaster reproduction), stems from prototypes of about 2900 BC. As Wachsmuth admits, 'even if this was executed at a later time, the knowledge which it embodies had long been preserved in the Mysteries'. The fact remains that there is a specific reference in the asterism map to 8 degrees of Aries, which relates to a sidereal system popular in Greece and Rome, not originated until 373 BC, which at least gives the earliest possible date for the structure. On the other hand, it is certain that the images and symbols are themselves derived from a very ancient tradition, which makes this zodiac one of the most fascinating astrological survivals from the past. See also ESNA ZODIACS.

Deneb   Fixed star, the zeta of Aquila, said to have derived its name from the

Arabic Al Dhanab (the tail). Of the nature of Mars conjunct Jupiter. See also DENEBOLA, for which this term is often confused.

Deneb Adige   Fixed star, the alpha of Cygnus, set in the tail of the Swan, the name derived from the Arabic for 'hen's tail'. Said to be of the nature of Mercury conjunct Venus – hence, it encourages ingenuity and mental versatility.

Deneb Algedi   Fixed star, the delta of Capricornus, set in the tail of the Goat, its name from the Arabic 'tail of goat'. Powell adopted this star to mark the termination of his sidereal Capricorn: see SIDEREAL ZODIAC. It is said by Ptolemy to be of the nature of Jupiter conjunct Saturn.

Deneb Kaitos   Fixed star, the beta of Cetus, set in the tail of the Whale, its name from the Arabian phrase meaning 'tail of the whale towards the south', and sometimes called Difda (the second frog) from the Arabic, which explains the curious Latin name Rana Secunda – see FOMALHAUT. It is said to give mental disturbances when emphasized in a chart, and Ptolemy likens its influence to that of Saturn.

Denebola   Fixed star, the beta of Leo, set in the Lion's tail, the name derived from the Arabic meaning 'the tail of the lion'. Powell used the star to terminate his sidereal Leo – see SIDEREAL ZODIAC. It is regarded as an unfortunate influence in a chart – as Allen says, 'thus opposed to REGULUS in character as in position'. Ptolemy links it with the influence of Saturn and Venus, and Simmonite to that of Uranus itself. Sometimes the star is wrongly called DENEB.

Depression   In a specialist sense, sometimes used as a synonym for the planetary FALL – see both DIGNITY and HYPSOMATA.

Descendant   As a strict technicality, the term should apply to the cusp of the 7th house, otherwise known as the Western Angle: however, it is often used to denote the whole arc of the house.

Descendant degree   The degree upon the DESCENDANT.

Descendant planet   Whilst one may reasonably think that this term applies to a planet on (or at least within orb of) the DESCENDANT, it is in fact applied to any planet following its CULMINATION – that is, when it is between the 9th house and the nadir.

Descending arc   This term, which may be applied to the degree of zodiac descending over the cusp of the 7th house, must be distinguished from the same theosophical term which is the equivalent of the 'shadowy arc', relating to the downward passage of life-waves into the physical realms.

descent of the soul   The hermetic tradition insists that the act of rebirth or incarnation is preceded by a descent of the soul through the planetary spheres. It is in the descent through the spheres that the incarnating spirit imbibes the influences proper to his being, and which are expressed in the chart at birth. The tradition is widespread, under many different terms, such as 'spherical descent', the 'path of rebirth', and so on, but the hermetic texts most often quoted are those attributed to Macrobius and Servius, derived from commentaries on Platonic and Virgilian texts respectively. The former is linked with an Orphic tradition, the latter with an Egyptian tradition, and they are accordingly dealt with under ORPHIC DESCENT OF THE SOUL and EGYPTIAN DESCENT OF THE SOUL. Several of the more important forms of ESOTERIC ASTROLOGY visualize a similar descent and rebirth – especially so the astrological forms described by Wachsmuth and expressed in certain lectures by Steiner. Some esoteric astrologers insist that the natal horoscope is itself to be interpreted as a record of the experiences undergone by the soul during its spherical descent.

Desertum   Name given to the 21st of the LUNAR MANSIONS in some medieval lists. Probably equated with the undistinguished stellar area along the ecliptic between the sigma of Sagittarius (Nunki) and DABITH in Capricornus.

Detriment   A specialist astrological term relating to a planet which is said to be in its Detriment, or in Detriment, when in a sign opposite to that which is its own sign or house. A planet in Detriment is weakened.

Devil   According to the astrologer Leo, Mars has been called the Devil. If this is so, then it is likely to have been connected with the old idea of NEGATIVE MARS, which has rule over the 8th house, which some astrologers link with post-mortem experiences. Inherent in this nomenclature is the diabolic dualism which the early Church fought to eradicate, for it seems that Saturn went under the *nom de guerre* of Satan. Whilst the difference between Diabolus (Devil) and Satanus (Satan) was recognized in early occult schools, it was largely lost in ecclesiastical literature.

Devore symbols   See INDIVIDUAL SYMBOLS.

dexter aspect An ancient term, probably derived by Ptolemy (who was unaware of the motion of the earth) and used in reference to aspects calculated against the order of the signs. The reverse computation is a SINISTER ASPECT. Traditionally, the dexter is more powerful, but the whole theory of aspects requires to be considered against the natures of the planets involved: the terms 'dexter' and 'sinister' are virtually archaic.

Dhanishtha The Sanskrit name for the constellation Aquarius – see KUMB-HAM.

Dhanus The Sanskrit name for zodiacal Sagittarius – but see also MULA.

Dhira Al Dhira, the name given to the 5th of the Arabic MANZILS, the Arabic word meaning 'forearm', as does the medieval equivalent Brachium.

diameter An archaic term for the aspect of OPPOSITION.

Diapason Ratio connected with the planetary system set out in the theory of the MUSIC OF THE SPHERES, and depending upon the supposed planetary distances. This is the octave, a ratio of two to one, and in the planetary diagrams it embraces the Sphere of the Moon up to the Stellatum itself, including thereby eight spheres.

diapente Ratio connected with the planetary system set out in the theory of the MUSIC OF THE SPHERES, and depending upon the supposed planetary distances. This is the interval of the fifth, or the sesquialtera, a ratio of two to three. In the planetary diagram this ratio embraces the Sphere of the Moon up to and including the solar sphere.

diaphanous heaven See NINTH HEAVEN.

diatessaron Ratio connected with the planetary system set out in the theory of the MUSIC OF THE SPHERES, and depending upon the supposed planetary distances. This is the interval of the fourth, a sesquitertia, a ratio of three to four. In the planetary diagrams this ratio embraces the Sphere of Mercury up to and including the solar sphere, incorporating three spheres.

dichotome Term now almost obsolete, from the Greek meaning 'cut in half', and used of a planet (especially of the Moon) which is half illumined – that is to say, in SQUARE aspect to the Sun.

Didemoi One of the several Greek names meaning 'twins', used to denote the sign and constellation GEMINI.

Dignity

Dido   In astrology, a name given to one of the HYPOTHETICAL PLANETS by Wemyss, said by him to be the ruler of zodiacal Virgo – a suggestion never widely accepted. It is possible that Wemyss had in mind the influence of Pluto, which (though not discovered by astronomers until 1930) was recognized by some astrologers as an extraneptunian influence some years before this. Wemyss gave his Dido a mean motion of about 1 degree per annum.

Didymoi   An alternative spelling for DIDEMOI.

Difda   A name for the fixed star DENEB KAITOS.

Dignity   Each planet is said to have its own dignity, which may be essential, accidental, or both: see therefore ACCIDENTAL DIGNITY. It seems that there was once greater precision in the use of the word than is now the case: originally, an essential dignity was manifest only through signs, while accidental dignities were manifest through houses, but the two have now been confused. A planet in its own sign is certainly said to be in its essential dignity, sometimes called a 'domal dignity'. When a planet is in the sign of its EXALTATION, it is also said to be dignified. Less certain, however, is a tradition that a planet posited in the same triplicity as that of which it is ruler may be said to be dignified. Mutual reception by Exaltation was sometimes regarded as an indication of dignity. The opposite of dignity is DEBILITY. When taken to extremes, the theory of dignities is extremely complex, especially in regard to accidental dignities. Wilson provides a fairly comprehensive list, and ascribes a system of specific values, of which the following are representative:

Planet in its house	5	Planet in mutual reception by house	5
Planet in Exaltation	4	Planet in mutual reception by Exaltation	4
Planet in own triplicity	3	Planet in own term	2
Planet in own Face	1		

Table 31 sets out the traditional dignities and debilities. The information in table 31 is based on the traditional Ptolemaic cosmoconception, which does not take into account the modern rulerships of the suprasaturnian planets, given in table 32. Some astrologers insist that Exaltation and Fall take place in specific degrees of the zodiac, as follows:

Exaltation: SU 19 Aries; MO 3 Taurus; ME 15 Virgo; VE 27 Pisces; MA 28 Capricorn; JU 15 Cancer; SA 21 Libra; DH 3 Gemini.

Planet	Rules	Detriment in	Exaltation in	Fall in
		TABLE 31		
SU	Leo	Aquarius	Aries	Libra
MO	Cancer	Capricorn	Taurus	Scorpio
ME	Gemini	Sagittarius	Aquarius	Leo
	Virgo	Pisces	(Virgo)	(Pisces)
VE	Taurus	Scorpio	Pisces	Virgo
	Libra	Aries		
MA	Aries	Libra	Capricorn	Cancer
	Scorpio	Taurus	(Virgo)	(Pisces)
JU	Pisces	Virgo	Cancer	Capricorn
	Sagittarius	Gemini	(Capricorn)	(Cancer)
SA	Capricorn	Cancer	Libra	Aries
	Aquarius	Leo		

The degree of Fall is that directly opposite in the zodiac to the degree of Exaltation. It must be observed that many authorities offer significant variations on these traditional dignities.

Planet	Rules	Detriment in	Exaltation in	Fall in
		TABLE 32		
UR	Aquarius	Leo	Scorpio	Taurus
NE	Pisces	Virgo	Cancer	Capricorn
PL	Scorpio	Taurus	Aries	Libra

Dione   Name given to one of the moons of Saturn – see also HYADES.

Dionysian Hierarchies   See CELESTIAL HIERARCHIES.

Dionysian period   See VICTORIAN PERIOD.

Dioscuri   Most usually the word applies to the GEMINI CONSTELLATION, but sometimes to the two stars in this asterism, CASTOR and POLLUX.

diosemaisis   An archaic term used to denote weather-forecasting from astrological, stellar and meteorological phenomena.

Dirah Fixed star (double), the mu of Gemini, set in the left foot of the Northern Twin, said to be of the nature of Mercury and Venus.

direct application The APPLICATION to aspect of planets when both are in direct motion along the ecliptic. See RETROGRADE APPLICATION.

directing See PRIMARY DIRECTIONS.

directional aspects See PRIMARY DIRECTIONS.

directions The art of making directions, as of interpreting them, is the most important part of PREDICTIVE ASTROLOGY, and is involved with studying the future configurations of planets and Angles in regard to a radical chart, with a view to learning something about the future of the native. The term is almost certainly derived from the idea of directing planets and Angles to some future position. There are actually three different forms of directions: two are involved with symbolic time systems (see PRIMARY SYSTEM and SECONDARY DIRECTIONS), while a third is concerned with the projection on to the radical chart of the actual movements of the planets in a non-symbolic way: see TRANSITS.

direct motion Planets moving, or appearing to move, in the direction of the signs are said to be in direct motion: see RETROGRADE.

disappear When a planet reaches the Nadir, it is sometimes said to disappear, and its influence is said to 'die'. An obsolete term.

discord The Point of Discord is one of the so-called Arabian PARS. If the degree of Mars in a natal chart is taken as the Ascendant, the adjusted position of Jupiter marks the Point of Discord.

disjunct signs A term used in Ptolemaic astrology, approximately the equivalent of DISSOCIATE SIGNS. A synonymous term was alien signs.

dispositor Under certain circumstances, a planet may be said to dispose of another, which really means that it superimposes its own influence upon that other planet. Such a disposing planet is called the dispositor. In fact, while the traditional basis for disposition may have been well-founded and simple, subsequent theories have rendered the subject very complex. In simple terms, the theory is that the ruling planet of the sign on the cusp of the house is the dispositor of a planet actually sited in that house. This indicates how unsatisfactory the term is, for the influence of the disposed planet is not removed entirely (that is, disposed of, in its modern connotation), but merely subjected in its influence to the dispositor. Thus, if Mercury disposes Jupiter, the influence of the latter is merged with the former, and does not disappear altogether. The example is chosen to raise a further point – that some authorities insist that Mercury itself may not be disposed.

dissociate aspects The original theory of ASPECTS was based on the supposed harmony and disharmony between the four elements: see ELEMENTAL ASPECTS and JAIMINI. Thus, two planets in trine aspect (120 degrees apart) were harmoniously united not so much by the magic of the number 120, or by any geometric proportion, as by the fact that they were manifesting in signs of the same triplicity. Planets in square aspect were inharmoniously linked by virtue of their being in triplicities inharmonious to each other. Since orbs are permitted in calculating the power of aspects, it sometimes happens that an aspect which fulfils the requirements of geometry or mathematics does not fulfil the requirements of the elemental relationship which lies at the base of aspectal theory. Thus, for example, two planets may be technically conjunct with an orb of 4 degrees, yet if these planets are near to a cusp, it is possible that they are manifesting through different elemental powers: while geometrically conjunct, they are not elementally conjunct. Clearly, such an aspect may in no way carry the full force of a normal aspect – if indeed it carries any aspectal force at all. Such an aspect is called a dissociate aspect. Unfortunately, the same term 'dissociate' was once used as an equivalent for the QUINCUNX, though this use is now discontinued.

dissociate signs Term applied to groups of signs which are either adjacent or five signs apart. The utility of the term appears to revolve around the important (though often overlooked) distinction between *DISSOCIATE ASPECTS and normal aspects.

distance A term derived from a modern survey of the influence which PLANETARY DISTANCES have on horoscopic factors. However, the idea was mooted, even in the late medieval astrology, when the conception of distance was applied to a model of our solar system very different from the present one. Few modern astrologers ascribe to the theory that planetary distances should condition the interpretation of planetary influences: it is significant that the largely experimental *COSMOPSYCHOGRAM, which once incorporated distance data (figure 25), does so no longer. The astrologer Schreiweis has done much work on the subject, and has come to the conclusion that the

distance

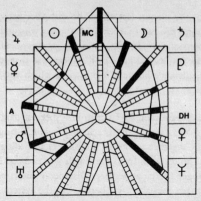

Figure 25: Cosmopsychograph with radiating distance values (after Ebertin).

influence of distance is measurable in terms of chart influences, expressed in terms of linear distances. As evidence of his research, he provides a general schema for the interpretation in terms of groups of planetary polarities (positive for a good aspect, negative for a malefic aspect). In each case, a planet close to the Earth tends to offer an intensification of the planetary nature. These pairs of contrasts are set out for close and far linear distances from the Earth, though information is not available for all the planets. See DISTANCE VALUES. The ancient astrological idea of distance (as, for example, represented in Dante's *Convivio*) was fundamental to certain concepts, but these distances were not concerned with planetary influences or powers (which were in any case regarded as being immaterial, and therefore not conditioned by spatial considerations), as with the theory of planetary harmonics, in the ancient sense of the musical tones. It was required of the theory of the *SPHERES that the maximum distance of any planet should correspond precisely with the minimum distance of the superior adjacent sphere, and it was of overriding importance that these distances should be harmonically related. The harmonics of the ancient *MUSIC OF THE SPHERES was initially linked with such distances, and not with planetary motions, periodicities or 'notes', as in the later astrological theory. That such distances were entirely chimerical (by modern standards, at least) is in a sense of no real importance: the conical shadow cast by the Earth into space (into which shadow we may look during the night-time) was measured according to such distances, for example. According to Alfraganus, who so fundamentally influenced Dante (and hence much modern thinking), this distance was 871,000 miles, which was in turn related to strict ratios for other planetary distances within the series of concentrics. We must not forget, however, that even the order of the planets (let alone the distances between their spheres) was disputed among the early Greek astronomers who influenced the promulgation of the PTOLEMAIC SYSTEM, and the best generalization which may be made about the traditional view of distances was that it was not so much the actual spatial distance between the planets which was important, as the ratios between them.

distance values A term used by Ebertin in respect of data incorporated into his COSMOPSYCHOGRAM, the inner circle of which originally set out in symbolic form the distance values of the planets according to a fixed system – see figure 25. The distance from the Earth was marked in terms of a scale value as 0 for apogee to 100 at perigee. Distance-value ephemerides for years after 1881 were made available by Ebertin, but the system appears no longer to be widely used.

diurnal Pertaining to the day, which is really pertaining to the periodicity of the axial rotation of the Earth, rather than to any concept of daylight. See DAY.

diurnal arc The variable arc through which any celestial body appears to pass from sunrise to sunset.

diurnal horary time See HORARY TIME.

diurnal horoscope A term sometimes used of a chart in which the Sun is above the horizon – hence cast for day-time. In a more specialist sense it is a term used for a figure designed to determine future events, the data for the figure being derived from the natal figure. The chart is erected as though the native were reborn each day at the same sidereal time, though the planets are inserted according to the changes recorded in the ephemeris for the future days in question.

diurnal house See DAY HOUSE and MASCULINE PLANETS.

diurnal planets These are the Sun and Jupiter, according to Ptolemy – but see NOCTURNAL PLANETS. In modern astrology a planet is sometimes called diurnal when it is in the day quadrants of a chart.

diurnal ruler Term applied to any planet that rules by day: see DAY HOUSES.

diurnal signs See POSITIVE SIGNS.

divine proportion See GOLDEN NUMBER.

Divine Year   One of the Years of the Gods in the Hindu system, regarded as being equal to 360 years of mortals. See YUGA.

Divorce Point   The Point of Divorce is a modern variant on the so-called Arabian PARS. The degree opposite the Ascendant is rotated to the Ascendant, and the adjusted natal Venus marks the point.

dodekatamorion chart   A somewhat confusing term (see DODEKA-TEMORION) derived from the Greek to denote a chart based on a division of the ecliptic into twelve equal parts. The majority of horoscopes are of this structure.

dodekatemorion   An ancient system of establishing degree points, used in Hellenistic astrology (though derived from the Babylonian), and now rarely used. The dodekatemorion of any point in the zodiac (Z) is 12 times the distance from Z of the arc between Z and the beginning of the sign in which it is located. Thus, the dodekatemorion of a planet located in 22 degrees Taurus is Aquarius 16 degrees: $(12 \times Z) = (12 \times 22) - 264$ degrees: $(264 + Z) = (264 + 22) = 286$ degrees – by conversion into zodiacal arc, 286 degrees equals 9 signs and 16 degrees. Therefore, the dodekatemorion is 16 Aquarius.

Dodonides   The group name of the constellations HYADES and PLEIADES, which make up the 14 nymphs of Dodona, the Grecian Mystery centre where, according to myth, Jupiter placed in care the infant Bacchus. They were also sometimes called the Atlantides, their father being Atlas.

dog days   A period originally computed from the heliacal rising of the dog star SIRIUS (the name probably from the Greek *serios*, scorching), the alpha of CANIS MAJOR, and extended for 40 days from that time. It was said to be the hottest time of the year, and ran through some of the weeks of July and August, from approximately 4 July. The tradition is, of course, Mediterranean in origin. In the esoteric tradition, a cycle of 40 days is involved with the descent of spirit into matter.

Dog of Orion   See CANIS MAJOR.

dog stars   Usually a reference to SIRIUS and PROCYON, but sometimes to the actual asterisms of CANIS MAJOR and CANIS MINOR.

dolphin   One of the names given to both the constellation and zodiacal sign

PISCES. When so used it is generally in reference to the esoteric nature of the sign, and then either to the era of Christianity, which falls under the Age of Pisces (see PRECESSION), or to the incarnate Christ. A dolphin was sometimes used as an image of Pisces, as for example in the zodiacal scheme on the ceiling of the arch in Merton College, Oxford (figure 26), though in this case

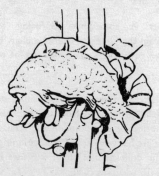

Figure 26: Dolphin as symbol for Pisces, from Fitzjames arch, Merton College, Oxford. The zodiacal sculptures are dated for 1497.

it is an heraldic evocation of the armorial bearings of Fitzjames, who was warden of the college at the end of the 15th century, when the zodiac was sculpted. See also DELPHINUS.

domal dignity   When a planet enters its own sign (that is, the sign of which it is ruler), then it is in one of its Essential Dignities, called the domal dignity. Sometimes such a planet is said to be 'domiciled' – for which a clumsy alternative 'domicilited' is also used. Unfortunately, the loose use of precise astrological terms in recent years has resulted in the term domiciled being applied in a non-specific sense, so that any planet is sometimes said to be 'domiciled' in any house, or indeed in any sign.

domicile   See DOMAL DIGNITY.

domicilium solis   Term for both the LEO CONSTELLATION and REGULUS.

domification   Specialist term meaning 'turning into houses', and relating to the procedure of applying one or other of the HOUSE SYSTEMS to the chart construction.

Dorothean astrology

dominical cycle See VICTORIAN PERIOD.

dominical letter The so-called 'Sunday letter', a term sometimes included in astrological lexicons, even though it is related only to an ancient method of computing the days of the week (and specifically the movable feasts) by means of letters of the alphabet. The term has nothing to do with astrology.

dominion A planet or a sign may properly be said to have dominion over some aspect of the material realm – for example, a sign may have dominion over a particular country (see CHOROGRAPHY) or over a precious stone, animal, and so on. See, however, RULE.

Dominions Or Dominations, the name given to the first of the CELESTIAL HIERARCHIES of the second rank, sometimes called the Kyriotetes, after the nomenclature proposed by Dionysius the Areopagite. They are associated with the Sphere of Jupiter. The Dominions are said by occultists to control the metamorphosis of matter, as the expression of spiritual events. Steiner calls them the Spirits of Wisdom.

Door of Birth A term used in esoteric astrology for the sign Cancer, ruled by the Archangel Gabriel – see GATE OF BIRTH.

Door of Death A term used in esoteric astrology for the sign Capricorn – see GATE OF DEATH.

Dorado Constellation, the head of which marks the south pole of the ecliptic, formed only in the early 17th century, the name apparently intended to refer to the coryphaena goldfish of the tropical seas.

Dorothean astrology A term used by the historian Pingree in his brief but excellent history of astrology, in reference to one of the two intellectually respectable traditions of Hellenistic astrology (the other being the PTOLEMAIC SYSTEM). The term refers to the writings of Dorotheus of Sidon (*c.* AD 75) whose principal later adherents were Firmicus Maternus (see MATHESEOS) and Hephaestio of Thebes, probably an Egyptian astrologer, of the 5th century. The refinements on the Hellenistic form of astrology for which the Dorothean is noteworthy is the system of SORTES, a specialist tradition relating to the PROROGATOR, and an early form of predictive astrology which has been called CONTINUOUS ASTROLOGY.

Dorsa Leonis   Name given by Simmonite to the fixed star VINDEMIATRIX, which he wrongly located in the sign Virgo.

Dorsum   Fixed star, the theta of Capricornus, set in the back of the Goat: an unfortunate influence linked with bites from venomous creatures.

doryphory   A curious survival from ancient forms of astrology, when horoscopes were often cast through direct observation of the heavens. In short, the term is really the equivalent of conjunction, though in a somewhat specialist sense. The doryphory of the Sun is any planet which rises shortly before it, as a sort of herald (the Greek term means something like 'spear bearer'), whilst the doryphory of the Moon is any planet which rises shortly after it, carrying its train, so to speak. Save in poetic language, the term is now obsolete.

double blind trial   See BLIND TRIAL.

double-bodied signs   See BICORPOREAL SIGNS.

Double Summer Time   See DAYLIGHT SAVING TIME.

Draco   Constellation circling the North Pole, from 63 to 81 degrees of the equator. Known even in ancient times as the Dragon, though under different names, such as Serphens, Python and Anguis: it was Sir (snake) to the Babylonians, whilst the Al Thuban of the Arabian astrologers was actually a translation of the Greek *dracon*. It was held in high esteem by the Egyptians, though they called it Hippopotamus. In the mythology derived from the Greek legends, this is the dragon which guarded the golden apples in the Hesperides. Astrologers maintain that it gives a brilliantly analytic mind, and much travel, but of course its very extent should make one wary of interpreting its effects as a whole – the fixed stars in the asterism which have proved of interest to astrologers are THUBAN and RASTABAN.

Draconic zodiac   A curious term originated (probably) by Fagan in reference to an interpretation of Babylonian imagery and mythology. The term is applied to a schema derived from the ideal sidereal longitude of the DRAGON'S HEAD in the 1st degree of Aries on the 1st Nisan of the so-called HYPSOMATIC YEAR, the positions of the traditional planets being measured from that node.

Draconitic period   The name given to the period of the Moon's motion from one of its nodes back to the same node (27.212 days). The name is derived from the ancient terminology which postulated a link between the Dragon and the Sphere of the Moon – see, for example, DRAGON'S HEAD.

Dragon See ATALIA and DRACO.

Dragon's Eyes See RASTABAN.

Dragon's Head The north node of the Moon, the point where this body crosses the ecliptic to begin its northward journey. The term, originally Caput Draconis in traditional astrology (though with several variants), was derived from the link made between the Sphere of the Moon and the celestial Dragon forces: the Dragon was imagined as being curled around the Earth, in symbol of the lunar sphere. It is said (with little evidence) that the eclipses were explained in terms of this celestial dragon swallowing the Sun and then regurgitating it. The Dragon's Head is undoubtedly a most powerful nodal point in the horoscope, though it is one usually made efficacious only by transits and progressions. It is said to have the equivalent nature of Jupiter and Venus, though some astrologers insist that when touched by directions it releases into the life of the native benefits due from karma. Some other authorities, such as Cornell, insist that it is of the nature of the Sun. For another, unrelated use of the term see ETTANIN.

Dragon's Tail The South Node of the Moon, the point where this body crosses the ecliptic to begin its southward journey. For the possible origin of the term, see DRAGON'S HEAD. This was the Cauda Draconis of traditional astrology, a powerful nodal point said to carry the influence equivalent to that of Saturn. Some astrologers insist that when touched by directions, it releases into the life of the native karmic consequences which present difficulties. In popular astrology it is regarded as a malignant force, equivalent to that of Saturn.

dry See PRINCIPLES and QUALITY.

dry signs According to the ancient Ptolemaic classification, these are Aries, Taurus, Leo, Virgo, Sagittarius and Capricorn: see PRINCIPLES.

Dschubba Fixed star, the delta of Scorpius, set in the head of the Scorpion, its Arabic name meaning 'brain of the scorpion'. Powell adopted this star to mark the commencement of his sidereal Scorpio – see SIDEREAL ZODIAC. Robson appears to confuse it with his Isidis. Ptolemy says that it is of the nature of Mars and Saturn, and it is reputed to cause immorality and malevolence.

dual signs   See BICORPOREAL SIGNS.

Dub   One of the several variant names for the constellation URSA MAJOR.

Dubhe   One of the seven stars forming the constellation URSA MAJOR, said by Ptolemy to be of the influence of Mars.

dumb signs   See MUTE SIGNS.

duodenary measure   An arc, or measure, used in SYMBOLIC DIRECTIONS, and so named by the astrologer Carter. It is based on the division of the zodiacal sign into twelfths, giving the equivalent of 2.5 degrees. This is the dwadeshamsa of the Hindu astrology. See DWAD.

Duo Pavones   See GEMINI CONSTELLATION.

dwad   A dwad is an arc of 2.5 degrees, obtained by the division of a zodiacal sign by twelve. As with the DECANS and FACES, the sequence of the dwads is accorded a planetary rule, which follows the traditional associations of the twelve signs, starting with the actual ruler of the relevant sign. Thus, Mars being the ruler of Aries is the ruler of the 1st dwad of Aries, followed by Venus, Mercury, Moon, and so on.

dwadeshamsa   In Hindu astrology, a term meaning 'one-twelfth', and applied both to the division of the entire circle of the zodiac into the twelve houses, and to the division of an individual sign into the series of DWAD. See also DUODENARY MEASURE.

Dwapara Yuga   The Bronze Age of the Hindu chronology, a periodicity of 864,000 mortal years. See, however, YUGA. The period given includes the twilights or sandhya and sandhyansa, which are 200 divine years each.

Dynamis   The beings of the CELESTIAL HIERARCHIES linked with the planet Mars, and the rulers of the Fifth Sphere. They are sometimes called the Powers or Mights, though the Greek term is actually cognate with the modern 'dynamic'. In Christian symbolism, these beings are often represented as subduing devils, which is clearly associated with the powerful duality of the Aries-Scorpio rulership (see NEGATIVE MARS), which visualizes the higher nature of Mars as linked with the Eagle, the lower nature with the Scorpion: for the higher to develop, the lower (demonic element) must be thrust down. Steiner calls them the Spirits of Motion.

Dysis   A Greek term for the Western Angle of the horoscope, the cusp of the SEVENTH HOUSE.

E

eagle   An astrological term with several connotations. It is the English
translation of the Latin AQUILA constellation. Sometimes the image of the
eagle is associated with the Evangelist St John, and hence with the
zodiacal sign Scorpio (see TETRAMORPH). Some astrologers insist that the
eagle was used in ancient times to stand in place of Scorpio itself, but there is
no evidence of this in surviving records: the rationale seems to be that since
Scorpio is the sign in which redemption takes place, then it alone of all the
twelve signs of the zodiac may be symbolized by two images. The image of
the earth-bound scorpion represents the unredeemed nature, while the image
of the aspiring eagle symbolizes the redeemed nature. The esoteric theories
underlying such astrological doctrine appear to support this view, which is
actually expressed in the imagery of Dante. The Eagle is also the attribute of
the god Jupiter, and is used as a symbol of the planet. See PHOENIX.

Ear of Corn   See SPICA.

Earth Chain   In the theosophical cosmoconception, the Earth is regarded as
a unit globe in a series of seven globes, each possessed of counterparts (see
GLOBES), called the Earth Chain. This Chain is classified according to a
convention of alphabetical letters which refer to equivalents in similar
septenary chains (see CHAIN):

A and G are 'lower mental' and invisible to ordinary sight.
B and F are 'astral' and invisible to ordinary sight.
C is the physical planet Mars (and its invisible counterparts).
D is the physical planet Earth (and its invisible counterparts).
E is the physical planet Mercury (and its invisible counterparts).

Inevitably, this schema has been reduced to a level of ordinary com-
prehensibility, with the result that the astrologer Leo blandly writes about
the Earth Scheme consisting of three physical planets – Mars, Earth and Mercury.

earth element   The earth element finds expression in the zodiac through the
three signs Taurus, Virgo and Capricorn (see EARTH SIGNS). It is the

element most deeply associated with the physical body and environment (in which connexion, see TRIPLICITY). Earth is esoterically an incarnating principle, which seeks to either draw or invite the immaterial into physical form: any incarnating spirit feels the weight of such form, with the result that the element is also associated with inertia, with physical and mental passivity, and so on, and is inclined to find pleasure in material well-being and physical comforts. The aridity of earth calls for the enlivening fluidity of water — which is to say that the earth element has a beneficial affinity with the WATER ELEMENT. See also MELANCHOLIC.

Earth Scheme   See EARTH CHAIN.

Earth Shine   Term applied to the dim light on the dark side of the Moon's crescent surface, caused by the reflection of sunlight from the Earth to the Moon.

earth signs   The three zodiacal signs Taurus, Virgo and Capricorn are expressions of the EARTH ELEMENT, sometimes called the 'earth asterisms' rather than the earth signs. Taurus is earth expressed through Fixity, Virgo is earth expressed through Mutability, and Capricorn is earth expressed through Cardinality. Therefore, by analogy with the earth of the physical realm, Taurus seeks to be fixed in one place, enjoying the rhythmic life of a conservative dependence upon environment; Virgo seeks to deal practically with the passing (ephemeral) things of the world, and often finds an orderly or even rhythmic service to others pleasurable; Capricorn is driven by an urge towards movement, so that fulfilment is usually translated as an upward aspiration, an orderly climbing of the social ladder, often intimately concerned with the idea of service to others. In modern astrological parlance, these three signs are the physical or rhythmic signs, and in the esoteric tradition they are associated with different manifestations of inhibited will forces. Since the earth element was regarded as the most 'earth-bound' and conservative of the four elements, the three earth signs were seen as being dominated by a sort of spiritual inertia (of which an aspect is a sense of inhibition, especially in regard to self-expression) which might be thrown off only by an effort of tremendous will. An earth type who succeeds in galvanizing his will forces becomes a most powerful personality, and is often deeply creative. See also EARTH TRIPLICITY.

earth trigon   See SIDEREAL MOON RHYTHMS.

earth triplicity   The earth triplicity consists of the three signs derived from the element of earth – Taurus, Virgo and Capricorn (see EARTH SIGNS). Taurus is visualized as the broad and fertile region of earth, the scene of productive agriculture; Virgo is visualized as a deep valley, pervaded by a degree of gloom and sadness; Capricorn is visualized as a high peak, impressive in its sense of serene age. Each of the three earth signs is restricted in its ability to express itself, and usually achieves self-expression as a result of a drive towards self-improvement. Since the earth triplicity is connected literally with the 'earth around' – that is with the environment – each of the signs reflects something of this connection: Taurus relates to the financial standing, Virgo to the 'control' of environment (often through occupation), while Capricorn relates to social standing, to 'how one is seen' as a physical being in the outer world.

earth zodiac   In his presentation of the system of SOLAR BIOLOGY, Butler postulates what he calls an 'earth zodiac' or 'inner zodiac', as a separate zodiacal aura, enclosing the Earth and the Moon, yet between the orbits of Venus and Mars (see figure 27). The ordinary tropical zodiac is visualized as enclosing the solar system in the normal way.

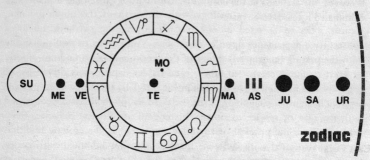

Figure 27: The earth zodiac in the spheres (after Butler).

East Angle   See EASTERN ANGLE.

Easter   The movable feast of Easter remains one of the few modern festivals in the Western world which is still determined by cosmic phenomena, connected as it is with the 84-year lunar–solar cycle. The Council of Nicaea (AD 325) attempted to resolve early controversy about the date on which the Resurrection of Christ should be celebrated by determining that it should fall on a specific day, rather than date (see DAYS OF THE WEEK), on the Sunday preceding the vernal equinox. Eventually a method of computing this movable feast day was developed to adopt the Sunday following the Paschal full Moon (that is, the full Moon subsequent to the vernal equinox, which was fixed calendrically as 21 March). An exception to this rule was that should the equinox also fall on a Sunday (or, indeed, should the day coincide with the Jewish Passover, when Christ was supposed to have been slain), the Easter should be put forward one week. The ecclesiastical terms Pascha or Pascale appear to have been derived from the Jewish name for the Passover (Pesach): the English name Easter appears to have been derived from a pagan festival in honour of the spring goddess Eastre.

Eastern Angle   An alternative name for the Ascendant, or for the cusp of the 1st house: it is associated with the sign Aries, and represents the symbolic point of sunrise in every chart, the point where the ego (or associated Mars, ruler of Aries) begins to assert itself. See also HOROSCOPE.

eastern houses   The eastern house is the FIRST HOUSE, but the term eastern houses refers (confusingly) to the houses in the eastern half of the figure – that is to the 1st, 2nd, 3rd, 10th, 11th, and 12th houses. Strictly speaking, however, the arc ends at the cusp of the 10th house, since once a planet has culminated it ceases to be eastern.

East Point   In astrology, this term is often wrongly used as synonymous for the ASCENDANT degree, while in fact the Ascendant measures the ecliptic, which rises due east only with the vernal or autumnal equinoxes. The term is much used, however, and may be regarded as a reference to the symbolism inherent in the stereographic projection of the chart, as the symbolic point of sunrise in each chart.

East Point system   A system of HOUSE DIVISION proposed by the astrologer Leo at the beginning of this century, for reasons which are not very clear. The system appears to be based on an equal division of the horizon being projected on to the ecliptic from the East Point, but the description is confused and incomplete: most certainly it is not a workable model, for the declination of a planet in itself is one of the most important determinants of its house position in the projection suggested. But see also ZENITH SYSTEM.

Ebertin mid-points   See MID-POINTS.

eccentric   An eccentric orbit is one formed about a centre which is itself in revolution around a second centre – see DEFERENT and PTOLEMAIC SYSTEM. The planets were sometimes called 'eccentrics', but the term is no longer used.

eclipse   See LUNAR ECLIPSE, OCCULTATION, SAROS CYCLE, SOLAR ECLIPSES and UMBRAL ECLIPSE.

ecliptic   The apparent path traced by the movement of the Sun around the Earth, measured against the backdrop of fixed stars. It is called the ecliptic because it is on this path that the solar and lunar eclipses take place. The plane projected by this ecliptic to the Celestial Sphere is inclined to the plane of the equator by 23 degrees and 27 minutes, decreasing by 48 seconds per century, probably to reach a minimum of 22 degrees and 54 minutes around AD 6600. The twelve signs of the zodiac are equal-arc divisions of the ecliptic into 30 degrees of arc each, based on the time division, reflected in both the solar and lunar calendars. The measurement of the fixed stars against the ecliptic is based on a spatial division. See, for example, SIDEREAL ZODIAC.

ecliptical inclination   Traditional astrology allowed for a maximum inclination of the planetary orbits to the ecliptic of between 7 and 9 degrees, which in some models of the zodiacal system defined the path or band of the zodiac (as opposed to the ecliptic itself). The modern figures for the maximum distances for each of the planets is set out in table 33.

—————————————— TABLE 33 ——————————————

Planet	Degrees	Minutes	Seconds	Planet	Degrees	Minutes	Seconds
MO	5	8	0	SA	2	29	25
ME	7	0	14	UR		46	23
VE	3	23	39	NE	1	46	28
MA	1	51	0	PL	17	8	34
JU	1	18	21				

ecliptic system   The name given to the system of projecting celestial co-ordinates by taking the ecliptic as the frame of reference. This is the system most used in astrology, and contrasts with the alt-azimuth system of navigation and the equatorial system of astronomers, for whom the celestial equator is the frame of reference. In regard to this model, see also GALACTIC CENTRE, PRECESSION and ZODIAC.

effeminate sign   For obscure reasons, Wilson terms Pisces an effeminate sign.

Egyptian calendar   Sometimes called the Alexandrian calendar, though the latter is really a fixed system, whilst the Egyptian calendar proper is a rotating system. In popular astrology there is much misunderstanding of the Egyptian calendrical system, which is complex in the extreme – the calendars of Lucky and Unlucky Days which survived in mutilated forms into medieval astrology (see EGYPTIAN DAYS) are different from the ordinary complex calendar, while both are different from the religious-ritual calendar and the funerary calendar. It is worth noting that the cycles for which Egyptian calendrical systems are famed in popular astrology are all of a relatively late development – the Sothic cycle is probably no older than the Antonines (but see SOTHIC CALENDAR), and the so-called Sadu cycle (30-year cycle) is not actually calendrical or astrological, but anniversarial, relating to jubilees. In simple terms, the year began on the first day of Thot (our 19 July), and consisted of 365 days, divided (quite reasonably in terms of the Nilotic inundations) into three seasons of four months, each of 30 days, with five epagomenal days at the end of the twelfth month, which only in much later times bore the Coptic names, in a variety of spellings:

1	Thot	4	Choisk	7	Phamenot	10	Payni
2	Paophi	5	Tybi	8	Pharmouti	11	Epiphi
3	Athyr	6	Mesheir	9	Pachon	12	Mesori

Egyptian constellations   See CONSTELLATION NAMES.

Egyptian days   A series of calendrical lists of 'fortunate' or 'unfortunate' days said with good reason to have originated from Egyptian astrological practices, but probably derived ultimately from Assyrian sources. The lists of good and bad days (good and bad for certain kinds of actions, etc.) are sometimes called 'horoscopes' in popular literature, but they are not related to astrology in this sense at all: it is likely that their original listings were derived from solar–lunar considerations which were eventually ossified and divorced from calendrical or astrological concepts. It seems that the original Egyptian days were each divided into three sections, or parts, each subject to cosmic scrutiny. A whole day might therefore be wholly good or bad, or a proportional admixture of good or bad: there are thus three different types of Egyptian days (rather than the so-called 'good' and 'bad' which have survived in popular literature), the admixtures representing a struggle between the good and bad principles. The prohibitions involve what appears to be primitive horary astrology, such as setting out on a journey, work, sexual congress, and so on. It is likely that the Egyptian days were influenced by the various festival days, and later contributed to the *fas* and *nefas* days of the Roman system. A bowdlerized traditon of the 'Egyptian days' survived into rudimental Sortes or elementary predictive methods well into medieval popular astrology and geomancy. The Egyptian days are sometimes called the Lucky and Unlucky Days.

Egyptian descent of the soul   In his commentary on a section from Virgil's *Aeneid*, Servius refers to a tradition linked with the descent of the soul through the planetary spheres, prior to rebirth, which postulates a progress inimical to the good of the soul (see, however, ORPHIC DESCENT OF THE SOUL). This image is (perhaps rightly) linked by some scholars with the Egyptian Mystery centres. Unlike the Orphic descent, which is descriptive of a septenary, Servius describes only five of the spheres, but in so doing he expresses the negative polarity of the planets in question, which may well link with the Ptolemaic records of the PLANETARY RULERSHIP, since the Sun and the Moon do not have negative or positive rule, like the other five planets, and it is indeed the luminaries which Servius leaves out. The account of the descent pictures the 'sluggishness' of Saturn, the 'desire for domination' of Jupiter, the 'passionateness' of Mars, the 'lustfulness' of Venus, and the 'cupidity' of Mercury. It is as though the pure soul is blinded by these influences, and is afterwards unable to use its own powers. The scholastic view of this curious 'spherical descent' is discussed by Mead, but the interesting thing is that virtually similar keywords are still used to represent the negative aspects of these planets in modern astrology.

Egyptian terms   See PTOLEMAIC TERMS.

Egypto-Babylonian zodiac   See HYPSOMATIC ZODIAC.

Eight Places   The Eight Places of Manilius have often been confused with an eightfold HOUSE SYSTEM, though it is likely that the places were merely derived by extrapolation from the Greek LOCI – at least, the astrologer Firmicus, who was the only other astrologer besides Manilius to mention the eight places, considers them to be derived in such a way. The eight places are obtained by a simple division of the four quadrants, and it would seem that the interpretation of the significance of the places (or arcs, as they should be called, even though originally they may have been points) is fundamentally influenced by the related angles. Manilius himself links the Angles with the planetary gods – the Horoscopos with Mercury, the Dysis with Pluto – through the term *Janua Ditis* (door of Dis) – the Midheaven with Venus, and the Imum Coeli with Saturn. This sequence appears to be linked with one of the many prevailing theories of the FOUR AGES. At all events, the eight places, conditioned by the angular influences, are listed as follows, in the modern order of houses:

1	Sedes Typhonis	1	Sedes Typhonis
2	Phoebe-Dea	6	Phoebus-Deus
3	DAIMON	7	Felix Fortuna
4	Porta Laboris	8	Porta Laboris

The historian Bouche-Leclercq sets out an admirable analysis of the probable meaning behind the system.

Eighth Hierarchy   See ARCHANGELS.

eighth house   The eighth of the twelve astrological HOUSES linked with the sign Scorpio and (in modern astrology) with the planet Pluto, and whilst popularly said to be the HOUSE OF DEATH, it is actually the house governing regeneration, of which the post-mortem experience is merely the extreme condition. It is called the Occult House because it has rule over hidden things – exoterically over legacies, but esoterically over karma, the final legacy. It is linked through Scorpio with the sexual parts – more properly with the generative system. In MUNDANE ASTROLOGY it rules the public income, especially from exports, the national debt, and interest rates, as well as important financial losses and gains.

Eighth Sphere   A much-misunderstood term, the real meaning of which may be grasped only from its context. In the PTOLEMAIC SYSTEM the term was used for the Sphere of the Fixed Stars (STELLATUM), though this designation changed as the model was adapted. In the early cosmoconceptive models the spheres were numbered from the centre outwards, though the actual numbers were merely designations, for the theory of the SPHERES was complex, and required a large number of interacting spheres within the designate concentrics. There were, for example, a number of elemental spheres encasing the geocentric Earth: this alone meant that even if we regard the spheres as numbered bands of interacting spheres, the Eighth Sphere 'stellatum' should really be numbered as the 'twelfth'. Now, in contrast to this, we find that when Copernicus introduced his own model (see COPERNICAN SYSTEM), he retained the theory of spheres (somewhat modified in the beginning, but totally rejected eventually), but numbered them inwards, towards the heliocentric centre. In the diagrammatic representation of this system, the 'stellatum' is now often numbered as the 'first sphere'. This rather tiresome distinction must be made here in order to explain why the term 'eighth sphere' is used in astrology for both the spheres of Heaven (Stellatum) and the spheres associated with Hell. In popular (occult) literature the Eighth Sphere is indeed equated with the Sphere of the Moon, which in the esoteric tradition is linked with the post-mortem purgatorial experiences. Allied to this, the Eighth Sphere has been called 'the planet of death' (the planet being the Moon, of course), and it is said to mark the sphere where lost souls find their dwelling (which, of course, makes it a very different place from purgatory). In fact, there is an extensive and highly potent literature dealing with the modern esoteric astrological conception of the Eighth Sphere, a literature spearheaded by Harrison, who examines with profound esoteric insight the mistaken notions that link this sphere with the Moon.

election   A term used to denote a chart cast according to the rules of ELECTIONAL ASTROLOGY.

electional astrology   A form of casting and interpreting charts to determine suitable times for commencing any specific activity, such as marriage, journeys, law-suits, and so on. The idea is that the astrologer seeks among the possible times practical to the querent the most satisfactory auspicious moment fitted to that particular endeavour. A chart erected for such a chosen moment is called the 'election', or the 'inceptional figure', and this is scrutinized according to the rules of HORORARY ASTROLOGY for success or failure. Electional astrology is usually practised alongside the radical and progressed natal horoscope of the querent. The rules for interpretation are complex, but follow the general drift of traditional genethliacal astrology, though with a few additional rules concerning DISPOSITORS and specialist rulerships maintained by planets and houses. See also INCEPTIONAL ASTROLOGY.

Electra   See PLEIADES.

Electrical Ascendant   See VERTEX.

elemental aspects The original theory upon which the doctrine of
*ASPECTS was based related to the supposed harmonies and lack of harmonies
(the SYMPATHY) between the elemental natures of the signs. Thus, planets in
trine were said to be operating in a harmonic sympathy because they each fell
into a sign of a particular element – for example, Mercury in Aries trine
Jupiter in Leo were both working through the fire element, and were thus in
unison. However, when an allowance is made for orb, this original elemental
nature may be ignored: an aspect may be noted on a geometric basis which
does not fulfil the requirements of the elemental basis. For example, Mercury
may be in 29 degrees of Aries, while Jupiter is in 2 degrees of Virgo: in
terms of the geometric theory of aspects, they are in trine (with an orb of 3
degrees), yet the aspect does not respect the ancient doctrine of elemental
accord, one planet being in a fire sign, the other in an earth sign. Such non-
elemental aspects are undoubtedly weaker in influence than elemental aspects,
and there is even good argument for discounting their influence altogether.

elementals A term properly applied to the classes of nature beings, the so-
called 'elementary spirits' or 'Sagani' of esoteric lore. It is important to
distinguish the term from ELEMENTS and from 'elementaries' (a word
properly applied to a class of disembodied spirits, or to the astral shells of
disembodied spirits). The four groups of elementals are often mentioned in
astrological literature – especially so in esoteric astrology – because they are
the expression on the natural plane (in the invisible realms which support the
natural plane) of the activity of the four elements, from which they take their
generic name. The GNOMES are the earth beings; the SYLPHS are the air
beings; the UNDINES are the water beings, and the SALAMANDERS are the fire
beings. It is erroneous to call these beings 'spirits' –see SOUL BEINGS.

elemental soul beings See SOUL BEINGS.

elementaries See ELEMENTALS.

elementary qualities A term used by Cornell to denote the ancient
QUALITIES: hot and cold, wet and dry.

elementorum See MACROCOSM.

elements The esoteric elements of the astrologers are four in number: earth,
water, air and fire, each manifesting through a different QUALITY three times
to make up the twelve divisions of the zodiac, as in table 34. Esoterically the
elements of the astrologers are five in number, for it is maintained in the

| | TABLE 34 | | |
Element	Cardinal	Fixed	Mutable
Fire	Aries	Leo	Sagittarius
Earth	Capricorn	Taurus	Virgo
Air	Libra	Aquarius	Gemini
Water	Cancer	Scorpio	Pisces

tradition that four are themselves restless, and even in enmity with each other, so that they are bound into a temporary pact (such as is required by the world of material forms) by a fifth element. This fifth element maintains the pact between sublunar things, and is often called the QUINTESSENCE, the secret AETHER of the ancients. One must observe that the occult elements are not what they seem, even in exoteric astrology. 'Let it be remembered,' says Blavatsky, 'that Fire, Water and Air, or the "Elements of Primary Creation" so-called, are not the compound Elements they are on Earth but nuomenal homogeneous Elements – the Spirits thereof.' It is widely recognized that the fifth element, the quintessence, will in the future become visible to human beings. Blavatsky takes this projection further however and speaks of two other elements, which will long remain invisible, rounding off the present tetrad as a septenary. This septenary is said to be the conditional modification of the one and only element, which is 'not the Aether', to quote Blavatsky once more, 'not even A'kasa but the Source of these'. For a full survey of the elements of astrology see AIR ELEMENT, AIR SIGNS, EARTH ELEMENT, EARTH SIGNS, FIRE ELEMENT, FIRE SIGNS, WATER ELEMENT and WATER SIGNS.

elevation   The altitude of a planet, nodal point or fixed star above the horizon. Sometimes the term is limited to the description of a horoscope, in which case it refers to the house-placings, the Medium Coeli being the most highly elevated of the cusps. The planet in the eastern diurnal arc nearest to this cusp is said to be 'elevated'. However, confusion proceeds from the fact that the term is also used loosely for any planet above the horizon. It is said in the ancient tradition that if either of the luminaries is elevated above a malefic, then the influence of the latter is mitigated.

eleventh house   The eleventh of the twelve astrological HOUSES, linked with the sign Aquarius and (in the modern systems) the planet Uranus, and associated with the native's ideals in relation to social life and evolution, as well as with the kind of acquaintances and friendships he or she will attract. The house is said to relate directly to the aspirations of the native, and for this reason it was called once the 'House of Hopes and Wishes', yet it does not have the same heart-felt drive of the FIFTH HOUSE in these matters, such impulses often arising more from ideals and intellectual attractions than from direct emotional involvement. The house indicates the areas which will open up the native to higher levels of consciousness – often involving connections with development groups or political-awareness groups – as the ruling Aquarius would suggest. Through the Aquarian connection it rules out the legs and ankles of the human body, and is esoterically connected with the entire circulatory system. In *MUNDANE ASTROLOGY, it has rule over the legislature, the stock exchange, public festivals, political aspirations and friendly societies.

Elgebar See RIGEL.

Elgeuze See ORION.

El Nath Fixed star (double), the beta of constellation Taurus, from the Arabian Al Natih (the butting one), located on the tip of the Northern Horn. Astrologers maintain that it brings eminence and good fortune, but its influence seems to depend upon the nature of the conjuncting planet.

elongation A term applied to the distance in degrees of arc of either of the inferior planets to the Sun. The maximum elongation of Mercury is 28 degrees and of Venus 46 degrees. The term is also used of the furthest distance which any planet may be from the Sun – see APHELION.

Elpheia See ALPHECCA and FIFTEEN STARS.

Elul See BABYLONIAN CALENDAR.

embolismic lunation A term often used as an equivalent for the EMBOLISM MONTH. The same term was used by the astrologer Placidus de Tito to denote a chart (really a series of charts) cast for the periodic lunar returns to the same arc relationship to the natal Sun position. See also SYNODICAL LUNATION.

embolismic month In some ancient CALENDAR systems an intercalary month is sometimes inserted in order to preserve a relationship between the days and the seasons: this is called the embolismic month.

Empyrean One of the ancient names for the TENTH SPHERE, though sometimes used (in post-medieval sources) to denote the extralunar spheres, and even as a synonym for heavens generally. See also MACROCOSM.

Enceladus Name given to one of the moons of Saturn.

enneal horoscope A chart derived from the NOVIENIC MOON, and used by the ancient Greeks – see NOVIENIC CHART.

enneatical Simmonite uses this Greek-derived term, *ennead* (ninth part) to apply to one of the climacterics, and records that it is supposed to bring 'a change of fortune'. It is also used of the ninth day of an illness, when a change of condition may be expected. On the ninth day, the Moon enters the orb of a beneficial trine to its radical position.

ens astrale See ASTRAL.

Ensis A nebula in the sword-sheath of the constellation ORION, said to be of the nature of Moon conjunct Mars, and to cause blindness and death when prominent in a figure.

epact A term with several related meanings. First, it is used to denote the number of days by which the solar year exceeds the lunar year. Secondly, it is used (though rarely in modern times) to denote any intercalated day or days. Thirdly, it is sometimes used wrongly in astrology to denote the age of the Moon on the first day of the year.

epanaphora A term derived from Greek astrology, as applied to a sign of the zodiac following (in the sequence of the signs) the Angles. Thus, if Aries is on the Midheaven, Taurus is epanaphora. The term was also used of the LOCI 2, 5, 8 and 11, which means that the term used in connection with houses was the equivalent of the later SUCCEDENT. The term has been wrongly defined by some modern astrologers, probably due to a misreading of Ptolemy: because of this, the term is sometimes used to denote arcs in which the planets are at their most influential position – the main 'epanaphoric arc' is the Midheaven, while only slightly less powerful is the Ascendant, the entire 30-degree arc immediately succedent to these cusps being regarded as the epanaphorics.

epanaphoric arc See EPANAPHORA.

ephemeral map A horoscope cast for purposes of HORARY ASTROLOGY.

ephemeral motion Term applied to the motions of the planets to distinguish from directional or progressed motions.

ephemerides The plural for EPHEMERIS.

ephemeris A table listing the positions of the planets and nodal points, usually in longitudes, latitudes and declinations, along with related astronomical data, usually including an ASPECTARIAN, and sometimes even tables of houses. Ephemerides have been in use from the beginning of personal horoscopy, but the first to be printed in the West was the *Kalendarium Novum* (a collection relating to the years 1474 to 1506) calculated by the German mathematician-astrologer Regiomontanus. Most ephemerides used by astrologers give their positions geocentrically, though modern astronomers' ephemerides list them heliocentrically. The majority of ephemerides prior to 1925 give information relative to noon on particular days, but the introduction of the concept of the astronomical day has resulted in many ephemerides adopting midnight as marking the time of their data. See COMPUTATION and TIME.

ephemeris time See TIME.

epicycle In the cosmoconception of the PTOLEMAIC SYSTEM, the epicycle is a small revolving circle, the circumference of which is centred upon the moving circumference of a DEFERENT circle. The theory of epicycles was originated mainly to support the Aristotelian dictum that extralunar bodies moved in perfect circles – a dictum openly contradicted by the appearance of things in a geocentric model. The theory appears to have been adopted by Ptolemy from Apollonius of Perga, who probably suggested the idea of epicycles in the 3rd century BC. The complex structures and models to which the epicycle theory gave rise permitted astrologers to combine the opposite motions of epicycles and deferents within a framework of velocities and radical magnitudes in such a way as to account for the (visually apparent) movement of planets in a remarkably accurate way. The theory of epicycles to some extent satisfied astrological demands, and was even adopted by Copernicus, though it eventually broke down under the weight of its own complexity.

epoch   Originally a term used to designate a point of reference in the calculation of dates, but more recently used to designate the PRE-NATAL EPOCH. The term is also used as a synonym for 'perodicities' or 'cycles' – see, for example, AGES. See also PLATONIC YEAR and PRECESSION.

epochal astrology   See PRE-NATAL EPOCH.

epochal chart   A horoscope figure designed to symbolically present in a chart information allegedly relating to the cosmic configuration attending the moment of conception, which is often mathematically derived from the natal chart (see PRE-NATAL ASTROLOGY). It is sometimes called the 'conception chart', or the PRE-NATAL CHART, and is generally projected geocentrically in accordance with the standard methods of HOUSE DIVISION. Those who favour ASTROSOPHY often use a heliocentric system, and an epochal chart which is sometimes called the 'pre-natal asterogram' – see ASTEROGRAM.

equal-division method   A system of HOUSE DIVISION which is to be distinguished from the MODUS EQUALIS (sometimes called by the same name). It was devised by the Australian astrologer Zariel in the 19th century, and sometimes called the Zariel Division. The cusps of the twelve houses are determined by the projection on to the ecliptic from the pole of the equator of twelve equal divisions of the horizon. As the astrologer Leo points out, the principle inherent in this system is tantamount to regarding the native as being born under the meridian of the birthplace, but at the equator, no matter where the place of birth. As a system it does not appear to have a wide acceptance.

equal house   See MODUS EQUALIS.

equal power   See BEHOLDING.

equation of time   A term usually applied to a set of tables showing the daily variation between the mean solar day and the true solar day: strictly speaking, the term should properly be applied to the variations so tabulated. The term is sometimes wrongly applied to the difference between mean time and sidereal time, which is properly CORRECTION OF TIME.

equations   See COSMIC PICTURES.

equator   The term equator is sometimes used to denote the imaginary circle dividing the globe of Earth into two equal parts, and sometimes to the plane of this circumference, projected into the Celestial Sphere, the equatorial plane. See also CELESTIAL EQUATOR and TERRESTRIAL EQUATOR.

equatorial chart A horoscope chart, or symbolic figure, erected to symbolize a plane through the equator of the Earth.

equatorial system The equatorial system is a model of the CELESTIAL SPHERE, which takes as a point of reference the diurnal rotation of the Earth. The 'vertical' co-ordinates are established by extending the North and South Poles into the Celestial Sphere, to create the corresponding CELESTIAL POLES. The equator is extended into the Celestial Sphere (hence called the CELESTIAL EQUATOR) to give the second co-ordinate. The equator is subject to a time division of hour circles, measured eastwards from the standard zero circle passing through Greenwich, and location along this circle is measured in hours and minutes of Right Ascension. The plane of the equator is regarded as a second zero, paralleled by parallels of declination, measured by Angular distance north (termed 'plus') or south (termed 'minus') of the equator. The points where the Sun's apparent orbit intersects the equator are called the 'equinoctial points', and the arc of greatest separation between the apparent path of the Sun and the equator are called the SOLSTICES.

equinoctial point See ECLIPTIC.

equinoctial precession See PRECESSION.

equinoctial signs These are Aries and Libra, which mark the spring and autumnal equinoxes, respectively.

equinox A term meaning literally 'equal night', and used of that time when the day and night are equal in length, which occurs when the Sun enters zodiacal Aries (see VERNAL EQUINOX) and Libra (see AUTUMNAL EQUINOX).

Equuleus Constellation, approximately 19 to 28 degrees Aquarius, and from 1 to 12 degrees north of the equator. The image is traditionally a figure of a horse or foal, said to represent either Celeris, the brother of Pegasus (near which the constellation is located) or another horse, Cyllarus. It is said to be a fortunate influence.

Equuleus Pictoris See PICTOR.

Eridanus Constellation, winding in a full 55-degree arc south of the equator beyond the confines of Pisces, Aries and Taurus, originally called by the Greeks Potamus (river). Its very extent makes its traditional associations with travel and with dangers at sea, questionable, though the power of the lucida ACHERNAR is beyond doubt. See also VIA LACTEA.

Eros   One of the SEVEN LOTS of Greek astrology, the Lot of love. The degree of Eros is the same distance from the Ascendant as is Venus to the DAIMON.

erratics   Some of the ancient astrologers used the term erratics as synonymous with planets, thus contrasting their movements with the orderly sweep of the fixed stars. The Greek word from which our term 'planets' was derived carried this sense of 'erratics', and the Babylonian for planets, *bibbu*, meant 'wild goats'. See also ECCENTRIC.

Esna zodiacs   The temple of Khum at Esna (modern Latopolis) in Egypt is famous for its bas-reliefs of a huge calendar and two so-called 'zodiacs', one of which is in the ceiling of the portico, the other in the ceiling of the inner shrine. Like the DENDERAH ZODIAC, these are constellation maps, containing planets. They run in two registers, the images of the zodiacal constellations being remarkably similar to the modern ones, which is scarcely surprising as it was such images that the Arabian astrologers transmitted to Europe by way of Greek and Roman astrology. Long believed to be of great antiquity, these maps have been dated from internal evidence, derived from the religious calendar of the temple, to 137 BC.

esoteric astrology   The term is derived from the idea that behind the outer form of ordinary astrology (exoteric astrology) there is an esoteric form which proceeds from an entirely spiritual realm of being. It has long been recognized by the more thoughtful astrologers and historians of astrology that much of the philosophy underlying their science properly belongs to the ancient wisdom once taught in the Mystery centres of Egypt and Greece, which has now become not only exoteric, but even badly mutilated in the process. It is equally apparent that the stream of esoteric knowledge is still available to those who are prepared to make the considerable effort required to reach into the spiritual realms from which it proceeds. Many astrologers have indeed made this effort, through the guidelines set out in the Neo-platonic, Rosicrucian and qabbalistic forms of astrology, with the result that it has been possible for them to formulate some of the ancient laws of esoteric astrology, and introduce into the exoteric art new and important truths. Drinkwater, who recognized seven streams of Western Mystery wisdom, quite rightly saw astrology as derived from ancient wisdom, but felt it prudent to add that 'astrology, though originally a great mystery religion in ancient Chaldea, is, in its present form, too fragmentary to be considered an independent Mystery school, and students of esoteric astrology are obliged to supplement it from other traditions, including theosophy itself'. It is apparent that the stream of esoteric knowledge which sprang from the ancient Mysteries is still available to those who are prepared to make the considerable effort to reach into the spiritual realms from which it proceeds. The theory of spiritual historicism upon which this account is based rests on concepts very different from those considered valid in academic circles, yet evidence for its truth is superabundant. The Neoplatonic astrology derived from gnosticism was essentially born of an esoteric stream designed to counter certain of the destructive elements inherent in orthodox theology. Through the medium of Arabian astrology, this esoteric antique astrology flowered in medieval astrological forms, many aspects of which were encapsulated in the symbolism of cathedrals and basilican churches. The 'new' astrology of the Renaissance, which so completely informed the arts and literature of that period, appears to have been born of an esotericism nourished in Florence, rooted in the new image of man derived from Neoplatonic sources, certain oriental ideas, and a spiritual cosmoconception which is now little understood. The esoteric stream seems to have gone into partial occultation during the period of the intellectual Enlightenment, and survives mainly in prolix alchemical and religious texts linked with Rosicrucianism, but in the decades around the turn of the present century there was a proliferation of different esoteric systems, some of which were syncretic and orientalizing rather than truly esoteric. The main difference between the esoteric astrology of modern times and the various exoteric forms is in the cosmogenetic and teleological aspect of the science – almost all esoteric astrological systems are founded on the premise that the cosmos is a living being, that the destiny of the solar system is intimately bound up with the destiny of humanity, and that human beings reincarnate periodically on to the earth. There is also a major difference between esoteric and exoteric forms in the view of the purpose behind astrology: the esoteric astrologers tend to see their studies involved with philosophy, with the study of cosmogenesis, and so on, and tend to make sparing use of actual personal horoscopes. Within most forms of even exoteric astrology there is usually a residue or accretion of esoteric lore which postulates the idea of ESOTERIC PLANETS. In many cases the esoteric nature of the planet is merely veiled by the exoteric planet (usually of the same name). The most notable example is the INTUITIONAL ASTROLOGY of Bailey, which in clairvoyant methodology was based on Blavatsky, and was deeply influenced by the Russian, yet succeeded in developing a strain of esotericism unique to itself. Other examples are found in the esoteric strain of qabbalistic astrology – especially in that linked with genuine Rosicrucianism, in the orientalizing astrology promulgated by Blavatsky, in the later ramifications of Thierens' esotericism, and in the Christianized esoteric astrology hinted at by Steiner and studied by many of his followers, including Vreede and Sucher. The most powerful influences of esoteric streams on the development of astrology have been through the writings and lectures of Steiner (see, for example, ASTROSOPHY and SIDEREAL MOON RHYTHMS), and the general renewal of interest in the stream of Paracelsian, Boehmian and Rosicrucian literature which followed on the esoteric writings of Blavatsky and her theosophy. Blavatsky had no follower sufficiently able to take her esoteric lines of thought further in the realm of astrology: the astrologer Leo (who wrote one of the earliest books purporting to deal with esoteric astrology) was himself no esotericist, and his writings, in so far as they continue theosophical lines of thought, are merely attempts to form correspondences between the orientalizing terminology of theosophy and traditional astrology, with some admixture of Hindu astrology. The chief contribution of this stream of theosophy to modern astrology, therefore, has not been in the esoteric realm so much as in its undoubted contribution to a reassessment of traditional astrology. The soi-disant ROSI-CRUCIAN ASTROLOGY of Heindel is actually derived from Blavatsky and Steiner, and must not be confused with the genuine esoteric stream of medieval Rosicrucianism. The esoteric elements contained in the astrological writings of Collin are derived ultimately from the esoteric system propounded by Gurdjieff and his follower Ouspensky, both of whom appear to have had scant respect for ordinary astrology.

esoteric Moon In her system of INTUITIONAL ASTROLOGY, the esotericist Bailey pictures the Moon of our Earth as a sort of veil, for the influence of planets which she variously identifies as Vulcan and Neptune. She gives no clear indication whether these two are to be regarded as hypotheticals, with specialist influences, or as disembodied influences of one or other of the hypotheticals described under VULCAN, and the traditional NEPTUNE. Bailey puts out many curious propositions, including, 'The Earth is a satellite of the Moon', and 'The Moon is a dead planet from which all the principles are gone.' Such statements are not in accord with the ancient esoteric tradition concerning the Moon, nor in accord with the traditions of ordinary astrology. See NON-SACRED PLANETS. The hypothetical LILITH has been called the 'esoteric Moon'.

esoteric planets The term has several applications, depending upon the

form of astrology under consideration. Within the cosmoconception of
INTUITIONAL ASTROLOGY, the term is used sometimes as synonymous with
SACRED PLANETS, and sometimes in a more exclusive sense. Within this
system, which traces the arising and manifestation of the SEVEN RAYS, the
sacred planets (called also the esoteric planets) are given the unorthodox
rulerships set out in table 35. For the 'rays' see individual entries, FIRST RAY,

─────────────────────── TABLE 35 ───────────────────────

Ray	Constellation	Orthodox planet	Esoteric planet
first	AR	Mars	Mercury
	LE	Sun	Sun
	CP	Saturn	Saturn
second	GE	Mercury	Venus
	VG	Mercury	Moon (veiling a planet)
	PI	Jupiter	Pluto
third	CN	Moon	Neptune
	LI	Venus	Uranus
	CP	Saturn	Saturn
fourth	TA	Venus	Vulcan
	SC	Mars	Mars
	SG	Jupiter	Earth
fifth	LE	Sun	Sun
	SG	Jupiter	Earth
	AQ	Uranus	Jupiter
sixth	VG	Mercury	Moon
	SG	Jupiter	Earth
	PI	Jupiter	Pluto
seventh	AR	Mars	Mercury
	CN	Moon	Neptune
	CP	Saturn	Saturn

etc. However, in the face of this list, Bailey does distinguish specific
esoteric planets by name: *ESOTERIC MOON, *ESOTERIC PLUTO and *ESO-
TERIC VULCAN, though it is evident that each of the orthodox planets has
also an esoteric side or nature. It is difficult to generalize the main different
views taken by practical astrologers (see ESOTERIC ASTROLOGY) of the
working of the esoteric planets, but the indications given by Bailey herself of
the fundamental difference between the 'sacred' and the NON-SACRED

esoteric Pluto

PLANETS may be extended to that between the esoteric and the exoteric. She says that it is the development of the individual being (the native) which marks the ability to respond (or not to respond) to the esoteric level of a given planet, or to its esoteric equivalent: an initiate, or highly developed human being, lives according to cosmic laws which are different to those which embrace the average person, for the more evolved human may respond to a higher range of vibrations marked out by, or heralded by, the esoteric planets. A careful reading of Blavatsky will show that each of the planets is viewed as having an esoteric nature veiled from ordinary understanding: it is this tacit esotericism which makes many of Blavatsky's statements appear to be so questionable to the ordinary reader, even though they are usually statements founded on a most profound grasp of esoteric cosmogenesis.

esoteric Pluto In her system of INTUITIONAL ASTROLOGY, which leans upon the esotericism of Blavatsky, Bailey pictures the Pluto of our solar system as a sort of veil, working through a rulership over Pisces. It is clear, however, that she adopts the standard interpretation of the destructive regenerative polarity of the planet. In view of this, the planet might more readily be termed the 'Bailey-Pluto', rather than 'esoteric Pluto', in accordance with the present tendency to distinguish by personalized suffix the various Plutos noted by modern astrologers.

esoteric Sun Blavatsky puts into words an idea which is implicit in several esoteric strains of astrological thought, and which is openly mentioned in the hermetic literature of the immediate post-Christian era – that the Sun of our solar system 'stands for, or veils, a hidden planet'. This veiling Sun is later called the 'esoteric Sun' in the INTUITIONAL ASTROLOGY of Bailey. The esoteric Sun is the image of divine intelligence and wisdom, and it is this which encourages Blavatsky to an etymology which derives the Latin *Sol* from *solus* meaning 'one alone', and linking the Greek *Helios* (Sun) with the meaning 'most high'. The triple Sun of esotericism is linked with the esoteric Sun, being a veil for the Trinity – the 'spiritual Sun' is God the Father; the 'heart of the Sun' is God the Son, and the 'physical Sun' is God the Holy Spirit. Blavatsky traces the relegation of the Sun to the status of 'planet', and the consequent loss of its esoteric and triple symbolism to the work of the Christian astrologers 'who had not been initiated'. However, if the esoteric documents and art-forms of the early Christians are interpreted correctly, it will be seen that the tradition of a spiritual sun, linked indeed with the Trinity and with the Logos, survived for well over a thousand years into our era. The central solar image of the SAN MINIATO ZODIAC, made during an exoterically geocentric period, points to this mystery, for example. While Blavatsky makes a distinction between the visible and spiritual Suns, she also claims that the esoteric Sun is 'the substitute for the invisible inter-Mercurial planet'.

esoteric Uranus Both Blavatsky and Bailey write of an esoteric Uranus: as the former says, 'Uranus is a modern name, but one thing is certain, the ancients had a mystery planet which they never named.' It is this esoteric Uranus which figures in the INTUITIONAL ASTROLOGY of Bailey, as one of the three synthesizing planets (the other two being Neptune and Saturn) through which Sirius influences our solar system. See also ESOTERIC ASTROLOGY.

esoteric Vulcan In her system of INTUITIONAL ASTROLOGY the esotericist Bailey has proposed an important role for her planet Vulcan, sometimes called an esoteric planet, at other times one of the *SACRED PLANETS. This Vulcan is said to be within the orbit of Mercury, and appears to have a crystallizing effect – at least, this is so in connection with its rule over Taurus in the UNORTHODOX ASTROLOGICAL RELATIONSHIPS. It is this same Vulcan which is said to be veiled by the present Moon – see ESOTERIC MOON.

essential dignities See DIGNITY.

estival quadrant See QUADRANTS.

estival signs See AESTIVAL SIGNS.

ethereal orbs See ETHEREAL PLANETS.

ethereal planets A term probably originated by the astrologer Harris and applied by him to a number of planets (invisible to all but advanced clairvoyants) said to be in orbit between Mercury and the Sun. This chain of planets he called the Sisterhood, and claimed they were *AROMAL PLANETS. It seems likely that the ethereal MELODIA was of this category – but see also HYPOTHETICAL PLANETS. Blavatsky insists that at least four (possibly six) of our own planets are 'ethereal orbs', by which she appears to mean that the spiritual planet is separate in space, if not also in time, from the physical orb: for example, she gives the name Sol to a transmercurial planet, nearer to the Sun than Mercury, which 'became invisible at the end of the Third Race', in Lemurian times, even before the epoch of Atlantis.

Etheric Race   See FIRST ROOT RACE.

etheric spheres   A specialist esoteric term probably coined by Thierens, relating to the future evolution of the spheres, and the consequent development of new planetary bodies which are (supposedly) present yet invisible to ordinary vision. Within this new development, Aries will unfold a planet called Pluto or OSIRIS; Taurus will unfold a planet called VESTA or Isis; Gemini will unfold a planet called HERMES or Mercurius, and Cancer will unfold a planet called BACCHUS or Horus. Thierens calls these 'spheres' because they include what must be visualized as the traditional spheres of Aries, Taurus, Gemini and Cancer. See also ASTRAL SPHERES.

Ettanin   Fixed star (double) the gamma of Draco, and a prominent star for the Egyptians who called it Isis or Taurt Isis, and for whom it marked the head of the Hippopotamus, as figured in the DENDERAH ZODIAC. Its rising through the central passages of the temples of Hathor at Denderah was said to be visible about 3500 BC. The name, with many variants (a standard being Eltanin) is from the Arabic Al Ras al Tinnin (the dragon's head) wrongly translated by Ebertin as meaning 'right eye of the dragon', and linked by him with the Martian influence. With a well-placed Saturn, however, it is said to be helpful for mental concentration and for 'esoteric studies'. Lying almost at the zenith of Greenwich, it is sometimes called the Zenith Star.

Euctemonian calendar   During the 5th century BC the Greek astronomer Euctemon defined a solar calendar in terms of solar months of (approximately) 30 days' duration, linked with the phenomena of equinoxes and solstices. Euctemon gave the twelve months of his calendar the same names as the zodiacal asterisms of the BABYLONIAN ZODIAC, or at least derived them from the corresponding images. Prior to the JULIAN CALENDAR reforms, this sequence of solar months began with the vernal equinox in the month of Aries (APRIL). As the historian Powell says, the Euctemonian calendar is still in vogue today, in popular Sun-sign astrology.

Eudaimon   A Greek-derived term – sometimes Eudamon or Eudaemon (meaning approximately 'good spiritual influence') – applied to the 11th house of a horoscope figure, which was regarded as a source of benefits, just as the 12th house Cacodemon was regarded as a source of difficulties.

Eudora   See HYADES.

Europa   One of the satellites of Jupiter.

Exaltations

Evangelists The Christian symbolism relating to the Four Evangelists is derived from pre-Christian astrological lore, as indicated by Guthrie among others. The Bull of St Luke is from the image of Taurus; the Lion of St Mark is from the image of Leo; the Angel of St Matthew (carrying a book or scroll) is from the image of Aquarius (carrying an urn). The Eagle of St John is the 'redeemed' image of Scorpio – see EAGLE. The cosmic imagery of the Evangelists is ultimately linked with the Hierarchies of the CHERUBIM and SERAPHIM.

Evestrum A Paracelsian term used to denote several aspects of the eternal substance of heaven. It is the ethereal and invisible substance (or, rather, agency) of the heavens, and the sidereal body, the starry body or astral body of man, which is itself built from that substance or agency.

Exaltations An Exaltation is a sign (or sometimes a degree) of the zodiac in

TABLE 36

Planet	Exaltation	Degree	Planet	Exaltation	Degree
SU	Aries	19	MO	Taurus	3
ME	Virgo	15	VE	Pisces	27
MA	Capricorn	28	JU	Cancer	15
SA	Libra	21			

which a planet is said to exert its most powerful characteristic influence. The traditional Exaltations of the planets were presented in Ptolemaic astrology in terms of signs only: a planet was said to be Exalted in a particular sign, the Ptolemaic order of which corresponds to the modern as set out in table 36. Ptolemy attempted a rationalization for these Exaltations, but is unconvincing

TABLE 37

Planet	Tropical (degrees)			Sidereal (degrees)		
UR	Scorpio	25	(24.9)	Sagittarius	9	(8.7)
NE	Aquarius	5	(4.6)	Aquarius	19	(18.4)
PL	Cancer	4	(3.9)	Cancer	18	(17.7)
Dragon's Head	Pisces	17	(16.7)	Aries	1	(0.5)

in his argument. The specific degrees of Exaltation have come down from very ancient sources: as the astrologer Fagan has shown, these Exaltation degrees are probably derived from the *HYPSOMATA. Following the computations of the hypsomatic degrees for what he called the HYPSOMATIC YEAR, Fagan provides the Exaltation degrees of the so-called new planets and the nodes as set out in table 37. It will be noted that in Hindu astrology the north node (RAHU) is Exalted in Taurus, and the south node (KETU) in Scorpio. See also DIGNITIES, but distinguish from the Hellenistic, and now archaic NATIVITY EXALTATION.

Exsusiai   See POWERS.

Ezrian house system   See HORIZONTAL SYSTEM.

F

face The term has been used in a variety of specialist astrological forms, resulting in much confusion. The only proper modern application of the term is to the division of the signs of the zodiac into 72 arcs, each of 5 degrees, so that each sign is divided into six equal-arc divisions. Each of these 5-degree arc faces has been accorded a planetary ruler, as well as specific associations (usually connected with the signs to which they relate): they are sometimes (more properly) called 'demi-decans' – but see DECAN. The incorrect use of the term has sprung from a twofold error. The term face has been used as the equivalent of the PROPER FACE in which connection it is usually misapplied due to Ptolemy's own obscure and inconclusive definition. Unaccountably, Wilson (quoting older sources) defines face as the equivalent of a decan, and gives traditional decanate readings for 10-degree arcs as applicable to the 5-degree arcs – he may have taken the word from Angelus.

facial types See APPEARANCE.

Facies Name given to the nebula in the 'face' of the figure of constellation Sagittarius. It has an evil reputation among astrologers, for when operative in a chart it causes blindness, accidents and violent death.

Faelis The constellation of the Cat, formed almost in joke by La Lande in 1804, but never widely adopted. Located between Antlia Pneumatica and Hydra.

Fair Star of the Waters See SIRIUS.

Fall A planet is said to be in its Fall when in the sign (or degree) opposite to that of its *EXALTATION (or Exaltation degree). In the ancient astrological system, a synonym for the term was 'prison', the idea being that a planet's influence was imprisoned in the sign opposite to its Exaltation.

falling star See METEOR.

familiarities The familiarities of planets were the aspects between them, the

elemental relationships in the original sense of aspectal theory. See FAMILIAR-
ITY.

familiarity   A term derived from medieval astrology, used as a synonym for
ASPECT – the familiarities were often lists of aspects in a given chart or in a
table of aspects. The term is the Latin equivalent of Ptolemy's *oikeiosis*
(kinship) which appears to embrace the idea of aspects, but was also used in a
much wider sense, of all planetary relationships.

Fargh al Mukdim   Arabic term (the fore-spout of the waterbucket, according
to Robson), the 24th of the Arabic MANZILS, located by the modern alpha
and beta of PEGASUS.

Fargh al Thani   Arabic term (the lower spout of the waterbucket, according
to Robson), 25th of the Arabic MANZILS, the medieval Haurens, located by
the modern alpha of ANDROMEDA.

fatality   For Point of Fatality, see HOUR GLASS.

father   The Point of Father is not to be confused with the House of Father.
The former is one of the so-called Arabian PARS – if the degree of Saturn in a
natal chart is revolved to the Ascendant, the adjusted position of the Sun
marks the Point of Father. The House of Father is a name given to the
TENTH HOUSE.

February   Now the second month (variable, 28 days, and 29 days as intercalary), the name being derived from the early Roman pre-JULIAN CALEN-
DAR, after a festival of expiation, a catharsis, the Latin *februatio* applying to
the ceremony of purification: the name Februata being applied to the
goddess Juno who was worshipped at the ceremonies. The festival of the
Lupercales was celebrated in the middle of this month in Roman times, and
as the imagery in the vast mosaic of the months from El Djem indicates, this
was involved with purification and catharsis (see Foucher). Esoterically the
month seems to be connected with the cleansing of the feminine element in
preparation for the beginning of spring. In the earlier European calendars,
this was the last month of the year.

feminine   See SECTA.

feminine planets   The schema of *PLANETARY RULERSHIPS recorded by

feral signs

Ptolemy involved the idea that planets had a dual nature (though this was not claimed of the luminaries), variously expressed in the dual rulerships over the zodiacal signs other than Cancer and Leo. For example, the Mercury which ruled Gemini was positive or masculine, while the same planet ruling Virgo was negative and feminine. Although this traditional idea has been widely misunderstood, it is still possible to find reference to such terms as 'feminine Mercury' or even 'feminine Mars', though it is more usual to find the expression NEGATIVE MARS, (the terms 'feminine', 'negative' and 'nocturnal' being interchangeable in this context). Quite a different tradition, connected with the old teaching of PRINCIPLES, has also emerged from the Ptolemaic stream of astrology. This maintains that certain planets are by their own intrinsic nature feminine: these are the Moon and Venus. Of the planets, Mercury alone partakes of both the male and female qualities, and is accordingly termed the ANDROGYNOUS PLANET.

feminine signs Ptolemy, reporting an ancient astrological tradition, and presenting a male argument, says that the even-numbered signs are feminine (hence, TA, CN, VG, SC, CP and PI), whereas the others are masculine. Having so defined the term, he then gives another quite different application for it, claiming that the sign on the Ascendant may be regarded as being masculine, the sequence of the zodiac in the chart marking the polarity of the sexes throughout the figure: in such a nomenclature, the sex of the signs changes according to the nature of the horoscope itself. An increasing misuse of the term in popular astrological circles is bent on confusing the signs with the IMAGES: it is supposed that the signs Virgo and Libra are feminine, presumably because the images (derived from the constellations) are traditionally those of women – however, in terms of the Ptolemaic definition, Libra is a masculine sign.

fera See LUPUS.

feral A somewhat unfortunate term, meaning 'pertaining to a wild beast', but in a specialist sense applied to the disposition of those with Leo on the Ascendant, or with the Ascendant degree located in the last 15-degree arc of Sagittarius. The name is clearly derived from the IMAGE, and has little practical value. The term is also used of those natives whose luminaries are so located, but with malefic planets on the Angles. Curiously, the Moon is also said to be feral when *VOID OF COURSE and to remain so whilst in her occupant sign. But see also FERAL SIGNS.

feral signs These are the BESTIAL signs – but see also FERAL.

fiducial The fiducial or fiduciary is a point (in astrology very often a fixed star) determined as a fixed position for a basis of comparative measurement. The term is derived ultimately from the Latin *fides* (faith) and refers to something in which one may place trust. The fixed star ALDEBARAN was one of the earliest of fiducials – see, for example, STELLA DOMINATRIX. See also AYANAMSA.

fifteen stars Many of the medieval star-lists pay special attention to 15 fixed stars, each of which was accorded a sigil, and a special place in relation to the lore of magical gems and talismans used for amuletic and therapeutic magic (for these sigils, see SIGILS – FIXED STARS). Contrary to what one might expect, the 15 stars were not evenly distributed throughout the heavens, nor were all of them of the greatest magnitude. The schema in figure 28 shows the distribution in approximate relation to the ecliptic (here the dotted line). It will be seen from this that three of the stars are derived from the ROYAL STARS of the ancient Persians (the missing star being FOMALHAUT). The following list gives the modern names, followed by the most frequently used of the medieval terms in brackets, the numbers referring to the schema in figure 28:

1 Aldebaran	6 Procyon	11 Polaris (Cauda Ursae)
2 Pleiades	7 Regulus (Cor Leonis)	12 Alphecca (Elpheia)
3 Algol	8 Algorab (Ala Corvi)	13 Antares (Cor Scorpii)
4 Capella (Hircus)	9 Spica	14 Wega (Vultur Cadens)
5 Sirius (Canis Major)	10 Arcturus (Alchameth)	15 Deneb Algedi

fifth element This is the *QUINTESSENCE, the secret etheric force which unites the four elements of earth, air, fire and water.

Fifth Heaven A term used synonymously with FIFTH SPHERE, which in the cosmoconception built around the PTOLEMAIC SYSTEM was the Sphere of Mars, ruled by the DYNAMIS.

Fifth Hierarchy See CELESTIAL HIERARCHIES.

fifth house The fifth of the twelve astrological HOUSES, linked with the nature of Leo, and with the Sun, representative of the creative power, organizational abilities, approach to pleasures, and the general talents of the native. It has been called the House of Children, and whilst this may

fifth ray

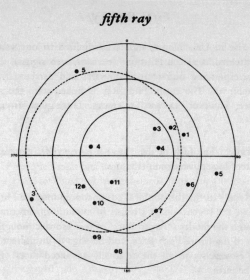

Figure 28: The fifteen stars of medieval lore (after Nowotny).

be taken quite literally, it must be seen as relating to a more general conception of 'offspring', such as works of art, creative concepts, and so on. The house has rule over the heart, loins and back of the human frame. In *MUNDANE ASTROLOGY it rules children, educational systems, sports, pleasures (of a public nature), the organizational principle, and creative arts.

fifth ray In the esoteric system of INTUITIONAL ASTROLOGY described by Bailey, the fifth of the *SEVEN RAYS is called the Ray of Concrete Science

--------- TABLE 38 ---------

Constellation	Orthodox planet	Esoteric Planet
LE	Sun	Sun
SG	Jupiter	Earth
AQ	Uranus	Jupiter

and is linked with the will to act: in another context, Bailey links it with the 'cosmic seed of liberation', which is an aspect of destruction. This Fifth Ray is said to arise in Ursa Major, and is transmitted to our solar system by means of three zodiacal constellations (perhaps even signs?), through three pairs of corresponding ORTHODOX PLANETS and ESOTERIC PLANETS, as shown in table 38. The entire Fifth Ray is linked with the sacred planet Venus – see, however, SACRED PLANETS. Distinguish from the FIFTH RAY TYPE.

Fifth Ray Type   The fifth of the *SEVEN RAY TYPES, associated with the planet Jupiter: distinguish from FIFTH RAY.

Fifth Root Race   In the theosophical cosmoconception, the Fifth Root Race is the fifth of the seven human streams of evolution, sometimes called the Aryan (though the modern 'Aryan Race' of non-esoteric thought is merely a subdivision of the entire Fifth Race, and belongs to a different conception), applied by esotericists to all the relatively advanced races of the present surface of the earth.

fifth sphere   The Sphere of Mars – see PTOLEMAIC SYSTEM.

figura coeli   A Latin term meaning 'celestial figure', used of the CHART.

figure   A synonym for CHART, probably originally the 'celestial figure'.

figurines   A term adopted by the German astrologer Meier in connection with his study of aspectal patterns. He distinguishes four groups of figurines, and 'conjunction' figurine, the 'harmonious' figurine, the 'polar' figurine and the 'polypolar'. Meier interprets these aspectal patterns in combination with what he calls the 'Ruler of Tension', which indicates the 'type of destiny'. The conjunction is said to exhibit little tension, the harmonious exhibits a gifted or creative personality, whilst the polar (which includes the polypolar) exhibits much tension in the life of the native.

fire element   The fire element finds expression in the zodiac through the three signs Aries, Leo and Sagittarius (see FIRE SIGNS). It is the element most deeply connected with spiritual activity (in which connection, see FIRE TRIPLICITY). Fire is esoterically an 'expressive' principle, the important aspect of this tradition revolving around the idea (more fully expounded in alchemical than in astrological texts) that fire, to operate fully in accord with its nature, needs an agency of control. Fire which is not so controlled is called in alchemical terminology 'common fire': it generates nothing but destruction. Therefore, when assessing the nature of a fire element in a chart, the serious astrologer is advised to look for the 'controlling factor' – that which gives useful direction to the creative and expressive potential which resides in the elemental force of fire – in this connection, see FIRE SIGNS. Within all the fire conditions known to astrologers, there is implicit (and indeed sometimes explicit) a destructive power. The consuming nature of fire calls for a constant supply of 'oxygen', which is expressed in the harmony the fire element bears with the *AIR ELEMENT, a harmony which may be reduced to fire being predatory at the expense of air. See CHOLERIC.

fire signs The three zodiacal signs Aries, Leo and Sagittarius are reflections on the higher plane of the FIRE ELEMENT. The expressive and creative power of fire works through Aries chiefly in regard to selfhood – the native seeks outlets which permit a growth of an independent image of self. The control of fire in this case is to be discerned in the opposite side of the zodiac, in the less selfish nature of Libra: the all-consuming drive towards egohood of the Arietan nature must be tempered or deflected by a development of an awareness of others, and of a sense of harmony or balance through relationships. The expressive nature of fire works through Leo mainly in regard to materiality – the native seeks to impress his own exuberant personality and love of life on the forms of matter, which is one reason why the Leonine nature is so often associated with artistic ability and creativity. The control of fire in this case is to be discerned in the nature of the opposite sign Aquarius, which offers the possibility of widening the scope of activity in the service of the community rather than the isolated self. The expressive nature of Sagittarian fire works mainly through the mental realm, through altruistic idealism and a profound (often practical) grasp of how the world of man may be extended into the mental plane – through travel or education. The destructive element enters into the basic duality of the sign, for uncontrolled fire feeds the passions of the lower duality of the sign, and this may consume the individual to a point where he is deflected from his real inner being or purpose. The clue to control in this case lies in the opposite polarity of Gemini, which offers a certain detachment from the passionate realm. The three zodiacal types inheriting the power and destructive nature of fire are often 'driven' to creativity as though by inner tensions – they are impulsive and iconoclastic in attitude, yet at the same time frequently full of originality and warmth.

fire triplicity These are the three signs derived from the element of fire, manifesting different aspects of the CHOLERIC temperament: Aries, Leo and Sagittarius. Fire may be visualized as a source of great heat, smouldering, ready to leap into high flame. Leo is a creative fire, sometimes radiating warmth as from a hearth, usually an illumination for others. Sagittarius may be seen as a firebrand moving through the air, ever in movement, dangerous when come to rest. Each of the fire signs is impulsive and dynamic, sometimes over-confident or even brash and exhibitionist. Since the fire triplicity is connected with the creative element of soul-life, each of the three reflect something of this world. Aries attempts to explore and leave his creative mark through romantic involvement with people and places – especially so through grandiose schemes. Leo is creative by imposing his own warmth on things of the earth – through the arts of painting and sculpture, for example. Sagittarius seeks to leave behind his fire-spark through the development of ideas.

firmament This name was originally applied to the Sphere of the Fixed Stars – the Stellatum in the PTOLEMAIC SYSTEM. However, in general modern use, the term is used in a general sense, for the so-called 'vault of heavens'.

firmamento chiericati See VICENZA CYCLE.

First Heaven Sometimes this term is used as being synonymous with 'First Sphere', which in the Ptolemaic cosmoconception was the Sphere of the Moon, esoterically the realm of Purgatory. In modern esotericism, however, the term is used of the lowest realm of the heavens (the so-called 'Devachan' in Sanskrit) into which the spirit passes after the purgatorial experience.

First Hierarchy See CELESTIAL HIERARCHIES.

first house The first of the HOUSES of the chart, linked with the ASCENDANT, with the sign Aries, and with the planet Mars: it is the representative of the selfhood of the native. Sometimes it is called the House of Fulfilment, the Point of Consciousness, and the House of Self, and it is generally regarded as the lens through which all the influences of the horoscope pour into the external world, through the guise of personality. It rules the personal and natural disposition of the native, denotes the head and face in the human frame, and is linked with the grandparents (the grandmother for the male native, the grandfather for the female – though both are actually important as spiritual progenitors). In *MUNDANE ASTROLOGY the house rules the country as a whole, the establishment or 'ruling majority', the defendant, and those in power.

first mover A translation of PRIMUM MOVENS.

first order cycle   When the point of conjunction between two planets repeats itself, the period elapsing between these two conjunctions is termed a first order recurrence cycle, or 'first order cycle'. Future cycles of this same first order will eventually make up any small discrepancy as to the original point of conjunction, so that eventually a series of first order cycles will result in a conjunction at the original starting point of the sequence. This cycle of first order cycles is called a 'second order recurrence cycle', or 'second order cycle'.

first point   The first point, without further qualification, usually refers to the first degree of Aries, to the commencement of the zodiacal circle of measurement: it is the so-called *ZERO DEGREE or 'zero point'. Since this point is almost always identified as 0 degrees of Aries, it has given rise to many misunderstandings in regard to the interpretation of a chart – see therefore ROUNDING OFF.

first ray   In the INTUITIONAL ASTROLOGY described by Bailey, the first of the *SEVEN RAYS is called the Ray of Will or the Ray of Power, and is linked with the will to incite and produce initiation or beginnings. This first ray is said to arise in Ursa Major, and is transmitted into our solar system by means of three constellations (perhaps even signs) through their related pairs of *ORTHODOX PLANETS and *ESOTERIC PLANETS, as set out in table 39. The

---
TABLE 39

Constellation	Orthodox planet	Esoteric planet
AR	Mars	Mercury
LE	Sun	Sun
CP	Saturn	Saturn

---

entire first ray is linked with the sacred planet ESOTERIC VULCAN – but see also SACRED PLANETS. Distinguish from FIRST RAY TYPE.

First Ray Type   The first of the *SEVEN RAY TYPES, associated with the planet Saturn. Distinguish from FIRST RAY.

First Root Race   In the theosophical cosmoconception, the First Root Race is the term used to cover the first of the SEVEN RACES, or streams of human evolution. The race, whilst human, had no dense bodies, and is sometimes called the Etheric Race.

First Sphere   The Sphere of the Moon – see, however, FIRST HEAVEN and EIGHTH SPHERE.

first station   The stationary point (see STATION) at which a planet becomes retrograde.

Fishes   The popular name for the zodiacal sign and constellation Pisces.

Fixed Cross   An alternative name for the FIXED QUADRUPLICITY of Taurus, Leo, Scorpio and Aquarius: see FIXITY. Not to be confused with the Major configuration of the GRAND FIXED CROSS. The esotericist Bailey traces a spiritual evolution in the sequence of the Fixed Cross, which is visualized as a power of transmutation.

fixed ecliptic chart   See CHART SYSTEMS.

Fixed quadruplicity   The Fixed quadruplicity is Taurus, Leo, Scorpio and Aquarius – see FIXITY.

Fixed signs   These are Taurus, Leo, Scorpio and Aquarius – see FIXITY.

fixed stars   A most important element of early personal astrology was the tradition of fixed-star influences in the form recorded by Ptolemy, and exploited and developed by the later Arabian astrologers. A literary tradition has been established (itself based on the Ptolemaic model) of equating the influence and power of the fixed stars with what have been called the 'planetary equivalents' – for example, an evil star might be equated with Saturn as a planetary equivalent, a benevolent star with Venus or with the Sun. However, as the studies preserved from Arabian astrology confirm, the influence of stars is involved with matters much deeper than merely such planetary equivalents, for it is clear that each star emits an influence personal to itself. Traditional astrologers insist that such influences – unique or otherwise – may be transmitted into the life of the native only when the stars are conjuncted by natal planets (or nodal points, Angles, and so on). The traditional view is that for a fixed star to be operative in a chart, then it must be exactly conjunct a planet or nodal point – however, some astrologers have adopted a system of orbs (see, for example, the outstanding MORINEAN ORBS). This allowance for orb has resulted in much confusion, since the fixed stars which figure in the astrological tradition are so numerous that if a wide orb is permitted then there is scarcely a horoscope figure which does not demonstrate a variety of complex 'fixed star' interpretations. If (quite sensibly) a precise conjunction is permitted for the influence to be operative, it becomes clear that fixed stars account for many things in charts which are otherwise quite inexplicable – especially so in connection with matters of genius and eccentricity of life-history and personality. For a general survey of fixed stars, in relation to astrological significance, Allen, Devore, Ebertin (Hoffmann), Powell and Robson appear to be indispensable. Even so, after allowing for radial motions, proper motions and precession – as well as for disagreement about geometrical variations in the conversion of Right Ascensions and longitudes – there is considerable disagreement among all authorities as to precise positions and influences. The longitudes given for each star in the table is that provided by Robson, subjected to correction for precession and for rounding off, amended to 1985. The asterisks indicate information not derived from this source. Ebertin provides a table (accurate for 1900, 1920, 1937 and 1950) in terms of longitudes for 73 stars, along with useful information and (somewhat dubious) case histories. Powell provides an accurate table of latitudes and sidereal longitudes for 105 stars in connection with the *SIDEREAL ZODIAC. For the general significance of the 105 stars listed below, see the entries under the individual star names.

Achernar (PI 15)	Acrux (SC 12)	Acubens (LE 14)
Aculeus (SG 26)	Acumen (SG 29)	Adhafera (LE 29)
Agena (SC 24)	Alamac (TA 14)*	Albireo (AQ 2)
Alcyone (TA 30)	Aldebaran (GE 10)	Algenib (AR 9)
Algenubi (LE 21)	Algol (TA 27)	Algorab (LB 14)
Alhecka (GE 25)	Alhena (CN 9)	Al Jabha (LE 28)
Almach (TA 14)	Alnilam (GE 24)	Alphard (LE 28)
Alphecca (SC 12)	Alpheratz (AR 14)	Alpherg (AR 27)
Altair (AQ 2)	Antares (SG 10)	Arcturus (LB 25)
Armus (AQ 13)	Ascella (CP 14)	Ascelli (LE 8–9)
Baten Kaitos (AR 22)	Bellatrix (GE 21)	Bentenash (VG 27)*
Betelgeuze (GE 29)	Canopus (CN 15)	Capella (GE 22)
Caphir (LB 10)	Castor (CN 20)	Castra (AQ 20)
Dabith (AQ 4)	Deneb (CP 20)	Deneb Kadige (PI 5)
Deneb algedi (AQ 24)	Deneb Kaitos (AR 2)*	Denebola (VG 22)
Dirah (CN 5)	Dorsum (AQ 14)	Dubhe (LE 15)*
El Nath (GE 23)	Fomalhaut (PI 4)	Giedi (AQ 4)
Graffias (SG 3)	Hamal (TA 8)	Han (SG 9)
Isidis (SG 3)	Khambalia (SC 7)	Labrum (VG 27)
Lesath (SG 24)	Manubrium (CP 15)	Markab (PI 24)
Markeb (VG 29)	Menkalina (GE 30)	Menkar (TA 14)
Mintaka (GE 23)	Mirach (AR 30)	Mirak (VG 15)*

Nashira (AQ 22)	Oculus (AQ 5)	Pelagus (CP 13)
Phact (GE 22)	Polaris (GE 29)	Polis (CP 3)
Pollux (CN 23)	Primum Hyadum (GE 6)	Princeps (SC 3)
Procyon (CN 26)	Propus (CN 19)	Rasalhague (SG 23)
Rastaban (SG 12)	Regulus (LE 30)	Rigel (GE 17)
Rigil Kentaurus (SC 30)*	Sabik (SG 18)	Sadalmelik (PI 3)
Sadalsuud (AQ 24)	Scheat (PI 29)	Seginus (LB 18)
Sharatan (TA 4)	Sinistra (SG 30)	Sirius (CN 14)
Skat (PI 9)	Spica (LB 24)	Tejat (CN 3)
Terebellum (CP 26)	Unukalhai (SC 22)	Vindemiatrix (LEB 10)
Wasat (CN 19)	Wega (CP 15)	Zaniah (LB 5)
Zavijava (VG 27)	Zosma (VG 11)	

Fixity   Fixity is the name applied to one of the three *QUANTITIES, which acts as a block to movement, and restricts the free flow of energies. The result is that the Fixed signs (Taurus, Leo, Scorpio and Aquarius) are frequently resistant to change, though fortunately blessed with a resourcefulness and persistence which usually makes them attractive and reliable powers for good. The four FIXED SIGNS of the zodiac gave rise to the esoteric symbolism of the EVANGELISTS. Sometimes the Fixed quadruplicity is called the FIXED CROSS.

Flames   In esoteric astrology the seven *SACRED PLANETS are sometimes called the Flames.

flexed   A term sometimes used as synonymous with MUTABLE.

Flora   Name given to one of the ASTEROIDS, discovered in 1847.

flying grype   See AQUILA.

Fomalhaut   Fixed star, the alpha of PISCES AUSTRALIS, set in the mouth of the Fish, the Arabian Fum al Hut meaning 'mouth of the fish'. This was one of the ROYAL STARS of the Persians (among whom it was called Hastorang), when it marked the winter solstice. It has many names, mainly variants on the Arabian Al Difid al Awwal (first frog). When rising or culminating in a chart it is said to give lasting honour, although the conjunction with planetary bodies tends to bring unfortunate strains, and intensification of energies, as well as secret affairs.

Fortunes

Foramen Fixed star (variable), the eta of ARGO NAVIS, which seems to take its modern name (Latin for 'opening') from the fact that it is set within the distinctive Keyhole Nebula. It is said to be of the nature of Jupiter with Saturn, and gives danger to the eyes. With the Sun in conjunction it promises danger from shipwreck.

foreground A specialist term used of planets near the Angles of the chart, which due to this proximity are regarded as being especially powerful, and are thus in the foreground of the chart. See also BACKGROUND and MIDDLEGROUND.

Forehead of Scorpio A term used by Ebertin to denote the fixed star GRAFFIAS. Ebertin treats this beta of Scorpius as equivalent to the influence of the delta Scorpius, which he calls Akrab.

forming When an aspect between two or more planets or nodal points is being built up, that aspect is said to be forming. When an aspect is forming, though still within a specific number of degrees of the precise aspect, it is said to be 'in forming orb'. The faster planet is described as applying to the given aspect: see APPLICATION.

Fornax The constellation Fornax Chemica, or Chymiae (alchemical furnace), formed by La Caille in 1762, located 10 to 13 degrees of Taurus, and between 26 to 40 degrees beyond the equator. It is said to bring fondness for scientific work.

fortified A term used of a planet which is strongly placed, especially by being well-aspected, either by ELEVATION, or by being in a sign of its own DIGNITY.

fortnight chart A term used by Bradley to denote half of the LUNAR RETURN chart.

Fortuna A frequently used short-form for PARS FORTUNA.

fortunate signs These are the signs Aries, Gemini, Leo, Sagittarius and Aquarius – see, however, SECTA.

Fortunes A collective term for Jupiter and Venus: the former is the Great Fortune, the latter the Lesser Fortune.

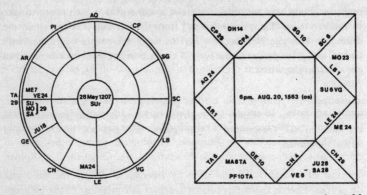

Figure 29 (left): Foundation chart for the SAN MINIATO ZODIAC, *cast for 28 May 1207, at sunrise (after Gettings). Figure 30 (right): Foundation chart cast for 6 pm, 20 August 1563 (after Taylor).*

foundation chart   Term used for a horoscope cast in relationship to the founding time or date of a building, or an important reconstruction. In some cases – especially in medieval astrology – such charts were cast in advance, in order to predetermine a suitable date for commencement of the structure, or to encapsulate in the building an important astrological or symbolic lore. Of such a kind was the chart (figure 29) cast for the SAN MINIATO ZODIAC, of which only the date (1207) now remains, though a modern recasting of the relevant figure shows that the symbolism within the church was involved with the orientation of the zodiac towards a satellitium in Taurus. In some cases, foundation charts have been cast in retrospect, in order to establish certain astrological presumptions, but in general the chart is derived from ELECTIONAL ASTROLOGY. Careful examination shows that there is no clear dividing line between what has been called 'calendrical orientation' and 'foundation astrology' in connection with ground plans (and even symbolism) for ancient temples, churches and even non-ecclesiastical buildings (see, for example, Taylor, Nissen and Schwaller de Lubicz). Figure 30 reproduces the horoscope cast for the laying of the foundation stone of a monastic church, recorded by Taylor: the Julian date (20 August 1563) was the feast day of St Bernard, but it is likely that the time was astrologically determined in the long term to include the beneficent conjunction of Jupiter with Saturn, and (in the short term) to place Mercury on the Descendant, thus linking the CHRONOCRATORS with the 'Nuntius' healer – the two most important considerations in relation to a monastic church. As Plutarch records, the foundation chart for Rome was a popular subject among astrologers, and several variants have survived – the lack of agreement is the probable cause of the several dating systems derived from the Roman calendar, of which some five different systems existed by the time of Christ (see FOUNDATION DATES). Such things have not changed significantly in modern times – there are at least nine different charts for the 'foundation' of the United States of America.

foundation dates   In ancient times, there were a variety of different systems for determining the beginning of calendrical or other dating systems. The only ones which are relevant to European astrological charts (other than in extremely specialist cases) are the Roman and the Christian. The Roman system is derived from the supposed founding date of Rome, determined by the equivalent of our 754 BC. This date was designated in Latin as AB URBE CONDITA – from the foundation of the City – often shortened to AUC. According to this system, the probable birth of Jesus (in our own system AD 4) would be given as 756 AUC. However, in respect of Roman horoscopes, see also CALENDRICAL SYSTEMS and KALENDS. The modern European dating system did not entirely replace the Roman until late into the medieval period. Shortly after AD 540 Dionysius Exiguus suggested the foundation date from ANNO DOMINI, relating to the birth of Jesus, which was gradually accepted, with the introduction of our now familiar abbreviation AD. Unfortunately there has been no certainty as to what this actual date was, and consequently correlations with the Roman foundation system have varied. The system was not widely adopted for over a century, however, and the correlate BC is of relatively modern adoption. Early Christian writers attempted pre-AUC and (later) AD dating with foundation dates relating to the supposed birth of Abraham, given as 2016 BC equivalent, and even in terms of the estimated date of the fall of Troy (given as 1183 BC in modern equivalent). Edwards and Schultz have made excellent attempts to marshal the material relevant to Christology and to certain aspects of astrology. There is some evidence to suggest that the early foundation dates were astrologically determined, as the survival of charts for AUC indicate.

Four Ages   In his rag-bag of esoteric and astrological speculations, the *Metamorphoses*, Ovid mentions four 'previous ages' to the one in which mankind presently lives (this being the 'fifth age'): he calls these ages the Gold, Silver, Bronze and Iron Ages, following a tradition linked with Hesiod, these being separated from the fifth age by a great flood. These four ages may be linked with the esoteric theory of cosmogenesis (see for example SUN PERIOD which corresponds to the Gold Age): there is also a correspondence with the Hindu theory of cycles – see YUGA. In early times an important analogy was established between the four world ages and the ages of man, the seasons, the elements and so on:

GOLD AGE   Fire   Youth
SILVER AGE Air   Adolescence
BRONZE AGE Water   Maturity
IRON AGE   Earth   Old Age

By analogy with the four elements, which marked a descent from the refined energy of light fire to the dark weight of earth, the Ages of Man were seen as a descent away from the Divine Fire into the dark prison of Earth. See also ORPHIC AGES.

four animals The astrological four animals are actually three animals united by a fourth human figure – the lion, bull and eagle linked with the winged human. These are the four FIXED SIGNS of the zodiac. In the esoteric astrological melothesia, the lion represents the feeling element in man (the heart); the bull represents the will-life, or physical element; the EAGLE represents the capacity for higher thinking, whilst the winged human represents the union of these three capacities in the human form. See TETRAMORPHS and FOUR ZOAS.

four elements The astrological four elements are actually five in number, the exoteric four being earth, air, fire and water, united by the fifth, which is the QUINTESSENCE – see ELEMENTS. The earth element was often personified in the image of a bull (linked with Taurus), the air element in the image of an

Figure 31. Personification of air and fire (after Nowotny).

eagle (see figure 31), the fire element with a lion (linked with Leo – see figure 31), and the water element was often personified in the image of a fish or sea-horse.

four-footed signs A descriptive term derived from the images associated with certain signs, which were in turn derived from their related constellation images. The signs are Aries (the Ram), Taurus (the Bull), Leo (the Lion), Sagittarius (the Centaur) and Capricorn (the Goat). They should be called by the alternative name of 'Animal Signs', since in many zodiacal images and constellation maps not all of these creatures are four-footed. For example, there are many pictures of two-footed Sagittarius, while Capricorn is scarcely anything other than a fish-tailed creature with two legs (the goat image being relatively modern).

four principles See PRINCIPLES and FOUR QUALITIES.

four qualities Sometimes called the 'four principles' to distinguish from the Qualities, these are the four states of being associated with the planets and zodiacal signs in early astrology: hot (*Calidium*); cold (*Frigidum*); humid or wet (*Humidum*); and dry (*Siccum*). See QUALITY.

four stars See ROYAL STARS.

fourteen stars The list of star-names set out under the medieval FIFTEEN STARS contains one asterism (the Pleiades), and the list is therefore sometimes more correctly referred to as the 'fourteen stars'.

Fourth Heaven Sometimes this term is used as synonymous with the Fourth Sphere, which in the cosmoconception built around the *PTOLEMAIC SYSTEM was the Sphere of the Sun, ruled by the EXSUSIAI. In the qabbalistic system, this Heaven is sometimes called Machonon, and in common with all esoteric systems, assigned rulership of the Archangel MICHAEL.

fourth house The fourth of the twelve astrological HOUSES, linked with the nature of Cancer, and with the planet Moon. It represents the domestic life of the native, and is linked with the early environment, especially with the mother (indeed, it was formerly called the House of the Mother), though esoterically it deals with the whole experience preceding birth (see GATE OF BIRTH). This house is a useful index of the imaginative faculties and moods, and relates to hidden things on account of which it has unfortunately been called the 'grave'. It is linked with the breasts and (some astrologers maintain) with the digestive organs, but it is essentially associated with the rib-cage, the protective element around the heart, and the protective attitude of kith and kin forms an important aspect of the fourth house associations. In MUNDANE ASTROLOGY it rules the opposition party, crops, buildings and the outcome of contests and litigation.

fourth ray   In the esoteric INTUITIONAL ASTROLOGY of Bailey, the fourth of the *SEVEN RAYS is called the Ray of Harmony through Conflict, and is linked with the will to harmonize or relate, and with the illumined will. The fourth ray is said to arise in Ursa Major, and is transmitted to our solar system by three constellations working through pairs of corresponding ORTHODOX PLANETS and ESOTERIC PLANETS, as set out in table 40. The

TABLE 40		
Constellation	*Orthodox planet*	*Esoteric planet*
TA	Venus	Vulcan
SC	Mars	Mars
SG	Jupiter	Earth

entire fourth ray is linked with the sacred planet *MERCURY – see, however, SACRED PLANETS. In connection with the rule of esoteric Vulcan over Taurus, see also ESOTERIC VULCAN. Distinguish from FOURTH RAY TYPE.

Fourth Ray Type   The fourth of the SEVEN RAY TYPES, associated with the planet Earth. Distinguish from FOURTH RAY.

Fourth Root Race   In the theosophical cosmoconception, the Fourth Root Race is the name given to the fourth of the SEVEN RACES stream of evolution. This race is sometimes called the 'Atlantean', and was said to occupy mainly the lost continent of Atlantis (also called Kusha), located approximately over the present Atlantic Ocean. The main continental area was destroyed in the Miocene period, but the final submergence of Poseidonis (to which Plato refers as ancient myth) did not occur until 9564 BC.

Fourth Sphere   The Sphere of the Sun – see PTOLEMAIC SYSTEM.

Four Watchers   See ROYAL STARS.

four zoas Term originated by the poet Blake, probably under the influence of the esotericist Boehm, from the Greek plural (*ζoa*), and derived from occult astrological lore. Blake follows the esoteric tradition in identifying the four beings with the fourfold nature of man – Tharmas is the body (probably a word derived from the Greek *Thumos*, the equivalent of the 'Etheric' – see AETHER); Urizen is Reason (thinking – probably a play on the Germanic *Ur* and the English 'horizon'); Luvah is the emotional life (probably a play on the word 'love'), while Urthona is 'imagination' (probably again the use of the Germanic 'Ur' and the English 'Earth-owner'). These four zoas correspond to the four fixed signs of the zodiac (with which the word 'zoa' is cognate) – Taurus, Scorpio, Leo and Aquarius, respectively – the four 'beasts' which figure in the TETRAMORPH: see FOUR ANIMALS.

fractional method A system of arc measurement used in SYMBOLIC DIRECTIONS, probably first described by the astrologer Carter, as an extension of the ONE DEGREE METHOD. The method is derived from a suggestion made by Carter that the arc between any two natal planets might be divided by a conventional numerical division to give significant directions. The radical arc he called the 'primary relationship', which was susceptible to divisions which he called 'secondary relationships'. Since in practice any division of the basic primary may be used, it is a system which even Carter admits may be 'easily ridiculed', and appears to have commanded little following in circles concerned with such projections.

framing A term derived from the interpretation of the French *encadrement* used by the astrologer Volguine, which has been wrongly equated with MIDPOINTS. An *encadrement* is a bracketing effect of two planets on another planet which is itself enclosed within the smaller arc of the ecliptic which they together define. The enclosed or framed planet is said to blend its influence with both of the bracketing planets. Angles also play a part in such framing, being regarded as the equivalent of planetary influences defining arcs. The idea (unlike the term) is by no means new, and is indeed explicit in the theory of aspects, though the bracket of arc is limited to fairly narrow orbs.

Frederici Honores Constellation named in honour of Frederick II of Prussia, formed by Bode from the 34 stars in the area between Cepheus, Andromeda, Cassiopeia and Cygnus. Now obsolete.

friendly planets Planets which have Exaltations in each other's signs, or which have communal triplicities, are said to be friendly to one another. Devore generalizes the term, however, by listing a number of relationships based on the conception of 'unfriendly planets'; Wilson (perhaps with good reason) suggests that the whole concept is nonsensical.

Frons Leonis Name given to the 10th of the LUNAR MANSIONS in some medieval lists. Probably located by reference to REGULUS.

fruitful signs Each of the WATER SIGNS are called fruitful. The term, now misunderstood, appears to have been used in the limited sense of being the opposite of *BARREN: the tradition now insists that such a water sign on the cusp of the 5th house is an indication of children.

fruit trigon See SIDEREAL MOON RHYTHMS.

frustration A specialist term properly limited to HORARY ASTROLOGY. It is applied to the action of an aspecting planet which intercedes into the effect of another aspect (being made to the body to which it is itself applying), and thus frustrates the promise of that other aspect. The full rule relating to frustration is a little more complex than this would suggest, however, for it may take place only when the frustrating aspect is already within orb: the interceding (swifter) planet which moves in to form an exact aspect before this orb may reach fulfilment is called the 'frustrator'.

frustrator See FRUSTRATION.

full Moon See PLANETARY PHASES.

function An equivalent term for KEYWORD ('keywords of function') applied in a specialist sense by the astrologer Pagan. Each of the signs of the zenith are accorded a function keyword which is included in a schema, and is said to modify the Sun-sign of the particular chart:

AR	Warrior or Pioneer	LB	Statesman or Manager
TA	Builder or Producer	SC	Governor or Inspector
GE	Artist or Inventor	SG	Sage or Counsellor
CN	Prophet or Teacher	CP	Priest or Ambassador
LE	King or President	AQ	Truthseeker or Scientist
VG	Craftsman or Critic	PI	Poet or Interpreter

future planets A few of those planets listed by certain modern astrologers as being among the HYPOTHETICAL PLANETS are, properly speaking, planets of future development, which have been called future planets. These are putative developments of other planets which, while at the moment doubtless in our present system (in the ordinary sense of the word) must be regarded as 'seeds' capable of growth and evolution. Among such future planets are AIDONIUS, BACCHUS, COELUS, MERCURIUS, NEREUS and VESTA, though Jupiter and Vulcan are names given to future incarnations of the present Earth – see, for example, JUPITER PERIOD and VULCAN PERIOD. These future planets are to be distinguished from the esoteric SACRED PLANETS. In some cases, the communality of names may be misleading: for example, VULCAN is a name given to a hypothetical planet, to a future planet and to a sacred planet.

G

Gabii altar A Roman altar of the Gabii family, now in the Louvre, which bears several interesting correspondences between signs, gods/goddesses and symbols, as follows:

AR	Minerva	Owl	LB	Vulcan	Bonnet
TA	Venus	Dove	SC	Mars	Wolf
GE	Apollo	Tripod	SG	Diana	Hound
CN	Hermes	Tortoise	CP	Vesta	Lamp
LE	Jupiter	Eagle	AQ	Juno	Peacock
VG	Ceres	Basket	PI	Neptune	Dolphin

Gabriel The Archangel accorded rule over the Sphere of the Moon, and over elemental water. Gabriel is the Angel of the Annunciation (see WATER ELEMENT), which derives from the esoteric cosmology giving him rule over the lunar sphere, the last of the planetary spheres through which the descending spirit passes on the way to incarnation – see, for example, DESCENT OF THE SOUL and GATE OF BIRTH. Gabriel is one of the *SECUNDADEIAN BEINGS.

Gados A name ('torch' in Greek) given by Ptolemy to the star ALDEBARAN.

galactic centre A term used to denote the centre of gravity for the entire matter of the GALAXY. The astrologer Landscheidt has provided an ephemeris for the galactic centre, and the related SOLAR APEX, which he calls the 'apex-ephemeris'. In 1980, the galactic centre was 26 degrees 35 minutes Sagittarius, and the solar apex was 2 degrees 10 minutes Capricornus. The progress of the former is about 1 minute per annum, the progress of the latter being approximately 8 minutes of arc per decade. See also GALACTIC POINT.

galactic point The point where the plane of the solar system cuts the plane of the GALAXY (see VIA LACTEA) is located by most astrologers on the axis of 15 degrees Sagittarius/Gemini: see, however, GALACTIC CENTRE. A planet on the former degree is in direct conjunction (alignment) with the galactic centre, while a body in the latter degree will be cut off (symbolically speaking) from a view of the galaxy by the interposed body of the Earth. The degrees are regarded as extremely critical, especially in mundane affairs: as the astrologer Davison has remarked, Mars (conjunct with Uranus) was in the latter degree at the exploding of the atomic bomb over Hiroshima.

galaxy A term derived from the Greek *gala* (milk), in reference to the milky appearance of the group of stars which mark out an irregular and luminous band, which is in fact a view into the 'island universe' of stars which contain our own solar system. This galaxy is the Via Lactea, and while the term is generally applied to our own stellar system, it is also applicable to other systems beyond our own. See also GALACTIC CENTRE.

Galen's method An ancient horary method of ASTRO-DIAGNOSIS.

gamalei Certain natural stones or gems, which, because of some powerful astrological influence, were said by medieval magicians to be magically efficacious. The artificial gamalei are those engraved with astrological, hermetic or magical sigils, towards talismanic ends. They are sometimes called 'gemetrei' and 'gamathei', but Paracelsus calls them 'gamahei', and says that they are 'stones graven according to the face of heaven'. See also BIRTH STONES and SEALS.

Ganymede Name given to one of the satellites of Jupiter.

Gate of Birth Name applied to the sign CANCER, not merely because of its mundane associations (see FOURTH HOUSE), but also because in the ancient THEMA (for the beginning of the world), Cancer was often shown on the Ascendant. In esoteric astrology, it is said that the incarnating soul enters the physical realm through the sphere of the Moon, which is, of course, associated with rule over Cancer, and also accounts for the term Gate of Birth. The esoteric basis for the term is most clearly set out in Macrobius in his description of the descent of the soul through the seven zones of the planetary spheres: see STARRY CUP. In her account of esoteric astrology, Bailey says that the sign of Cancer has been recognized down the ages as the 'doorway into life of those who must know death'.

Gate of Death Name applied to the sign CAPRICORN, sometimes to the SPHERE OF SATURN. In esoteric astrology it is said that the soul in its postmortem experience aspires to travel to the SPHERE OF SATURN, which marks the boundaries of time, and it is almost certainly this association which gives rise to the term. Equally, since Cancer is associated with the GATE OF BIRTH, it is also probable that, by virtue of being the extreme polarity of this sign, Capricorn was by extension associated with death, as the Throne of that Planet which marked the end of time. See, however, CROCODILE. In her account of esoteric astrology, Bailey says that the sign Capricorn has been recognized down the ages as the 'doorway into life of those who know not death'.

Gavel Point   The Point of Gavel, sometimes called the Part of Pluto, and even the Point of Organization, is one of the modern additions to the series of so-called Arabian PARS. If a chart is revolved so that the degree occupied by the Sun is transferred to the Ascendant, then the position of Pluto in this revolved arrangement is said to mark the Gavel Point within the houses.

geminated   Virtually an obsolete synonym for 'double', apparently restricted to horary astrology.

Gemini   The third sign of the zodiac, and one of the constellations. It corresponds as a zodiacal sign neither in location nor extent with the constellation of the same name – see GEMINI CONSTELLATION. The modern sigil for Gemini II is said by some to be a drawing of the human twin holding hands with the mortal twin (see CASTOR and POLLUX). Some see the significance of the sigil as relating to the square area inside the rectilinear structure, for the rule of Gemini over the human lungs points to the permanent rhythmic relationship which the human being holds to the outer world, constantly taking in its air and oxygen, so that the inner becomes the outer, the outer the inner. The esoteric astrologers place great emphasis on the numerical relationships of 72, which may be discerned through this exchange – see Wachsmuth. Gemini is of the air element, and of the Mutable Quality, the influence being mental, intellectual and versatile: the Geminian exhibits a strong need to relate to others, and is at worst histrionic. This 'airy' nature of the sign is expressed in the many keywords attached to it by modern astrologers: versatile, idealistic, communicative, imitative, inventive, alert, inquisitive – in a word, all those qualities which may be associated with an air nature expressing itself with a view to establishing communication with the world. In excess, the Geminian nature may be described in terms which express the tendency to superficiality of all air types, the keywords being: restless, impatient, unstable, superficial, lacking in concentration, inconsistent and diffused. The type tends to remain (at worst) childish and immature, and (at best) retains its youthful mentality and physical appearance. Gemini is ruled by the planet Mercury and offers no traditional

DIGNITY, though some authorities (following the Arabian astrologers) mark Gemini as the Exaltation of the DRAGON'S HEAD.

Gemini constellation Zodiacal constellation, the extent of which is set out in figure 32. The Latin name Gemini (twins) merely follows a tradition already ancient in classical times, along with the many variants (Didymoi to the Greeks, Dioscuri to the Romans). According to ancient mythology, the twins were sons of Leda, hence the poetic 'Ledaean Stars' of Milton – yet their personalized names are almost legion: Gemini Lacones, the Spartan Twins, for example, though they are perhaps more famous as CASTOR and POLLUX, less so as the pairs of Apollo and Hercules, Theseus and Piritheus. The image for the constellation appears to have been two giants originally, but in medieval times (probably under the influence of Romance literature) the image became male and female, as in the 13th-century image of the SAN MINIATO ZODIAC (figure 32). Since the Arabian

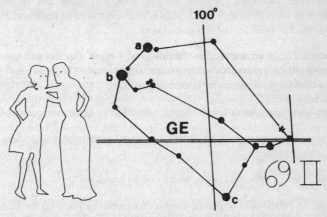

Figure 32: Image of Gemini from the 13th-century SAN MINIATO ZODIAC, *and the constellation from the* MODERN ZODIAC *of Delporte.*

constellation artists present the pair as peacocks, a medieval title 'Duo Pavones', sometimes occurs. Kircher, in tortuous arguments, claims that the Egyptians regarded the twins as Horus the Elder and Horus the Younger, though this does not appear to be accurate, yet it links perfectly with an esoteric tradition outlined by Steiner. The fixed star alpha is Castor the mortal twin, while the beta is Pollux, the immortal: see also ALHENA, DIRAH, PROPUS and WASAT.

Gemma One of the early names (probably late Latin) for the fixed star

ALPHECCA, which Ebertin wrongly translates as 'jewel' and relates to the 'crown' in which it is placed. The name was actually used in the sense of 'bud', in reference to the unopened blossom of the floral crown, as it was pictured in early texts.

gems See BIRTH STONES.

general astrology A term used by the historian Pingree in his division of the forms of astrology into four distinct groups (the others being genethliacal, catarchic and interrogatory). By 'general astrology' he intends to denote that form relating the situation in the heavens at a particular moment (such as eclipses, the vernal equinox, and so on) to events affecting broad classes of people, or the entire world. The term is approximately the equivalent of MUNDANE ASTROLOGY.

genethliacal astrology A term derived from the Greek and applied in general to natal astrology, which deals with the casting and interpretation of personal charts cast for the moment of birth.

genethliaci An archaic term for 'astrologers' – those who cast and interpret 'genitures', or practice GENETHLIACAL ASTROLOGY. The term may have been used for the first time by Isidore of Seville.

geniculator An ancient Greek name for the HERCULES CONSTELLATION.

geniture Term used both to denote a personal horoscope and (incorrectly) the interpretation of the same.

genius See AGATHOS DAIMON, which may be translated 'good genius'.

geoarc A subdivision of the daily orbit of a particular point on the Earth: a measurement relating to the *HOUSE SYSTEMS, and used to distinguish measurement by * HELIARC.

geocentric Term applied to a measurement or image of the solar system viewed as though the Earth itself were the central point of celestial motions. In general terms, when astrologers refer to the 'geocentric model' they mean the *PTOLEMAIC SYSTEM. Most modern systems of astrology (and virtually all ancient astrological systems) are geocentric. However, in modern times there has been a move to introduce a HELIOCENTRIC ASTROLOGY.

geocentric nodes The only geocentric nodal axis to receive much attention in traditional astrology is that of the Moon – see DRAGON'S HEAD and

DRAGON'S TAIL – the axis of these nodes being the points where the plane of the ecliptic is intersected by the TERRESTRIAL ECLIPTIC, which is itself defined by the path of the Moon. The geocentric nodes of the other planets are rarely expressed by a single axis (but see HELIOCENTRIC NODES), though in recent years ephemerides of geocentric planetary nodes have been made available, and the astrologer Dobyns has related the significance of nodal interpretation to traditional astrological concepts, including SYNASTRY.

geographical latitude The Angular distance, north or south, of the Earth's equator. Distinguish from LATITUDE, which in an astrological context is very often CELESTIAL LATITUDE, save in reference to place of birth.

geographical longitude The Angular distance between the meridian of a given place and the Prime Meridian of Greenwich, expressed in degrees east and west of Greenwich.

geographic astrology For a consideration of the zodiacal rulership accorded geographic locations, see CHOROGRAPHY.

geomancy Geomancy in its simplest form is really a manner of making predictions by using stones (or indeed any inorganic materiality of the earth, such as twigs) in order to construct a figure which is then subject to intepretation by a series of fixed rules. These fixed rules were originally unrelated to astrological traditions, and were limited to the consideration of the interactions of 16 geomantic figures. The simple geomantic method was rendered more complex by permitting an influx of astrological ideas, such as the doctrines of elements, aspects, planetary rulerships, houses and signs, into a divinatory system to which they did not properly belong. As a predictive method, geomancy is therefore only peripherally connected with astrology, and then only because the early practitioners of the art in medieval Europe chose to introduce a system of recording the results of divination within a chart which resembled the horoscope figure and which (eventually) attached to its own system a series of associations derived from astrological lore. Table 41 sets out the traditional associations. Of late, due to an ignorance of historical terminologies, the word 'geomancy' has also been misused, and applied to the study of telluric forces, the so-called 'ley-lines', which have nothing to do with the predictive art of geomancy. By definition, geomancy looks down to the earth, while astrology looks up to the stars, so that the recently coined term 'astro-geomancy' is something of a misnomer.

Georgium Sidus The name given to the newly discovered URANUS by

TABLE 41

Name of geomantic figure	Element	Planet	Sign
Via – Way	Water	Moon	Leo
Populus – People	Water	Moon	Capricorn
Coniunctio – Joining	Air	Mercury	Virgo
Carcer – Prison	Earth	Saturn	Pisces
Fortuna Major – Greater good	Earth	Sun	Aquarius
Fortuna Minor – Less good	Fire	Sun	Taurus
Acquisitio – Gain	Air	Jupiter	Aries
Amissio – Loss	Fire	Venus	Libra
Loetitia – Joy	Air	Jupiter	Taurus
Tristitia – Sadness	Earth	Saturn	Scorpio
Puella – Girl	Water	Venus	Libra
Puer – Boy	Fire	Mars	Aries
Albus – White	Water	Mercury	Cancer
Rubeus – Red	Fire	Mars	Gemini
Caput – Head	Earth	Caput Draconis	Virgo
Cauda – Tail	Fire	Cauda Draconis	Sagittarius

Herschel in 1781, the Star of George in honour of the reigning George III. It would appear that the same Uranus was used by Bode about two years later, though astrologers (who had for some years suspected the existence of such a planet) had called it Ouranos long before.

geo-sophic   See THREE PROTOTYPES.

Gestalt-horoscopy   One of the names given to the astrological system developed by the German astrologer Dr Walter A. Koch.

Ghafr   Al Ghafr (the covering), name of the 13th of the Arabian MANZILS.

Giant   See ORION.

Gibbor   See ORION.

gibbous   A term most usually applied to the Moon in one of its PHASES, but applicable equally to both Mercury and Venus. The Moon is said to be gibbous when both of its sides are convex, shortly after completion of the square aspect. See, however, PLANETARY PHASES.

Giedi   See ALGEDI.

Girtab   See SAGITTARIUS CONSTELLATION.

giver of life   See HYLEG.

Glastonbury Zodiac   Name given to a supposed earth-zodiac, contained within a circle of about 9-mile diameter, with Butleigh (Somerset) at the centre, and Glastonbury Tor to north-north west of the circle. The figures were originally traced out by Maltwood (after whom the zodiac is sometimes named) in the Twenties. In the 16th century, the scholar John Dee had mentioned the existence of a zodiac in or around Glastonbury, but his descriptions (noted in his diaries – see DEACON) are vague, and there is no evidence that this so-called 'Dee Zodiac' corresponds to the Glastonbury Zodiac. The images (which do not in every case correspond to either zodiacal or constellational images) are traced out in landscape contours, roads, earthworks, rivers, pools, and other 'natural' formations. Sometimes called the Somerset Giants, their antiquity is (unconvincingly) claimed by Harwood Steele to have been derived from initiate knowledge brought to Britain by Sumer-Chaldean priests. The supposed zodiac, whatever it is, is certainly not Chaldean. The tracing of these figures by Maltwood was involved with her study of the Arthurian and Grail legends, the zodiac itself being the Round Table, and the circle of figures which her imagination projected into Ordnance Survey maps was adapted to the Grail cycle – Virgo becoming Guinevere, Aries Gawain, and so on. Many specialists, including the present writer, deny that there are any figures at all – certainly, if the projections are regarded as figures in any sense, then they do not relate to any zodiac, antique or modern, nor to any known constellational figure. A serious historian cannot but be inclined to see the zodiac as entirely a product of the human imagination: the figures are probably nothing other than autoscopic projections. From a strictly historical point of view, the figures imagined in this vast circle could not have the antiquity ascribed to them. For example, for historical reasons, Capricorn (zodiacal or constellational) could not have been imaged as a goat, no more than Cancer would have been a ship: the antiquity claimed for the zodiac would suggest that Libra and Scorpio would not have been imaged as Maltwood presents them. Modifications made by Caine to Capricorn, Scorpio and Libra are equally unconvincing, on strictly historical grounds.

globe period   In the theosophical cosmoconception, the period of time during which any given GLOBE in a CHAIN is fully active (which is to say, supportive of the main stream of life) is called a globe period. The passage of such 'life support' through all the seven globes of a particular chain is called a 'round', a septenary of rounds make up one 'chain period', and seven 'chain periods' (that is, 343 globe periods, or 49 rounds) make up what is called a single SCHEME OF EVOLUTION.

globes In the theosophical cosmoconception, the Earth and the planets are called globes. However, each individual globe is said to possess a counterpart, often described in terms of graduations of finer substance than ordinary matter. A physical globe, such as our own Earth, for example, is regarded as the dense centre of seven interpenetrating worlds, each occupying the same space, yet all of different qualities of materiality. Such counterparts are approximately analogous to the invisible subtle sheaths which are said in occult circles to surround and interpenetrate the body of man (the so-called 'astral', 'lower mental', 'higher mental', 'buddhic' and so on – specialist Sanskrit terms not defined in the present text). Not all the globes are regarded as being physically rooted, however. For example, the globes of our EARTH CHAIN consist of three physical globes, along with two astral and two mental globes. The astral globe is said to possess higher corresponding bodies, something akin to the higher sheaths of the angelic realm in esoteric lore.

Globus Aerostaticus An obsolete constellation – see AETHERIUS.

Glutiens Name given to the 23rd of the LUNAR MANSIONS on some medieval lists. Probably identified by the fixed star Nashira, the gamma of Capricornus.

glyphs A term often wrongly used as being synonymous with SIGILS, and perhaps the most misused of all astrological specialist terms: even the careful Blavatsky misuses it. The word 'glyph' has a proper application to 'relief symbols' in sculptural and architectural forms, but not to written forms. Some glyphs have undoubtedly become sigils. For example, some of the bas-relief symbols on Egyptian tombs (properly glyphs) have become sigils, by virtue of being translated to the written form on papyrus or paper: the modern sigil for the Sun ⊙ was once precisely such a glyph as the Egyptian determinate for 'god' and for 'time', it became properly the sigil for the Sun in the 15th century AD, when it was adopted as a symbol by Italian astrologers. In this sense, therefore, the bas-relief images on the DENDERAH ZODIAC may be called glyphs, but *IMAGES are not *ipso facto* glyphs.

gnomes The class of *ELEMENTALS linked with the Earth. They have many different names in popular lore, but two types of gnomes mentioned in esoteric lore are the Diemeae (the kind which live in large stones) and the

Durdales (tree spirits) – in certain of the astrological *CALENDRIA MAGICA, they are often called Pigmei.

Goat Popular name for the zodiacal sign CAPRICORN, and used also of Capricornus. Originally, in the Babylonian and Egyptian zodiacs and constellation images, the sign and constellation were represented by an image of a goat-fish (see Figure 10).

Goh system Abbreviated name (*Geburtsortshäuser*) given for the House system introduced by the German astrologer Koch in which cusps are calculated for the place of birth. Tables are available according to this GOH System for all latitudes between 0 degrees and 60 degrees, for each successive degree passing over the Midheaven. See KOCHIAN SYSTEM.

Golden Age See FOUR AGES and KRITA YUGA.

Golden Fish See PISCIS AUSTRALIS.

golden number In astrology the term is applied to a method for determining the calendrical fall of the METONIC CYCLE by application to a table of such cycles. In a non-astrological context the same term is also used of a ratio, the Divine Proportion of the ancients, which results from the division of a line in such a way that the smaller part is in the same proportion to the greater part as that greater part is to the whole.

Gonasi Term used by Ptolemy for the HERCULES CONSTELLATION.

Good Daemon Old name for the ELEVENTH HOUSE.

Goods Point The Point of Goods is one of the so-called Arabian PARS. If the Lord of the 2nd house is taken as the Ascendant of a figure, then the new cusp of the 2nd house marks the Point.

Gorgonifer See PERSEUS.

gradial transit A term used first by the astrologer Sepharial in reference to the idea that a progressed planet usually projects an ever-increasing arc (the gradial transit arc) due to progression. This arc is presumed to remain sensitive to major transits during the life of the native.

Graffias Fixed star (triple), the beta of Scorpius, set in the head of the scorpion, and hence sometimes called Frons Scorpii. The probable etymology of the term, from the Greek *graphaios* (crab) may explain why the scorpion in early astrology is sometimes actually pictured as a crab, even though the image was linked with zodiacal and constellational Cancer: Allen records that St Augustine believed the desert scorpion was generated from the aquatic crab. Under the modern name Akrab, Powell adopted this star to mark the termination of his sidereal Libra – see SIDEREAL ZODIAC. It has an evil reputation among astrologers, and was said by Ptolemy to be of a nature equivalent to Mars conjunct Saturn.

grand climacteric According to some astrologers, the 63rd year of life – but see CLIMACTERICS.

Grand Climactic See CHRONOGRATORS.

Grand Cross An aspectal pattern involving at least four planets which are so placed as to form two axes of opposition which intersect at right angles. If there are no orbs involved, then each of the planets will be in a sign of a different element. The Grand Cross is an extension of the so-called COSMIC CROSS. It is said to bring considerable tension into the life of the native, often with creative results. See MAJOR CONFIGURATION.

Grand Fixed Cross Name given to a MAJOR CONFIGURATION in which two pairs of oppositions are placed in all four of the Fixed quadruplicities – that is, in Taurus, Leo, Scorpio and Aquarius. Distinguish from FIXED CROSS and GRAND CROSS.

Grand Trine When two planets are in trine to each other, and both separately trined by a third planet, the whole configuration is said to be a Grand Trine: it is a *MAJOR CONFIGURATION. Provided there is not a disjunct trine involved in this figure, each of the planets must be in the same triplicity, and the aspect is quite rightly said to be highly beneficial.

Graphiel In medieval occultist literature, Graphiel is the name of the Intelligency of Mars. Agrippa accorded him the number 325.

Grave A term sometimes used for the *FOURTH HOUSE.

great chain of being See CHAIN OF BEING.

Great Dragon A term with a variety of applications in occult, esoteric and astrological contexts. In astrology the term is sometimes used in reference to the path of the Moon (or indeed to the Sphere of the Moon) which, in its intersection of the ecliptic, gives the nodes, themselves named after the image of an encircling dragon (see DRAGON'S HEAD), and often depicted as a scaled dragon. The term is also sometimes used in astrological contexts of the constellation DRACO: but see also ATALIA. In alchemy the reference is often to time, as the great destroyer, frequently imaged as the Ouroboros dragon, an image often found in esoteric astrological contexts. In occult thought, which touches upon modern esoteric astrology, the name is sometimes given to the dark adversary Ahriman, who is often depicted as a scaled dragon under the feet of Michael. In spiritualist circles it is a term (first used by Davis) for a mysterious portion of the heavenly Summerland, inhabited by the morally deficient beings called Daskka. While apparently relating to very different ideas within these several contexts, there is actually a unifying principle in this term, for the Sphere of the Moon is traditionally linked with the diabolic post-mortem existence – but see EIGHTH SPHERE.

greater benefic   See BENEFICS.

Greater Cloud   See NUBECULAE MAGELLANI.

greater luminary   An ancient term for the Sun, in comparison with the lesser luminary which is the Moon.

Great Ilech   The hidden VIRTUE in medicine, derived from planets or stars.

Great Nebula   Name given to a nebula in the constellation ANDROMEDA, said by astrologers to be of an evil nature, threatening sight, and promising a violent death. It is sometimes called VERTEX, the Queen of the Nebulae and Little Cloud.

Great Year   In medieval manuscripts, Great Year is a term used to denote the period of time in which all the planets will return to a given fiducial – there is little agreement as to how long this period is, however: at one extreme, William of Conches gives 49,000 years, whilst at the other extreme Petarius gives 350,635 years. This vast cycle has been confused with the PLATONIC CYCLE of 25,920 years, which is sometimes called the Great Year.

Greek planets   The names of the Greek planets are: SU Helios; MO Selene; ME Stilbon, Hermes; VE Phosphoros, Hesperus, Aphrodite; MA Pyroeis, Ares; JU Phaethon, Zeus; SA Phainon, Kronos.

Greek zodiac   The historian Powell makes a convincing case for calling the tropical zodiac the Greek zodiac. It is in effect a zodiac devised by Greek astronomer-astrologers according to solar principles not inherent in the earlier *BABYLONIAN ZODIAC. The division is not of the zodiacal asterisms, as in earlier zodiacs, but of the ecliptic itself, which is a solar co-ordinate. The division is twelvefold, in equal 30-degree arcs, in a system which had been used before but which was unique in that the first arc was conceived as being linked not with a fixed star fiducial but with solar phenomena. This solar phenomenon was defined by the *EUCTEMONIAN CALENDAR, and the whole system of the Greek zodiac, using names derived by Euctemon from the Babylonian zodiacal/constellational system, was probably formulated by the Greek Hipparchus in the 2nd century BC. This appears to have been the first zodiac technically free of the fiducials of fixed stars. Inevitably, the very freedom of the system has given rise to the problems arising from *PRECESSION, which in turn raises the question of determining the correct *AYANAMSA. See also HELLENISTIC ZODIAC. The ancient Greek names for the signs of the zodiac (with variant spellings) were:

AR	Krios	TA	Tauros	GE	Didymoi
CN	Karkinos	LE	Leon	VG	Parthenos
LB	Chelai (Zugos)	SC	Scorpios	SG	Toxotes
CP	Aigokeros	AQ	Hydroxous	PI	Ichthys

Greenwich Mean Time   See TIME.

Greenwich Sidereal Time   See TIME.

Gregorian calendar   The modern European and American calendrical system, named after the reform of the *JULIAN CALENDAR initiated under Pope Gregory XIII, and based on the work of the astrologer-physician Aloysius Lilius and a Jesuit named Claivius. The vernal equinox had regressed by a time equivalent of ten days since the Council of Nicaea of 325 (see EASTER), and so by proclamation of a Bull of 1582 the Pope annulled ten days, thereby adjusting the calendar so that the equinox was back to 21 March. The length of the solar day was corrected, and the year made to commence on 1 January, while the present system of leap years was established. The 97 leap years which occur in the course of 400 years allows an error of time which is only 26 seconds short of celestial measurement. From an astrological point of view the changes subsequent to this Bull have led to much complication. The Gregorian system was immediately adopted in Spain (and colonies), Portugal and parts of Italy (then not unified). With some adjustment it was adopted also in the Low Countries and most of France. However, it was not adopted in England until 1752, following an Act passed in 1751, resulting in the famous 'suppression' of eleven days. Germany and Sweden had already adopted the system in 1700. The Russians adopted it only in 1918, and the Greek Orthodox countries (variously) between 1916 and 1923, the Turks being the last to adopt it in 1926. Unfortunately, the correct dating (and hence computation of horoscopes) cannot always be ascertained from the material given here, since some localities did not adopt the changes immediately, and in some cases the years were not always begun in Gregorian terms. In pre-Conquest England, the year began on 25 December, but after the Conquest, 1 January was adopted, though later changed to 25 March following general medieval practice, related to the Feast of the Annunciation. In 1752, and for all later years, it began on 1 January.

Griffin See PHOENIX.

grouping A modern term used nominally as an equivalent of the technical SATELLITIUM, but in general use applied to the general configuration of planets in a chart.

Grus Constellation formed and named by Bayer in 1604, though from stars included in most earlier maps within the Southern Fish. The name means 'crane', which is highly appropriate, as Allen remarks, since the crane was claimed to be the symbol of the astrologer in Egypt (at least, in the widely accepted medieval and Renaissance interpretation of Egyptian symbolism). However, in England Bayer's term was actually translated as meaning 'bittern' or 'flamingo'.

grype This term is usually a reference to the Falling Grype, a medieval equivalent of Vultur Cadens, one of the important FIFTEEN STARS, now called WEGA.

guarded A little-used term applied in a specialist sense to one or more planets which, while elevated in the chart, are guarded on the eastern side by the Sun, and on the western side by the Moon.

Guardians of the Heavens See ROYAL STARS.

H

Hades   Name given to one of the HYPOTHETICAL PLANETS in the URANIAN ASTROLOGY. This transneptunian is included in an ephemeris, and is said to relate to lowness, dirt, antiquity and secrets.

Hagiel   Name given by Agrippa (quoting ancient qabbalistic sources) to the INTELLIGENCY of Venus. He records the magical number of Hagiel as 49, which is a typographical error, and does not follow the normal procedure of correspondences in relation to the *MAGIC SQUARE of Venus.

Hagith   One of the names for HAGIEL.

Hakah   Al Hak'ah (the white spot), the 3rd of the Arabian MANZILS.

half moon   See PLANETARY PHASES.

half-sums   See MID-POINTS.

Halley's Comet   See COMETS.

Halyian system   A method of *HOUSE DIVISION closely based on the 10th-century ALBATEGNIUS SYSTEM, whereby the projected arc was defined by the passage of two hours of Right Ascension for each house, measured from the Ascendant. The system is ascribed to the Arabian astrologer Albohazen Haly, who lived in the 11th century, and is sometimes called the Albohazen system.

Hamal   Fixed star, the alpha of constellation Aries, set in the forehead of the Ram, the name from the Arabic Al Hamal (the sheep). As might be expected from its placing, the star is said to cause violence and cruelty: Ptolemy likened it to the nature of Mars conjunct Saturn. Penrose has calculated that several of the ancient Greek temples were orientated to this star, which is perhaps not surprising since the constellation of Aries was seen by the ancient Greeks as one of the symbols for Zeus.

Hamal   Al Hamal, the Arabic term for Pisces.

Hamburg school   A name often applied to the unconventional system of URANIAN ASTROLOGY, on the grounds that Witte and Sieggruen established in Hamburg the school which developed and promulgated their ideas.

Han   Fixed star, the zeta of OPHIUCHUS, in the left knee of the figure, the name apparently from the Chinese for the feudal state of Han. It is said to be of the nature of Venus conjunct Saturn, bringing troubles and disgrace.

Hanael   One of the names given to the spiritual being ruling the Sphere of Saturn.

Hanah   Al Han'ah (the brand), the 4th of the Arabian MANZILS.

Hand of Ophiuchus   Modern name for the fixed star epsilon of Ophiuchus, apparently identified with the Nitach-bat (the Man of Death) in the Euphratean system. It is of an evil disposition.

harmonic chart   Name given to the chart designed to present data according to the system of HARMONICS proposed by Addey. The positions of the planets and related points are set out on a calibrated circle. The harmonic format which arises from the insertion of 90, 72, 60 degrees (etc.) permits the astrologer to note particular harmonics. See ARC TRANSFORM CHART.

harmonics   A term applied to an entirely new form of astrological research and interpretation, based on statistical methods originally (at least) connected with numerological theory. The theory to which the now extensive literature of harmonics relates is largely derived from the researches of the astrologer Addey and his later co-workers, stemming from a seemingly indefatigable application of statistically arranged material to computerization. Addey found that when large numbers of birth data were submitted to specialist statistical analysis, planetary and nodal positions exhibited wave-like fluctuations which were not accounted for in the usual methods of astrological interpretation. Subsequent exploration of such analysis showed that numerical harmonic relationships seemed to apply in various groups of charts, and Addey came to the conclusion that all integral divisions (that is, harmonics) of the 360-degree circle were susceptible to interpretation in terms of specialist meanings, and were therefore capable of forming the basis for a new approach to astrological studies. Catalogues for the harmonics which fall at and within each of the 360 degrees of the circle are now available, designed to facilitate the use of a potential 16,000 harmonics and their equivalents in this specialized form of chart interpretation. Although the exploration of the statistical method now related to harmonics was pioneered by the Swiss astrologer Krafft in the 1930s, his methods were later seen to result in invalid conclusions. The main impulse for the study came from Addey, who was of the opinion that the significance of harmonics lay in the numerological importance underlying them: in later times he questioned the validity of the higher harmonics – but see, for example, SUBSISTENCES. Whatever the underlying reasons for the significance of the harmonics, a system of interpretation has been built around the intervals established by simple divisions in terms of whole numbers. For example, the division by four is seen as relating to 'difficulty, effort and achievement' – corresponding, one notes, to the basic fourfold division of the square aspect. The division by five, linked with the 'mind, discrimination and the human perception', appears to rest on a somewhat personalized view of the spiritual and material nature of man, however, and finds no correspondence with traditional theory. The series established by the division of the interval of nine is said to point to the various stages of completion and fruition in the life of man – 'birth, death and rebirth'. From a practical point of view, the difficulties with harmonics (for those not wholly taken up with this new type of astrology) are the very number of possible intervals available, and the actual model of man which is supposedly portrayed or revealed in the harmonic system. It must, however, be recognized that the use of the statistical approach to large groups of horoscopes does appear to have established its own form of astrology and astrological data. For example, when Seymours subjected to such 'group analysis' the charts of 200 centenarians, then the ordinary standards of astrological doctrines appeared to have no really valid application. In contrast, research by Addey into the relationship between harmonics and longevity (involving 972 nonagenarians) established a convincing 108th harmonic (9×12) relating to the number 9, itself linked with the life-cycle. However, it is worth observing that this in no way diminishes traditional astrology, which is itself designed to relate only to the unique individual chart, and has never (until comparatively recent times, in the work of Krafft, Gauquelin, and so on) been regarded as susceptible to statistical interpolation and analysis – the very uniqueness of the individual chart seeming to prohibit such an approach. It is therefore important to recognize that what the statistically based method of harmonics establishes is the groundwork of a new approach to astrology, which does not invalidate the traditional forms. Harmonics is rooted in a statistical approach to data which is foreign to the spirit of traditional astrology: harmonics links certain data derived from astrological concepts to the field of waves, the advantage being that wave-patterns are themselves readily subjected to analysis by the computerized techniques beloved by the practitioners of harmonics. Harmonics certainly expresses more of the modern attitude to the cosmos (an attitude which is distinctly materialistic, in its striving for much-vaunted scientific objectivity) rather than to the individual chart-reading, which is the prime object of traditional astrology. Such points must be made, if only because some modern popular literature suggests that harmonics is really nothing more than a modern equivalent of the ancient aspectal theory – which is simply not the case. The aim and symbolic fruits of harmonic analysis involve a view of astrology very different from that expressed in the theory of aspects, and indeed embrace many ideas rooted in traditional astrology. There has been a move, arising mainly from the practitioners of harmonics, to claim that the system is reflected in the much earlier divisions of Hindu astrology: however, many of the ancient methods of divisions in the Hindu system may be traced to adaptations made by Indian astrologers to the sign divisions, SORTES and LOCI of Hellenistic astrology which formed the basis of Hindu exoteric astrology (see, for example, LORD OF THE GREEKS), which is alien to the spirit of harmonics. Much more work must be done, especially on the history of the evident adaptation of Hellenistic methods to the peculiar socio-cosmological beliefs of the Indian astrologers, before such links with the modern harmonics may be made with any real validity.

harmonies See KEPLERIAN HARMONICS and PLANETARY TONAL INTERVALS.

harmony of the spheres See MUSIC OF THE SPHERES.

Hasalangue See OPHIUCHUS.

Hasmodai Name given by Agrippa (quoting ancient qabbalistic sources) to the Daemon of the Moon, for whom he gives the magical number 369 which is the linear addition of the lunar MAGIC SQUARE. See also LUNAR INTELLIGENCY.

Hasta Name (the hand) given to the 11th of the Hindu NAKSHATRAS.

Hastorang See ROYAL STARS.

Hauriens Primus Name given to the 26th of the LUNAR MANSIONS in some medieval lists. Probably identified by the fixed star Markab.

Hauriens Secundus   Name given to the 27th of the LUNAR MANSIONS in some medieval lists. Probably identified by the fixed star Algenib.

Hayz   A term applied to a minor DIGNITY derived from the Arabian astrology, and limited to use in horary astrology. When a masculine diurnal planet is above the horizon in a diurnal chart, then it is said to be 'hayz'. When a feminine nocturnal planet is below the Earth in a nocturnal chart, then it is said to be 'hayz'. The placing was said to be mildly beneficial.

Heart   In a specialist sense, one of the Arabian PARS, the Part of Heart, sometimes called the Part of Venus or the Point of Love. If a chart is revolved so that the degree occupied by the Sun is transferred to the Ascendant, then the position of Venus in this new arrangement will mark the Heart within the houses.

Heart of Hydra   See ALPHARD.

Heart of the Royal Lion   See REGULUS.

heart of the Sun   See CAZIMI and ESOTERIC SUN.

Heart of the Zodiac   A term (rarely) used in modern times to denote the zodiacal and constellational Leo.

Heaven   See EMPYREAN, SPHERES and TENTH SPHERE.

Hebe   Name given to one of the ASTEROIDS, discovered in 1847.

He-goat   A term used by some modern astrologers for both the zodiacal sign Capricorn and Capricornus.

Heka   Fixed star (double) the lamda of Orion, the name derived from the Arabic Al Hak'ah (white spot): it is considered to be an unfortunate influence. See also HAKAH.

heliacal rising   The first rising of a star after its period of invisibility due to conjunction with the Sun.

heliacal setting   The last setting of a star prior to its period of invisibility due to conjunction with the Sun.

heliocentric astrology

heliacal visibility See HELIACAL RISING, HELIACAL SETTING and MOR-
INEAN ORBS.

heliarc A subdivision of the annual orbit of the Earth: a measurement
relating to sign divisions, and used to distinguish from measurements made
in terms of GEOARC.

heliarc figure A name for a chart cast employing the simple subdivisions of
the SOLAR HOUSES.

heliocentric astrology Heliocentric astrology is founded on the calculation
and interpretation of charts which symbolically present data from a solar
rather than from a geocentric standpoint: the heliocentric figure charts the
planetary positions from a viewpoint of an observer in the centre of the Sun.
Special heliocentric ephemerides have been designed for astrological use, and
since the heliocentric motions of planets and nodal points is more uniform
than the geocentric equivalents, such tables lend themselves to ease of
calculation. Since the zodiac proper is based upon a point calculated in
geocentric terms, heliocentric astrology tends to postulate and rely upon a
constellational zodiac, though there appears to be no precise consensus as to
the limits of the twelve asterisms: but see SIDEREAL ZODIAC which is widely
used by modern heliocentric astrologers, especially those concerned with
Astrosophy. The heliocentric chart differs considerably from the geocentric
equivalent, as the Sun itself is absent (in the sense that the Earth is absent
from a geocentric chart), and there is no mundane system of houses, no
Ascendant, and no Moon (as this is incorporated in the orb of the Earth).
Thus, the three fundamental influences of the geocentric chart – Sun, Moon
and Ascendant – are missing. Heliocentric astrologers place great emphasis
on planetary nodes, and the axes arising from these, as well as upon
heliocentric aspects. The very fact that a different range of aspects is available
to Venus and Mercury has perforce involved new interpretations foreign to
traditional astrology (though the 16th-century astrologer Morin offered such
interpretations on an academic basis). It must, however, be realized that the
projection of an aspectarian upon a heliocentric chart in no way follows the
traditional rules or philosophy of astrology, since aspects were literally
involved with the idea of influences (in the sense of 'in-flowing' forces) of a
JAIMINIC elemental nature into the matrix of the Earth, which in turn
presupposed a zodiac (see ELEMENTAL ASPECTS). While heliocentric charts
are championed by some modern astrologers, the non-telluric measure implies
an element quite foreign to the individualizing tendency proper to traditional
chart interpretation. Some interesting, if highly speculative, work has been

done by Sucher (influenced by Vreede and Steiner) in connection with heliocentric charts, relating to historical periods and epochs (see WORLD CYCLE OF MUNDANE EVENTS), though some of the historical data employed is of dubious value. Such historical interpretation is aided by the fact that heliocentric nodes advance with a fair regularity at about one degree in a century, and thus afford a most convenient time-conversion. To date, most of the studies (for example, those of Erlewine and Thorburn) of chart comparisons between geocentric and heliocentric systems, appear to have been conducted using too few cases to support valid inference. Some useful findings, especially in connection with surveys of nodes, seen as both points and axes, and in connection with heliocentric aspects, have been ably presented by Dean. Heliocentric astrology is to such an extent a modern phenomenon that when the traditionalist Leo attempted to define it at the beginning of the present century, he made a large number of significant errors in his text, including the idea that it had 'sprung up in America in modern times'. The major problem is that, aside from one or two learned astrologers, the various heliocentric systems are being promulgated by students who have an insufficient knowledge of the rules and implications underlying the vast corpus of doctrines relating to traditional astrology, with all its esoteric and exoteric undertones and implications. A really satisfactory account and study of heliocentric astrology is still awaited.

heliocentric chart A chart designed to stereographically project the planets against the stars from an imaginary point at the centre of the Sun, and used in HELIOCENTRIC ASTROLOGY. Such a projection has little relationship to the many different methods of conventional chart projection, and is in most cases a projection of the ecliptic. While there is a remarkable tendency in some astrological circles to interpret heliocentric charts from a point of view of traditions established for geocentric charts, much attention is paid to the planetary nodes and ingresses. Heliocentric charts are often used in AS- TROSOPHY (see ASTEROGRAM), and much work has been done on the heliocentric interfaces by Erlewine.

heliocentric nodes Since the planetary nodes advance at approximately 1 degree per century, it is possible to set down the positions of the axes in degrees valid for a long period of time. Such mean heliocentric nodes as are listed in table 43 are those generally used in heliocentric astrology, expressed in whole degrees (but see ZERO DEGREE). These whole degrees are valid for the mid-point of our present century, when only Jupiter's node was near to entry into the next degree (being 9.9 degrees in 1950, which will place it in

TABLE 42

Planet	Node	Planet	Node	Planet	Node
ME	18	VE	17	MA	20
JU	10	SA	24	UR	14
NE	12	PL	20 (but see below)		

the 11th degree after 1960). The information for Pluto relates to the true node, and may therefore be + or − 1 degree.

heliocentric system A model of the solar system which places the Sun as a fixed point at the centre, and uses this as the basic co-ordinate to which all cosmic measurements are related. In astrology, this is contrasted with the GEOCENTRIC SYSTEM, which in its modern form is much adapted from the original Copernican model, and is heliocentric. The idea of a heliocentric system was known to the ancient Greeks, as for example in the model proposed in outline during the 3rd century BC by Aristarchus of Samos. In setting out the form of his own PTOLEMAIC SYSTEM, Ptolemy mentions the Aristarcheian model, but appears to think that it does not give a satisfactory explanation of celestial phenomena. Most modern astrology is founded upon a geocentric view of the solar system, though all information used in the calculation of nativities is now ultimately derived from the heliocentric system. See also HELIOCENTRIC ASTROLOGY.

Helios The Greek astrological term for the Sun – but see APOLLO.

Hellenistic zodiac The term is used to denote the Babylonian-based astrological computation schema of the Hellenistic period derived from tables brought from the conquered Babylon by Callisthenes (331 BC): it is Hellenistic, therefore, rather than Greek (see GREEK ZODIAC). The *AYANAMSA favoured was that calculated according to the tables of Naburiannu, which gave a 10-degree arc, though it seems that the Greeks incorporated it into their schema without being aware of its significance. Fagan claims that this was indeed the zodiacal schema of such well-known astrologers as Manilius, Firmicus Maternus, Vettius Valens and Manetho. He further insists that the planetary positions of the so-called *DENDERAH ZODIAC are cast for the Neomenia of 17 April, AD 17, according to the Hellenistic zodiac – a point which is much contested by other historians. See also GREEK ZODIAC.

hemisphere Half of a circle of a sphere. In relationship to the symbolic representation of the Celestial Sphere which is now called the horoscope figure, the visible hemisphere is usually called the DIURNAL ARC, while the obscured hemisphere is called the NOCTURNAL ARC. The eastern hemisphere, ascending from the cusp of the Imum Coeli to the Midheaven, is called the oriental arc, while the Western hemisphere, descending from the Midheaven to the Imum Coeli is called the 'occidental arc', or hemisphere.

Hercules Name given by the astrologer Wemyss to one of the HYPOTHETI-CAL PLANETS, said by him to be the ruler of Leo. He gives the planet a revolution of approximately 654 years. See also POLLUX.

Hercules constellation Constellation, located approximately 28 degrees Libra to 2 degrees Capricorn, and from 5 to 53 degrees north of the equator. Allen quite rightly records this as one of the oldest sky figures, though it was known by quite a different name to the early Greeks, who saw it variously as a kneeling figure (hence *Engonasi*, the Kneeler). The original Engonasi was seen as a wide variety of mythical heroes, of which Hercules was merely a later type, and over 80 different names for the figure have been recorded, most of them linked with heroes. Another name has been translated as 'Phantom', though the Greek *Eidolon* really meant something beyond merely 'ghostly', perhaps more like 'astral form'. This latter derivation reminds us that it was considered by astrologers to give dangerous passions, such as are related to unbridled astral forces. Ptolemy likens the whole constellation to the nature of Mercury, yet curiously enough its influence is supposed to promote tenacity, fixity of aim, and general strength of character.

Hermean zodiac A term used by Massey to denote the zodiac of popular syncretic esotericism, in which the twelve signs are associated with ancient gods and mythologies – the HERMES ZODIAC, or HERMETIC ZODIAC. For example, in this zodiac Pisces is figured as Ichthon (Ichton), and with the female goddess who brought forth the young Sun-god as a fish. One version of the Hermean zodiac is that used by Kircher (see figure 22 under COSMIC PANTHEONS), which contains several important hermetic associations – such as Amun for Aries, Isis for Virgo, Typhon for Scorpio, Canopus for Aquarius and Ichthon for Pisces.

Hermes Name given to Mercury – but see GREEK PLANETS. It is also the name given to one of the HYPOTHETICAL PLANETS in the system proposed by the Dutch astrologer Ram, accorded rule over zodiacal Gemini. This transplutonian is included in an ephemeris, and claimed to relate to 'spiritual ideality and occult ability'. The same name was also used of a hypothetical by Thierens, in relation to his *ETHERIC SPHERE, as the future ruler of Gemini. Thierens gives the alternative (confusing) name 'Mercury' for this future embodiment.

Hermes zodiac   See HERMEAN ZODIAC.

hermetic Mercury   The mercurial forces linked with planetary Mercury in hermetic lore is something different from that associated with ordinary Mercury – that is, with the Mercury of the non-esoteric astrological forms. It is termed the Permanent Water, the Vitalizing Spirit, and is linked with the quintessential forces of the world: often the magical symbol for the SEAL OF SOLOMON is drawn with the sigil for Mercury in the central space, to link it with the quintessence, and so placed it is no longer the ordinary exoteric Mercury. Many of the esoteric names for this planet suggest the idea of 'liquid', a symbolic form for the quintessence of etheric forces – the Blessed Water, Virtuous Water, Philosopher's Vinegar, Dew of Heaven, Virgin's Milk, and so on. Esoterically, the power of this hermetic Mercury is as unifier in the human being of the polarity of thinking (imaged as Salt) and the will-life (imaged as Sulphur). Sometimes the thinking element in man is symbolized by positive Mars, and the will-life by negative Mars. Without the control of hermetic Mercury, these polarities would ruin the life of man.

hermetic rule   A term used for the so-called TRUTINE OF HERMES.

hermetic Sortes   Sometimes called the 'seven Sortes' (to distinguish from the 'twelve Sortes', as the TWELVE PLACES of Manilius were sometimes called), these are the ancient equivalents of the PARS, attributed to the Trismegistian literature. The distinction made between the hermetic sortes and the standard sortes or PARS rests in the method of computation. The Alexandrian Paul tells us that the hermetic *klipoi* (*pars*) are to be calculated from the Ascendant (but see also SORTES), in the day horoscope by direction of the signs, and in the night horoscope by inverse direction against the signs, according to the rules set out in table 43.

hermetic zodiac   See HERMEAN ZODIAC.

Herschel   See URANUS.

Hesperus   One of the names given to the planet Venus, as the evening star. See also LUCIFER.

—————————————— TABLE 43 ——————————————

Sort	Planet	Arc between
Nemesis	Saturn	SA and Fortuna
Victory	Jupiter	JU and Daimon
Courage	Mars	MA and Fortuna
Daimon (genius)	Sun	See DAIMON
Love	Venus	VE and Daimon
Necessity	Mercury	ME and Fortuna
Fortune	Moon	See PART OF FORTUNE

hexagon An ancient term for the SEXTILE aspect, presumably derived from the fact that the division of the 60-degree arc into the zodiacal circle produces points which may be inscribed within a hexagon.

hidden planets In the esotoric developed by Bailey, it is claimed that there are many hidden planets – around 70 or so – in our solar system. However, in her INTUITIONAL ASTROLOGY, only the three interlinked systems of five 'Non-sacred Planets', eight 'Orthodox Planets' and seven 'Sacred Planets' are genererally considered. See also AROMAL PLANETS.

hiding places A term derived from Babylonian astrology, and explained by the modern astrologer Fagan as referring to those locations (arcs) in the heavens where the stars disappear and then re-appear in their heliacal phenomena.

hiemal signs The hiemal or hyemal signs (from the Latin *hiems*, winter) are Capricorn, Aquarius and Pisces.

Hierarchies See CELESTIAL HIERARCHIES.

hieroglyphic monad See MONAS HIEROGLYPHICA.

Hieroz-Bonatis directions Term used of a system of PRIMARY DIRECTIONS, described by the astrologer Hieroz, relating to the calculation of successive charts from a radical, whereby the fiducial is the Sun's radical place, each subsequent chart being derived from the Sun's return to that position. The system was described by Hieroz in 1941, and only ten years later did he discover that a similar technique (see BONATIS DIRECTIONS) had been proposed by Bonatus, as a result of which he changed the term 'Hieroz directions' to 'Hieroz-Bonatis directions', to accommodate the medieval source.

Hieroz directions See HIEROZ-BONATIS DIRECTIONS.

Hindu zodiac From very ancient times the Hindu astrologers have distinguished between the tropical and sidereal zodiacs, wisely providing different names for the two systems. The Sanskrit terms for the two have been variously Anglicized, but table 44 sets out the correspondences given in the

	TABLE 44	
Western name	*Hindu tropical*	*Hindu sidereal*
Aries	Mesham – Maish	Acvini
Taurus	Rishabham – Vrishab	Krttika – Krittika
Gemini	Mithunam – Mithun	Mrgacirsha – Mrigasitas
Cancer	Karkatakam – Karka	Punarvasu
Leo	Simham – Simha	Magha
Virgo	Kanya	Phalguni
Libra	Tula – Tulam	Citra
Scorpio (Scorpius)	Vrishikam – Vrishchik	Vicakha
Sagittarius	Dhanus – Dhanu	Mula
Capricorn(us)	Makaram – Makar	Ashada – Ashadha
Aquarius	Kumbham – Kumbh	Dhanishta – Cravishta
Pisces	Meenam – Meena – Minam	Bhadzapada – Bhadrapada

esoteric treatments of Blavatsky, Row and Thierens, along with main variant spellings. It will be observed that many of the names used for the sidereal zodiac are inevitably re-duplicated in the lunar asterisms of the Hindu *NAKSHATRAS.

Hippos See PEGASUS.

Hircus See CAPELLA and FIFTEEN STARS.

Hismael Name given by Agrippa (quoting ancient qabbalistic sources) to

the Daemon of Jupiter, for whom he gives the magical number 136 – see
IOPHIEL.

historical astrology   Garacaeus distinguishes four types of astrology – the
ordinary genethliacal, horary, astro-meteorology (which he terms '*De Mutationibus Aeris*'), and a fourth, '*De Regnorum Mutationibus*', which Wemyss
equates with the 'rise and fall of Kingdoms and Religions', as a sort of
'historical astrology'. Clumsy as the term is, there appears to be no available
alternative to describe this important branch (which certainly must not be
confused with MUNDANE ASTROLOGY, which is perhaps what Garacaeus had
in mind). The SECUNDADEIAN BEINGS of the occultist Trithemius were
derived from such historical astrology. See also HISTORIC ASTROLOGY.

historic astrology   What Ptolemy called CATHOLIC ASTROLOGY.

Hitschler method   Name given to a method, devised by the Swiss astrologer
Hitschler, by which allocation is made of every known chemical element or
compound to a special degree of the zodiac, the list being prepared for
therapeutic purposes.

hoarse signs   One of the curious terms derived from the natures of the
images attached to the twelve signs of the zodiac, and relating to Aries,
Taurus, Leo and Capricorn. The name is said to be linked with the noises
which are made by the animals associated with the corresponding images –
though what sort of noise the goat-fish of Capricorn makes is anyone's guess.
Although Wilson follows tradition and insists that those born under these
signs will have a certain roughness or harshness of voice (those under
Capricorn apparently 'speak with a kind of whistling sound'), the fact is that
Taurus (through its ancient rule over the throat and larynx) often produces a
most beautiful and harmonious voice. See MUTE SIGNS.

homodromi   A term derived from the Greek, meaning 'fellow runners', and
applied to the two planets Mercury and Venus, which by virtue of being
inferior planets may never appear to be very far from the Sun in terms of
zodiacal arc. The outer limit for Mercury is approximately 28 degrees, and
that for Venus approximately 76 degrees.

homonymous error   A designation used by Fagan to apply to the confusion
of a tropical sign with a sidereal sign of the same name, sigil or symbol. The
term is a useful one for purposes of discussion and reference, though Fagan's
own 'solution' to the error is questionable. Fagan suggests that the tropical
and sidereal were coterminous in AD 221, yet the issue is not quite so simple

as he implies. The fiducial, the nature of the sidereal zodiac (in the extent of its demarcations), and so on, make all speculation as to a possible temporary identity of the two extremely dubious. A further confusion in the homonymous error is the assumption that the coincidence of two fiducials may be taken to imply an extensive coterminity of the two sequences, and thus an identity of structure.

horary astrology The astrological art of interpreting specific questions in terms of a chart erected for the moment when the question is formulated. Horary astrology has its own specialist rules, though most of them are related in a commonsense way with those of GENETHLIACAL ASTROLOGY. The astrologer Lilly, who (if his autobiography is to be believed) had a wide practice in horary astrology, recommended that the figure be cast in respect of propositions made and the 'doubts of the querent will be instantly resolved'. It would seem that the horary figure represented an autoscope, a symbolic centre for concentrating the intuitive faculties. The interesting thing about the method is that it places the querent himself within the framework of the conditions surrounding the question, for he or she is represented by a planetary significator, according to clearly formulated rules. The basic rules of the art are too complex to be set out usefully in the present work: Wilson sacrificed a fair proportion of his book on general astrological terms to an admittedly incomplete analysis of the horary rules and terminology. In a sense, these rules reflect the basic tenets of astrology, though there is a simplistic and highly symbol-conscious approach, involving different time-scales, with planets regarded not as so many sources of differentiated energies and spiritual qualities, as symbols of conditions relating to the questions in hand: the planets become promittors in a horary chart. The study of the many horary charts and recorded interpretations noted in English libraries by Thomas make a fascinating background to the investigation of social conditions in relation to astrology. The most frequent questions are about theft, law-suits, partnerships, the safety of goods and persons in sea voyages, imprisonment, lost objects and animals. Sometimes the querent asks about death (mainly about the deaths of others), which should properly be determined from a progressed genethliacal chart.

horary time Horary time is measured in equal arcs marking $\frac{1}{12}$ of the diurnal or nocturnal arc of a star. When the days are long, then the diurnal horary time will be longer than the nocturnal horary time, and the corresponding 'hours' will be shorter.

horas Term used in connection with the classification of signs into two halves of 15 degrees. The solar horas are the positive halves, the lunar horas the negative halves. It is said that the horas of the positive signs (AR, GE, LE, LB, SG, AQ) are solar in the first half, lunar in the second half. In the negative signs (TA, CN, VG, SC, CP, PI), the first half is lunar, the latter half solar. According to Cornell, the horas of the negative signs are called 'magnetic' and 'magneto-electric' horas respectively. The malefic planets in the solar horas are said to be powerful, though weak in the lunar horas: the benefic planets are stronger in the lunar, weaker in the solar.

horimea An obsolete term, apparently used of the HYLEG planet when it has passed the Midheaven of a chart.

horizon chart A horoscope, or symbolic figure, erected to symbolize a planet through the horizon – but see HORIZON SYSTEM.

horizon system A method of HOUSE DIVISION based on the projection of the division of the horizon into twelve arcs of 30 degrees, starting from the meridianal circle. The cusps obtained by this projection were said to be the centre of the houses. The method is attributed to the 12th-century Jewish astrologer Ibn Ezra, and is sometimes called the Ezrian house system.

horizontal system The horizontal system, when applied to the CELESTIAL SPHERE, takes as a point of reference any specific location on the Earth. The point directly overhead is called the zenith, the point immediately below, the nadir. The plane at right angles to these is called the horizon, and this (within the projection of the sphere) is regarded as having infinite extension. This horizon plane is then regarded as being divided into the 360 degrees of equal arc. Through each of these degrees run as many vertical circles, centred upon the mid-point between zenith and nadir, and joining these two points. Parallels of altitude are visualized as parallel to the plane of horizon. Measurements taken from this Celestial Sphere describe precisely a location in terms of horizon plane and parallels of altitude.

Horlogium Oscillatorium Constellation, Pendulum Clock, introduced as an image by La Caille in 1752, and (according to Robson) inducing a 'well-stored mind and a fondness for history'. It was for no apparent reason later refigured and called Horoscope.

horlogia See ZODIACAL CLOCKS.

horoplanetary domification A term constructed by the astrolger Hieroz to denote the *PLACIDEAN SYSTEM, on the grounds that Hieroz believed that this system was not actually proposed by Placidus (correct, as it happens) but by Ptolemy (incorrect, as it happens). He was on safer grounds in tracing the system to Scaliger, who mentioned it in his commentary on Manilius.

horoscope   In modern astrology the word 'horoscope' is used to denote the symbolic figure of the heavens, the CHART, sometimes indeed the 'horoscope figure' or the 'horoscope chart'. However, until the 18th century the word was frequently used as synonymous not with such a chart but with the Ascendant degree. This 18th-century use was etymologically correct, for the Greek astrology (from which much of the modern astrology derived) termed the Ascendant the 'Horoscopos' (translated into Latin as '*Horoscopus*'), apparently from two Greek terms meaning 'watching from a high vantage point'. In spite of the objections of such people as Blundeville, the word was by the 18th century used more and more to denote the chart (see CHART SYSTEMS), so that eventually 'horoscope' and 'chart' became synonymous. Ptolemy, who is one of the first to even pretend to set down the rules of how the Horoscopos is determined, is infuriatingly vague, and seems to refer to a method which does not correspond to anything used now. According to Ptolemy, the degree of the zodiac ascending is to be measured according to the method of Ascensions (as the Greek *anaphoron pragmateias* is translated), but there is little discussion of the vexing problem of CLIMATA. Wilson defined the term in its original sense, but this is now almost lost, save to a few historians used to dealing with ancient charts and readings. See CHART SYSTEMS, and for a meaning related to constellations HORLOGIUM OSCILLATORIUM.

horoscope chart   See CHART SYSTEMS.

Horoscopos   See HOROSCOPE.

Horus   Name given to one of the HYPOTHETICAL PLANETS by Thierens in relation to his esoteric view of the *ETHERIC SPHERE, the future ruler of zodiacal Cancer. The hypothetical was also called Bacchus. See ORION.

hot   See FOUR QUALITIES and PRINCIPLES.

hour glass   In a specialist sense, the Hour Glass is one of the so-called Arabian PARS, the Part of Saturn, sometimes called the Point or Part of Fatality, or the Point of Love of Brethren. If a chart is revolved so that the degree occupied by the Sun is transferred to the Ascendant, then the position of Saturn in this new arrangement will mark the Hour Glass within the houses.

hours   See PLANETARY HOURS and HORAS.

house cusp   See CUSP.

house division   In computing an astrological chart, as a necessary preliminary to interpretation, the astrologer usually incorporates three quite different systems of co-ordinates into the symbolic figure. One of these is the system of house division, a grid which fixes diurnal motion. This grid may be inserted by means of a variety of different geometric or symbolic systems, all of which are aimed at dividing the heaven into (usually twelve) mundane areas called 'houses', which are marked by cusps. In almost all methods, each quadrant of the Celestial Sphere is trisected by one way or another, to give the cusps which are themselves symbolically related to the zodiacal signs. Three separate methods have been distinguished. The first is by direct division of the ecliptic, one of the earliest of which is the *MODUS EQUALIS, the equal-house method, familiar to Hellenistic, if not to ancient Greek astrologers, and an entirely symbolic method of dividing the heavens. The second is that involved with division by projecting on to the ecliptic, one of the earliest of which was the *CAMPANEAN SYSTEM, which projected from the Prime Vertical. The third, which involves trisecting the semi-arcs by time-based divisions, is the oldest of the non-symbolic systems, being devised by the Arabic astrologer Haly in the 11th century – see HALYIAN SYSTEM. The basic principles involved in the construction of the various house systems, as projective methods, are set out under different headings. However, the two diagrams in figure 33 point to one of these principles, in demonstrating simplistically the difference between the constructive method

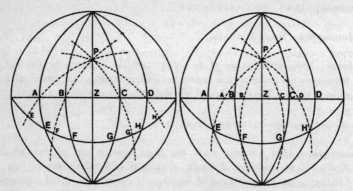

Figure 33: Comparative method of house division. (Left) The Modus Rationalis of Regiomontanus and (right) the Campanean system.

of the *Modus Rationalis* of Regiomontanus and that of Campanus. In the Regiomontanean system (left) it will be noted that the polar arcs (P) cut the equator (EFGH) at the same points as do the house circles (the vertical arcs), but they do not cut the Prime Vertical (ABCD) at the same points as these house circles. In the Campanean system (right), it will be noted that the polar arcs (P) cut the Prime Vertical (ABCD) at the same point as the house circles, but they do not cut the equator (EFGH) at the same points as these house circles. These two diagrams illustrate clearly that such house systems are based on a solution to a problem which does not exist for constructions involving horoscopes cast at the equator, for then the Prime Vertical (ABCD) and the equator (EFGH) are identical. It is only when we wish to determine the house divisions for a place other than the equator that we must decide which of these two circles are to be divided equally, by (in effect) choosing a particular house system in tabular form. In fact, a large number of house divisions have been evolved in order to facilitate computation of the cusps intermediate to the four Angles, and the relevant information for most systems has been made available in a variety of TABLES OF HOUSES, which enable astrologers to insert cusps in their charts without resorting to complex calculations, and even without understanding the geometry or philosophy involved in what they are doing. Essentially, the quadrant must be trisected by either time considerations or by spatial considerations – see HOUSES. About a dozen house systems are worth mention in modern times, though the American astrologer Munkasey has recorded 20 different systems described in extant literature, and has published tables for seven different methods. In addition to those already noted, the following list gives all the systems designed to establish a twelvefold grid division to symbolize diurnal motions: ALBATEGNIUS SYSTEM, ALCABITIUS SYSTEM, EAST POINT SYSTEM, EQUAL DIVISION METHOD, HALYIAN SYSTEM, HORIZON SYSTEM, KOCHIAN SYSTEM, MERIDIANAL SYSTEM, MORINEAN SYSTEM, NATURAL GRADUATION SYSTEM, PLACIDEAN SYSTEM, REGIOMONTANEAN SYSTEM, PORPHYRIAN SYSTEM, TOPOCENTRIC SYSTEM and ZENITH SYSTEM. However, other divisions, involving a grid of eight houses, have been recorded (see LAVAGNINI SYSTEM and OCTOPODOS). The 2nd-century AD astrologer Manilius divided the mundane belt into eight houses of equal arc, instead of the usual twelve – but see the TWELVE PLACES of Manilius, which are not house divisions at all. The so-called '8th-house system of Firmicus' is merely the first eight houses of the normal twelve divisions. Dean mentions systems of ten, 24 and 48 cusp divisions, as well as systems applied clockwise, instead of the usual widdershins. See also the entry under URANIAN ASTROLOGY which touches upon symbolic methods sometimes confused with house division. The ultimate purpose in computing and house cusps is to locate the areas of specific mundane influence revealed by the twelve houses: however, there is no general conformity in regard to the interpretation of the significance of the cusps themselves, for according to some systems the cusps mark the centre of the houses (as Campanus insisted they should) while in others they are said to mark the boundaries. According to the widely accepted doctrine of ANGLES (which provide the basic grid around which the house divisions are made), the cusps should be regarded as marking the centre of the individual houses, on the grounds that the Ascendant is identical with the cusp of the 1st house. The general outline of traditional interpretation relating to the twelve houses is set out under TWELVE HOUSES.

House of Children   See FIFTH HOUSE.

House of Death   Traditional astrology uses this unfortunate term for the EIGHTH HOUSE of the horoscope figure, because of its link with the negative Mars of Scorpio. The matter dealt with in the 8th house is related more to the regeneration principle than to death (which is merely the ultimate experience of regeneration). A more recent term, the House of Karma, referring to the idea that karma and death are interlinked, is perhaps more suitable, provided that karma is itself seen as a redemptive process.

House of Hopes and Wishes   Traditional term for the ELEVENTH HOUSE of the horoscope figure. As with so many of the older terms, it contains an interesting kernel of truth, for the house is esoterically linked with the way in which the native is related to the realm of ideals, to the ancient 'world of archetypes'. This relationship is often expressed on the mundane level in terms of hopes and wishes, and perhaps explains the tradition which maintains that a malefic planet in this house may be regarded as an indication that the native will feel a sense of dissatisfaction with his or her life – in that their hopes and wishes may be continually frustrated.

House of Karma   See HOUSE OF DEATH.

House of Money   See SECOND HOUSE.

House of One's Own Undoing   Traditional term for the TWELFTH HOUSE of the horoscope figure. Like many of the ancient terms, this does encapsulate a truth when the working of the house is visualized from an external point of view. The association with zodiacal Pisces (and in modern astrology with the nebulous Neptune) means that spiritual matters generally come to a head in this house – which is one reason why it is called a 'terminal house'. The mundane division therefore represents a dissolving process, that important process of spiritual re-orientation (even of withdrawal from the material plane) which, from the outer point of view, often gives the impression of being the result of self-inflicted injury: hence it is sometimes called the House of Sorrows.

House of Self   See FIRST HOUSE.

House of Short Journeys   See THIRD HOUSE.

House of Sorrows   Traditional term for the TWELFTH HOUSE – see HOUSE OF ONE'S OWN UNDOING.

House of Father   Traditional term for the TENTH HOUSE, usually explained in terms of social conditions in the late medieval period (when the term was coined), for in those days the father was usually instrumental in determining the career of the native, or at least set a standard in relation to the social standing of the native. Esoterically, however, the term is linked with the idea of 'Old Father Time', the ruler (Saturn) of the associate Capricorn, which is associated with the Sphere of Saturn. In the ancient cosmoconception, this sphere marked the end of time – beyond the Sphere of Saturn there was that timeless condition which was originally termed 'eternity', which is now often understood as meaning 'infinite extension of time'.

House of the Mother   See FOURTH HOUSE.

House of the Public   See ELEVENTH HOUSE.

house ruler   This term is misleading, for a house is not ruled by a planet. A sign is ruled by a planet. Since a sign is associated with its relevant house, however (for example Aries with the 1st house), the idea has arisen that the planetary ruler of the sign may also rule the related house. Properly speaking, a planet is associated with a particular house. However, when a term is so widespread in its misuse, it might be said to have acquired its meaning almost by default. The early astrologers who observed the growth of the misuse attempted to stem it by making a distinction whereby the planet was LORD of the house, and RULER of a sign. Unfortunately, the two terms are now regarded as being virtually interchangeable by modern astrologers.

houses   The traditional horoscope figure is divided into twelve arcs, which are symbolically presented as being equal either in a spatial system or in a time system (see CHART SYSTEMS). This division is superimposed upon the projected Celestial Spheres, with the symbolic horizon line (usually) marking the cusps of the 1st and 7th houses (the East Point and West Point, respectively). The 10th house cusp marks the symbolic nadir, called the IMUM COELI. The remaining eight house cusps are disposed according to one of the many different systems of *HOUSE DIVISION. Each of the individual houses has many different names, derived usually from traditional astrology (for example, the 8th house is sometimes called the HOUSE OF DEATH, the Occult House, the House of Karma, and so on): usually, however, the houses are listed by number: the sequences of houses for the various chart systems of the quadrated or circular type are given under CHART SYSTEMS. A large number of different terms and classifications have survived from ancient astrology into modern use – the most powerful houses are the ANGULAR (1st, 4th, 7th and 10th); less powerful are the SUCCEDENT (2nd, 5th, 8th and 11th), while the weakest houses are the CADENT (3rd, 6th, 9th and 12th). See also EASTERN HOUSES, OCCIDENTAL HOUSES, ORIENTAL HOUSES, PSYCHIC HOUSES, RELATIVE HOUSES, TEMPORAL HOUSES, and INTERCEPTED HOUSE.

house systems See HOUSE DIVISION.

humane While the text of Ptolemy appears to take it for granted that the so-called HUMAN SIGNS are humane, he distinguishes Libra (the image for which was a pair of scales) as humane because of the influence exerted by the sign.

human signs These are Gemini, Virgo and Aquarius, which are presumably so called because their images are in the form of humans – the twins, a woman holding an ear of corn or a flower, and a man pouring the water urn: but see also HUMANE. The first half of Sagittarius is traditionally included in the list, since it is imaged as half-human (usually a centaur). The whole term is used in contrast to the BESTIAL SIGNS.

humid See FOUR QUALITIES.

humours A conception of a tetrad of bodily fluids which were recognized as influencing the disposition (later the 'humour') of a person, through excess or deficiency. The humours were Phlegm, Black Bile, Yellow Bile, and Blood. These humours were dominated by a theory of *perisomata* (Greek word meaning approximately 'concerning humours') which was involved with the idea that certain excesses or dificiencies produced illness – a man might be said to be 'out of humour' through lack of a bodily fluid. This *perisomata* theory eventually led to the development of the conception of the

TEMPERAMENTS which were associated with each of the four humours and elements, their names being derived from the Greek, as in table 45.

Humour	Greek equivalent	Temperament	Element
Phlegma	Phlegma	Phlegmatic	Water
Black Bile	Melina Chole	Melancholic	Earth
Yellow Bile	Chole Nxathe	Choleric	Fire
Blood		Sanguine	Air

TABLE 45

hunting the Moon A phrase used of a powerful and difficult aspect: Moon square Saturn. It has long been regarded as one of the most difficult of aspects, for the malignant rays of Saturn play unimpeded on the sensitive Moon, that representative of the personality and the anima (in its Jungian sense), so that male and female, age and youth, are in tension, and the form-building of Saturn is in disharmony with the diffusive form-releasing of the Moon. The aspect as a radical consideration is difficult, but for reasons connected with the cycles of the Moon and Saturn, it becomes an influence which persists in transits and progressions through the life of the individual, in that the two cycles roughly coincide. By progression, the Moon takes approximately 28 years to complete its cycle, whilst Saturn takes almost 30 years in direct motion to return to its radical position. This means that roughly every seven years there is a major square, opposition or conjunction (of progressed and radicals) which calls into play the baleful influence of the original square aspect. These progressed aspects may be reinforced by the transits of Saturn. It has been observed that when the Moon is applying to the square, the influence becomes stronger as life proceeds. When the Moon is separating from the radical square, the influence, while strong in childhood (depending upon the orb of separation), gradually becomes weaker, and after the first 28 years is diminished. The aspect was sometimes termed 'chasing the Moon'.

hurtful signs The signs Aries, Taurus, Cancer, Scorpio and Capricorn are so called, presumably because of the natures of their IMAGES – the ram, bull, crab, scorpion and 'goat', respectively.

Husbandry The Point of Husbandry, sometimes called the Point of Fortune in Husbandry is one of the so-called Arabian PARS in a chart. If the degree of Venus in a natal chart is taken to the Ascendant, the adjusted position of Saturn marks the Point.

Hut   Al Hut, the Arabic term for Pisces.

Hyades   An asterism set in the forehead of the constellation Taurus, one of the most lovely groups in the night sky. In the Greek mythology they were the daughters of Atlas and Aethera, and therefore half-sisters of the PLEIADES, with whom they made up the DODONIDES. Pliny, who called them the Parilicium, described them as the 'violent and troublesome' stars which bring tempests, which explains one of their many appellations, 'moist daughters'. Ptolemy says that the group are of the nature of Mercury conjunct Saturn: they have an evil influence, inducing violence, poisoning and a difficult life. See also PRIMUM HYADUM.

Hydra   A winding constellation, the 'water snake', located from about 5 degrees of Leo through to 23 degrees of Scorpio, and from 10 to 36 degrees south of the equator. As the name suggests, astrologers supposed that it threatens shipping, and gives a passionate nature. It is possible that the constellation was linked with that of Scorpius, since the early Greek sources name it Nepa, apparently a word of African origin, meaning 'land crab', the related word 'Nepas' meaning 'scorpion'. Allen points to its evil reputation, and records a (dubious) story that the asterism was seen as a dragon, said to symbolize the winding course of the Moon itself. He mentions also that when a comet was originated near Hydra, then it was believed that poison was being scattered through the world, but these fanciful ideas are now associated with Draco. An astrologically important star in this asterism is ALPHARD. Distinguish from HYDRUS.

Hydra's Heart   See ALPHARD.

Hydroxous   Greek term, meaning 'water power' used to designate the zodiacal sign and constellation Aquarius.

Hydrus   A long and slender constellation, the 'snake' not to be confused with HYDRA, the head touching the polar stars, the tail almost reaching ACHERNAR. It was originated as a separate constellation in 1604 by Bayer, and has gained a reputation among astrologers for 'danger of poisoning'.

Hygeia   Name given to one of the asteroids, discovered in 1849.

hyleg A term derived from the Arabian astrological system, and used of a planet which (by virtue of its position in a special mundane arc) is regarded as the 'giver or sustainer of life', the APHETA. The mundane arcs in which a planet may be hyleg are listed under HYLEGIACAL PLACES. A planet which is so hyleg is said to have influence of a beneficial kind on the length of a native's life. When such an aphetic planet progresses to an unfavourable aspect, or to a conjunction with an ANARETIC PLACE, then the life of the native is said to be threatened. The simplicity of the above description disguises the fact that the rules for determining hyleg are among the most complex in astrological interpretation, as the indications set out by Wilson show.

hylegiacal places In modern astrology these are usually called the 'principal places', a specialist term used to denote mundane arcs in which the two luminaries are said to exercise the most beneficial influence within a chart, being in such arcs the promoters of life. These places, properly the 'hylegiacal arcs', are the 1st, 7th, 9th and 11th houses in popular lore, but Ptolemy located the hylegiacal arcs as being 5 degrees above the Ascendant to 25 degrees below, with a similar arc-coordination for the other houses listed above. The English astrologer Lilly called the degrees occupied by the Ascendant, Midheaven, Sun, Moon and Fortune the hylegiacal places, though he was certainly confusing these with the places of the MODERATORS – even so, his confusion has persisted in some astrological circles, so that the hylegiacals are sometimes misunderstood. See also APHETIC PLACES and HYLEG.

hylegiacal arc See HYLEGIACAL PLACES.

hyper astrology See HYPER-CHART.

Hyperborean See SECOND ROOT RACE.

hyper-chart A term used in modern astrology (as, for example, by Dean) to describe a chart which would contain within its symbolic configurations all astrological data, such as Angles, houses, aspects, declinations, mid-points, co-Ascendants, lunations, perihelia, interfaces, asteroids, hypotheticals, and so on. The interpretation of such a chart ('hyper-astrology' as it has been called) would be beyond the grasp of an astrologer. Dean has suggested that a reasonable synthesis could be obtained within ten years by a team of a thousand astrologers working around the clock. The fact is that by mere intellectual extrapolation and irresponsible symbolism one might establish almost any system of interpretation which ultimately will be impractical – one might, for example, quite easily propose a hyper-medicine, just as useless as hyper-astrology.

Hyperion   Name given to one of the moons of Saturn.

hypogeon   The Greek term meaning 'under the earth' is often interpreted in astrological circles as being the Hellenistic equivalent of the IMUM COELI, but this is not correct. The early Greek astrologers, as well as the later Romans such as Ptolemy, defined the 10th house and the axial-correspondent 4th house by simply counting signs from their Horoscopos 1st house, which meant that they did not distinguish the Medium Coeli or the Imum Coeli in the same way as later astrologers. The term therefore originally simply meant '4th house'.

hypothetical Moon   See LILITH.

hypothetical planets   Some astrologers have claimed the existence of a number of planets in our own solar system which are neither visible nor recognized by science, but which have been located (largely) by such psychic means as clairvoyance. Some astrologers have postulated the existence of over 1,000 such planets, called AROMAL PLANETS, while Blavatsky claims that there are six ETHEREAL PLANETS. At least 25 of the more widely used hypotheticals have been accorded ephemerides by various astrologers, while information regarding position, revolution and supposed influences has been furnished for some 35 such bodies. The astrologer Dean, who appears to work with statistical methods and with an analytical thoroughness which scarcely informed traditional astrological doctrine, has come to the conclusion that the hypotheticals are entirely without credibility. However, at least one hypothetical originally perceived by clairvoyant means was later recognized as a genuine non-hypothetical: see PAGAN-PLUTO. A complete list of the names of hypotheticals would actually include many asteroids, and the hypothetical moons of certain planets, such as Jupiter: but see NON-PLANETS. The 31 names of important hypotheticals given here (in reference to specific sections in this present work) includes only such as may in theory be termed 'planets', and indeed only such as may be regarded of importance in the recent history of astrology. Unfortunately, since the work on hypotheticals has never been co-ordinated, the list includes some names that denote several hypotheticals which have been given the same name by different astrologers. See therefore the entries under the following names:

Admetos	Apollon	Cupido	Demeter	Dido
Hades	Hercules	Hermes	Horus	Isis
Jason	Kronos	La Croix	Lilith	Lion
Melodia	Midas	Minos	Moraya	Osiris
Ov	Pan	Persephone	Pluto	Polyhymnia
Poseidon	Transpluto	Vulcan	Vulkanus	Wemyss-Pluto
Zeus				

A few specialist astrologers refer to some planets as hypotheticals when they are not really hypotheticals at all. A whole group of planets frequently mentioned in esoteric literature are related to the future evolution of our present solar system, and names are used to denote the 'evolved state' of modern planets which are regarded as being just as involved in the evolution as is the human race: see FUTURE PLANETS, LUCIFER and SACRED PLANETS. The esoteric planets of Bailey's INTUITIONAL ASTROLOGY also belong to a category of their own, and are scarcely hypotheticals.

hypotheticals   In astrology, usually the HYPOTHETICAL PLANETS.

hypsomata   The modern teaching concerning the EXALTATIONS of the planets in certain zodiacal signs, from which the theory of DIGNITIES is derived, appears to have rested on an ancient tradition of hypsomata or 'degree Exaltations'. The astrologer Fagan has attempted to show that these hypsomata relate to an important chart (more properly an adjusted schema) cast for the year 786–785 BC for the latitude of Babylon (figure 34), related to heliacal settings, and the use of the fixed star SPICA (then held to be in 29 degrees of Virgo) as fiducial of the zodiac. Table 46, which is derived from Fagan's own researches, compares the hypsomata for this year (with mean difference calculations) in order to show that the luminaries and five planets were in their hypsomatic longitudes and heliacal positions during that one year, the positions of the luminaries being those of 4 April 786 BC (the 1st Nisan – see BABYLONIAN ZODIAC). The degrees given for the new planets and their nodes are extrapolated by Fagan from the 786–785 BC chart. A modern hypsomata, really an Exaltation degree for the node, is given as 3 degrees of Gemini by the Arabian astrologer Albiruni (11th century AD), which was repeated by the 17th-century astrologer Lilly, and by this means was adopted into the Western astrological tradition. The hypsomata of Fagan do correspond to the Hellenistic Exaltation degrees recorded by Neugebauer, yet it is worth recording that other systems and variations have survived from ancient times. For example, the Roman encyclopaedist Pliny quotes the Exaltation degrees given above, with the variations:

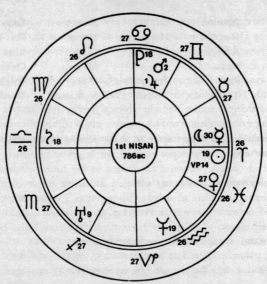

Figure 34. Horoscope for the 1st Nisan of 786 BC *which gave the European hypsomata (after Fagan).*

| ──────── TABLE 46 ──────── |||
Planet	Hypsomata	786–785 BC	Difference
SU	AR 19	AR 18.8	+ 0.2
MO	TA 3	AR 29.4	+ 3.6
ME	VG 15	VG 16.4	− 1.4
VE	PI 27	PI 26.7	+ 0.3
MA	CP 28	AQ 0.8	− 2.8
JU	CN 15	CN 15.3	− 0.3
SA	LB 21	LB 20.6	+ 0.4
UR	SC 25		
NE	AQ 5		
PL	CN 4	(Dragon's Head in PI 17)	

SU – AR   29   MO – TA   19   SA – LB   20

Albiruni gives quite different figures, which he claims to have from Hindu astrology. The depressions (or FALL) of the planets were accorded correspond-ing degrees by Hellenistic astrologers, in direct opposition to the hypsomata, but this tradition does not appear to have survived in Western astrology.

hypsomatic ayanamsa The modern astrologer Fagan has applied his theory of hypsomatic Exaltations (see HYPSOMATA) to a study of the difference between the sidereal zodiac and the tropical zodiac (see AYANAMSA), to derive a mean value ayanamsa of 13.8 degrees, which would place the sidereal longitude of the Vernal Point for 786 BC in Aries 13.8 degrees. See, however, HYPSOMATIC YEAR.

hypsomatic degrees See EXALTATIONS.

hypsomatics A modern misuse of HYPSOMATA.

hypsomatic year A term derived by Fagan from his researches into the true basis of the tradition underlying the 'Exaltation degrees' and the origins of the modern zodiac. It is the year 786–785 BC – see HYPSOMATA.

hypsomatic zodiac A modern name (used especially by the astrologer Fagan) for the zodiac of Fixed signs to distinguish this from the tropical zodiac. It is also called the 'Egypto-Babylonian zodiac', and is based on a twelvefold 30-degree arc, the initial point being located in the 1st degree of constellation Taurus, determined by the ancient fiducial star ALDEBARAN. See HYPSOMATA.

I

Iapetus   Name given to one of the moons of Saturn.

iatromathematics   A term used for that division of astrology concerned with medicine – see MEDICAL ASTROLOGY.

IAU zodiac   Abbreviated form for the 'International Astronomical Union Zodiac' – see MODERN ZODIAC.

Ichthys   Greek name for the zodiacal sign and constellation Pisces.

Ides   See KALENDS.

Idle   The astrologer Wilson classifies zodiacal Pisces as an idle sign, though it is actually no less active (all things being equal in the given chart) than any of the other water signs. For another curious Wilsonian view, see EFFEMINATE SIGN.

Iklil al Jabhah   Name given to the 15th of the Arabian MANZILS (the crown of the forehead).

Ilech   A term from the Paracelsian thesaurus. The Great Ilech is the 'star of medicine', which is to say that it is the hidden principle *VIRTUE behind the medicinal element: it is the healing power (derived from the planets and stars) which we assimilate when we take medicine. The Supernatural Ilech is defined by Waite as 'the supercelestial conjunction and union of the stars of the firmament with the stars of things below'. It is, in other words, the name given to that subtle connecton which, the ancients insisted, links all sublunary things with the celestial realm. Crude Ilech is said to be a composition of the three alchemical principles of Salt, Sulphur and Mercury which represent (in the microcosm of man) the Thinking, Willing and Feeling life, respectively. The Paracelsian term is linked with the ILIASTER.

iliaster   A term from medieval esoteric astrological lore (linked with alchemy), apparently originated by Paracelsus from the Greek word *hyle* (matter) and *astrum* (star). It is applied to that spiritual force in matter which strives towards perfection, and towards the building of forms. It works against the *CAGASTER.

image In connection with astrological and esoteric symbolism one must note that the term 'image' is cognate with the word 'magic' – see SYMBOL. Specifically in connection with astrology, the word may properly be applied to the pictorial symbols associated with the zodiacal constellations and signs. While it is taken for granted nowadays that an image of a constellation (such as the Ram of Aries – figure 35) is to be linked with the zodiacal sign of the

Figure 35: Images of the constellations – Aries and Sagittarius.

same name, the evidence is that the images were originally associated only with the constellations (being derived ultimately from the Babylonian imagery, and continued in the planispheres so wrongly termed zodiacs – see, for example, DENDERAH ZODIAC). The method of symbolizing the signs is properly by means of SIGILS, while the method of symbolizing constellations is properly by means of images: thus, figure 35 is an image of Aries, whilst the form ♈ is the sigil for Aries. In modern times the two systems are so interlinked for the differences to be of concern only to academic purists – but see HOMONYMOUS ERROR. Many of the amulets derived from astrological symbolism were even in medieval times called 'imagines'. Paracelsus defines the 'science' of the image as that of representing the properties of heaven and 'impressing them' into the material realm. In the occult tradition, the man-made (or more appositely, 'magic-made') image is far more powerful than a natural SIGNATURE, for 'like virtue is not found in any herbs' as Paracelsus says. See also GAMALEI.

imaginary motion   A term used in connection with *SYMBOLIC DIRECTIONS to distinguish the symbolic arc motions imposed upon radical planets from the *APPARENT MOTION of planets.

immersion   A term normally applied to either the Sun or Moon when entering the eclipse, though sometimes the term is used also of an occultation.

Immum Coeli   A technically incorrect spelling of the Latin IMUM COELI.

impeded   A planet which is afflicted by an unfortunate planet, such as Mars, Saturn or Uranus, is said to be impeded or impedited. The Moon is said to be impeded when in the aspects of conjunction, square or opposition to these malefics, but also when so aligned with the Sun.

impedited   An archaic term used by Simmonite to signify bodies afflicted by evil stars. But see also IMPEDED.

imperfect signs   These are Leo, Virgo and Pisces, so called because (according to Wilson, at least) 'those born under them, and with bad aspects, are deformed in some way or other'. These are also called the 'Broken Signs' and 'Mutilated Signs'. To these Hall adds Cancer and Capricorn.

Imum Coeli   A Latin term ('the lowest part of the heavens'), often wrongly applied to the cusp of the 4th house, which in many chart systems falls on the northern ANGLE of the horoscope figure: see MIDHEAVEN. The IC (as it is often abbreviated) is sometimes called the Anti-Midheaven.

inceptional astrology   A term suggested by the astrologer Carter to distinguish two branches of astrology concerned with 'beginnings'. An inceptional chart is connected with ELECTIONAL ASTROLOGY but is not manipulative. Most Electional Charts are cast in order to ensure favourable astrological influences for the undertaking of an enterprise – for example, in determining the appropriate moment to lay the foundation stone for a building (see FOUNDATION CHART). An inceptional chart is non-manipulative, or it is not concerned with determining a 'favourable' moment, but merely records the astrological conditions at the beginning of an enterprise (often with a view to studying the outcome of that 'beginning'). A historical inceptional chart is that cast by Flamsteed for the laying of the foundation stone of Greenwich Observatory in 1675, which has been reproduced in facsimile (figure 36) by Pearce. The aptness of this chart might suggest that it was perhaps a

inconjunct

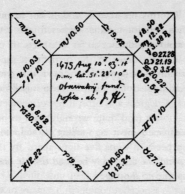

Figure 36: Inceptional chart for the foundation of Greenwich Observatory, cast by Flamsteed for 3.15 pm, 10 August 1675 (after Pearce).

Foundation Chart, for Flamsteed would be quite capable of calculating a propitious moment, and 3.15 pm is certainly an odd 'accidental time' to lay a stone – it is perhaps a nominal inceptional chart, so disguised to deflect the wrath of his peers who had scant respect for elections. An example of a simple inceptional chart is that cast for the launching of the liner *Titanic* (from which the disaster was noted in advance of the event – see figure 18 under CHART SYSTEMS) or indeed several of the charts cast for space ship launches, the times for which are determined by quite other considerations than astrological factors: see, for example, SELENOCENTRIC CHART.

inceptional chart A term used of a horoscope chart cast for purposes of INCEPTIONAL ASTROLOGY, to be distinguished from an ELECTION CHART.

inclination An astrologically specialist term used of a planetary movement towards a position in a chart other than the natal position: it is most frequently used in connection with progressions (that is, within the symbolism of the chart) but in fact all planets have inclination. The term is also used in regard to the plane of the orbit of a planet which subtends an angle (the inclination) to the plane of the orbit of any other planet. Most usually this degree of inclination is related to the ecliptic, which marks the orbit of the earth. But see also OBLIQUITY.

inconjunct This term is frequently misused in modern astrology. It should properly be used as an equivalent or synonym for DISJUNCT, but it has now several meanings due to general misapplication. Devore says that a planet is inconjunct when it forms no aspects, and is not in mutual disposition to another planet. This definition really applies to the disjunct aspect, which may be said to be void of influence so far as aspects are concerned. The term was also applied in early times to the QUINCUNX aspect of 150 degrees, which (for all the theory underlying DISSOCIATE ASPECTS) is widely regarded as an aspect of some power. The term is further used not of aspects (which are always a matter of degrees and orbs, if not of geometry) but of signs and houses, which are loosely (and quite wrongly) said to be aspect relations to each other: thus, an 'inconjunct sign-aspect' is supposed to describe some sort of relationship between signs five-signs apart. By this reasoning, every sign is inconjunct, and the term may be used only if one believes that signs and houses may aspect one another.

increase   When a planet has come to the Ascendant, it is said to 'commence' and 'increases' for the entire semi-arc. For Part of Increase, see POMEGRAN-ATE.

increase in number   See INCREASING IN MOTION.

increasing in motion   Any planet accelerates as it reaches its apogee. When the Moon is thus SWIFT IN MOTION it is regarded as more fortunate than usual. A more ancient term was 'increase in number'.

increasing light   Any planet, after conjunction with the Sun, increases in light, from a geocentric standpoint. An increasing Moon is a waxing Moon.

indication cycle   Sometimes wrongly defined as a cycle of 15 years, this is a time system (said to have been recognized by the Chaldeans) determined by segmenting the zodiacal circle by 15 degrees, allowing one degree for a year. It is not strictly speaking a cycle at all, but merely a chronological system: the name was derived from the Roman practice of using the period as suitable for the reassessment of taxes.

individual degrees   Devore provides a list of DEGREE SYMBOLS linked with the constellational zodiac, which he calls 'individual degrees', and which incorporates material relating to influence from fixed stars, nebulae, clusters, planetary nodes, Exaltations (and Falls), as well as a great deal of material relating to the aspectarian for the period between 1940 and 1946. Some of the degree interpretations are from 'ancient authorities', but since the aspectarian covers such a limited span, and certain of the nodal points are subject to precession, the overall validity of the entire list as a degree symbol system is

questionable. Since the information was published by Devore, they have
been called (misleadingly) the 'Devore symbols'.

Indus Constellation formed by Bayer between Grus and Pavo. Robson
records that it is supposed to give a penetrating mind, and an interest in
orientalism, mysticism and sport.

infantine quadrant See QUADRANTS.

Inferior Midheaven An old term for the IMUM COELI.

inferior planets Originally this was a term applied to the planets 'below'
(inferior to) the Sun, in the concentrics of the SPHERES of the PTOLEMAIC
SYSTEM – hence, the inferiors were Venus and Mercury. The change in the
meaning of the word in an ordinary sense has now made it an unfortunate
term, and it is usually more fitting to call them the 'lower planets', and even
the 'inner planets' – both terms also derived from the Ptolemaic model.

influence In a general sense, the term means 'to have an effect on something',
but the etymology of the word, which is from the Latin *in-fluere* (to flow
into), indicates that in origin this word was once directly linked with
astrological concepts, for it was held that the macrocosm flowed into the
microcosm. The word is now defined in its astrological sense by the *Oxford
English Dictionary* as 'the supposed flowing from the stars of an ethereal fluid
acting upon the character and destiny of men, and affecting sublunary things
generally'. Such a definition is exceedingly dated, however: the modern
astrologer will see such 'supposed' influences proceeding from other sources
than merely the stars – for example, from the zodiac (which is not to be
confused with asterisms), from the planets, from Angular relationships
between planets, from nodes, and a multitude of other cosmic phenomena or
supposed conditions. Also, few modern astrologers would be prepared to
acquiesce to the idea of 'ethereal fluid' as being a transmitter of such
influences. There is actually no more satisfactory theory of influences than
that expressed in the tradition of *SYMPATHIES which underlies the doctrine
of the CHAIN OF BEING – though admittedly this is descriptive, rather than
anything explanatory. A related term from Paracelsian astrology is ILECH.
See also SYNCHRONICITY.

infortune major A term used for the planet Saturn.

infortune minor A term used for the planet Mars.

infortunes   The two planets Mars and Saturn are traditionally the infortunes, the infortune minor and the infortune major, respectively. To this traditional classification, some astrologers add Uranus. See MALEFICS.

ingress   The technical term used to denote the entry of any planet, fixed star or nodal point, into a sign or quadrant of the zodiac.

inheritance   The Point of Inheritance, sometimes called the Point of Magistery and Possessions, is one of the so-called Arabian PARS. If the degree of Saturn in a natal chart is taken as the Ascendant, then the adjusted position of the Moon in the houses marks the Point.

initial chronocrators   A term used originally in connection with an Hellenistic astrological system of assessing age rulerships, described by Bouche-Leclercq. Only the luminaries may be the initial chronocrators: the Sun in a diurnal horoscope, the Moon in a nocturnal chart. This chronocrator stands guard over the first period of age, a period of ten years and nine months, or 129 months in all. This periodicity is derived from the ancient PLANETARY PERIODS, for the sum of these (MO 25, ME 20, VE 8, SU 19, MA 15, JU 12 and SA 30) gives 129, which may be read as either 'months' or 'years' according to the ancient system of numerical correspondences.

initiating signs   These are Aries, Cancer, Libra and Capricorn.

inner planets   See INFERIOR PLANETS.

intellectual trinity   A term used by Butler in his SOLAR BIOLOGY in respect of the zodiacal Aries, Taurus and Gemini.

intellectual zodiac   One of Blavatsky's terms for the tropical zodiac, used to distinguish this from the constellational zodiac, which she calls the NATURAL ZODIAC. The term 'intellectual' is being used in its ancient sense, as something related to the spiritual realm, and thus distinguished from the natural world of created forms.

Intelligency   A term used to denote a class of spirits, originally the Intelligencies, who regulated the movement and life of the celestial SPHERES. However, the function appears to have been forgotten by about the 17th century, so that the term was (quite wrongly) equated with intelligence, and even those who recognized that the term was designate of a class of spirits attached to the group an indeterminate activity and series of rulerships.

Agrippa, still aware of the actual function of these beings as 'spherical regulators', gives the names shown in table 47, along with a series of sigils and magical squares for the Intelligencies of the planets (by which he meant 'planetary spheres', and the corresponding Daemonia (see DAIMONS). It is

————————————————— TABLE 47 —————————————————

Planetary spheres	Intelligencies	Daemonium
Saturn	Agiel	Zazel
Jupiter	Iophiel	Hismael
Mars	Graphiel	Barbazel
Sun	Nachiel	Sorath
Venus	Hagiel	Kedemel
Mercury	Tiriel	Taphthartharach
Moon	Malcha*	Hasmodai

* In full, Malcha is 'Malcha hetharsisim hed beruah Schehakim'.

likely that the idea of the ensouled spheres entered into astrological lore by way of Aristotle himself, as indeed was the idea that these were the *motori*, or drivers of the spheres. The nomenclature is, however, Hebraic.

intensities A term used in connection with the measurements for the aspects. The intensities of aspects may be measured by several rule-of-thumb methods. Carter, for example, ranks the aspects in the order of decreasing strength: conjunction, opposition, square, trine (sextile/semi-sextile/sesquiquadrate), and so on. This approximately corresponds to the interpretation of the two different groups of aspects described by Ptolemy, the MAJOR ASPECTS being stronger, the MINOR ASPECTS weaker. Morin (who in any case sought to schematize everything in a highly personal manner) relates the opposition as stronger than the square, the trine stronger than the sextile, on the (dubious) grounds that each is twice the other. Doane, in connection with his *ASTRODYNES, offers a well-researched system of weighting, by which the numerical values assigned to the aspects are increased by about 25 per cent if the luminaries are involved: this idea is far from new, however, though the weighting is special to Doane. Most disagreements among authorities as to the ratios of intensities seem to involve the strengths of the minor aspects.

intercepted house See INTERCEPTED SIGN.

intercepted sign A term which, though frequently used, is something of a misnomer. It is generally supposed to refer to a sign which, due to the variations in time of Right Ascension, is not posited on one of the cusps in the symbolic representation of the mundane system. In fact it is really an intercepted house – some insist that planets within such a house are given emphasis, while others insist that there is a general weakening. Quantitatively speaking, however, the house covers a much wider area of influence than would a normal mundane arc, and one would therefore expect the influence to be more considerable than normal. Certainly there is a statistical propensity for such houses to contain more planets than other houses.

interface A term used in connection with heliocentric astrology, relating to nodes in orbital planes other than that of the ecliptic. Such nodal planes are termed interfaces.

International Astronomical Union zodiac See MODERN ZODIAC.

interpolation The act of inserting within a specific order of things, especially within a series of numbers. In astrology, the term is sometimes used of the process of determining the position of a planet between two known positions (usually the two positions listed for a diurnal cycle in an ephemeris). Again, the cusps of the intermediate houses, between the four Angles, are interpolated, according to the methods of *HOUSE DIVISIONS.

interpretation Term often used of the reading of an astrological figure according to the interacting factors seen within the chart, both from a radical point of view, and from a point of view of transits and progressions.

interrogatory astrology A term used by the historian Pingree in his division of the forms of astrology into four distinct groups (the others being genethlialogy, general astrology and katarchic astrology). Interrogatory astrology is that concerned with answering specific questions on the basis of the situation in the heavens at the time of making the query: it is therefore the equivalent of HORARY ASTROLOGY.

interval A term more appropriate than the traditional word 'aspect' to describe the application of number symbolism involved in *HARMONICS to chart figuration. Each interval has its own significance: in the system of harmonics, the intervals are based on 179 harmonic numbers, yet the significance of many of these is still to be established beyond reasonable doubt.

However, the meanings of the intervals relating to 4, 5, 6, 7 and 8 appear to be agreed among specialists.

intuitional astrology   In her study of ESOTERIC ASTROLOGY, which in its principles and application is far removed from orthodox astrology, the esotericist Bailey suggests that it is her 'intuitional astrology' which must eventually supersede what is today called 'astrology'. The esoteric system proposed by Bailey, which is partly an extension of esoteric principles formulated by Blavatsky, involves the *NON-SACRED PLANETS, the *ORTHODOX PLANETS, the *SACRED PLANETS, the *SEVEN RAYS and the *THREE CROSSES, among other specialist concerns and recondite terminologies. The term 'intuitional astrology' is that proposed by Bailey herself. It is termed 'intuitional' to distinguish it from other formulated esoteric astrological systems, as it appears to have originated largely by means of clairvoyance. By laymen the system has been variously called 'Tibetan astrology' and 'Lucian astrology', derived from the fact that the works of Bailey are linked with the teachings of 'a Tibetan', and are published by the Lucis Trust.

invariable plane   A central plane of the solar system, said to incline at approximately 1 degree and 35 minutes to the ecliptic. In astrology it has direct relevance to HELIOCENTRIC ASTROLOGY, and to the calculative justification for many of the Great Ages which are related to the planetary perihelions involving cycles of vast duration.

inversion   When the extent of an aspect is subtracted from an arc of 180 degrees, the result is another aspect: each is an inversion of the other.

Io   Name given to one of the satellites of Jupiter.

Iophiel   Name given by Agrippa (quoting ancient qabbalistic sources) to the INTELLIGENCY of Jupiter. He gives the magical number of Iophiel as 136, which is four times the linear sum of the *MAGIC SQUARE of Jupiter (34).

Iris   Name given to one of the asteroids discovered in 1847.

Iron Age   As a specialist occult term, derived from Hindu chronology, and from modern esotericism – see KALI YUGA and FOUR AGES.

Ishtar   One of the ancient Babylonian names for a goddess regarded as the equivalent of Venus. But see also VIRGO CONSTELLATION.

Isidis   Name given by Robson to a fixed star, probably the delta of Scorpius, DSCHUBBA, set in the head of the Scorpion, not in the right claw.

Isis   Name given to three different HYPOTHETICAL PLANETS. The first was that identified by the astrologer Sutcliffe, who claims it to be extraneptunian, with a periodicity of 360 years. The second Isis was that proposed by Thierens, as the future ruler of Taurus (not properly speaking a hypothetical at all – see ETHERIC SPHERES). The third is that listed by the astrologer Jayne, as a hypothetical transplutonian. Additionally, the name Isis is linked with the constellations: the Egyptian goddess of this name was the pre-Christian prototype of the Virgin Mary, and inevitably esoteric astrology has linked the Egyptian name with both the constellation and zodiacal sign Virgo, thereby echoing an ancient association, since Eratosthenes (among others) called the constellation Isis.

Ithuriel   The name given by modern Rosicrucians to the Archangel who acts as ambassador from the planet Uranus. In earlier traditions the name is used of one of the Cherubim (as, for example, by Milton in *Paradise Lost*), while in the popular demonological tradition it is used of the spirit of Mars.

Izards   The second series of creations of Ormuzd, the Zoroastrian being of light. They are 28 in number, and are said to watch over the virtue of the world, acting as interpreters of human prayer. The Izards are often linked with the lunar cycle of 28 days, and with the 28 LUNAR MANSIONS.

J

Jabha   Al Jabha, fixed star of the 3rd magnitude, set in the mane of the Lion, though the term is from the Arabic Al Jeb'ha (the forehead). Not a beneficial influence, it is said to carry the equivalent nature of Mercury conjunct Saturn, and to be conducive to violence.

Jabha   'Al Jabhah' (the forehead), name of the 8th of the Arabian MANZILS, in reference to the head of the Lion (the fiducial is the alpha of Leo).

Jadiyy   Al Jadiyy, the Arabic name for Capricorn.

Jael   In some forms of ceremonial magic, the Governor of zodiacal Libra.

Jaimini   A Sanskrit term applied to one of two Hindu astrological methods of computing aspects according to elemental influences (that is, from sign to sign, rather than by geometric considerations – see ELEMENTAL ASPECTS). This method (which is like the Parasara Hindu method) is similar to that described by Ptolemy in his *Tetrabiblos*. See also TAIJAK.

Jaiminic   A term adapted from Hindu astrology (see JAIMINI) in application to the use of aspects viewed from a point of view of the elemental natures, to distinguish from aspects considered exclusively as geometric relationships. It was Ptolemy who introduced into astrological literature the concept equivalent to the jaiminic aspects.

January   Named as the first month of 31 days in the GREGORIAN CALENDAR, after the god Janus (who was two-faced, looking to the past and the future, and thus an excellent symbol for a cycle which has neither beginning nor end). Esoterically Janus was one of the numina which rendered sacred all entrances – the points of intersections between past and future – and he was thus associated with the ruler of time, Saturn. Long before the Julian reforms, and for some four centuries afterwards, the esoteric significance of this month was expressed in the festival of the Compitalia, a survival in the cities of rural Lares lustral sacrifices, which may be associated with Aquarius.

Janus Name given to one of the *HYPOTHETICAL PLANETS by Wemyss, who accorded it rule over zodiacal Sagittarius, and a revolution of 45 years, in an orbit between Saturn and Uranus. Since the planet was first announced, circa 1912, little evidence for its influence has been adduced.

Japetus One of the moons of Saturn.

Jason One of the four hypothetical planets designated by Maurice Wemyss. This Jason is proposed as ruler of Sagittarius, with Dido as ruler of Virgo, Hercules as ruler of Leo, and the hypothetical Pluto as ruler of Cancer. In the astrological literature, the Jason of Greek mythology (after whom the hypothetical is named) is often mentioned: this hero was the leader of the Argonauts (see ARGONAUTS) who brought the Golden Fleece from Colchis.

Jauza Al Jauzah. See ORION.

Jawza Al Jawza, the Arabic name for Gemini. See also ZAWZAHR.

Jeuze See ORION.

Johndro locality chart A horoscope chart devised by the astrologer Johndro in which the angles are calculated according to the locality of the birth place, the so-called LOCALITY ANGLES.

Jonas theory A term used specifically to denote an ovulation cycle theory, which the Czechoslovak psychiatrist Jonas claims to be astrologically determined. This ovulation cycle (which does not correspond to the normal ovulation cycle) is said to reach its peak each time the transits of Sun and Moon reach their monthly radical Angular relationships in the female chart, allowing an orb of 15 degrees. Unfortunately, scrutiny of his statistical method, and indeed some of his astrological computations, have led to serious doubts as to the validity of the theory. Dean quite rightly comes to the conclusion that the majority of independent workers have not confirmed Jonas's findings. Jonas has also broadcast at least one other theory relating to the sexuality of the child, but it is in connection with the former theory that his name is now widely known.

Jones patterns A term used of a system of classifying planetary groupings in personal horoscopes as an aid to rapid interpretation of underlying psychological tendencies: it is named after its originator, the American astrologer Marc Edmund Jones. Seven different patterns have been described, and their supposed influences outlined: see BOWL, BUCKET, BUNDLE, LOCOMOTIVE, SEE-SAW, SPLASH and SPLAY.

joy   This is now almost an obsolete term in its specialist astrological sense, not because it fails to carry a useful and distinctive connotation but because of the confusion which surrounds it in the astrological literature. Some astrologers have substituted the term 'joy' for the planetary rulerships over signs, as a synonym indeed for what Ptolemy called the 'thrones'. Thus, the Sun in Leo was in its joy – though Ptolemy (and others) would insist that he was merely on its throne. Other astrologers describe the joys as mundane affairs, as follows:

MO in 3rd house	ME in 1st house	VE in 5th house
SU in 10th house	MA in 6th house	JU in 11th house
SA in 12th house		

Ptolemy had a very different view of the joys, however, yet as is so often the case he is not very clear in his definition of the term. It would seem that he intended a planet to receive joy from another planet when (though not itself in any dignity) it was 'harmoniously linked' (whether by aspect or not is hardly clear) with a planet which was in a dignity. In this case the dignity appears to have become a joy by means of transmission of sympathy. Wilson attempts a general definition on much these lines, and proceeds to a series of joys which are so prolix that they have the effect of reducing astrology to a level which few astrologers might follow. Devore simplifies enormously the idea of a joy by suggesting that it is experienced by a planet when another is placed in one of the former's essential dignities – thus returning nominally to something verging on the Ptolemaic concept. The term does not actually call either for total rejection from the astrological lexicon, or for reform. After setting out a brief though incomplete survey of the various definitions of 'joy', Devore suggests sensibly that the joy should be restricted to mundane considerations. It has long been held by some astrologers that the joys should be an expression of delight felt by planets of a similar nature when they are posited in each other's houses. The pairs of planets are Sun and Jupiter, Saturn and Mercury, Venus and Moon. By such standards, the Sun may be said to have his joy in the 9th house.

Jubana   Al Jubana (the claws), the 14th of the Arabian MANZILS, a throwback to the Greek constellation map which placed the claws of Scorpius where the fiducial alpha of Libra is now located.

Julian calendar   Important calendrical system used from 45 BC to AD 1582 in most parts of the civilized European world, and in some parts even until the present century (see GREGORIAN CALENDAR). It forms the basis of all Roman and medieval astrological computation and horoscope data. The system is named after Julius Caesar who ordered the reforms of the ancient Roman calendar, the work of adjustment being done mainly by Sosigines of Alexandria and Marcus Favius. The year contained 365 days, with an intercalary every fourth year, an adjustment which permitted a close relationship with the tropical year, and involved an error of only 11 minutes – which itself eventually called for the reforms of Gregorius, which ousted the Julian system. The Julian was more than merely an adjustment of the earlier Roman system, but the transference to a solar-based method of computation, away from the lunar system of its predecessor. After a small necessary adjustment (required due to a misapplication of one of Sosigines's rules) in 8 BC, the months of unequal days, with February as the one susceptible to intercalary adjustment, were those which we have inherited in our own Gregorian system. Indeed, the fifth month gave us July in memory of Julius Caesar himself, while August was derived by Augustus when he ordered the adjustment of 8 BC. The date of primal reform was followed by the famous 'Year of Confusion' (46 BC), when in order to adjust for precession, two months were inserted into the year.

Julian day   See JULIAN PERIOD.

Julian period   A period introduced by Scaliger to provide a continuous measure of time in cycles of great extent. The period was a combination of the solar cycle of 19 years, the lunar cycle of 28 years, and the indication cycle of 15 years. Scaliger artificially linked the beginning of this cycle with the first day of the year of the Julian calendar, and began the cycle in 4713 BC. One benefit proceeding from the Julian period is that it has been possible to establish an easily manipulated scale of historical references, the Julian Day, which is merely determined in terms of daily aggregates from the beginning of the cycle. The Julian Day is rarely used in astrology.

July   The seventh month (with 31 days) of the GREGORIAN CALENDAR, its name being derived from that given to the original Roman month *Quintilius* (the fifth month of the JULIAN CALENDAR) in honour of Julius Caesar (born this month), who had initiated the fundamental calendrical reforms.

June   The sixth month (with 30 days) of the GREGORIAN CALENDAR, its name probably being derived from that of the Roman goddess Juno, who was patron of womanhood (called Pronuba) – this month being favoured for marriage (even though the Junoesque 'Matronalia' was celebrated on the 1st March). Juno's Grecian prototype, Hera, was certainly worshipped as the embodiment of the fecund Earth, as many of the rituals of the Samian Hera indicate: the imagery associated with the month is pre-Christian, and (as Foucher indicates) is far from understood.

Juno   Name given to one of the *ASTEROIDS, discovered in 1804, and linked by Dobyns with the sign Libra, along with a co-ruler, the asteroid PALLAS. The same name has been given to one of the moons of Jupiter.

Jupiter   The planet in orbit between Mars and Saturn, ruler of the zodiacal Sagittarius. Jupiter is the giant planet of our system, with an equatorial diameter eleven times that of the Earth, and with twelve satellites (four visible with good binoculars), its rotational speed variable. In traditional forms of astrology, prior to the discovery of Neptune, Jupiter was said to have rule over the zodiacal Pisces (see DAY HOUSES): modern astrologers generally give the throne of Pisces to Neptune, however. Jupiter is Exalted in Cancer, and has its Fall in Capricorn (but see DIGNITY). The sigil for Jupiter ♃ is said to represent the semi-circle of spiritual potential lifting the heavy cross of materiality, which certainly symbolizes the uplifting nature of the planet, though this sigil is actually post-medieval in origin (see SIGILS – PLANETS). Jupiter represents the profound side of the mental life: it stands for the speculative thought of the native, and for the spiritually expansive side of his nature. In particular Jupiter offers an index of the moral nature of the native, and of the degree to which he will respond enthusiastically to life and its responsibilities. The responsibilities themselves may come from Saturn, but how they are shouldered depends very much upon Jupiter. When Jupiter is beneficially emphasized in the chart (through dignity or aspect) it gives a generous, optimistic, loyal, popular and generally successful nature. On the other hand, an unfortunately located Jupiter tends towards uncontrolled expansion, and frequently an inability to accept limitations – hence, self-indulgence, prodigality and recklessness ensue. Traditionally Jupiter has rule over the thighs and liver in the human body (see MELOTHESIC MAN), though modern astrologers place under its rule the posterior pituitary gland, and the working of the blood: its action seems to be involved with the regulation of blood supply and breathing, which is the index of health in the human body. The traditional description of the Jupiterian type is of one who has a well-formed, though large (even plump) body, with a tendency to develop a paunch in later years. The women are particularly beautiful: the forehead is noticeably large, the hair dark and wavy, and in the male there is a tendency for it to recede in later life. The eyes are usually commanding, though also kindly, the whole bearing a union of the dignified avuncular-paternal. The Jupiterian usually finds expression on the physical plane through activities demanding careful thought, speculation, growth and steady expansion – nowadays this is especially so in the realm of law, industry and government. While the Jupiterians are usually capable in money matters, they do tend to be prodigal with personal finance, yet with the special ability to handle and control the money of others. They delight in a rich and varied life (especially one involving travel or exploration), and are usually in great social demand. They participate easily in public functions, and are of a philanthropic nature, often with a fine feeling for music, and a love for philosophy which brings them into constant contact with creative people.

Jupiterian Chain   See JUPITER SCHEME.

Jupiter period   The name given in esoteric astrology to a future condition (embodiment) of the earth, when humanity (having learned the lessons relating to objective consciousness in the present cycles) will attain to self-directed picture consciousness. Distinguish from JUPITER SCHEME.

Jupiter–Saturn cycles   Jupiter conjuncts with Saturn every 19.859 years, at an advance of some 123 degrees. The first order cycle is 59.5779 years (usually rounded off to 60 years), and a second order is 794.372 years (usually rounded off to 795 years). See CHRONOCRATORS.

Jupiter Scheme   The Jupiter Scheme or Jupiterian Chain are terms derived specifically from the theosophical cosmoconception linked with the *SCHEME OF EVOLUTION, and are not to be confused with the JUPITER PERIOD. It is claimed that while Jupiter is not yet inhabited, its moons are, and the evolutionary development of this CHAIN will reach a very high level.

jyestha   The 16th of the Hindu NAKSHATRAS (the oldest).

K

Kabbalistic astrology See QABBALISTIC ASTROLOGY.

Kalb Al Kaib (the heart), the 16th of the Arabian MANZILS, applying to the heart of the Scorpion, as in the medieval lunar mansion Cor Scorpionis.

Kalends In the ancient Roman calendrical system there were no weeks, and months were divided according to three points of measurement for each month called Kalends, Nones and Ides, names surviving from an ancient lunar calendar (for example, the 'Idus' was the full Moon). These were used in astrological data and horoscopes well into late medieval times, surviving even into collections of genitures of the so-called NOTABLE NATIVITIES type. The Kalends were the first day of the month. The Nones were variable, being usually the fifth day, though in March, May, July and October, it was the seventh day. The Ides were also variable, being usually the 13th day, but in the four above months it was the 15th. Days were calculated before these points, inclusively, so that (for example) 29 May in our (New Style) system would be the fourth Kalends, written in horoscopic abbreviation as 'a.d.iv Kal. Jun.' – the abbreviation 'a.d.' meaning *ante diem*, the whole phrase reading '*ante diem iv Kalendiae Junius*'. The day immediately prior to a reckoning point was indicated by the Latin *pridie*, usually shortened to 'prid': our 4 January would be written 'prid.Non.Jan.'. The seven-day week system appears to have been first of all popularized by Roman astrologers (no doubt because of the involvement with predictive techniques linked with lunar rhythm), but not made official until the time of Constantine.

kali yuga A Sanskrit term which has been Europeanized mainly through the influence of theosophy, and translated as meaning Iron Age, Black Age or Dark Age. The term often crops up in esoteric astrology because it is widely acknowledged that we now live in this age, which the Hindus claim began in 3102 BC. It is said by some occultists to be a period of 360,000 years, but others insist that the age has already terminated (1880). See YUGA, which sets out the additional periods of twilights for the Kali Yuga.

Kallistos One of the several important variant names for URSA MAJOR.

kalpa In the Hindu chronology, a kalpa is a 'day of Brahma', a period of 4,320 million years. The 'night of Brahma' is of the same length. One 'year of Brahma' (a solar kalpal in modern parlance) is made up of 360 such pairs, while a century of such years makes up an entire cycle, the mahakalpa, of 311,040 trillion years. Unfortunately, it is apparent that the word 'kalpa' is used indiscriminately in reference to ages. Purucker records that a kalpa is also a mahamatava, a period equivalent to a whole lifetime of the globes in a PLANETARY CHAIN.

Kandra The Sanskrit term for the ASCENDANT, as used in Hindu astrology.

Kanya The Sanskrit name for zodiacal Virgo – but see also PHALGUNI.

Karkatakam The Sanskrit name for zodiacal Cancer – but see also PUNAR-VASU.

Karkinos One of several Greek names (meaning 'crab') used to denote the zodiacal and constellational Cancer. It is sometimes written 'Carcinos'.

karma A term derived from the Sanskrit root *kri* (to make or to do), and meaning approximately action, though linked with the idea of consequence of actions, being the 11th Nidana (chain of causation) of orthodox Buddhism. In Western occultism it is applied to the law of Karma, the unfolding of destiny through repeated earth lives, in which merit and demerit are reflected in life conditions, events and inner attitudes. In applications to astrology it is increasingly used in esoteric circles, and many astrologers regard the natal chart as an impress of the particular karma which the incarnating ego (that is, the native) has undertaken to resolve in this particular lifetime. Some astrologers see the horoscope as reflecting the whole of accumulated karma: the astrologer Robson writes, 'the horoscope is the result of a series of past lives ... and by a study of the map of the trend of former lives may be discovered.' Some emphasis is placed on karma in ASTROSOPHY, subsequent to the work of Steiner.

Karoubs See OPHANIM.

Katababazon Although almost archaic now, the term was once used widely as a synonym for the DRAGON'S TAIL, derived from the Arabization of a Greek term. There are many variant spellings and sounds, as for example Catahibazon and Katahibazon.

Keplerian harmonics

Katahibazon   See KATABABAZON.

Katakam   A Sanskrit term for zodiacal Cancer, as used in Hindu astrology.

katarchic astrology   A term derived from the Greek (*Katarche*, beginning) and used to refer to that form of astrology concerned with the study of influences arising from any given moment deemed to mark the beginning of an enterprise or undertaking: it is in effect HORARY ASTROLOGY.

Katotyche   A Greek term meaning approximately 'ill fortune', and used to denote the 'Sixth House'.

Katune   A term used in MAYAN ASTROLOGY to denote a period of 20 × 360 days. Volguine links the Katune with the Jupiter–Saturn cycle, but in reality it appears to be a system of dating.

Kemedel   Name given by Agrippa (quoting ancient qabbalistic sources) to the Daemon of Venus, for whom he gives the magical number 157, without doubt a misprint for 175, the value of the Venusian MAGIC SQUARE. But see HAGIEL.

Kephziel   One of the several variant names for the ruler of Saturn.

Keplerian aspects   A name given to a group of aspects linked with the QUINTILES, originally proposed by the astronomer Kepler. These are the BI-QUINTILE, DECILE, QUINDECILE, QUINTILE, SEMI-DECILE and IN-CONJUNCT. Kepler personally used the major and minor aspects of traditional astrology in his horoscope work, but evolved the others on numerological grounds.

Keplerian harmonics   The equation made by Kepler in his *Harmonices Mundi* of 1619, between the celestial motions of the solar system and musical intervals, has never been systematically integrated into astrological theory, mainly because the Keplerian harmonics are heliocentric, while traditional astrology has been mainly rooted in geocentric cosmoconceptions. In effect, Kepler breaks away from the astrological tradition (see MUSIC OF THE SPHERES) by rejecting the archaic notion of the spheres as a basis for harmonics, and by rejecting even planetary distances or orbits, as these were no longer applicable within the extension of the COPERNICAN SYSTEM proposed by Kepler in his famous laws. Instead, Kepler established a new harmonic series in terms of the heliocentric diurnal Angles of the planets, in their passages through aphelion and perihelion, which he showed demonstrated a complete system of musical intervals. The series, set out in an excellent article by Haase, may be digested as follows. Saturn is connected with the major third, with the twelve tones and the minor third. Jupiter is connected with the octave, the three octaves and the perfect fifth. Mars is connected with the minor third (plus two octaves and plus one octave). The earth itself is connected with the diatonic semitone, the perfect fifth and the major sixth. Venus is connected with the chromatic semitone, the minor sixth and the double octave. Mercury is connected with the minor third (plus one octave) and the major sixth. This, as Haase says, is a 'surprisingly simple structure of the harmonies of our planetary system'. See also PLANETARY TONAL INTERVALS.

kernal   See COSMOPSYCHOGRAM.

Ketu   The Sanskrit term for the DRAGON'S HEAD, which is accorded much importance in Hindu astrology. For the Hindu 'Cauda', see RAHU.

key cycle   A system of calculating progressions, designed to rectify (minute) errors implicit in the normal method. The system is an adaptation of SOLAR RETURN calculations, though adjustable to any time or space relating to the Earth.

keynotes   An equivalent term for KEYWORDS, applied in a special sense by the astrologer Pagan to characteristic aims of the twelve signs:

AR Hope	CN Patience	LB Beauty	CP Reverence
TA Peace	LE Glory	SC Justice	AQ Truth
GE Joy	VG Purity	SG Wisdom	PI Love

keys   A term translated from the astrology of the French occultist Christian, in reference to his 'Keys of Magism', a series of 570 fixed readings linked with the planets, signs, houses, aspects, and so on, derived from a syncretic interpretation of the astrological doctrines of Firmicus Maternus, Junctinus of Florence and Pelusius. Included in this somewhat confused schema are the so-called 'keys of the annual horoscope', which represent 159 fixed readings (merely extended KEYWORDS) supposedly intended to facilitate comparison of the progressed aspects to the natal chart.

keys of the annual horoscope   See KEYS.

keywords In modern astrology a procedure has been established of portray-
ing the qualities of planets and signs (as well as other interpretative factors)
in terms of single words or short aphorisms, each designed to capture the
essence of the specific influences: these have been called keywords. The
system was designed originally for teaching purposes, but has been used of
late (especially by Sunley) as the basis for investigating relationships between
traditional astrological forms, particularly that between the tropical and
sidereal zodiac. One of the most perceptive systematic portrayals of planetary
influence based on an extension of the keyword system is that given by
Davison, but a fine synoptical single-keyword series, in which essentials are
reduced to one word for signs, houses (Hs) and planets, is that given by
Jones:

AR Aspiration	LE Assurance	SG Administration
TA Integration	VG Assimilation	CP Discrimination
GE Vivification	LB Equivalence	AQ Loyalty
CN Expansion	SC Creativity	PI Sympathy

1st Hs Identity	2nd Hs Possession	3rd Hs Convenience
4th Hs Integration	5th Hs Offspring	6th Hs Obligation
7th Hs Complementation	8th Hs Ideality	9th Hs Consciousness
10th Hs Honour	11th Hs Expectation	12th Hs Confinement

SU Purpose	JU Enthusiasm	SA Sensitiveness
MA Initiative	PL Obsession	VE Acquisitiveness
NE Obligation	ME Mentality	UR Independence
MO Feeling		

See also FUNCTION, KEYNOTE and WATCHWORD.

Khambalia One of the star-names for the lamda of constellation Virgo,
which has gained a reputation in astrological textbooks not because of its
size, but (probably) because of its proximity to the ecliptic, and because it
marks one of the LUNAR MANSIONS. The name is said to be from the Coptic
meaning 'crooked claw': it is said to be equivalent to Mercury conjunct Mars,
and therefore the bringer of violence: the native will have a changeable and
argumentative nature.

Khatti See ASTROLABE.

Kidinnu tables   See AYANAMSA.

killing place   See ANARETIC PLACES.

kinship   See FAMILIARITY.

See BABYLONIAN CALENDAR.

Kneeler   See ancient name for the HERCULES CONSTELLATION.

known data method   A term used to denote a method of checking astrological traditions and speculations (and even formulating new astrological hypotheses) by the empirical method of checking horoscope data (radical or progressed) against the known case histories of natives.

Kochab   Fixed star, the beta of URSA MINOR, which Ebertin claims brings suicide when with an afflicted Sun. Allen suggests that during the millennium prior to our era, Kochab may have been one of the pole stars.

Koch houses   See KOCHIAN SYSYEM.

Kochian system   A method of HOUSE DIVISION devised by the German astrologer Koch, based on the individual calculation of the cusps of intermediate houses for the place of birth. It is one of the methods which trisects the diurnal semi-arc by time division. This method produces a grid which has no valid application in polar regions, and TABLES OF HOUSES are available only to 60 degrees of latitude, yet even so it rapidly gained support among astrologers after its introduction in 1962. 'One wonders,' writes Genuit, 'why, over hundreds of years, nobody has hit upon this correct and obvious idea.' The method is sometimes called the GOH system (derived from the German *Geburtsorts Häusertabellen*) and sometimes the birthplace system.

Koilon   A term derived from the ancient Greek *koilos* (hollow), and used by esotericists to denote 'root matter'. Koilon is invisible to all but the most highly developed of clairvoyants.

Kolisko effect   A term used to denote the effect of (sidereal) planetary aspects on certain chemical reactions, recorded as litmus pictures (the so-called MORPHOCHROMATOGRAMS). Kolisko, following the suggestions of Steiner, established over several years' experimentation that the precipitation of mineral-salt solutions (such as argentum nitricum, ferrum sulfuricum and

Kore

plumbum nitricum) left filter paper patterns which were conditioned by planetary conjunctions and oppositions, thus largely vindicating in a vividly pictorial manner Steiner's claim that 'so long as substances are in a solid state, they are subject to the forces of the Earth, but as soon as they enter the liquid state, the planetary forces come into play'. Kolisko demonstrated that the metals were affected by their corresponding 'traditional' planets (that is, silver by Moon events, iron by Martian events, lead by Saturnine events, and so on). Figure 37 gives a visual comparison of the patterns left by such

SU CONJ. SA
21.11.1926

Figure 37: Precipitations of lead solution on litmus paper (left) under normal conditions, compared with (right) during a conjunction between Sun and Saturn, in 1926 — see below (after Kolisko).

precipitations, derived from a large number of photographs published by Kolisko and her followers. In connection with this figure, which compares a non-aspectual morphochromatogram with one demonstrating a superior conjunction of Saturn and Sun, Kolisko writes: 'under normal circumstances a picture similar to plate A was to be expected. But what did we see? Instead of heavy, massive forms an utter blank. An invisible hand had blotted out the working of the lead in the solution. And whose was the invisible hand? It was the Sun. The Sun had stood before the planet Saturn, and here below on the earth the lead could not manifest its activity. When the Stars speak, man must stand still in silent awe.' Fyfe has continued the basic research technique of Kolisko, more particularly in regard to lunar phenomena, as well as in regard to plant saps: see CAPILLARY DYNAMOLYSIS.

Kore Greek term, sometimes Romanized as koure, of Attic origin, meaning 'maiden', and often used to denote the zodiacal and constellational

Virgo. She was especially called the Staxeodos Koure, 'the wheat-bearing maiden'.

Koronis   See HYADES.

Kosmobiologie   See COSMOBIOLOGY.

kosmos   See MACROCOSM.

krasis   See MUSIC OF THE SPHERES and TEMPERAMENTS.

Krios   The Greek term (meaning 'ram') used to denote the zodiacal and constellational Aries.

Krita Yuga   The Hindu Golden Age, a periodicity of 1,728,000 years – sometimes called the Satya Yuga. This period includes the twilights sandhya and sandhyansa of 400 divine years each. See also YUGA.

Krittika   The Sanskrit name for constellation Taurus: but see RISHABHAM. Krittika is also the name given to the 1st of the Hindu NAKSHATRAS.

Kronos   One of the ancient terms for the planet Saturn, but not originally the Greek god-equivalent of Saturn. Kronos was in Greek mythology the youngest of the giant Titans, the son of Uranus and Gea, and was later identified with Saturnus by the Romans – presumably through his ancient association with agriculture (Gea being the Earth goddess). By the time that the astrological tradition was being set down in book form, Kronos and Saturnus were both used for the same planet: Ptolemy (writing in Greek) generally refers to the planet as Kronos. The same name (Kronos) is given to one of the several HYPOTHETICAL PLANETS used in the system of URANIAN ASTROLOGY, and is said to have a meaning linked with authority.

Kuja   The Sanskrit word for the planet Mars, as used in Hindu astrology.

Kumbham   The Sanskrit name for zodiacal Aquarius: but see also DHANISH-THA.

Kurri   See ASTROLABE.

Kursi   The shoulder of the *ASTROLABE which carries the swivel ring. This

Kyriotetes

shoulder projection, sometimes called the Throne, was often elaborately decorated in Islamic astrolabes.

Kyriotetes   See DOMINIONS and CELESTIAL HIERARCHIES.

L

Labrum Fixed star set in the constellation CRATER. It is supposed to combine the natures of Mercury and Venus, and (presumably because of its symbolic name as Holy Grail) it is said when rising on the Ascendant to give ecclesiastical preferment.

La Croix Name given by the astrologer Charubel to an extraneptunian planet, one of the HYPOTHETICAL PLANETS, which he perceived clairvoyantly some years prior to 1897, and for which he claimed a revolution period of just over 340 years. He gave it the alternative (though ambiguous) name Mundane Cross, and claimed for it the 'higher properties of Mars'. It has been observed that the revolution of this hypothetical is almost the same as that named ISIS. Later astrologers assigned La Croix to the day house of Capricorn, the night house of Aries, and exalted it in Sagittarius, though it has not been widely adopted by modern astrologers.

Lady of Heaven One of the translations of a Sumerian term for Venus.

Lagnam A Sanskrit term for the ASCENDANT, as used in Hindu astrology.

lamb A term sometimes used of zodiacal or constellational Aries, derived from the TALISMANIC MAGIC of the medieval period, when there was a move (for example, in the writings of Arnaldus de Villanova) to integrate the pagan elements of astrology with Christian magic.

Lamb of God A name sometimes used of zodiacal or constellational Aries – but see also LAMB.

lame degrees See AZIMENE DEGREES and MUTILATED DEGREES.

latitude Within the limits of astrological use, the term is applied to a system of co-ordinate measurements from the ecliptic, north or south, towards either of the poles.

Lavagnini system A method of HOUSE DIVISION based on the attempt to apply the equal-house symbolic method (see MODUS EQUALIS) in such a way that both Ascendant and Midheaven may be integrated for all latitudes. It is an eight-point system, each of the points reflecting the combination of the four Angles and the pairs of zenith/nadir and east/west points. The 1st house of the grid is determined by bisecting the Angle between the Ascendant and the east point. The 10th house is determined by the bisection of the Angles between the zenith and the MC. Over this grid the four Angles are inserted, and these cusps are regarded as marking the centre of influence of the houses. The system is attributed to the Italian astrologer Lavagnini.

leading planet   See LOCOMOTIVE.

leading signs   These are Aries, Cancer, Libra and Capricorn.

leaf trigon   See SIDEREAL MOON RHYTHMS.

Ledaean stars   See GEMINI CONSTELLATION.

lemniscate   A closed figure, having a resemblance to a figure 8, the term being derived from the Greek *lemniscos* (ribbon). It is relevant to astrology since certain esoteric systems (especially those of Steiner and Vreede – but see also Schultz) insist that planets do not move in elliptical orbits (as the developed COPERNICAN SYSTEM has taught), but in lemniscatory patterns.

Lemuria   A name given in esoteric circles to the long-lost culture and continent occupied by the THIRD ROOT RACE. In its occult sense, the term is not to be confused with the Roman festival held in mid-May to appease ghosts (the Latin *lemures*, meaning approximately 'shades of the departed').

Leo   The fifth sign of the zodiac, and one of the constellations. It corresponds as a zodiacal sign neither in location nor extent with the constellation of the same name – see LEO CONSTELLATION. The modern sigil for Leo ♌ is said by some to be a corruption of the initial of the Greek term for the asterism: however, the present form of the sigil appears to derive from a late medieval form which is not like the lamda, and the significance remains hidden, even though many explanations have been given by various astrologers – see SIGILS – ZODIAC. Leo is of the fire element, and of the Fixed Quality, the influence being creative, self-reliant, enthusiastic, warm-hearted and positive. As with all fire types, there is a strong element of selfishness, but the Leonine is rarely insensitive to the needs of others. The warm and creative outlook of Leo is expressed in many keywords attached to the sign by

modern astrologers: self-expressive, dignified, inspirational, exuberant, magnanimous, flamboyant (sometimes theatrical), hospitable and altruistic – in a
word all those qualities which may be associated with a fire nature expressing
itself without fear, and with a wish to be creative. In excess the Leonine
nature may be described in terms which express its underlying selfishness, the
keywords being: vain, demanding, predatory, imperious, dogmatic, self-
satisfied, ostentatious and militant. Leo is ruled by the Sun, and the sign
marks the detriment of Saturn, while some astrologers say that it marks the
fall of Mercury.

Leo constellation   Constellation linked in name, if not in influence, with the
zodiacal LEO, located as indicated in figure 38 (though it is likely that the
figure was much smaller in ancient times). Figure 38 sets out the modern

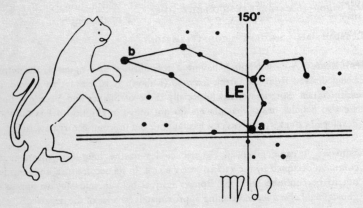

Figure 38: The image for Leo, from the 13th-century SAN MINIATO ZODIAC, *and
the constellation from the* MODERN ZODIAC *of Delporte.*

asterism according to Delporte, derived from the MODERN ZODIAC. Even
from early times (see BABYLONIAN ZODIAC) the asterism was named Lion,
and its image was that of a lion – that in figure 38 is from the medieval SAN
MINIATO ZODIAC. In classical times it was associated with the Nemean lion
of the Hercules cycle. Ptolemy describes the influence of the constellation in
terms of the smaller asterisms of which the image is composed, but it is said
to convey the Leonine virtues of nobility and courage – for specific influences,
see the related stars ADHAFERA, DENEBOLA, REGULUS and ZOSMA.

Leo Minor   Constellation, 'the lesser lion', formed by Hevelius from a number of stars between the constellations Leo and URSUS. It appears to be of a nature approximate to that of the asterism Leo, bringing generosity and nobility, combined with courage.

Lepus   Constellation located approximately 5 degrees Gemini to 3 degrees Cancer, and from 14 to 25 degrees south of the equator, 'the hare' from very early times. In some legends this hare was placed near to the hunter ORION, a nearby asterism: in Egyptian astrology the asterism appears to have been the Boat of Osiris, and thus also linked with the equivalent of Orion. Ptolemy gives it a nature equivalent to Mercury with Saturn, and the tradition insists that it gives timidity, though a quick wit.

Lesath   Fixed star, the upsilon of Scorpius, set in the tail of the Scorpion, its name from the Arabic Al Les'ah (the sting). It is said to bring danger and malevolence when emphatic in a chart, and to be of the nature of Mercury conjunct Mars.

lesser benefic   See BENEFICS.

Lesser Cloud   See NUBECULAE MAGELLANI.

Libra   The seventh sign of the zodiac, which corresponds neither in location nor extent with the constellation of the same name (see LIBRA CONSTELLA-TION). The sigil for Libra ♎ is said by some astrologers to be a vestigial drawing of a pair of scales, with which the sign and constellation has been associated from the very earliest times, the Babylonian name for the asterism being Zibanitu (scales). However, one of the more frequently used of the Egyptian hieroglyphics for the sign shows that the sigil is derived from a picture of the Sun setting over the Earth ⌂ an appropriate image for a sign which is now associated with the Descendant, that symbolic place of sunset. Libra is of the air element, and of the Cardinal Quality, its influence being harmonious, elegant, orderly, comparative, peaceful, changeable and helpful. The nature of Libra is manifest in human beings in terms of the many keywords used by modern astrologers: gentle, artistic, sensitive, helpful, peaceful, spiritualized, delicate, perceptive, affectionate – in a word all the qualities which may be associated with an air type expressing itself with a delicate awareness of others. In excess, the Libran nature may be described in terms which express an impractical nature, as well as a lazy streak, the keywords being: untidy, easily persuaded, 'lost in the clouds', unable to cope, and so on. The Libran says 'I want to be carried' more often than others do, and she may do this gently or imperiously, depending upon which side of the nature is being called into activity. Libra is ruled by the planet Venus, marks the Exaltation of Saturn and the Fall of Pluto.

Libra constellation   The constellation of Libra covers an arc of just under 21 degrees, and it would appear that at one time it was considered to be merely an extension of the adjacent Scorpius, of which it was the *chelae* (claws): figure 39 sets out the form of the asterism according to the MODERN

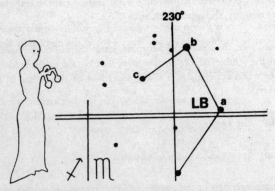

Figure 39: The image for Libra, from the 13th-century SAN MINIATO ZODIAC, *and the constellation, from the* MODERN ZODIAC *of Delporte.*

ZODIAC, however. In the occult tradition, the original significance of this constellation was intimately connected with the two signs Scorpio and Virgo, but it is clear that the asterism was defined over three thousand years ago: it was the Babylonian Zibanitu (scales), and inevitably it was later identified with Astraea, the holder of the balance which weighs the deeds of men, though of course this goddess was linked with Virgo. It is perhaps no accident that the image for Libra in the medieval zodiac (figure 39) is distinguished from that of Virgo merely by the attribute of the scales. Early names which still appear in medieval manuscripts are Zichos, and the Latin Jugum (yoke or beam). The fixed stars in the asterism which have importance in the astrological tradition are NORTH SCALE and SOUTH SCALE, originally the North Claw and the South Claw, respectively.

Life Point   The Point of Life is one of the so-called Arabian PARS. If the degree of the syzygy preceding the birth is used to denote a new Ascendant in the chart, then the position of the natal Moon in that new arrangement marks the Point of Life.

Lightning Flash   In a specialist sense, one of the modern additions to the Arabian PARS, the Part of Uranus, sometimes called the Point of Catastrophe. If a chart is so revolved that the degree occupied by the Sun is transferred to the Ascendant, then the position of the radical Uranus in this new arrangement will mark the Lightning Flash. The same term is also used (albeit in translation) of Boehme's 'schrack', linked with the operation of the Sun in his SEVEN PROPERTIES.

light of time   This is a term used to denote the Sun in the day-time (or in the day-time chart), the Moon in the night-time (or in the night-time chart).

light planets   These are the Moon, Mercury and Venus.

lights   The Sun and Moon are the lights (also the luminaries): the Sun is the greater light, the Moon the lesser light. Confusingly, the Moon and the two inferiors are also sometimes called the 'light planets'.

light trigon   See SIDEREAL MOON RHYTHMS.

Lilith   Name given to a planet generally described as one of the HYPOTHETICAL PLANETS, though it is really a 'hypothetical Moon' of the Earth, invisible to ordinary sight save (it is said) when its body passes between the Earth and the Sun. Lilith is called the Dark Moon by Goldstein-Jacobson, who has provided an ephemeris of positions, and gives her a revolution of 126 years. As the choice of name would suggest, Lilith is regarded as a sinister and malevolent planet, bringing temptations and betrayals, and ruling such matters as abortion and stillbirth. The word Lilith appears to have been derived from the Assyrian 'lilitu', a semi-human succubus whose name was introduced into Rabbinic literature as the second wife of Adam. From this ancient liaison were born devils, spirits and 'lilin'. Some occultists call Lilith the 'first wife' of Adam, however.

limiting data   See ADJUSTED CALCULATION DATE.

line of advantage   Name given to an imaginary line conceived as running between the cusps of the third decans of the 3rd and 9th houses of a chart. When the DRAGON'S HEAD is to the east of this line, it is judged more favourably than otherwise.

Linon   See PISCES CONSTELLATION.

Lion   One of the modern names for zodiacal and constellational LEO. It is also a name given to one of the HYPOTHETICAL PLANETS, listed by the astrologer Jayne as a transplutonian.

Lion's Heart   See REGULUS.

Lion's Tail   See DENEB.

little world   See MICROCOSM.

Loadstar   See POLARIS.

local Angles   A term derived from the work of Johndro, designed to establish a workable chart projection called the JOHNDRO LOCALITY CHART. The locality Midheaven is the local Midheaven of the birth place added to the longitude of the Sun at birth. The locality Ascendant is derived from the above, for the latitude of birth, in the normal way.

locality Ascendant   See LOCALITY ANGLES.

locality Midheaven   See LOCALITY ANGLES.

local mean time   See TIME.

local meridian   The meridian passing through the position of the observer on any part of the Earth's surface.

local sidereal time   See TIME.

loci   A term derived from the Hellenistic astrological *topoi* (places), relating approximately to a system of equal-house divisions of 30-degree arcs linked to the Angles rather than to the ecliptic. The historian Neugebauer records a second variant of the standard equal-house system, which locates the beginning of the loci 5 degrees before the Angles, though still in the twelvefold divisions of 30 arcs. The order of the loci follows the standard order of modern *HOUSE DIVISION. The twelve loci which are linked with a sequence beginning with the Ascendant are the equivalent of the 1st house and are said to govern the following life conditions:

lord

1 Life, spirit and 'breath' (pneuma).
2 Livelihood, property, partnerships, relations with women, business, profit from inheritance. It is called Gate of Hades.
3 Brothers, living abroad, royalty, wealth, friends (also, 'slaves') – sometimes called 'Goddess'.
4 Parents, life in the temple, reputation, children, and so on.
5 Children, friendship, accomplishment and (curiously) marriage – sometimes called 'good fortune'.
6 Slaves, illness, enmity, infirmity – sometimes called 'bad fortune'.
7 Marriage.
8 Death, trial, penalty, loss and weakness.
9 Travel, friendship, benefit from kings, manifestations of gods, soothsaying – sometimes called 'God'.
10 Careers and honours, reputation, accomplishment, children, wife.
11 Friends, hopes, gifts, children and accomplishments (also, freed persons) – sometimes called 'good daemon'.
12 Enmity, foreign country, slaves, illness, dangers, court trials, infirmity, death – sometimes called 'bad daemon'.

Neugebauer records that Vettius Valens mentions the 8th locus is counted not from the Ascendant, but from the Part of Fortune, though this seems to savour of the system of LOTS, the Hellenistic equivalent of PARS.

Locomotive   Term applied to one of the *JONES PATTERNS, in which the distribution of the planets is such that a space of at least 120 degrees in the figure is opposed by a more or less evenly distributed 240 degrees of 'occupied' space. The planet furthest east (read in a clockwise direction) is said to be the LEADING PLANET, and gives an especially powerful practical drive (in terms of the planet's house). The entire Locomotive distribution is said to give an executive ability, and at worst leads to a ruthless self-seeking.

Long Ascension   The signs of Long Ascension are those in the zodiacal arc from Cancer to Sagittarius, inclusive.

longitude   In a strictly astrological sense, the term is used of a co-ordinate of measurement of the zodiac, in degrees of arc, starting at the beginning of Aries: thus, 53 degrees of longitude is 23 degrees of Taurus.

lord   This term is used widely, and often inaccurately in modern astrology. Properly it should apply to a planet which, by domal placing, is the most powerful in a given chart. However, see LORD OF A HOUSE, LORD OF THE HOROSCOPE and LORDSHIP.

Lord of Action   An archaic term, originally appearing in English from the translations of Ptolemy. As so often with Ptolemaic astrology, the rules for determining the Lord of Action are complicated, and in fact lead in many cases to two Lords of Action being apposite in the same chart. In its simplest form, the Lord is that planet which makes its morning appearance closest to the Sun (and/or) nearest to the Midheaven, in the latter case when in (beneficial?) aspect to the Sun. The rules need not concern us here, however, for the archaic term is of value only in so far as it points to an ancient tradition by which the 'action' (that is, the profession, and professional outlook, of natives) may be determined, other than by the simplistic 'Sun-Sign' or 'Ascendant' or 'Midheaven' interpretation. The traditional rule of planets over professions is adhered to (for example, if Mercury is the Lord, then his subjects will be writers, businessmen, mathematicians, teachers, bankers, soothsayers, astrologers, and in general those who perform their functions by means of documents). It is interesting that among the Mercurial professions Ptolemy included 'sacrificers'.

lord of a house   Rather than being the ruler of a house, a planet should be deemed the lord of a house (see RULE). The term was introduced by late medieval astrologers, intent on removing confusion in terminologies. A planet may be said to be in its THRONE when in the sign it rules, whereas it is lord of the corresponding house. In fact, the lord of the house is the planetary ruler of the sign which occupies the cusp of that house. This term must be distinguished from LORD OF THE HOROSCOPE.

Lord of Karma   A term used by the esotericist Bailey of Saturn, the planet which imposes retribution, and which demands payment of all debts.

Lord of the Geniture   The ALMUTEN – but see also LORD OF THE HORO-SCOPE.

Lord of the Greeks   Name given to a scholar, 'Yavanesvara', who translated into Sanskrit a 2nd century AD Grecian (probably Alexandrian) popular handbook on astrology, of which even the title is no longer known, since the original has been lost, surviving only in this translation, and in the versification made by an Indian in the 3rd century AD. This, together with another lost translation used by Satya, has formed the root of Indian astrology. Inevitably, the basic Hellenistic socio-cosmic views were adapted to Indian social conditions and philosophy, specifically to the Indian caste systems, doctrines of reincarnation, and so on. Such was also united to the horary system of the Hindu NAKSHATRAS. The adaptation of the several Hellenistic

SORTES and other divisions (such as the OCTOPODOS) to the Hindu system has given rise to the notion that there is or was a correspondence between the Hindu system and modern HARMONICS, though the bases of division are quite different.

Lord of the Horoscope   The European equivalent of the ALMUTEN. It is maintained that any of the planets may be so placed in the radical chart as to be deemed the most powerful influence in that chart, in which case it is called the Lord or Ruler. Carter suggested that such a 'Lord' stands for the native himself, while the rest of the chart stands for his external life – but this plays no part in the traditional view. The Lord or Ruler is properly the determinant of the term often applied by astrologers to a nativity, in a simplistic manner – such as Martian, Neptunian, and so on. This mode of descriptive speech indicates that a Lord may be regarded as of about the same importance as the Ascendant, since the astrologer will speak of persons in terms of signs (that is, as an Arietan or Leonine) in reference to the Ascendant sign. Strictly speaking, in traditional astrology at least, the term 'Lord of the Horoscope' was used of a planet emphasized by its domal position, while the 'Ruler of the Horoscope' was used of a planet emphasized by its sign-position, or by special aspectal conditions. However, in recent years, the two terms have been merged, and it is now difficult to separate domal strengths from other dignities.

Lord of the Nativity   The ALMUTEN: see LORD OF THE HOROSCOPE.

Lord of the Triangle   A descriptive phrase, which might just as well be 'Lord of Triplicity', and related to an ancient system of rulerships which have disappeared from modern astrology, though it is regarded as being important by Ptolemy, and is sometimes adduced in astrological argument, as for example in connection with the vexing question of the arc rulerships of the TERMS. The Lord of the Triangle is determined as follows:

The FIRE TRIPLICITY is ruled by the Sun by day, by Jupiter at night.
The EARTH TRIPLICITY is ruled by Venus by day, by Moon at night.
The AIR TRIPLICITY is ruled by Saturn by day, by Mercury at night.
The WATER TRIPLICITY is ruled by Venus by day, by Moon at night, though this last pair are regarded as 'co-workers' with the dominant ruler Mars.

Lord of the Year   A name given to a planet which has most dignity in a SOLAR REVOLUTION CHART, or in an INGRESS FIGURE.

Lord of Triplicity   See LORD OF THE TRIANGLE.

Lordship   Properly speaking, the relationship which a planet may be said to have with its associate house is one of lordship. A planet may be said to be Lord of a particular house – even though in popular astrology a planet is often said to have rule over a particular house. Thus, Mars may be said to be Lord of the 1st house, and Ruler of Aries.

Lords of Flame   See SATURN PERIOD.

Lords of Form   The name given to one of the TWELVE CREATIVE HIER-ARCHIES, associated with zodiacal Scorpio, who according to the esoteric cosmogenesis are especially concerned with the evolution of man as a spiritual being during the present Earth period. They are in the present stage of evolution linked with the EXSUSIAI, and therefore with the Sphere of the Sun.

Lords of Harmony   See CHERUBIM.

Lords of Individuality   The name given to one of the TWELVE CREATIVE HIERARCHIES, associated with the zodiacal Libra, who according to the esoteric cosmogenesis prepared the Desire Body of man during the MOON PERIOD.

Lords of Love   See SERAPHIM.

Lords of Mind   The name given to one of the TWELVE CREATIVE HIER-ARCHIES, associated with zodiacal Sagittarius, who according to the esoteric cosmogenesis formed the equivalent of humanity during the SATURN PERIOD. They are in the present stage of evolution linked with the ARCHAI.

Lords of Will   Esoteric name for the THRONES.

Lords of Wisdom   The name given to one of the TWELVE CREATIVE HIERARCHIES, associated with zodiacal Virgo, who according to the esoteric cosmogenesis prepared the vital body (Etheric Body) of men during the SUN PERIOD.

Lot of Daimon   One of the most important of the SEVEN LOTS of Greek astrology, and formerly held as of equal importance to the PART OF FORTUNE. The Daimon of the Greeks is not the 'demon' of medieval lore, and the Lot is a beneficial one. The degree of the Daimon is the same distance from the Ascendant as the arc between the Moon and Sun (figure 40); the sigil for the

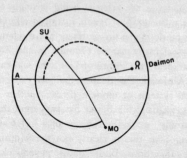

Figure 40: Construction of the Lot of Daimon (after Bouche-Leclercq).

daimon used in this diagram is that used by late Greek astrologers. From the Part of Daimon are determined the Lot of Love (EROS) and the Lot of Victory – see SORTES.

Lot of Eros   The ancient equivalent of the Part of Venus: see HERMETIC SORTES.

Lot of Necessity   An ancient Part of Mercury: see HERMETIC SORTES.

Lots   See SEVEN LOTS.

Love   For Part of Love, see EROS.

Lowell-Pluto   A name suggested for the present PLUTO shortly after it had been discovered by Lowell. At that time, however, more than one astrologer (most notably, Wemyss and Pagan) had already named two of the HYPOTHETI-CAL PLANETS 'Pluto', and the new name was proposed to avoid confusion – see, however, PAGAN-PLUTO and WEMYSS-PLUTO.

Lower Midheaven   A misleading variant used for the IMUM COELI.

Lucian astrology   A misnomer, sometimes 'Lucis astrology', for the esoteric system proposed by Bailey as her INTUITIONAL ASTROLOGY. The term is derived from the Lucis Trust, which publishes the works of Bailey.

Lucida Corona Scorpionis   Late medieval term for the fixed star CAPELLA.

Lucifer In astrology the name sometimes given to the planet Venus when it appears before the Sun (see HESPERUS) – also called the Morning Star. The same term is also sometimes used in different senses in contexts which touch upon astrology. Lucifer is the celestial being wrongly equated with Satan (probably due to a misreading of Isaiah 14:12): at all events, the name has been adopted in esoteric circles to represent the modern equivalent of the being of the Sun (originally named Ormudz in the Zoroastrian dualism) opposed by the 'darkness', the Prince of Lies, who was called Angrimayu, in modern occultism Ahriman. This dualism has been carried into several of the esoteric astrological systems of modern times. Another use of the term is linked with the ASTEROIDS, for Butler has suggested that these (which he rightly calls 'planetoids') mark the remains of a lost planet, which he calls Lucifer or Morning Star: this is clearly not the Morning Star of Venus. In support of his suggestions, Butler quotes several clairvoyants (such as Swedenborg and Davis), but the idea appears to be essentially biblical in origin. Although the name Lucifer is not used in esoteric lore in such a connection, it is held in many occult traditions that the planetoids are indeed a detritus of a lost planet.

Lucky and Unlucky Days See EGYPTIAN DAYS.

Luminaries The Sun and Moon are often called the luminaries in traditional astrology. It is quite usual for modern astrologers to explain this term by saying that the Moon is a luminary because it lights up the night, yet the fact is that the ancients believed that all planets emitted some degree of light in addition to that reflected by the Sun.

luminosity The luminosity of a star (which is different from its MAGNITUDE) is its absolute stellar magnitude – see STELLAR MAGNITUDE.

Luna The Latin name for the Moon.

lunar astrology The facility with which the ever-changing lunar disc might be studied against the night sky in Mediterranean lands meant that the earliest forms of astrology were based on material derived from lunar phenomena and measurements. Some of this lunar material survives in the LUNAR MANSIONS, and so on, but it is perhaps imaginative of the astrologer Fagan to claim that astrology was in ancient times 'essentially lunar' – calendrical systems may have been, yet there is little evidence from the surviving Greco-Roman traditions (see, for example, BABYLONIAN ZODIAC

lunar eclipse

and GREEK ZODIAC) that astrology was 'essentially lunar'. However, see LUNAR STATIONS and LUNAR ZODIAC. In modern times, particularly in the realm of ASTROPHY, there has been a renewed interest in the equivalent of lunar astrology, in connection with bio-dynamic farming and related studies – see for example, SIDEREAL MOON RHYTHMS.

lunar body A term used in contrast with the SOLAR BODY by the esoteric astrologer Thierens intent on establishing the relationships of spheres in what he regards as the future evolution of the solar system. The lunar body is made up of the ASTRAL SPHERES of Aquarius and Pisces, and the ETHERIC SPHERES of Aries to Cancer, inclusive. The solar body is made up of the spheres of the superior planets and the Spheres of Venus and Mercury. See, however, FUTURE PLANETS.

lunar cycles The Moon completes a cycle once every month. During this time it forms four powerful aspects with the Sun – the conjunction, opposition and two squares. Each of these four Angular relationships induces a monthly rhythm in the life of individuals, the nature, intensity and timing of which will depend very much upon the positions of the natal Moon and Sun. It is for this reason that the number 4 has been associated with the Moon, which in turn explains why some writers insist that there is a four-year cycle of the Moon, which is not the case: see the LUNAR NODE CYCLE and REVOLUTION PERIODS. When this quaternary is translated into years by the methods of progression, it is expressed in periodicities of seven years, totally 28 years in a full cycle. See also AGES OF MAN, SAROS CYCLE and SIDEREAL MOON RHYTHMS.

lunar eclipse When the orb of the Earth interposes itself between the light of the Sun and the Moon, the phenomenon is called a lunar eclipse. The Moon is usually itself visible, but with the appearance of a dark object against a dimly illumined backdrop of sky. Clearly, a lunar eclipse may take place only when the two luminaries are in the aspect of opposition. Experience has shown that eclipse leaves a sort of 'degree influence' which is only made operational later by transits or progressions. The ancient astrological rule held that the effect of a lunar eclipse would last the equivalent of one month for every hour of eclipse, timed from the beginning of the occultation. Eclipses (see also SOLAR ECLIPSE) in a chart are regarded as malignant if the degrees involved are conjunct the radical Sun, Moon, Ascendant, Midheaven, any malefic, or any of these progressed (to the time of the eclipse). Esoteric astrology insists that lunar eclipses mark the entry of evil forces into the Sphere of the Earth. See also ATALIA.

lunar gods   The term is usually a theosophical equivalent for the Pitris, 'the ancestors of previous human races', according to occultism. In Western terminology, however, the lunar gods are really the ANGELS.

lunar horas   See HORAS.

lunar Intelligency   Following ancient qabbalistic sources, Agrippa records the Hebraic *Schedbarschemoth Schartathan* as relating to both the 'daemon of the lunar demons' and the 'intelligency of the lunar intelligencies' (see, however, INTELLIGENCY), and accords them the number 3321, which is the multiple of the lunar number 9 and the linear sum of the MAGIC SQUARE of the Moon, which is 369.

lunar mansions   An ancient method of dividing the ecliptic belt into arcs of 12 degrees and 51 minutes, corresponding approximately to the daily mean motion of the Moon through the ecliptic. By such divisions, 28 segments, called 'lunar mansions' are obtained (the word 'mansions' appears to be derived from the Arabian 'manzils'). Each of these mansions is named, and has been subject to astrological interpretation along the lines of a LUNAR ZODIAC. Since each of the mansions is said to emit an influence, the planets located in the individual signs transmit these influences in accordance with their own natures, and thus determine the natural disposition of sublunary creatures. In modern times, interpretation of the mansions appears to be limited to a consideration of the Moon itself, however. There are now three major extant lunar systems of mansions used in astrological circles – that derived from the Arabian system, called MANZILS the Hindu system of NAKSHATRAS and the Chinese SIEU, the names for which are set out in table 48, as given by Robson: the fiducial is given for the Arabian system, there being (slight) variations between the systems. It was the Arabian system which most influenced the development of Western astrology until the Hindu system was popularized by theosophy at the beginning of this century.

lunar month   The official lunar month is 29.531 days, which marks the cycle in which the Moon returns to its original position in relation to the Sun. It is the sidereal month (that interval between two successive conjunctions of Sun and Moon) of 27.322 days which is used by the majority of astrologers in regard to calculations influencing progressions: this is the cycle reflected in the lunar periodicity of 28.

lunar node cycle   The nodes of the Moon (see DRAGON'S HEAD and DRAGON'S TAIL) which move gradually around the ecliptic in contrary

TABLE 48

Chief fiducials	Manzils	Nakshatras	Sieus
1 Eta Tauri	Al Thurayya	Krittika	Mao
2 Alpha Tauri	Al Dabaran	Rohini	Pi
3 Lambda Orionis	Al Hak'ah	Mrigasiras	Tsee
4 Gamma Geminorum	Al Han'ah	Ardra	Shen
5 Alpha Geminorum	Al Dhira	Punarvarsu	Tsing
6 44 M Cancri	Al Nathrah	Pushya	Kwei
7 Lambda Leonis	Al Tarf	Aslesha	Lieu
8 Alpha Leonis	Al Jabhah	Magha	Sing
9 Delta Leonis	Al Zubrah	Purva Phalguni	Chang
10 Beta Leonis	Al Sarfah	Uttara Phalguni	Yen
11 Beta Virginis	Al Awwa	Hasta	Tchin
12 Alpha Virginis	Al Simak	Citra	Kio
13 Iota Virginis	Al Ghafr	Svati	Kang
14 Alpha Librae	Al Jubana	Visakha	Ti
15 Beta Scorpii	Iklil al Jabhah	Anuradha	Fang
16 Alpha Scorpii	Al Kalb	Jyestha	Sin
17 Lambda Scorpii	Al Shaulah	Mula	Wei
18 Zeta Sagittarii	Al Na'am	Purva Ashadha	Ki
19 Pi Sagittarii	Al Baldah	Uttara Ashadha	Tow
20 Alpha Capricorni	Al Sa'd al Dhabih	Abhijit	Nieu
21 Mu Aquarii	Al Sa'd al Bula	Sranava	Mo
22 Beta Aquarii	Al Sa'd al Su'ud	Sravishta	Heu
23 Alpha Aquarii	Al Sa'd al Ahbiyah	Catabhishaj	Shih
24 Alpha Pegasi	Al Fargh al Mukdim	Purva Bhadra-Pada	Shih
25 Alpha Andromeda	Al Fargh al Thani	Uttara Bhadra-Pada	Peih
26 Beta Andromeda	Al Batn al Hut	Revati	Goei
27 Beta Arietis	Al Sharatain	Asvini	Leu
28 Delta Arietis	Al Butain	Bharani	Oei

direction to the movement of the planets, complete a cycle in approximately 18 years and 7 months, which is used as an important periodicity in astrology, especially in regard to the CLIMACTERICS of human life.

lunar return A name applied to a chart cast for the time at which the Moon returns to its radical position in a previous chart (most usually the natal chart of the subject). It may be measured in terms of the tropical or sidereal zodiac, though the astrologer Bradley favours the latter in his own comprehensive survey of what he calls SOLUNAR RETURNS. He insists that the lunar return chart, with its periodicity of approximately 27.3 days (the sidereal month) is basically a health chart. See also SENNIGHT CHART, SOLAR RETURN and SYNODICAL RETURNS.

lunar revolutions   See SYNODICAL RETURNS.

lunar rhythms   See SIDEREAL MOON RHYTHMS.

lunar semi-circle   Wilson defines this term as meaning 'from Aquarius to Cancer, both included'.

lunar stations   Often used as a synonym for the LUNAR MANSIONS, even though it is clear from Arabic documents that the stations were points marked by fiducials, while the mansions (or MANZILS) were constellational arcs (sometimes described as 'asterisms'). The names of the stations, as well as the related fiducials (given here in terms of the modern stars, rather than the Arabic names) are listed under LUNAR MANSIONS. The list is derived from the Arabic astrologer Al Biruni.

Alsharatan	(beta and gamma Arietis)
Albutain	(epsilon, delta and pi Arietis)
Althurayya	(Pleiades)
Aldabaran	(alpha Tauri)
Alhak'a	(lambda and phi, Orionis)
Alhan'a	(gamma and zeta Geminorum)
Aldhira	(alpha and beta Geminorum)
Anathra	(Praesepe, Aselli)
Altarf	(lambda Leonis, zeta Cancri)
Aljabha	(zeta, gamma, eta and alpha Leonis)
Alzubra	(delta and theta Leonis)
Alsarfa	(beta Leonis)
Al'awwa	(beta, eta, gamma, delta and epsilon Virginis)
Alsimak Al'a'zal	(alpha Virginis)
Alghafr	(iota, kappa and lambda Virginis)
Alzubana	(alpha and beta Librae)
Al'iklil	(beta, delta and pi Scorpii)
Alkalb	(alpha Scorpii)

Alshaula	(lambda and upsilon Scorpii)
Alna'a'im	(delta, epsilon and gamma Sagittarii)
Albalda	(this is the 'desert' without fiducial – see BALDAH)
Sa'd'Aldahbih	(alpha and beta Capricorni)
Sa'd-Bula	(mu, nu and epsilon Aquarii)
Sa'd Alsu'ud	(beta Aquarii)
Sa'd-Al'akhibiya	(gamma, zeta, pi, and eta Aquarii)
Alfargh Al'awwal	(alpha and beta Pegasi)
Alfargh Althani	(gamma Pegasi and alpha Andromedae)
Batn-Alhut	(beta Andromedae)

lunar year A lunar year is a cycle of twelve lunar months, which gives 354 days, some 11¼ days shorter than the SOLAR YEAR.

lunar zodiac It is clear from such ancient systems as the LUNAR MANSIONS that there was once a widely used system of fiducials of a constellational nature linked with the Moon. All the surviving lunar mansions have now been codified into arcs of equal extent, yet surviving traditions, and one or two interesting medieval terminologies, indicate that the equivalent of a lunar sidereal zodiac did exist in early times. In the schema set out in figure 41, derived from information furnished mainly by Nowotny in regard to medieval star lists, it will be seen that the 28 divisions are marked by fiducials presenting unequal divisions. Several of the names indicate specific stars as fiducials, while one (the Desertum of 21) relates to an empty space between fiducials. Again, certain fiducials appear to have been determined by orientation lines imagined between two stars (presumably capable of projection), thus to allow for variations in the lunar path (take for example the line between Porrima and Vindemiatrix of 13). It is possible that the sigils derived from Agrippa for a sidereal zodiac are probably ultimately from a tradition of a lunar zodiac, though neither sigils nor zodiac appear to have survived into late medieval astrology. The application of the lunar sidereal zodiac to astrology is not attested, save in the codifications of the lunar mansions (much used in HORARY ASTROLOGY). The names given in table 49 are linked with the numbers (given in the table under REF for 'reference') in the corresponding asterisms of figure 41.

lunaticus An ancient term used of a human being suffering from certain forms of insanity associated with the influence of the Moon. Paracelsus appears to have used the word alongside saturninus (i.e. 'under the influence of Saturn') for people whose strong brain activity stifled the life of the soul,

Ref.	Medieval name	Fiducial
		TABLE 49
1	Cornua arietis	Mesarthim
2	Venter arietis	Delta or epsilon of Aries
3	Caput tauri	Pleiades
4	Cor tauri	Aldebaran
5	Caput canis validi	Lambda of Orion
6	Sidus parvum lucis magnae	Alhena
7	Bracchium	Between Castor and Pollux
8	Nebulosa	Praesepe – but see NEBULOSA
9	Oculus leonis	Alterf
10	Frons leonis	Regulus
11	Capillus leonis	Zosma
12	Cauda leonis	Denebola
13	Canis	Vindemiatrix
14	Spica	Spica
15	Cooperta	Chi to phi of Virgo
16	Cornua scorpionis	Zubenelgenubi to Zubenelschamali
17	Corona super caput	Akrab
18	Cor scorpionis	Antares
19	Cauda scorpionis	Shaula
20	Trabs	Between delta Scorpius and phi Sagittarius
21	Desertum	See DESERTUM
22	Pastor et aries	Giedi
23	Glutiens	Between epsilon and gamma of Aquarius
24	Sidus fortunae	Sadalsuud
25	Papilio	Mu to gamma of Aquarius
26	Hauriens primus	Between Scheat and Markab
27	Hauriens secundus	Between Sirrah and Algenib
28	Piscis	Between Mirach & phi of Pisces

creating thereby a permanent duality, resulting in a rigidity of outlook and unpleasant psychic characteristics. The term 'lunaticus' is therefore used by Paracelsus because it is the lunar part of the human being (the soul life) which is being oppressed by the rigidity and fixity of Saturn.

lunation Term applied generally in reference to a new Moon, specifically to

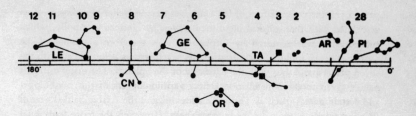

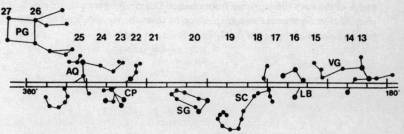

Figure 41: Lunar zodiac of unequal arcs, constructed from a medieval list of 'mansions' (constructed from several sources – nomenclature from medieval sources quoted by Nowotny).

the exact conjunction of the Moon with the Sun. It is also sometimes used as an equivalent term for the SYNODIC MONTH, which is really a lunation period.

lunation period   See SYNODIC MONTH.

lune   The portion of the surface of a sphere contained within two great semi-circles of the celestial sphere. Also, the French word for Moon.

lupus   The Latin term Lupus (Wolf) is often used to designate the ancient constellation formerly called Wild Animal – the 'Bestia' of Vitruvius, the 'Fera' of Germanicus, and so on. In the 12th century constellation figures at Sacra di San Michele, in the Val di Susa, the word 'Bestia' is used, yet in the earlier Arabic texts it appears as a Panther, Al Fahd. The asterism is set south of Libra and Scorpio, to the east of Centaur. It is often figured in association with this latter constellation image.

luxurious sign   Aries is said to be a luxurious sign because of its temperamental disposition towards luxury. The classification is misleading, however, as all the fire signs are given to luxury, Leo far more than Aries.

Lyra

Lyra   Lyra, the constellation of the Harp, set next to stellar Hercules, is often associated with that musical instrument given by Hermes to Apollo, which subsequently was acquired by Orpheus – it is therefore a symbol of the artistic-spiritual impulse in civilization, its seven strings being linked with the seven planets, and thus with the harmony of the Spheres. The many different names given to the constellation reflect various aspects of this mythology, and certain names (such as Testudo, Musculus and the Arabic Salibak) recall that some early harps were made from shells. However, the more traditional image of the harp has persisted from classical times – for example, we see the constellation represented as a harp on coins from the sacred island of Delos (the traditional birthplace of Apollo).

M

Ma'at See WEGA.

Machon One of the qabbalistic names for the FIFTH HEAVEN.

macro-anthropos See THREE PROTOTYPES.

macrocosm Term derived from a Greek compound meaning 'great arrangement', and applied to the ordered systems of the celestial realms, bodies,
spheres and the related levels of consciousness. The macrocosm is sometimes
called the Kosmos (in contrast with COSMOS) in occult circles. Any complete
entity which contains the working of all the laws within its ambient macrocosm is called a microcosm or 'little world'. Man is himself such a microcosm and is indeed said to be the archetypcal microcosm. The study of the
relationship which holds between the microcosm and macrocosm forms
the basis of astrology. Occultists divide the macrocosm into three parts.
There is the Empyrean, which corresponds in the microcosm to the intellectual part of man (usually termed *ratio*, reason), located in the head; the
Aethereum, which corresponds to the vital faculties, located in the heart, and
the Elementorum, or the natural faculties, located in the human stomach.
These correspond in the astro-alchemical systems with Salt, Mercury and
Sulphur, respectively. In the astrological cosmoconception, these correspond
sequentially to the realm of the fixed stars, the planetary spheres, and the
Sphere of Earth. Within the esoteric tradition, man is a microcosm only
while living in a physical body. Before birth, and after death, he is integrated
into the macrocosm.

macrocosmic chart A term derived in modern astrological circles to denote
the standard configuration of the schematic heavens with Aries on the
Ascendant, Taurus on the cusp of the 2nd house, and so on, without
particular reference to space or time: it is the ideal chart, the schematic chart.
Every geocentric chart – be it natal, horary, progressed or otherwise – is
really a variation, set in space and time, of the macrocosmic chart.

Magellan's Clouds See NUBECULAE MAGELLANI.

magha The 8th of the Hindu NAKSHATRAS, the mighty one. The same Sanskrit term is used for the constellation Leo – but see also SIMHA.

magi A term sometimes used as meaning astrologers – but see MAGIAN.

magian A name given to certain ancient priests connected with the worship of a fire-god, in Assyrian, Babylonian and Persian cults. Blavatsky links them with the Sanskrit term *maga*, which is used for 'priests of the Sun', mentioned in the *Vishnu Purana*. The same Sanskrit root has given us such words as 'magic', 'magician', 'image', 'imagination', and so on. Since the WISE MEN who sought out Jesus by means of the so-called STAR OF BETHLEHEM have been called Magians it is clear that they were more than mere astrologers: they were initiates.

Magian Star See STAR OF BETHLEHEM.

magical calendars See CALENDARIA MAGICA.

magical jewels See BIRTH STONES.

magical stones See BIRTH STONES.

magic square Name given to an amuletic device used in the medieval magical tradition, and linked with the series of numbers ascribed to the planets, or to the spiritual rulers of the planets. Figure 42 is the traditional

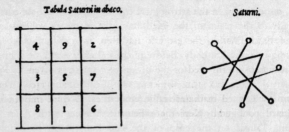

Tabula Saturni in abaco.

Saturni.

Figure 42: The magic square for Saturn, and derived sigil (after Agrippa).

magical square of Saturn: in each case, the lines add up to 15, which is the number of Saturn. A particular relevance to astrology is that the series of sigils associated with the planets (the so-called 'signacula' or 'characteres') the

Intelligencies and the Daemons of the planets, were derived from a graphic overlay of these magic squares. In the sample of Saturn (figure 42), three separate lines are traced between the nine figures in the sequence 123, 46 and 987: the triple-lined figure (sigil) which results is associated with the planet Saturn. Each of the traditional planets was accorded such a magical square, with a corresponding INTELLIGENCY and DAEMON, each ascribed traditional numbers which were derived from the linear additions of the magic squares (and multiples thereof). These numerical values are listed under the names of the relevant spirits – see therefore: (Intelligencies) AGIEL, GRAPHIEL, HAGIEL, IOPHIEL, LUNAR INTELLIGENCY, NACHIEL and TIRIEL; (Daemons) BARZABEL, HASMODAI, HISMAEL, KEDEMEL, SORATH, TAPHTHARTHARATH and ZAZEL.

magnetic Ascendant   See VERTEX.

magnetic signs   Scorpio is said to be the magnetic sign, though sometimes the negative signs in general are termed magnetic. Cornell, however, rightly associates the term with the *HORAS of the negative signs, the so-called 'magnetic' or 'magnetic-electrical' horas, which are half-signs.

magnetic workboard   Name given to a magnetized drawing surface designed to facilitate the study of the Ebertin COSMOGRAM: the workboard includes calculation discs, plastic dials, and the so-called 'graphic 45-degree ephemeris', intended to facilitate the reading of annual configurations.

magnitude   See STELLAR MAGNITUDE.

magus   A term sometimes used as meaning 'astrologer' – but see MAGIAN.

Maia   See PLEIADES.

major aspects   A term used of the group of *ASPECTS given by Ptolemy – the CONJUNCTION, SEXTILE, SQUARE, TRINE and OPPOSITION. The MINOR ASPECTS are derived mathematically from these. A third important group, more recently derived by Kepler, are the QUINTILES.

major configuration   A modern term used to denote a number of important aspectal patterns, which include the GRAND CROSS, the GRAND TRINE, the MYSTIC TRINE. the MYSTIC RECTANGLE, the Double BI-QUINTILE, the Double QUINCUNX and the T-SQUARE. These configurations are said to focus energies in a particularly powerful way.

major planets These are Mars, Jupiter and Saturn, though some authorities include only the last two, and designate the three as SUPERIOR PLANETS.

Makaram The Sanskrit name for zodiacal Capricorn (though the word actually means crocodile): see also ASHADHA. Makara is a frequent alternative in Western astrological texts.

malefic aspects Certain ASPECTS are said to be harmful or malefic to the native. Especially so are the SQUARE and SEMI-SQUARE. These are contrasted with the benefics, which contribute to the harmonious working between planets. However, it is widely recognized that the malefic aspects may well create tensions (through the planets involved) yet in so doing may present a challenge to the native, who, by efforts and wholesome exercise of will, may resolve the tensions, and so concentrate the energies as to render them creative forces. The ancient term is therefore quite reasonably regarded by many modern astrologers as misleading.

Maltwood zodiac See GLASTONBURY ZODIAC.

Manger See PRAESAEPE.

Manilius measure See MINOR MEASURE.

Manilius places See TWELVE PLACES.

mansions See LUNAR MANSIONS.

mantle See COSMOPSYCHOGRAM.

Manu See PLANETARY CHAIN LOGOS.

Manubrium A cluster of stars, often designated as the omicron of Sagittarius, set in the face of the archer. It is said to cause blindness, and is associated with fire and explosions.

manzils The Arabian term for the LUNAR MANSIONS, from which the European word 'mansions' appears to have been derived. The 28 asterisms of the manzils are recorded (in the nomenclature preserved by Robson) for each of the 12-degree 51-minute arcs, under LUNAR MANSIONS – though see also LUNAR STATIONS and LUNAR ZODIAC.

Mars

map Usually 'map of the heavens', another word for the HOROSCOPE in its modern sense. See therefore CHART.

March Named after the god Mars (the Initiator) in the GREGORIAN CALENDAR, with 31 days. Following ancient patterns, the year began in England on 25 March (the Feast of the Annunciation) well into the 18th century – but see GREGORIAN CALENDAR. Esoterically, Mars and March are linked with beginnings, and with the descent of the soul into matter. It might therefore be said that March pierces into the opening of APRIL, that the *Spiritus Mundi* might delve more deeply into the *maya* of MAY.

Marduk One of the Chaldean names for the planet Jupiter.

Markab Fixed star, the alpha of PEGASUS, now in the wing of the flying horse, though the Arabian name originally meant 'saddle'. Ptolemy ascribes to it the nature of Mercury conjunct Mars, and it is said to portend danger from cuts, stabs and fevers: it is one of those many Arabian stars which threaten a violent death.

Markeb Fixed star, the kappa of ARGO NAVIS, set in the buckler of the ship, and of the same Arabian derivation as MARKAB, with which it is often confused. It is said to bring profitable journeys, and through its traditional link with Jupiter and Saturn it brings breadth of knowledge and educational work.

Marriage Point The Point of Marriage is the name given to one of the so-called Arabian PARS. If the degree of Venus in a natal chart is taken to the Ascendant, then the cusp of the 7th House is said to mark the Point.

Mars The planet in orbit between the Sun and Jupiter, the planet which more clearly resembles the Earth in regard to physical factors than any of the other planets – though it has a diameter just over half that of the Earth. Mars has two satellites, Phobos and Deimos. In modern astrology Mars is the Ruler of zodiacal Aries (but see also NEGATIVE MARS), with its detriment in Libra. Some astrologers say that it is Exalted in Capricorn, with its corresponding Fall in Cancer. Mars represents the physical side of the native's life, and is an index of his energy, endurance, and so on – of his ability to carry projects through, and of the active side of his temperament. Mars is directly linked with the sex drive, in which it co-operates with the planet Venus, the former being active, the latter passive. When Mars is well emphasized in a chart, the native is courageous, enterprising, confident, active and proud – in particular the planet bestows great energy, with an ability to construct or destroy, and consequent on this latter urge there is often a general lack of refinement on the manifestation of the sign in which it is posited. A badly placed Mars, or one badly aspected, tends towards disruption and explosions: the native is rough, reckless, destructive and self-centred. The sigil for Mars ♂ is said to represent the cross of matter (the fourfold elemental life) weighing down the circle of spirit. It is in effect the reverse of Venus in its graphic origin. Traditionally Mars has dominion over the private parts in the human being, and, as it is essentially the male planet, it has particular rule over the penis – though its action is also clearly connected with the adrenals (see PLANETARY MELOTHESIA). The planet is a useful index of how quickly the physical energies may be called into play as a result of disturbances on the physical or astral plane. The Martian type usually finds expression on the physical plane through activities demanding the exertion of physical energy, and especially in connection with activities which give rapid results. All forms of physical and manual labour fall under the rule of Mars, so that at one end of the spectrum there is building and construction work, at the other end, surgery, driving and athletics. All work involving the use of iron and steel, sharp tools and weapons of war are connected with Mars, so that military establishments, slaughterhouses, waste-disposal units, mortuaries, and so on are under its dominion. The colours of Mars are the reds, especially the more 'violent' scarlets and carmines.

masculine See SECTA.

masculine degrees A term applied to certain zodiacal degrees. It is said that should such a degree mark the Ascendant of a female chart, then she will have a masculine appearance. The list furnished by Cornell is:

AR	8	15	30		LE	5	15	30		SG	2	12	30
TA	11	21	30		VG	12	30			CP	11	30	
GE	16	26			LB	5	20	30		AQ	5	21	27
CN	2	10	23	30	SC	4	17	30		PI	10	23	30

masculine planets With the exception of the LUMINARIES, each of the planets is regarded in traditional astrology as being possessed of masculine/feminine or positive/negative polarities, which were probably derived from the PLANETARY RULERSHIP described by Ptolemy. For relevant information, see also FEMININE PLANETS. Quite a different tradition, connected with the old tradition of PRINCIPLES, has emerged from the Ptolemaic tradition. This maintains that certain planets are (by their own intrinsic nature) masculine:

these are the Sun and the three superior planets Mars, Jupiter and Saturn. Ptolemy is reporting rules which were already ancient in his day when he says that these planets are masculine, that Mercury is 'common to both genders', while the Moon and Venus are feminine. He mixes (even confuses) a further strain of planetary sexuality from a different stream of astrology, however, when he lists a number of rules relating to variable sexuality. Matutine planets are masculine, while vespertine planets are feminine; oriental planets are masculine, occidental planets feminine. Cardan insists that any planet within the six-sign arc of the Sun (following the zodiacal order) are feminine, while all the rest are masculine. Wilson, after noting how complex the theory of planetary sexuality had become by his day, observes: 'Modern astrologers, too, with their usual sagacity, have rendered the confusion more confused.' In modern astrology, which is fortunately something different from that known to Wilson, the sexuality of planets has retained its complexity and importance mainly in regard to HORARY ASTROLOGY – otherwise, it is generally recognized as reasonable to treat planetary genders in terms of the gods and goddesses associated with the planets and spheres in classical terms – an exception might well be Neptune, which is really a feminine influence.

masculine signs   In the traditional schema, the signs in alternate order, starting with Aries, and ending with Aquarius, are masculine. See SECTA.

Mater   See ASTROLABE.

maternal trinity   A term used by Butler in his SOLAR BIOLOGY in respect of the zodiacal signs Cancer, Leo and Virgo; the ancient 'Summer Signs'.

mathematicians   Sometimes 'mathematicals', a term used as synonymous with astrologers in early treatises. Without doubt the word is derived from the early association between mathematics and astrological practices: the QUADRIPARTIUM of Ptolemy was in Greek the *Mathematices Tetrabiblon*, the 'Four Books of Mathematics'. By the 18th century, however, the term was often used in a pejorative sense.

Matheseos   The term is usually a reference to the books on astrological lore by Julius Firmicus Maternus, an astrologer who lived in the early part of the 4th century AD. The title of a good edition is given under Firmicus in the Bibliography, though the short title is *Matheseos libri VIII*. The text, often described as relating to Roman astrology, is actually based on a confusing series of astrological streams, partly dependent on DOROTHEAN ASTROLOGY, upon the hermetic literature associated with such sources as the so-called NECHEPSO, and upon Neoplatonic sources.

Mathesis Universalis A term used by the historian Haase in connection with his study of KEPLERIAN HARMONICS, denoting an attitude of mind expressed in the hope that 'with the help of mathematics everything could be grasped and explained'. Mathesis Universalis is concerned with a clear scientific picture or proof, and is literally materialistic in that it is quantitative, and while allowing us to 'grasp only one aspect of the world' (as Haase puts it) 'gives the erroneous impression that this picture is complete.' Haase comes to the conclusion that the great Kepler uses the methods of Mathesis Universalis to describe what the contrasting PANSOPHIA means.

mature quadrant See QUADRANTS.

matutine Literally pertaining to the morning, and used sometimes of planets which are above the horizon with the rising of the Sun, at other times of planets which arise with the Sun.

matutine culmination Term used by Ptolemy for one of his VISIBLE ASPECTS.

matutine setting Term used by Ptolemy for one of his VISIBLE ASPECTS.

maxima A term for the Grand Climactic of 800 years: see CHRONO-CRATORS.

maximum altitude See CULMINATION.

May The fifth month of the GREGORIAN CALENDAR, with a length of 31 days, named, as it is often claimed, after Maia, the daughter of Atlas and Pleione, and hence one of the sisters of the PLEIADES, which constellation rose in May. Maia was identified with her namesake Rhea-Cyubele, the Maia-Majestas, the fruitful mother, and is therefore linked with the fecundity of the Earth in this month – but, for an esoteric sequence, see MARCH. Esoterically one cannot ignore the ancient Sanskrit root *maya*, popularly translated as meaning 'illusion', though ultimately connected with the image-making faculty of humanity (*maya*, 'magic', 'image' and 'imagination' are semantically connected). It is probably this connection which explains the popular imagery of the Roman calendar (see, for example, Foucher, who deals with the Thrysdus calendrical mosaic, itself indicative of esoteric astrological nuances) which singles out the creative god Mercury, patron of the arts, to whom the 15th day of the month was sacred. For the early Roman calendar, in which May was the third month, see JULIAN CALENDAR.

Mayan astrology Virtually nothing is known about Mayan astrology, although, as Volguine points out, many of the pre-Columbian buildings were astrological in design – for example, the famous seven-terraced pyramid of Chichen-Itza, orientated to the cardinal points, with the step symbolism of 365, corresponds to the year. What we do know of the Mayan system suggests an extraordinary grasp of cycles, and numerical symbolism. Among the cycles was the 13-day periodicity, said to form a sign: 20 such signs gave a cycle of 260 days, called a Tonalamatl. There is also a 20-day periodicity, designated in the calendar by symbols from the natural world – animals, reptiles, wind, rain, etc. (see table below). Eighteen of these periods (to which were added five or six 'vacant' days, 'days without a name' called Xma-Kaba-Kin) formed a solar year. There was a cycle of 20 solar years, or 7,200 days: see also KATUNE. Volguine, apparently basing his information on Gentry, insists that the system of secondary directions (on a DAY FOR A YEAR symbolism) was practised in pre-Columbian America; however, the evidence is more than dubious. There is recorded a Venusian year – the synodic period of Venus, the mean duration of her four phases – of 584 days (several names were given to this planet, including Noh-Eh (big star), Chac Eh (red star), Zaztal Eh (brilliant star), and so on. The Mayan manuscript in Dresden records five such Venus revolutions. Linked with the Venusian years is a cycle of eight solar years (five Venusians). Another cycle is that of 13 Venusian years. Another cycle, of 18 solar years, is linked with symbols significantly close to those used for similar cycles in China. In the year-counts were several long periodicities, including 394 years, 600 years and 7,616 years. Acosta gives the following translations for the names of the zodiac:

AR	Splendour of the Lamb	TA	Powerful, Brilliant, Inflamed Male
GE	United Stars		
LE	Return of the Lance of the Lion	CN	Sleeping Snake
		VG	Divine Mother
LB	Ladder	SC	?
SG	?	CP	Ardent Goat or Horned One
AQ	Epoch of the Waters	PI	?

The Mexican Tonalamatl, the Book of Days (a sequence of 260 days), in conjunction with the solar year, is the foundation for the calendar. The Tonalamatl is reckoned in terms of 20 distinct periods, each with symbols and hieroglyphics. The names of the days are recorded by Hastings in table 50 (for the Mayan) and Volguine, supposedly for the Mexican, though there is a strong Toltec element in the list.

————————————————— TABLE 50 —————————————————

Mayan day	Hastings	Volguine	20-day Cycle	
1 Imix	Female Breast (?)	Crocodile	Pop	Mat
2 Ik	Wind	Jaguar	Uo	Frog
3 Akbal	Night	Star	Zip	Error
4 Kan	Copious (?)	Flower	Zo'tz	Bat
5 Chicchan	Biting Snake	Reed	Tzec	Scorpion
6 Cimi	Death	Death	Xul	End
7 Manik	That which hurries	Rain	Yaxkin	Renewal
8 Lamat	?	Liana Creeper	Mol	Assemble
9 Muluc	That heaped	Serpent	Ch'en	Alone
10 Oc	?	Flint	Yax	Fresh
11 Chuen	Monkey	Monkey	Zac	White
12 Eb	Row of teeth	Lizard	Ceh	Slave
13 Been	Worn out	Movement	Mac	Obstruct
14 Ix	?	Dog	Mankin	Sun (?)
15 Men	Maker	House	Muan	Bird
16 Cib	Perfumery	Vulture	Pax	Drum
17 Caban	That exuded (?)	Water	Kayab	Song
18 E'tznab	Hard (?)	Wind	Cumku	Explosion
19 Cauac	Storm	Eagle		
20 Ahau	King, Sun	Rabbit		

---

mean motion   The averaged motion of a body during an orbit, based on the accepted fiction that the planet will move at a uniform rate around the Sun – the truth being rather that the motion varies in relation to the ratio of its distance from the centre. The mean daily motions, of value mainly in respect to rough-and-ready progressional work, are set out in table 51.

mean precession   A cycle of 25,694.8 years – see, however, PRECESSION.

mean solar day   The standard measurement of the so-called 'clock time', derived from the average of the annual series of the TRUE SOLAR DAY sequences – see TIME.

Planet	Deg.	Min.	Sec.	Planet	Deg.	Min.	Sec.	Planet	Deg.	Min.	Sec.
MO	13	10	35	ME	4	6	0	VE	1	36	0
MA	0	31	0	JU	0	4	59	SA	0	2	1
NE	0	0	24	UR	0	0	42	PL	0	0	14

mean solar time   See TIME.

mean time   A term applied to the method of converting the theoretical MEAN MOTION of the Earth into intervals of equal and regular duration – see TIME.

measure of Naibod   An arc used in connection with SYMBOLIC DIRECTIONS, based on the mean daily solar increment arc, which is given as 59 degrees and 8 minutes, named after the German astrologer Valentine Naibod, one of the greatest 16th-century authorities on Arabic astrology.

measure of time   A technical term relating to the system of measuring the time of an event by conversion to arc of direction applied to particular charts. Several different systems are used: the astrologer Naibod suggests that the arc of Right Ascension should be converted into 1 year, 5 days and 8 hours for each degree. The Ptolemaic DAY FOR A YEAR system is a measure of time in which one degree is symbolically equated with a year of time: this progressional conversion is widely used in modern times. The method recommended by Placidus de Tito involves adding the Right Ascension of the natal Sun to the arc of direction, thus converting to Right Ascension, and allowing the standard progressional technique of a Day for a Year.

media   A term used for the 200-year cycle of the CHRONOCRATORS.

medical astrology   The branch of astrology which deals with the study and application of particular rules of casting and interpreting in regard to questions of health. In the past it was common for astrologers and doctors to combine natal astrology and HORARY ASTROLOGY towards this end, though there is a whole corpus of tradition attached specifically to the interpretation of purely medical considerations of a chart. The modern realm is partly represented by the Viennese Feerhov, who worked especially with astro-medical concepts linked with the alchemical allocation of human types, and the American astrologer Cornell, who, having found the astrological tradition both sparse and wanting, made a thorough record of astrological cases, which he subsequently published in encyclopaedic form. Much work has been done by Carey and Perry in relating the twelve biochemic remedies, the so-called 'Salts of Salvation' to the zodiacal signs and their physio-chemical allocations: see TWELVE CELL-SALTS. Most of the major modern astrological systems have their subdivisions relating to medical astrology, ranging from the fixed-diagnosis system of Wemyss to the (unsystematized) link between astrology and homeopathy of Steiner.

Medium Coeli A term from the Latin meaning 'middle of the skies', the equivalent term for Midheaven, sometimes confused with the 10th house cusp.

Medusa's Head See ALGOL.

Meenam The Sanskrit name for zodiacal Pisces – but see BHADZAPADA.

Megrez One of the seven fixed stars forming the constellation of URSA MAJOR, said by Ptolemy to be of the influence equivalent to that of Mars.

Meier figurines See FIGURINES.

melancholic One of the four TEMPERAMENTS, derived from an excess of Earth element in the psychological make-up of the personality. The melancholic temperament is conservative, withdrawn and quite stable, and is associated with the following keywords: practical, reliable, delighting in the physical and in the rhythmic, possessive, conservative, reserved, utilitarian, steadfast and hostile to novelty. Faults in the temperament (extreme melancholia) arise from a lack of fluidity, and excessive concern with materiality and self – under these conditions the type becomes commonplace, egotistical, narrow, unadventurous and miserly. See also HUMOURS.

melancholic quadrant See QUADRANTS.

melancholic signs Term used of the EARTH TRIPLICITY.

Melodia Name given to one of the ETHEREAL PLANETS, perceived clairvoyantly by the astrologer Harris, and said to have a circumference of 50,000 miles. It is very likely that this Melodia is the same as the ov described by

melothesic man

Charubel, which has a solar revolution of 297 years. Melodia was said to be in 18 degrees Aquarius in November 1905, which should locate it in the 1980s in the last decan of Taurus. The historical importance of Melodia from an astrological point of view is that it has been claimed to be the planet which rules the destiny of the United States of America. It has been accorded rule over Taurus (of which it is the day ruler) and Sagittarius (of which it is the night ruler), and is said to have a beneficial nature.

melothesic man   Term derived from the Ptolemaic astrological system for the zodiacal man, that image of man linked with the twelve zodiacal signs. The earliest melothesic descriptions go back to the first records of astrology, but the earliest melothesic image did not emerge in Europe until about the 12th century. Ptolemy was merely reporting an old tradition when he linked the planets and the signs to the human frame (see PLANETARY MELOTHESIA), and it is likely that West and Toonder are not being too imaginative in showing an ancient melothesia in early Egyptian architectural zodiacal images. The traditional associations visualize zodiacal man as a straightened circle, with the zodiacal sequence running from head to foot, as in figure 43. The

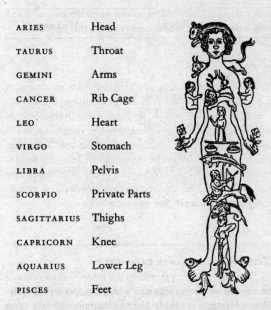

ARIES	Head
TAURUS	Throat
GEMINI	Arms
CANCER	Rib Cage
LEO	Heart
VIRGO	Stomach
LIBRA	Pelvis
SCORPIO	Private Parts
SAGITTARIUS	Thighs
CAPRICORN	Knee
AQUARIUS	Lower Leg
PISCES	Feet

Figure 43: Melothesic man (from a 15th-century Shepherd's Calendar).

traditional sequence gives AR over the head (mentation); TA over the throat (larynx and speech); GE over the 'dualities' of arms and shoulders (externally) and the lungs (internally), and the process of breathing; CN over the breasts and rib-cage (the 'protection' of the heart); LE over the heart itself, seen as the centre of the emotional life; VG over the stomach and womb (processes of elimination); LB over the small of the back and the kidneys; SC over the sexual organs (the generative and regenerative organs, as they are termed esoterically); SG over the hips and thighs, the liver and the entire hepatic system; CP over the skeletal frame, and the knee in particular, as well as the skin (which holds the frame in place, so to speak); AQ over the shin and ankle, and (internally) over the circulatory system; PI over the feet and the lymphatic system. There are many modern adaptations of these melothesic associations, and the bodily parts and functions have been extended to include rulership over the endocrine glands by Collin, as well as the so-called tissue salts (see TWELVE CELL-SALTS). Steiner offers a melothesic image which links man with the esoteric view of the tripartite union of Spirit, Soul and Body, in a series of septenaries. The Spirit of man (the 'Thinking' of table 52), discriminating ego, includes the head and

		TABLE 52	
Sign	**Thinking**	**Feeling**	**Willing**
AR	Upright position		
TA	Direction forwards		
GE	Symmetry	Head and feet	
CN		Breast enclosure	
LE		Interior – heart	
VG		Second interior	Kidneys, solar plexus
LB		Balance	Balance
SC		Reproduction	Reproduction
SG	Upper arm	Thigh	Thigh
CP	Elbow		Knee
AQ	Lower arm		Leg
PI	Hands		Feet

arms; the Soul life (the 'Feeling' of table 52) reflects the emotional working of man (what Steiner elsewhere calls the 'middle man'), and includes a recapitulation of the head down to the thighs; the Body life, (the 'Willing' of table 52) is expressed in a septenary which rules the lower limbs, the moving parts of man, centred on the power-centre of the will life, which is (esoterically) linked with the sexual centre. The threefold schema of septenaries is set out in table 52. This twelvefold system, when linked with the triadic image of man as a thinking, feeling and willing being, is an esoteric extension of the traditional melothesia. See also BIOLOGICAL CORRESPONDENCES and PLANETARY MELOTHESIA.

Menkalina Fixed star, the beta of AURIGA, set in the right shoulder of the figure (the Arabian term Al Mankib means 'the shoulder'). It is said to bring disgrace and danger of a violent death.

Menkar Fixed star, the alpha of CETUS, set in the jaws of the monster (the Arabic Al Minhar means, however, 'the nose'). Ptolemy says that it is of the nature of Saturn, and it is supposed to bring disease and ruin.

Menkib See SCHEAT.

Mensa Constellation formed by La Caille in 1752, originally as Mons Mensae (Table Mountain), after the African namesake – though later called simply 'Table'. It is said to give an ambitious nature – early difficulties in life are overcome with the passage of time.

Mercurius A name given by Thierens to a future development of his own HERMES as Mercury-Vulcan (see, however, VULCAN PERIOD). It is said to be a principle of transmission. The traditional Mercury is sometimes called Mercurius in late medieval texts, especially in astro-alchemical contexts, in which planetary virtues and metals were rarely distinguished.

Mercury The planet in orbit nearest to the Sun – though in the traditional astrology founded on the pre-Copernican models (see, for example, PTOLE-MAIC SYSTEM) it was thought to be in orbit between the Moon and Venus, and is often so depicted in diagrams of the spheres. It is the smallest of the planets, with a diameter of less than half that of the Earth. It has rule over Gemini and Virgo, with corresponding Detriments in Sagittarius and Pisces. Some astrologers say that it is Exalted in Aquarius, with its Fall in Leo. Mercury represents the ability of the native to communicate with others: in traditional astrology, it ruled human speech. Mercury is also an index of mentality, though it is concerned with the details, with short-term reactions, with memory and with day-to-day problems, rather than with profound metaphysical thought, which is really the domain of Jupiter. When Mercury is emphasized in a chart, it is usually an indication of a quickness of mind and expression, as well as of an alert attention: the subject will be fluent in thought and talk, though probably also of a changeable disposition. A badly placed Mercury, or one badly aspected, tends towards exaggeration of speech – in lying, sarcasm or excessive volubility. The sigil for Mercury ☿ has been interpreted in many different ways, but most astrologers see in it a union of the three forms of crescent, circle and cross, as representative of the ability of Mercury to bring things together and to unify: this is the exoteric level of looking at the esoteric tradition connected with the planet in astro-alchemical circles – see ALCHEMICAL MERCURY. Traditionally, Mercury rules the hands and arms of the human body, as well as the lungs: it is linked also by some modern astrologers with the thyroid gland – but see PLANETARY MELO-THESIA. Mercury is a useful index of how the nervous energies will flow and manifest, and it is also probable that the planet has rule over perception. The Mercurial type usually finds expression on the physical plane in work requiring rapid communications, writing, or as a middle-man, salesman, and so on. He likes to exercise his wits, or even craftiness, and enjoys the manipulating of slower types, even though they often infuriate him: he often has an imitative faculty. There is usually a wish to use verbal expressions, through speech, or through literature – for this reason the type is frequently involved with journalism, publishing, travel agencies, teaching, and so on. He is at his best when dealing with other people in a manipulative or helpful way – especially in situations where financial transactions take place. Mercury is associated with mixtures of colours – with tartans, polka dots, harlequins, and so on: the true Mercurian will wear a coat of many colours, and is not especially tasteful in his choice of colours, even though he may dress fashionably. The underlying urge of Mercury is to take, rather than to give, and so the planet often assumes (chameleon-like) the influences of those planets with which he is most clearly and strongly related in a particular chart – so, incidentally, it is with the Mercurian personality, which is sensitive to environment, and will tend to pick up and imitate 'influences' from those around.

meridian The meridian circle is that which passes through the zenith, the celestial poles, and the north and south points of the horizon.

meridian time See TIME.

meridianal culmination Term used by Ptolemy for one of his VISIBLE ASPECTS.

meridianal quadrant See QUADRANTS.

meridianal setting   Term used by Ptolemy for one of his VISIBLE ASPECTS.

meridianal subsolar   Term used by Ptolemy for one of his VISIBLE ASPECTS.

meridian distance   A measurement of distance between any point on the celestial equator and the nadir.

meridional   Sometimes used as an equivalent term for southern.

merope   See PLEIADES.

Mesarthim   Fixed star (double), the gamma of Aries, and sometimes called 'the first star of Aries' (and hence the 'first star' in the constellational zodiac). The origin of the name appears to have been lost. See SHARATAN.

Mesham   The Sanskrit name for zodiacal Aries – but see also ACVINI.

meteors   A meteor, or falling star, was in popular lore regarded as a sign of death, though according to Neoplatonic lore, it was a sign of conception (see, however, PERSONAL STAR). Eusebius of Alexandria records both beliefs. However, the fact that millions of 'shooting stars' fall to earth each day (30 million a day, so it is said, and these only such as may be seen by the human eye) means that such phenomena cannot be usefully integrated into chart interpretation. In spite of this, attempts have been made to relate the periodic meteor swarms, such as the Aquarids (end of July, in Aquarius) and the Taurids (20 November, from near Aldebaran), with genethliacal astrology. The fact that such swarms appear to radiate from a single point in the heavens is actually a sign that the Earth is passing over the point at which its own orbit intersects that of the meteors (which are like the Earth itself in orbit around the Sun), yet this origin-point has been used by some astrologers as the basis for interpretation.

meteorological astrology   See ASTRO-METEOROLOGY.

Metis   Name given to one of the asteroids, discovered in 1848.

Metonic cycle   A cycle of 19 years, named after the Athenian astronomer Meton who first formulated its true nature. The cycle marks the return of a cycle of solar eclipses to a specific degree of the zodiac. Distinguish from the SAROS CYCLE, and see also GOLDEN NUMBER.

Metonic return   Term sometimes used of a completion of the METONIC CYCLE.

Michael   The ARCHANGEL linked with the rule of the Sphere of the Sun, and said by some to be the leader of the Archangels: Trithemius makes him the leader of the SECUNDATION BEINGS. The name Michael is properly pronounced 'Mikhaiel'. In the esoteric Christian tradition, Michael is the guardian of the newly dead, on whose behalf he will fight the demonic hordes. He is linked in the astrological lore with rule over the element of fire, and is often depicted with the attributes of a golden sword and a pair of scales (derived from Egyptian symbolism, for the weighing of souls). See also FOURTH HEAVEN. For traditional symbolism, see SIGILS – PLANETARY SPIRITS.

micro-aspects   A modern term for aspects within an extreme angle of 7.5 degrees. Within the traditional view the term has no meaning, since aspects were supposed to express elemental relationships – see ELEMENTAL ASPECTS.

microcosm   Term from the Greek meaning 'little world' or 'small arrangement', and usually applied in astrological contexts to man, who is the small arrangement of the cosmos – see MACROCOSM. Technically, any complete entity which contains within it a summary of all the working of the laws within the macrocosm is a microcosm. In terms of the occult theory of cosmogenesis, Man is defined as the vehicle which spans the gulf between the highest and lowest manifestations of a particular world system. By such definition, each man is a model of the entire universe, and contains within him (either in developed form, or in potential) each and every manifestation to which the universe may be subject. For example, in the triadic world of the present age, the solar force represents the creative energies of the highest order, and these find a link with the brain and heart of man. The lunar and earthly forces, representing the lowest energies working from beneath man (sometimes indeed termed the demonic) find a centre in the sexual sphere – thus man is 'inhabited' by the (solar) angels and by the (telluric) demons, and is the unifier of these two different beings, through which he learns to exercise consciousness and love. This image is preserved in a vast number of ancient images depicting universal or microcosmic man, from the theological image of the Crucified Christ between the Sun and Moon, to the occult images popularized by Fludd: the solar image always represents the angelic nature, and the lunar image the demonic nature. A glance at the microcosmic man described by Fludd will indicate something of the triadic nature of the micro-macrocosm. Man is pictured within the ambient of the macrocosm, as a threefold being inhabiting a physical body. The lowest part, the Regio

Elementaris, is quite mortal, a shade which returns after physical death to the realm of darkness – it is of the matter of Earth, of the four elements. This is the realm charged with the faculty of perceiving colours, dimensions and materiality – the word 'matter' and the Sanskrit *maya* (illusion) are cognate. The highest spirit, the Regio Intellectus, is linked with the stars, and presents the clarity of selfhood: it governs the intellectual life of man, being objective in its view, capable of developed wisdom and justice. Connecting these two is the 'spirit of life', the Orbis Solis seu Cordis (orb of the Sun or of the Heart). The function of this third part is to remain within the true divine light within man, and to resist error: it is linked with the emotional life, detects spiritual similitudes and likenesses, and echoes the realm of the planets. No great difficulty would be required to link this triadic model with the triadic astro-alchemical model of the Paracelsian Crude Ilech (which is really another name for the microcosm). The Crude Ilech consists of Salt (the thinking part, which is linked in the alchemical model with the lunar forces), Sulphur (the will part of man, linked in the alchemical model with the Mars forces, and with sexuality), and the mediator, Mercury (the emotional part, in the seat of the heart). This triadic relationship between the microcosm and the ambient macrocosm is sometimes dispensed with in occult and astrological literature, and a septenary relationship (directed to the planetary forms) is established as an alternative model of man. It is maintained in occult circles that any valid spiritual model of the microcosm (and hence of the macrocosm) must be at once a triad and a septenary. The theory of the nature of the microcosm, in its specific relationship to the macrocosm, actually constitutes the bulk of modern occult studies, as, for example, developed by such esotericists as Blavatsky, Steiner and Bailey.

microcosmic man See PLANETARY MELOTHESIA.

Microscopium Small constellation formed by La Caille in 1752, north of Capricornus. It is said to give a methodical and meticulous nature.

Midas Name given to one of the HYPOTHETICAL PLANETS, said to be in transplutonian orbit when listed by Jayne.

middleground A specialist term used of planets which are near the cusp of the succedent houses, and, by virtue of this placing, less effective than they would be when on the Angles themselves. Such planets are said to be in the middleground of the chart. See also BACKGROUND and FOREGROUND.

Midheaven The Midheaven is a term properly applied only to the culminating degree of the ecliptic. Sometimes a symbolic stereographic projection of HOUSE DIVISION results in the cusp of the 10th house not corresponding with this culminating degree (as, for example, in the MODUS EQUALIS), so the 10th house cusp should not be called the Midheaven, as it so often is in popular astrology. The method for correctly calculating by trigonometry the relationships of the Midheaven to the Ascendant (Horoscope) was known to Ptolemy, though the method was hampered by the prevalent theory of the CLIMATA. The Midheaven is often called the Medium Coeli (Latin for 'middle of the heavens'), frequently abbreviated to MC.

Midheaven axis Term used of the vertical axis which in the horoscope figure corresponds to the axis of the Midheaven and the Imum Coeli.

midnight mark A term denoting the equivalent time of midnight (Greenwich Time) in regard to the place for which the chart is calculated.

mid-points A term applied to modern astrological practice by the German astrologer Ebertin (see COSMOBIOLOGY), relating to the supposed significance of the half-sum of two other (usually planetary) factors. The mid-point is regarded as susceptible to influences by transit or progression by planets and other nodal points – see COSMOGRAM. Ebertin traces the use of half-sums (as he also terms the mid-points) to a method of rectification used by the 13th-century astrologer Bonatus, but there appears to be some misunderstanding of what Bonatus had in mind. Mid-points were used in Europe prior to the work of such astrologers as Ebertin and Koch, even though modern astrologers often speak of the 'Ebertin mid-point system'.

Mights See DYNAMIS.

Milky Way See VIA LACTEA.

Mimas Name given to one of the moons of Saturn.

Minam One of the Sanskrit terms for zodiacal Pisces, as used in Hindu astrology. Sometimes the term is given in the form Minas.

Minimis One of the terms used to denote the 20-year CHRONOCRATORS cycle.

minimum altitude See CULMINATION.

minor aspects See MAJOR ASPECTS.

minor measure The minor measure of Manilius is a system of directing the horoscope figure on the basis of 72 arcs, each house representing a period of six years, each quadrant 18 years. The number 72 was regarded as being of profound esoteric significance (see, for example, TETRACTYS), for it linked man (through the pulse beat) with the zodiac (through the precessional rate, which retrogrades the Vernal Point 1 degree in 72 years). However, the fact is that the astrologer Manilius (contemporaneous with Christ) did not recognize such a rate of precession (then believed to be about 1 degree every century), so that the numerical symbolism is directly linked with the magical theory of early gnostic speculation. The system is really a device for projecting a time-grid (one-sixth of a house being the equivalent of a year – a sort of 'face for a year' method) to measure the supposed events in terms of time, the natures of the events being revealed by progressed aspects to the radicals. So far as the dating is concerned, the system is entirely symbolic, and deviates from standard progressional techniques.

Minos Name given to one of the HYPOTHETICAL PLANETS, recorded by the astrologer Jayne (in 1962) as in Aquarius, and claimed to correspond to a planet sighted in 1850.

Minotaur See CENTAURUS. Sometimes the SAGITTARIUS CONSTELLATION is called Minotaurus.

Mintaka Fixed star (double), the delta of ORION, set in the belt of the giant (the name from the Arabic Al Mintakah for belt) the first powerful star visible in the rising portion of this asterism. Depending upon the planet or nodal point which touches off its influence, the star is said to bring good fortune, developing the better side of the planet concerned.

minutary horoscope A term applied to a horoscope symbolism linked with the solar radix method of progression, based on the symbolic representation of the sidereal measurement of the Earth's rotation.

Mira Fixed star (variable), the omicron of CETUS, sometimes called the Stella Mira (wonderful star), and in the 16th century the first recorded example of a variable. It is the Collum Ceti (neck of the whale) which Ebertin wrongly translated as 'tail of the whale'. He claims that it is of the nature of Saturn and Jupiter, yet at the same time is far from beneficial, for under unhelpful connections it brings failures, enmity and even suicide.

Mirach   Fixed star, the beta of ANDROMEDA, set in the girdle of this figure, the name probably from the Arabian Mi'zar (girdle), which has given very many variants, such as Mirac, Mirax and (confusingly) Mizar; sometimes it is called the Zona Andromedae or Cingulum in reference to the belt of Andromeda. It is said to give great physical beauty, a brilliant mind, and good fortune in marriage: it may indeed be regarded as another Venus in the skies when emphasized in a chart, though it is particularly subject to disturbances from malefics.

Mirak   One of the seven fixed stars forming URSA MAJOR, which Ptolemy equates with the influence of Mars – distinguish from the 'Mirac' of MIRACH.

mirror of heaven   A poetic (and indeed misleading) synonym for CHART.

Mithraic Bull   See TAURUS CONSTELLATION.

Mithraic zodiac   As the cult of Mithras was deeply involved with the Mysteries, little is known for sure about the unique features of the astrology practised by its adherents, but many astrological symbols and even entire zodiacs associated with the cult have survived. A representative zodiac is that from the Mithraeum on the island of Ponza, studied by Vermaseren and Beck. It is a stucco ceiling zodiac in three concentric zones, the outermost of which carries the twelve images presented clockwise (set in unequal arcs, which may or may not point to a constellational map, though Beck suggests a connection with the distinction between 'solar and nocturnal houses' to account for the variations). In the middle zone is a snake, which Vermaseren sees as DRACO, but which Beck suggests may be the 'first ancient representation of the eclipsing dragon' (see ATALIA). In the central zone are the images of Ursa Major and Ursa Minor. Certain Mithraic bas-reliefs depicting the slaughter of the bull contain zodiacal images – one of the best known is the marble found in London with the tauroctonous Mithras surrounded by a circular anti-clockwise zodiac of twelve images, with the personifications of the winds, Sun and Moon on the outer edge. There are also several interesting images of serpentine leontocephalous figures (sometimes called the Aeon, or Zervan Akarana, or the Mithraic Kronos) into which have been inset images of the zodiacal series.

Mithunam   The Sanskrit name for zodiacal Gemini – but see MRGACIRSHA.

mixed application   This term is used to denote the *APPLICATION to aspect of one planet to another planet when one of the two is in retrograde motion. To be distinguished from DIRECT APPLICATION and RETROGRADE APPLICATION.

Miyan See REGULUS.

Mizan Al Mizan, the Arabic name for Libra.

Mizar One of the seven fixed stars forming URSA MAJOR, said by Ptolemy to be of the influence equivalent to that of Mars. Not to be confused with the variant Mizar of MIRACH.

mode See QUALITIES.

moderators A defunct term applied in Ptolemaic astrology to the Sun, Moon, Ascendant, Midheaven and Pars Fortunae, with the understanding that aspects from these condition the influence of other planets. Since this is now quite sensibly regarded as the basic tenet of aspectal theory (that indeed all planets act as moderators in this sense) the term is no longer valid. The astrologer Wilson used the term in a different sense, claiming that the moderators were so named because they each had their 'own mode of operating on the native', which is a fairly meaningless conception within the context of astrology. Wilson dismissed the Pars Fortunae as a moderator, for he personally had come to the conclusion that it could have no effect in nativities. The astrology Lilly so misunderstood the moderators as to call them the 'five hylegiacal places' (see HYLEG), and as a result the term has been misused in this sense by later astrologers.

modern astronomical zodiac See MODERN ZODIAC.

modern planets The so-called modern planets are those which have been discovered since the development of advanced optical instruments, and added to the planetary schema of seven planets used by the ancients. These planets are URANUS, NEPTUNE and PLUTO, though some astrologers also include the ASTEROIDS. See NEW PLANETS.

Modern Zodiac The Modern Zodiac, sometimes called the Modern Astronomical Zodiac, was defined during the 1928 conference of the International Astronomical Union (hence, it is sometimes also called the IAU Zodiac). Strictly speaking it is not a zodiac, but a constellational grid definition. It was published by Delporte in 1930, and within the present text

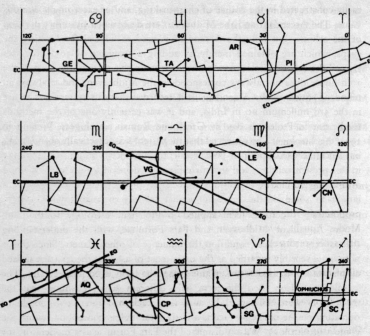

Figure 44: The Modern Astronomical Zodiac based on the computations of Delporte in 1930, defined by the International Astronomical Union in 1928.

his divisions are set out in the tabulations provided by Powell under SIDEREAL CORRESPONDENCES. The sequence of asterisms, with their curious grids, are given in figure 44 – but see also CONSTELLATION MAPS. The definition of the fiducial zero-point for this zodiac of unequal asterisms is uniquely defined in mathematical terms, in a complex system of arcs of constant Right Ascensions and declination for the equinox of 1975. However, as Powell and others have pointed out, within the system proposed, some of the constellations actually overlap, and Ophiuchus straddles the ecliptic (as may be seen in figure 44) even though this constellation is not normally taken into account in any zodiac proper. See SIDEREAL ZODIAC.

modes See QUALITIES.

Modus Equalis A popular system of HOUSE DIVISION, sometimes called the equal-house system. It is an entirely symbolic method of dividing the ecliptic, easily constructed in the course of chart-making, and of exceedingly dubious value. The Ascendant and the Midheaven are determined by direct division of the ecliptic, but only the Ascendant is used as fiducial for locating the cusps, which are obtained merely by adding a sequence of 30-degree arcs around the ecliptic circle from the Ascendant. In this system, the cusp of the 10th house does not always correspond to the Midheaven, and is always in mundane square to the Ascendant. It is recorded that this system was in use in the 3rd millenium BC in India, and it was certainly one of the methods used prior to Ptolemy's day by Greek and Roman astrologers. Writing in 1904, the astrologer Leo observed that the system was 'practically abandoned', but since the rise of popular (and frequently superficial) notions of astrology in this present century, and the corresponding demand for an 'easy' method of drawing up charts, the Modus Equalis has proliferated, and is now even taught in some of the more influential schools of astrology. A system proposed by the Australian astrologer Zariel has been confused with the Modus Equalis, even though it is not a symbolic method – see EQUAL-DIVISION METHOD.

modus rationalis   See REGIOMONTANEAN SYSTEM

moist   See PRINCIPLES.

Moist Daughters   See HYADES.

moist signs   According to the Ptolemaic astrology, these are Cancer, Scorpio and Pisces, to which some later astrologers add Gemini, Libra and Aquarius. The classification is little used today – but see SECTA.

moisture   An archaic term in its ancient specialist astrological sense. It is linked with the natures of the planets, which were said to be dry, wet, warm and cold, in accordance with the doctrine of the four Qualities – see, however, QUALITY. The wet condition (moisture) was said to increase in planets which were matutine. The Moon, however, increases in moisture in her first quarter, when it is nocturnal and in winter time.

monad   In its original Greek etymology, the term 'monad' was applied to the basic indivisible unit, and it was adopted into astrological and occult terminology with a similar meaning, though with a special application to the realm of spirit, rather than matter. The 'human monad' is in modern occultism that indestructible element within each individual human, which continues through the cycles of reincarnation. See also MONAS HIEROGLYPHICA.

Monas Hieroglyphica   A name given by Dee to a sigil sometimes called the Monad. The sigil itself is deceptively simple ☿ consisting basically of the standard sigil for Mercury inserted into the open sigil for Aries. The importance of the sigil lies not so much in its form as in the erudite and deeply esoteric argument used by Dee to explain its structural significance in his book (the *Monas Hieroglyphica* of 1564). Never adequately translated from the Latin, the text is a repository of esoteric innuendoes which makes it one of the more remarkable occultist products of that century, and certainly the most important attempt to construct around a single sigil a theory of graphic symbolism linked with the esoteric and astrological tradition.

Monoceros   Constellation on the celestial equator, parallel to the first 15 degrees of Leo, 13 to 23 degrees south of the equator. Probably formed in the early 16th century, perhaps as the image of a horse, perhaps even as the 'unicorn', as it was called by Bartschius. Robson records that it is supposed to bring an ambitious nature, with a love of travel and change.

Monomoiria   A term derived from Hellenistic astrology, though probably related to earlier astrological forms, and linked with the tradition that each degree of the ecliptic is associated with a planetary influence. This theory must be differentiated from that pertaining to SYMBOLIC DEGREES, however, though it perhaps did influence this latter tradition to some extent. A number of systems for locating the particular individual degree-rulerships exist, but the one recorded by Vettius Valens is perhaps the most influential. In this system, five columns of degrees are ranged against linear arrangements of the seven planets, following the normal descending sequence from Saturn to Moon, but with each range adjusted by one sequence, so that the upper range reads in the same order as the first vertical column. In the degrees of the Monomoiria listed by Neugebauer and Bouche-Leclercq, the concept of 'zero degree' is employed, which within such a context is misleading: the tabulation in table 53 is accordingly adjusted to the standard rounding-off process required by practical astrology. The system presumes of course that the degree influence is sparked off only when that degree is tenanted by a planet.

Mons Maenalus   Small constellation at the feet of the asterism BOOTIS, formed by Hevelius at the end of the 17th century. It is said by astrologers to give pride and great ability, though with some destructiveness.

Mons Mensae   See MENSA.

month   The basic lunar period, reconciled into sequences of twelve by

Moon

TABLE 53

Degree					Tenant Planet: Monomoiria						
---	---	---	---	---	SA	JU	MA	SU	VE	ME	MO
1	8	15	22	29	SA	JU	MA	SU	VE	ME	MO
2	9	16	23		JU	MA	SU	VE	ME	MO	SA
3	10	17	24		MA	SU	VE	ME	MO	SA	JU
4	11	18	25		SU	VE	ME	MO	SA	JU	MA
5	12	19	26		VE	ME	MO	SA	JU	MA	SU
6	13	20	27		ME	MO	SA	JU	MA	SU	VE
7	14	21	28		MO	SA	JU	MA	SU	VE	ME

reference to the solar cycle of one year – see CALENDARS. Each of the names of the months has (or, more accurately, had) some esoteric relevance to astrological lore. See therefore each of the months listed by name, JANUARY, etc.

monthly directions See MONTHLY SERIES.

monthly series One of four measures for symbolically computing time, used in SECONDARY DIRECTIONS, and based on the idea that the true solar day should be regarded as the equivalent of a month. The full term is 'monthly series of secondary directions', and (misleadingly) 'monthly directions'.

Moon The satellite of the Earth, called in astrology a planet. It has rule over zodiacal Cancer and is Exalted in Taurus, with its corresponding Fall in Scorpio. In a chart, the Moon represents the imaginative, reflective side of the native, and is justifiably linked with the subconscious element in the modern image of man. It is an index of the receptive, withdrawn, secluded part of the person. When the Moon is emphasized in a chart, it is usually an indication of a highly sensitive, impressionable and changeable nature, often linked with a personality in some way associated with the past, or with childhood. A badly placed Moon, or one badly aspected, tends to give rise to a personality which is hypersensitive, untidy, withdrawn, morbidly concerned with self, and strangely subject to misfortune. The sigil for the Moon ☽ represents this body as a reflection of the Sun, the idea being that what is important in the sigil is not so much the lighted crescent reflecting the light of the consciousness, as the part of the lunar globe which remains in darkness – that which is not seen in nature or in sigil: the sigil is therefore a most apt symbol of the subconscious and hidden part of man. Traditionally, the Moon is given rule over the breasts in the body, though in some systems both the womb and the lymph glands are associated with this planet – see, however, PLANETARY MELOTHESIA. By association with its throne Cancer, it has rule over the human rib-cage, over that protecting form wherein the heart rests secure. In the chart, the position of the Moon often gives a useful index of how fluids are working through the body. Lunar types seek positions in public life, and are very often strangely subject to fluctuations in popularity (or to frequent changes in jobs or locations). They generally find much happiness in the quiet retirement of home life, and sometimes they will involve themselves with work involving liquids, such as sailing, washing or drug-dispensing, while there is also a tendency for them to protect or look after people who are themselves dealing with liquids of some kind, or who are in need of special nurturing. The colours associated with the Moon are white, pearl, pale blues, silvers and iridescents: the metal of the Moon is silver, but some astrologers now link it also with aluminium. In connection with lunar cycles, see EMBOLISMIC LUNATION, LUNATION, LUNAR MONTH and SAROS CYCLE

Moon Chain A term applied in the theosophical cosmoconception to the Third Chain of the evolution of the Earth. The planet D was the equivalent of the Moon in the present EARTH CHAIN, and the present physical body of the Moon is visualized as a disintegrated crust, much diminished in size, and on its way to destruction. This theosophical teaching is much opposed by several other esoteric doctrines – see, for example, EIGHTH SPHERE. The most highly developed beings to emerge from this Chain are called the Sons of Twilight, the Sons of the Moon, and sometimes the 'Barhishads'.

Moon-forces See SUN-FORCES.

Moon period A term which has nothing to do with the cycles of the Moon, or the SIDEREAL MOON RHYTHMS. As a specialist term, it is derived from the modern esoteric astrology, and relates to cosmogenesis. The Moon period is said to represent the third of the planetary evolutionary states, following on the first (SATURN PERIOD) and the second (SUN PERIOD). Initially it was a condition of moisture, which was said to densify, though it did carry spiritual elements from previous stages of cosmogenesis, so that while the ancient Moon body is called Water, the equivalent of its atmosphere was called Fire-fog. This Moon period, which is actually only most tenuously linked with the present Moon of our system, is sometimes called the Old Moon. The period of the evolutionary state was said to have been overseen by the Hierarchy of Seraphim.

Moon revolution　Term derived from modern esoteric astrology, relating to cosmogenesis. It is said that the path of the evolution of our solar system, with its periods of activity and sleep, must recapitulate the entire sequence of evolution. Since the third stage of the cosmogenesis of our present system took place in the MOON PERIOD, the term 'Moon revolution' pertains to the third of these recapitulations. See also REVOLUTION PERIODS.

Moon rhythms　See SIDEREAL MOON RHYTHMS.

Moraya　Name given to one of the HYPOTHETICAL PLANETS, said to be transplutonian when listed by Jayne.

Morinean orbs　These are ORBS applied to the planets and fixed stars by the French astrologer Morin de Villefranche, determined by him as a result of direct observations of the skies, and linked with the heliacal visibility of the bodies. It takes about 72 minutes after sunset for the stars to become visible on the horizon in Europe. This period was regarded by Morin as the arc equivalent of the orb of the Sun. By relating the heliacal visibility of the known planets to the sunset arc, the orbs of the known planets were determined. The degrees of orb given by Morin are:

SA 7　JU 8　MA 6.5　SU 18　VE 13　ME 8　MO 12

Using exactly the same method, he determined the so-called 'orb of stars', according to brilliance, linked with the conception of 'magnitude':

1st mag. − 6　2nd mag. − 5　3rd mag. − 4　4th mag. − 3　and so on.

Morinean system　One of the systems of HOUSE DIVISION based on projection of equal-arc divisions of the equator on to the ecliptic from the pole of the ecliptic. It is a method proposed by the 17th-century astrologer Morin de Villefranche, and is sometimes called the Morinus system (after his Latinized name), and sometimes the Rational and Universal Method.

Morinus system　See MORINEAN SYSTEM.

Morning Star　In esoteric lore, one of the names given to the *PENTAGRAM, and in exoteric lore to Venus, which draws a pentagram in the skies in her conjunctions with the Sun (for further details, see Schultz). This form is exoterically linked with Venus, and esoterically with what historians call the *orans* gesture, which is itself an imitation of the five points of the pentagram (echoing the two feet, the two outstretched arms, and the head). Strictly speaking, it is not an *orans* (praying) gesture so much as an image of the departed soul, symbolized in the ASTRAL plane, at the beginning (in the morning) of the soul's post-mortem existence – hence the link with the 'morning star' Venus.

morphochromatogram   A term introduced by Kolisko to denote the litmus pictures (see figure 37, under KOLISKO EFFECT) made in the study of fluid dynamics related to planetary aspects.

mother   The Point of Mother is one of the so-called Arabian PARS in a chart. If the degree of Venus in a natal chart is revolved to the Ascendant, the adjusted position of the Moon marks the Point of Mother. The 4th house is sometimes called the House of the Mother.

Motori   See INTELLIGENCY.

movable signs   A term applied to the Cardinal signs, which should scarcely be described in the passive mode, in that they are really the 'moving signs', the initiatory triad of the twelve.

movable zodiac   One of Blavatsky's several terms for the constellational zodiac, her own 'Natural Zodiac'.

moving signs   These are Aries, Cancer, Libra and Capricorn, the Cardinal signs – see CARDINALITY.

Mrgacirsha   The Sanskrit name for constellation Gemini – but see MITHU-NAM.

Mrigasiras   The 3rd of the Hindu NAKSHATRAS (the head of the stag).

Muhurta   See MUHURTASASTRA.

Muhurtasastra   A Sanskrit term used for the practice of determining the right moment for an activity – in other words, for ELECTORAL ASTROLOGY, which in the Hindu system is dependent upon lunar factors, and makes considerable use of the NAKSHATRAS. Sometimes MUHURTA.

Mula   The Sanskrit name for constellation Sagittarius – but see also DHANUS. The same term is also used of the 17th of the Hindu NAKSHATRAS (the root).

mul-apin tables The two mul-apin tablets, dated to the 8th century BC (though incorporating material at least 500 years older) are astronomical Babylonian texts listing the rising of 18 stars and constellations in the zodiacal belt, in terms of a schematic year of twelve months. The Babylonian term means 'plough star', and refers to the asterism TRIANGULUM and certain contiguous stars. This combination of schematic year, along with the later introduction of the axial fiducial of the fixed stars ALDEBARAN and ANTARES, probably gave rise to the BABYLONIAN ZODIAC which was (according to Powell) devised as an alternative system to that of the NORMAL STARS.

Mulier Sedis See CASSIOPEIA.

mundane aspects These are *ASPECTS measured without regard to the zodiac, but only in terms of subdivisions of the semi-arcs, and in terms of houses. Thus, planets separated by two houses are in mundane sextile, though they need not be in zodiacal sextile, and separated by 60 degrees.

mundane astrology Astrology limited to casting and interpreting specialist charts directed towards the examination of national trends – a subdivision of what has been called HISTORICAL ASTROLOGY. The basic charts and system of interpreting in terms of mundane astrology are different from those used in natal astrology, for special emphasis is placed on such things as solstices, lunations, eclipses and planetary conjunctions. The term is rather confusing in view of the different application of the word 'mundane' to the house systems. See WORLD CYCLE OF MUNDANE EVENTS.

Mundane Cross See LA CROIX.

mundane house Since all houses are part of the mundane figure, the term is merely prolix. A mundane house is merely a house.

mundane parallel See PARALLEL IN MUNDO.

Mundi Templum See ARA.

mundoscope A name applied to a chart calculated in such a way as to present a planet through the Prime Vertical for a given moment in time – this time being the ingress of the Sun or Moon into a certain sidereal sign. The chart is rarely used, but was favoured by Fagan in his researches into the sidereal zodiacs of the ancients, and adopted by the astrologer Bradley.

Muses In the Renaissance chart of musical correspondences given by Gafurius are a number of associations between the Spheres (PLANETS), the Muses and MODES (these relate to the octave species of the Great Perfect System), but the correspondences given by Gafurius (table 54) do not appear to be the same as those established by modern musicologists.

TABLE 54			
Sphere	*Muse*	*Muse association*	*Mode*
Stellatum	Urania	Astronomy	Hypermixolydian
Saturn	Polyhymnia	Mimic arts	Mixolydian
Jupiter	Euterpe	Flautic art	Lydian
Mars	Erato	Lyric poetry (hymns)	Phrygian
Sun	Melpomene	Tragedy	Dorian
Venus	Terpsichore	Lyric poetry (dance)	Hypolydian
Mercury	Calliope	Heroic epic	Hypophrygian
Moon	Clio	History	Hypodorian

Musca Australis Constellation formed in the 17th century and variously called Apis (bee), Musca (fly), or Musca Indica (Indian fly), as well as the Southern Fly of the given term. It is located south of the Cross, and from 66 to 74 degrees south of the equator, east of the CHAMAELEON. It is said to give a capricious, changeable and industrious nature.

Musca Borealis Constellation, the Northern Fly, over the back of constellation Aries, probably formed by Habrecht in the 17th century, and sometimes (confusingly) called Apis – see MUSCA AUSTRALIS. It is sometimes called Vespa (wasp), or even Beelzebub after the Lord of the Flies. Robson says that it gives a practical, pleasure-loving and industrious nature, with some sarcasm and spitefulness.

Musca Indica See MUSCA AUSTRALIS.

Musculus See LYRA.

musical astrology See also MUSES.

music of the spheres An idea of a planetary music which has been attributed wrongly to Pythagoras (the conception of planetary spheres itself had not been developed before the days of Eudoxos). Pythagoras did, however, lay the basis for the idea of planetary ratios (which were later used in the 'music' or 'harmony of the spheres') when he showed that intervals of the scale could be expressed in simple ratios: the intervals known to the ancients were the fourth, fifth and octave. The discovery appears to have resulted from a lengthy search for a way to introduce a 'blend' (the Greek *krasis*) into the pairs of opposites which were at the time believed to underlie the basic nature of the material realm. Pythagoras applied these ratios to the supposed distances of the planets – an attempt which has been likened to an anticipation of Bode's series. The *harmonia* which was said to reign through all things as a result of such *krasis* extended well beyond the planets and music, for it coloured all created things. In relation specifically to music, Pythagoras maintained that since the ratios of the scales extend into the heavens, then the whole of the substellar realm should echo to a heavenly music. Several explanations have since been given as to why we in the sublunar sphere do not hear this music: some say that we are deafened by its perpetual sound, while others maintain that we do hear it, if only in our sleep, when our spirits are not so tightly bound to the body. Modern esotericism insists that the music may be heard only by initiates under normal circumstances, but that it becomes part of the post-mortem experience for every other human. Pythagoras was of the opinion that we did not hear it because our souls were out of *krasis*, out of harmony with the stellar music. When in the post-Eudoxian times the concept of the planetary spheres was developed, the original Pythagorean idea was extended and popularly called the music of the spheres: it was thus passed into the Neoplatonic tradition which elaborated on the basic ideas (see, for example, PLANETARY TONAL INTERVALS). The architect Alberti, and such musicologists as Gafurus, drew a detailed picture of the planetary relationships and corresponding tones and intervals, which so profoundly influenced humanistic art and literature in respect to the notion of planetary harmonies, and which in fact still reverberate in many of our modern conceptions of ratios and planetary natures. Alberti adopted many of the Greek ratios, and unions of ratios, which have subsequently been linked with the diagrams of the planetary spheres evinced in pre-Copernican systems: see DIAPASON, DIAPENTE, DIATESSARON and TONUS. In modern times Collin in particular has introduced a sophisticated concept of the harmonic relationships between planets, based no longer on distances (as is the Pythagorean model) but on revolutions and revolution periodicities, which he links to the tonic scale. He shows that in theory 36 octaves separate human time from what he calls 'solar time', and the same interval separates the vibrations of human music from the vibrations of the planetary motions.

Musattah

The Rosicrucian school, which has so deeply influenced the development of esoteric astrology, maintains that humans are not permitted to hear the music itself, but may hear copies through ordinary mortal music, and thus not lose the longing for the celestial world beyond the material realm. As Fludd (who believes that the music is produced from the impinging of the moving Sun upon the paths of the planets) puts it, 'music is a faint tradition of the angelic state.' The nine MUSES of Hesiod have been associated by occultists with the celestial music: these were the daughters of Zeus and Mnemosyne (whose name means 'remembrance') and so we are inclined to find in classical literature a link with the astrological teaching which would have our souls 'recalling' the celestial world. See also CELESTIAL HARMONICS and KEPLERIAN HARMONICS.

Musattah   See ASTROLABE.

Mutable Cross   In traditional astrology the Mutable Cross is that which is formed by Gemini, Virgo, Sagittarius and Pisces. This cross has been widely developed by the esotericist Bailey in her system of astrology, and is associated with the 'crisis of incarnation, relating to evolution, to the kingdom of nature, to planetary initiation, body, form and personality': she says that it is the cross of changing and absorbing experience, and visualizes in it an evolutionary sequence in reverse movement from Pisces to Gemini, through which the seeds of Piscean intuition flower into 'dreams of life' in Gemini. See also CARDINAL CROSS and FIXED CROSS.

Mutable signs   These are Gemini, Virgo, Sagittarius and Pisces.

Mutability   The term applied to one of the three QUALITIES (see CARDINALITY and FIXITY) which acts as a unifier of the impulsiveness of the former with the rigidity of the latter, and may even be said to carry a quality common to both. Indeed, the Mutable signs (Gemini, Virgo, Sagittarius and Pisces) are sometimes called the COMMON SIGNS. The Mutable nature is that of changeability and adaptability, but under pressure it will give rise to instability.

mute signs   The water triplicity are called the mute signs because it has been said that when Mercury (which rules the human voice) is afflicted in these signs, there is a liability to dumbness or to impediment of speech. The idea may be derived from the images for these signs – the crab of Cancer, the scorpion of Scorpio and the fishes of Pisces emit no audible sound.

Mystic Rectangle

mutilated degrees Tradition holds that some of the zodiacal degrees afflict the native with lameness if they are emphasized in the chart in any significant way – for example, by the Moon or the ruler of the Ascendant occupying them. These are called the mutilated degrees, even though they are properly speaking the 'mutilating degrees'. As the traditional list below will indicate, they are in some cases arcs, and it is quite possible that the original groups were linked with stellar influences, yet no precessed ratios give a satisfactory relationship with a constellation map, and so the tradition is hard to fathom:

TA 6–10	CN 9–15	LE 18–25	SC 18–19
SG 1 7–8 18–19	CP 26–9	AQ 18–19.	

mutual application When two planets applying to aspect move towards each other (one being retrograde), they are said to be in mutual application.

mutual aspects Aspects formed between two moving bodies at a specific time are said to be mutual: such aspects are formed between transitting planets, as opposed to those formed between such a planet and a natal planet or nodal point, or a progressed planet. The aspects listed from a natal chart prior to the listing of progressions or transits are mutual aspects. The term is used to distinguish such Angular relationships from those which are directional, progressed or transitory.

mutual reception Two planets which are posited in the essential dignities corresponding to the other are said to be in mutual reception – see therefore DIGNITY. Some astrologers maintain that the so-called 'domal dignities' may also give rise to mutual reception.

mystical planets See SACRED PLANETS. The astrologer Libra discussed the planets linked with the future development of Aries, Taurus and Gemini as being 'still invisible on higher planes', and called these the 'invisible mystery planets'. Libra links this future unfoldment of Aries with the Hierarchy of the physical plane, Taurus with the Hierarchy of the astral plane, and Gemini with the Hierarchy of the mental plane.

Mystic Rectangle A modern term applied to a configuration of aspects in which a pair of trines connect with a pair of sextiles in such a way that oppositions fall across the four corners. The term, which appears to have originated in French astrological circles, is not directly associated with mystical tendencies within the chart – though, as Brau puts it, the Mystic Rectangle 'combines the awareness of the opposition, the understanding of the trine, and the productivity of the sextile'. One wonders, however, if the term has arisen mainly because such a configuration is found in the chart of Blavatsky? The Mystic Rectangle is a MAJOR CONFIGURATION.

Mystic Tetrad   See TETRACTYS.

Mystic Trine   A term sometimes used to denote a pair (or more) of convergent trines in a single chart. In terms of the elemental theory which lies behind the doctrine of aspects, such a trine implies that each of the signs of one triplicity is given some prominence by virtue of its being occupied by a planet. For example, SA 15 LE, SU 15 AR, JU 15 Sg, would be a Mystic Trine in fire. The term is, in effect, an alternative for GRAND TRINE, but an emphasis is placed on the *effect* of such a powerful aspect, which is to bring into prominence the spiritual (or mystic) side of the planets and signs concerned. Sometimes, the term is restricted to denoting a Grand Trine in the water triplicity.

N

Na'am Al Na'am, the 18th of the Arabian MANZILS (the Ostriches).

Naburiannu tables See AYANAMSA.

Nachiel Name given by Agrippa (quoting qabbalistic sources) to the IN-TELLIGENCY of the Sun, to whom he ascribes the magical number 111, the linear sum of the MAGIC SQUARE of the Sun.

Nadi astrology See NADIGRANTHAMS.

nadigranthams General name given in Hindu astrology to collections of palm-leaf manuscripts used in consultative astrology. The origin of these nadigranthams is unknown, and the tendency of popularist literature is to give them a hoary and quite improbable antiquity – some claim the authorship of Brahma, others of Vyasa. It is often said that the palm-leaves contain horoscopes, but certainly such are not the main matter in those palm-leaf books examined by specialists in Sanskrit. The theosophist Row consulted an expert in nadigranthams, and his report indicates that it was a farrago of nonsense unrelated to astrological practice. Unaccountably, Brau makes of the nadigranthams a system of astrology which he calls 'Nadi astrology', in which the astrologers are supposed to have delineated horoscopes cast ('sometimes centuries before') the appearance of their clients. Though popularly believed to be true, such accounts of Indian astrology are really based on fantasy.

nadir The lowest point below the Earth, relative to any given position on the Earth – it is the opposite polarity of the ZENITH. Just as the zenith should not be confused with the 10th house, so the nadir should not be confused with the 4th house.

Naibodic arc See TABLE OF NAIBOD.

Naibod's measure of time See MEASURE OF TIME.

nakshatras the Hindu equivalent of the *LUNAR MANSIONS. While there are the usual 28 mansions or nakshatras in the Hindu system, only 27 of these are used in ordinary astrology (see ABHIJIT). The arcs consist of 13 degrees and 20 minutes, and begin at ASVINI, marked by the beta of constellation Aries. Their names, and their relationships to Arabian and Chinese lunar systems, is set out under LUNAR MANSIONS.

Nanar   One of several Chaldean names for the Moon.

Nannak   One of several Chaldean names for the Moon.

Naronic measure   A measure of arc used in SYMBOLIC DIRECTIONS, based on a division of the zodiacal circle by 600 (see NAROS CYCLE), which gives arcs of $\frac{2}{3}$ of a degree. It is claimed by Sepharial that this progression unit is a useful definition of the periods of 'depression and expansion' in life, while the astrologer Carter regards the measure as being of the utmost value in progressional work.

Naros cycle   A name given to a periodicity or cycle, sometimes called the Naronic cycle or the Neros cycle, claimed in esoteric circles to be one of the Mystery secrets, but exoterically said to be a period of 600 years. Blavatsky claims that there were three kinds of Naros cycle – the 'greater', the 'middle' and the 'lesser', and that only the last corresponds to the cycle of 600 years. One of the esoteric significances of the cycle of 600 is found in the numerical values and images ascribed to the Hebraic *Tau* and *Resh*, which letters of the alphabet are associated with the cycle. The image of *Resh* is a circle, while the sigil for *Tau* is a cross: the encircled Tau (or the encircled cross) is the basis for the symbolic form of the horoscope chart – yet the numerical values of *Resh* (200) and *Tau* (400) add up to 600, thus uniting the circle and cross in the idea of the Naronic cycle.

Nashira   Fixed star, the gamma of Capricornus, set in the tail of the Goat, the name said to be derived from the Arabic Al Sa'd al Nashirah (the fortunate one). Ptolemy says it is of the nature of Saturn and Jupiter, redemptive of evil, though tending to bring danger from wild beasts.

natal astrology   That branch of astrology which is concerned with the casting and interpreting of birth charts – the Latin *natus* meaning 'birth'. It is GENETHLIACAL ASTROLOGY.

Nathrah   Al Nathrah (the crib), the 6th of the Arabian MANZILS.

native   The subject for whom a horoscope (the NATIVITY) has been cast.

nativity In a specialist sense, the term used for the chart cast for a particular birth (from the Latin *natus*, birth): in modern terminology, the equivalent is HOROSCOPE. In some pre-19th-century astrological texts the term 'nativity' was used specifically to mean 'ascendant degree', no doubt in reference to the original etymology of the word HOROSCOPE.

nativity exaltation The historian Neugebauer records the term 'Exaltation (of the Nativity)' in regard to a Hellenistic method of determining a nativity exaltation which is to be distinguished from the planetary EXALTATIONS, even though it is now merely of historic interest. Reference is made to the astrologer Vettius Valens, who gives the exaltation sign of a given nativity in terms of the position of either the Sun or Moon in a chart. One simply counts the number of signs from the day horoscope Sun to Aries (the exaltation sign of the Sun), or from the night horoscope Moon to Taurus (the exaltation sign of the Moon), and then takes the same distance (counting in signs) from the Ascendant to mark the nativity exaltation.

natural astrology A term originated (seemingly by Brau) to denote both astronomy and ASTROMETEOROLOGY, the latter the equivalent of METEORO-LOGICAL ASTROLOGY. The term 'natural astrology' appears to be derived from translation of a French term which does not have an English equivalent.

natural graduation system A method of HOUSE DIVISION invented by the astrologer Colin Evans, and essentially an improvement on the PORPHYRIAN SYSTEM. The method assumes that the time divisions involved should be based on a gradual, though continuous, increase. A survey of the method is presented in the Evans edition of *Waite's Compendium of Natal Astrology*.

natural zodiac One of Blavatsky's terms for the CONSTELLATIONAL ZODIAC, which is rightly defined as a 'succession of constellations', and distinguished from her INTELLECTUAL ZODIAC. Unfortunately, the astrologer Heindel succeeds in confusing her zodiacs, and defines the Natural Zodiac as the 'twelve (sic) constellations'.

natus An archaic (Latin) term for the natal chart.

navamsa The Hindu term meaning 'ninth part', most usually the 'ninth part of a sign'. Due to a misunderstanding, the same term has been used by modern Western astrologers to mean the equivalent of a nine-house system (see NOVIENIC CHART), and has also been wrongly used as applicable to

HARMONICS. The astrologer Harvey has attempted to show that the navamsa chart is really a diagram of planetary positions in terms of the ninth harmonic. See also NOVENARY MEASURE.

navamsa-dwadasamsa   A term derived from Hindu astrology, meaning 'a (9 × 12) division'. It is important as an astrological division, for the 108th part of sidereal zodiacal arc (hence an arc of 16 minutes, 40 seconds) on the Ascendant is regarded as an indicator of the sex in pre-natal calculations – see EPOCHAL CHART. The problem with the method is that no fiducial is available, and no agreed AYANAMSA, so the application of the arc is of dubious theoretical value. The epochal method of the astrologer Bailey does not use the navamsa-dwadasamsa but correlates by means of what are called SEX POINTS which are supposed to be based on the Hindu NAKSHATRAS, incorporating the asterism ABHIJIT to give 28 divisions of the ecliptic.

Nebo   A Chaldean god, sometimes equated with the planet Mercury.

nebula   A star cluster in which the individual lights of stars are merged into an indistinguishable cloud (the Greek *nephos* means cloud, the derived medieval Latin *nebulatus*, cloudy). Astrologers regard the supposed influences from some nebulae as worthy of consideration in chart interpretation – see FIXED STARS.

Nebulosa   Name given to the 8th of the lunar mansions in some of the medieval lists – but see table 48 in LUNAR MANSIONS. The mansion was probably located by Praesaepe, which was the only universally recognized nebula prior to the invention of the telescope. It was accordingly called Nubilium (cloudy sky), Nebula or Nebulosa in Pectore Cancri (clouds in the breast of Cancer). See LUNAR ZODIAC.

Necessity   Name given to one of the SEVEN LOTS of the Greek astrologers, the Lot of Anagke (necessity). The degree is the same distance from the Ascendant as is Mercury from the PART OF FORTUNE.

Nechepso   See PETOSIRIS.

negative Mars   In the traditional system of planetary rulerships over the signs, Mars was assigned to Scorpio: in accordance with this schema (see PLANETARY RULERSHIPS and SECTA), the Scorpionic Mars was said to be feminine, nocturnal and negative, in contrast to the masculine, diurnal and POSITIVE MARS which ruled Aries. In post-medieval astrology the basis for this use of the term was largely forgotten, and there developed an idea of a sort of 'negative Mars' which was in some ways different from planetary Mars. It is for this reason that one frequently finds reference to negative Mars in contexts where there is no mention of other planets as being expressed in negative or positive equivalents. The fact is that there is a negative Mercury, a negative Venus, and so on. Usually the term simply means 'Mars as ruler of Scorpio', though perhaps the word 'negative' does inadvertently point to something of the need for redemption so characteristic of the sign. It must be understood, however, that even a redeemed Scorpionic nature is ruled still by negative Mars: but see EAGLE.

negative planets   See FEMININE PLANETS and NEGATIVE MARS.

negative signs   These are the alternate signs from Taurus to Pisces, inclusive. But see FEMININE SIGNS, for which negative signs is really an alternative term. See also SECTA.

Nemesis   Name given to one of the SEVEN LOTS of Greek astrology, Nemesis being (approximately) the goddess of retribution. The degree of this lot is at the same distance from the Ascendant as Saturn from Fortuna.

Neomenia   Literally the 'new Moon' from the Greek. In ancient times the new moons of spring and autumn were calendrical fiducials, the first marking the first day of the ecclesiastical year, the second marking the first day of the civil year. As the astrologer Fagan points out, the term 'neomenia' in antiquity had not quite the same meaning as today: in remote times the first day of the neomenia was that day (commencing at sunset) when the thin crescent was first visible in the western sky immediately after sunset. In ancient times the neomenia was regarded as being inauspicious.

Neptune   Planet in orbit between Uranus and Pluto, discovered in 1846 by Galle in Berlin, following a suggestion of the Frenchman Leverrier, though it had been observed (if mistaken for a star) by Lalande as early as 1795. Since its discovery, Neptune has displaced Jupiter from its traditional rulership, and has been accorded rule over the zodiacal sign Pisces. Astrologers see it as a dissolving influence: it bestows nebulousness and confusion, even deceptiveness when working under pressure. When working through its beneficent side, however, it induces an imaginative, inspirational, idealistic and artistic nature. The link with the sea is more than nominal (Neptune being the Roman equivalent of Poseidon, the god of the seas), for it is generally interpreted as a symbol of the ease of contact which the planet facilitates with the 'sea' of the unconscious: but see also NEPTUNUS. The sigil for Neptune Ψ is popularly explained as being derived from the trident of . Neptune: but see NEW PLANETS.

Neptune–Pluto cycles   The synodic period for these two planets is 492.328 years, which (with a minute error of less than 3 degrees) gives the first order cycle. The second order cycle is a vast 72,372.25 years, involving 147 synodics.

Neptune Scheme   The Neptune Scheme or the Neptunian Chain are terms derived specifically from the theosophical cosmoconception linked with the SCHEME OF EVOLUTION. It is claimed that the three planets in this Chain include Neptune, and two others beyond its orbit, perhaps even Pluto, which was unrecognized when the schema was originated. Blavatsky claimed that Neptune is not in our own solar system, so that the formulation of this scheme is derived from her followers (probably influenced by the teachings of Besant and Leadbeater).

Neptuni Sidus   See PISCES CONSTELLATION.

Neptunus   The name given to NEPTUNE by the esoteric astrologer Thierens. He departs little from the exoteric tradition in seeing the influence of the planet (which he also designates as the Grecian Poseidon) as the principle of absorption ruling the 'subconscious mind or dream-thinking of psychoanalysis' which will 'gather what is rejected by the wakeful or constructive mind: it may cause a morbid passivity, which leaves the mind open to become the playground for more or less "animal" elementals or influences from the astral world, and which finally very often leads to idiocy or other phenomena of mental instability'.

Nereus   A name given by the esoteric astrologer Thierens to a future development of his NEPTUNUS, as a transformation of the modern Jupiter in his ASTRAL SPHERE. It is said to represent the principle of absorption.

Nergal   A Chaldean name, probably for a personification of Mars as 'the giant king of war'. The same name is used in Hebraic astrology for Mars, though seemingly of the darker side of his nature.

Neros   See NAROS.

nervous signs   Term used as a synonym for AIR SIGNS.

new Moon   See NEOMENIA.

new planets   This is the curious term often applied to the three planets of the solar system which have been added to the traditional seven of the ancient astrological model – in order of discovery URANUS, NEPTUNE and PLUTO. Much has been written and speculated about the functions of these planets in astrological terms, and a digest is given under the respective headings. In regard to a treatment of the planets as representative of a group influence, the ideas of Walther are perhaps worthy of mention. Walther links the discovery of the new planets with the evolution of modern consciousness, to an extension of our understanding, to new technology, and so on. Yet, he does not regard the three as being divorced from the traditional seven, so much as being 'higher vibrations' of these. He compares the 84-year revolution of Uranus with the 84-day revolution of Mercury, for example, and suggests that Uranus rules those qualities which distinguish 'us' from 'others' – that which is outstanding, extraordinary or original in a native: the opposite of the peculiar and original is the trivial, the everyday. Neptune reflects Jupiter much the same way as Uranus does Mercury: Neptune urges towards self-sacrifice, and service to others. Walther says indeed that this is the planet expressive of the 'principle of the Love of Christ', and visualizes the sigil ♆ as the 'bowl of sacrifice'. The opposite of unlimited openness and sacrifice is egocentricity, seclusion in oneself. Pluto was visualized as a reflection of traditional Mars, as a 'higher octave of the red planet': when it is emphasized in a horoscope it may be taken as an indicator of creative faculties. The opposite of creative (or even destructive forces) of Pluto is 'paralysis of the instinct'. Walther remarks that while Neptune has a revolution approximately twice that of Uranus, Pluto has a revolution three times that of Uranus.

New Style   Term used in reference to the so-called new chronology derived from the GREGORIAN CALENDAR, to distinguish it from the earlier Julian system: in astrological charts the term is usually abbreviated to NS (or n.s.). See also OLD STYLE.

New Venus   See TYCHO'S STAR.

night home   See DAY HOUSE.

night house   Each of the traditional planets (excluding the luminaries) has an associate night house, a sort of mundane nocturnal lordship which is perfectly in accordance with the ancient model of planetary rulerships (see SECTA). The traditional term 'night house' is a misnomer, for it is a sign, and not a house, which is being denoted. The night house of Saturn is Capricorn, of Jupiter Pisces, of Mars Scorpio, of Venus Taurus, and of Mercury Virgo. These are also the negative planets (see NEGATIVE MARS). The concept is linked with the idea that each of the non-luminaries have a negative and positive side to their natures, finding special expression in the pairs of signs allocated to them. See also DAY HOUSE.

night of Bramha See KALPA.

night sign See NIGHT HOUSE.

Nine Orders One of the names for the CELESTIAL HIERARCHIES.

nine sacred animals See CALENDARIA MAGICA.

ninety-degree circle A method of astrological charting, linked with the COSMOGRAM of the Ebertin *COSMOLOGY. The method was developed to combine the normal 360-degree system of chart projection with a 90-degree system in order to facilitate the study of PLANETARY PATTERNS and the study of transits and directions. It was used in connection with a Ninety-Degree Dial, a separate calibration designed to aid rapid assessment of the patterns.

Ninib Name derived from Chaldean sources, and generally equated with the planet Saturn – though originally it was the name of a god.

Ninth Heaven At one time the name given to the PRIMUM MOVENS, the Primum Mobile of the PTOLEMAIC SYSTEM, that sphere above the eight spheres which gave the diurnal revolution to all the others: sometimes it was called the Crystalline Heaven, the diaphanous or transparent heaven (even though all the other spheres were themselves said to be both diaphanous and transparent). See, however, TENTH HEAVEN. The Ninth Heaven was also called the Ninth Sphere, but unfortunately this same term was also later applied to the Earth itself, as some cosmoconceptions reckoned the concentrics in descending order from the periphery, with the Primum Movens (the old Ninth Heaven) as the first.

Ninth Hierarchy See ANGELS.

ninth house The ninth of the astrological HOUSES, linked with the nature of Sagittarius, and associated with higher mentality and education. It was formerly referred to as the House of Religion, in the days when theology was of great popular significance, and less divorced from the life-energies of philosophy. It has rule over long journeys – originally over those which could not be made in a single day, for which some preparation was required, though it was also concerned with the idea of mental travel, with speculation and deeper thought. In the melothesic man it has rule over the thighs and hams. In MUNDANE ASTROLOGY it rules higher education, universities, transportation, immigration and long-distance communication (though not television or radio).

Ninth Sphere See NINTH HEAVEN.

Nisan See BABYLONIAN CALENDAR.

Noctua Name given to an obsolete constellation visualized in the tail of HYDRA, called Night Owl.

nocturnal arc The variable arc through which the Sun appears to pass from sunset to sunrise. It is, of course, the arc left when the DIURNAL ARC is subtracted from 360 degrees, and is conditioned both by space and time: in space by latitude, in time by the relationship between Sun and solstices.

nocturnal horary time See HORARY TIME.

nocturnal house See NIGHT HOUSE.

nocturnal planet A planet is said to be nocturnal when it is above the horizon in a nocturnal nativity – when, in other words, it is in the night sky. The term is also used in a specialist sense recorded by Ptolemy, who tells us that the Moon and Venus are nocturnal, the Sun and Jupiter diurnal, while Mercury is of the nature of both day and night (diurnal when matutine, and nocturnal as an evening star). The definition is then somewhat clouded by the introduction of an idea which has been misunderstood by generations of astrologers: Ptolemy assigns Saturn, a cold planet, to the warmth of day, where its influence is mitigated – it is therefore neither a nocturnal nor diurnal planet, but a sort of 'adopted-diurnal'. Similarly, Mars (which is a dry planet) is mitigated in action by the moisture of the night: it is therefore neither nocturnal nor diurnal, but 'adopted-nocturnal'. Later authorities, including Wilson, have misread Ptolemy, resulting in Mars being listed as a nocturnal planet, and so on.

nocturnal signs   See NEGATIVE SIGNS.

nodal   As a general term, nodal applies to any sensitive degree (or arc) on the ecliptic – for example, the Ascendant may be described as a nodal point. More specifically, the nodes are the points (hence the sensitive degrees) at which the orbits of planets (or other celestial phenomena, including even meteors) cut the plane of the ecliptic. When in an astrological context the term 'node' is used, without further qualification, the reference is usually to one of the NODES of the Moon.

nodal horoscope   A system of symbolically projecting the heavens for astrological purposes, based on the CHOISNARD CHART, but locating the ascending node of the Moon on the Ascendant. It is a system proposed by the astrologer Froger, whose own researches have indicated that the mean node rather than the true node of the Moon should be adopted.

nodes   In relation to planets, the nodes are the points in which the orbit of the planet intersects the ecliptic. The point at which the planet crosses the ecliptic into the northern latitude is called its north node, while the point where it crosses into the southern latitude is called the south node. In traditional astrology the only planetary nodes which are accorded any importance are those of the Moon, called the DRAGON'S HEAD, and the DRAGON'S TAIL, both of which regress approximately 3 minutes of arc each day. In some modern forms of esoteric astrology, and in all modern forms of heliocentric astrology, the nodes of the planets are of great importance (indeed they are the seminal forces in a heliocentric chart).

Nodus   See PISCES CONSTELLATION.

Nodus Coelestis   See ALRESCHA.

Nodus Piscium   See ALRESCHA.

nonagen   An aspect of 40 degrees – See NONAGON.

nonagon   Defined by Simmonite as an aspect of 40 degrees, which divides the heavens into nine equal arcs. Rudhyar calls the same aspect the 'Nonagen'.

nonagesimal   A name sometimes given to that point of the ecliptic which is 90 degrees from the Ascendant in the diurnal arc. This point should not be confused with the MIDHEAVEN – in the MODUS EQUALIS system of symbolic house division, the nonagesimal marks the 10th house cusp. It is sometimes misleadingly called the zenith projection.

non-elemental aspects See ELEMENTAL ASPECTS.

Nones See KALENDS.

non-planets An unfortunate term, since strictly speaking it relates to everything in the cosmos which is not a planet, but used by Dean to include a whole range of celestial phenomena only tenuously connected. These are specifically given as COMETS, ECLIPSES, FIXED STARS, HELIOCENTRIC NODES, HYPOTHETICAL PLANETS, NODES and PARS. The classification appears to be based on the (mistaken) notion that each of these is involved with influencing degrees or arcs of degrees of the zodiac with natures analogous to those of the planets, even while they are not themselves planets. Within such a definition, many other factors should have been included, such as the FUTURE PLANETS, the SACRED PLANETS, and so on.

non-sacred planets The esotericist Bailey distinguishes five non-sacred planets (to be distinguished from her SACRED PLANETS), which she links with the etheric centres of the human spine. The five non-sacreds include the Earth (which is said to be on the way to becoming a sacred planet), the Moon (connected with the spleen and thymus gland, and often said to 'veil' another planet – see therefore ESOTERIC MOON), Mars, which functions in connection with the sacral centre of Man, Pluto which is connected with the solar plexus, and the Sun (which is also said to be veiled by another planet – see therefore ESOTERIC SUN), linked with the parathyroids, and acting as a mediator between the higher and lower organs. The non-sacreds are not to be confused with the ORTHODOX PLANETS of Bailey's system.

non-sacreds Another term for the NON-SACRED PLANETS.

noon date See ADJUSTED CALCULATION DATE.

noon mark A term used to denote the equivalent time of noon (Greenwich Time) in regard to the place for which the chart calculations are applicable.

noon-point method A term originated by the astrologer Leo, 1912, to describe a simple method of RECTIFICATION, based on the fact that if the noon date for a particular figure is known, then so is the exact Ascendant.

The method appears to be entirely unresearched prior to publication, and is based on the chimerical application of the noon date into the life of the native, at a time which is supposed to characterize maximum astrological intensity, which is registered either by the native or by some astrologer interested in his chart. The noon date used in progressions is properly called the ADJUSTED CALCULATION DATE.

Norma   Constellation located to the north of the asterism TRIANGULUM, sometimes itself called the Southern Triangle, sometimes Quadra Euclidis (Euclid's Square). Astrologers accord it a reputation for rectitude, and an interest in geometry, etc.

normal stars   A term introduced by the historian Epping in reference to 31 stars which were sufficiently prominent in the Babylonian zodiacal belt to be used as fiducials. The 'belt' was some 10 degrees north, and 7.5 degrees south, of the ecliptic. In the early cuneiform texts of Babylonian astrology, planetary positions were given in reference to these normal stars. See also MUL-APIN TABLES.

Northern Angle   The cusp of the 4th house is called the Northern Angle, and is sometimes (wrongly) identified with the IMUM COELI.

Northern Claw   *See* ZUBENESCHAMALI.

Northern Cross   See CYGNUS.

Northern Crown   See CORONA BOREALIS.

Northern Fly   See MUSCA BOREALIS.

northern quadrant   See QUADRANTS.

northern signs   These are the six signs from Aries to Virgo inclusive – probably so called because in the standard chart (the MACROCOSMIC CHART) they fall below the horizon in the nocturnal and northern arc. They are sometimes called the boreal signs.

North Horn   See SHARATAN.

north node   See NODES.

North Point   The cusp of the NORTHERN ANGLE.

North Scale   See ZUBENESCHAMALI.

notable nativities   A term now used generally to denote any collection of horoscopes, or related data, but originally derived from the title of a book *A Thousand and One Notable Nativities*, published under the name of the astrologer Leo, comprising birth data and horoscopes derived from a large number of astrological sources prior to the first decade of our century. While a useful reference list (along with a second volume of related data), especially in regard to the quoted sources, it must be used only with caution, since some of the ROUNDING OFF is not accurate, and not all the primal sources are beyond suspicion. Many similar collections were published in previous centuries, most notably those given by Gauricus in 1542 (figure 45), those

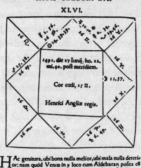

Figure 45: Horoscope of Henry VII of England, cast for 27 June 1491, at 10.40 am – the central inscription is inaccurate – after Cardan.

by Junctinus in 1583, the *Collectio Geniturarum* of Gadbury in 1662, and the many figures printed copperplate by Sibly in 1789. An up-to-date equivalent collection (which is certainly more accurate) is in the Sabian symbols of Jones.

nova   Usually defined generally as 'a new star', the nova is something of a misnomer: as a phenomenon, it is a sudden and often very great increase in light (and presumably in release of energies) of an extant star. Sometimes a nova is popularly described as 'an exploding star', but in fact the sudden burst of light and energy does not destroy the body, so that a nova will gradually return to being an ordinary star once again. Perhaps a nova might more usually be defined as a temporary brilliance in a star. The observed changes do, however, suggest that incredible energies are involved: the nova in the Andromeda Nebula of 1885 is reported to have emitted more light in six days than our Sun will emit in a million years! The nova observed in Cassiopeia by Tycho Brahe in 1572 (it remained visible in daylight for about 18 months) is often said to have been the first to be recorded, but there are several indications of earlier ones being observed (if misunderstood). The importance of the Cassiopeia nova was the profound effect it had on men's thinking about the nature of the cosmos: few things shook the emotional security men felt in the PTOLEMAIC SYSTEM more dramatically than the 1572 nova. It is likely that the influence of a nova was seen in terms similar to those of the adjacent or involved fixed stars. One wonders to what extent the dire tradition attached to Cassiopeia influenced Tycho Brahe, when he described the nova in that constellation as of an evil influence, an omen of wars, sedition, captivity, and deaths of princes, the destruction of cities, together with drought and fiery meteors in the air, bringing pestilence and venomous snakes? The causes of novae are unknown, and the attempts of astrologers to account for them within their system, to invest them with prophetic meaning, are usually quite pathetic. For example, the Nova Aquilae of 1918, which increased its brightness by 40,000 times in four days, was linked by Deluce with 'the year in which an American Expeditionary Force on foreign soil turned the tide of World War I' – a chauvinistic parochialism which is typical of this kind of astrology.

November The eleventh month (30 days) of our calendar, its name being derived from the Roman system prior to the introduction of the JULIAN CALENDAR reforms, when it was the ninth month (the Latin *novem* means nine). The esoteric significance of the month is linked with this original number, with its magical power as 3 × 3 marking the completion of two interrelated cycles, as a necessary prelude to the enactment of a new cycle which will begin in the following month. There is a possibility that this magical connotation was linked with the change of the time of Christmas to December (originally Christmas was celebrated in January). Certainly the imagery of the important Thrysdus (El Djem) mosaic calendar (see Foucher) suggests a connection with the Egyptian cult of Isis and Anubis, which in the later Christian imagery developed into Marian and Christophorian equivalents: in the November segment of this mosaic Anubis is dog-headed, a caduceus-bearing Mercury prototype of St Christopher.

novenary measure   An arc, or measure, used in SYMBOLIC DIRECTIONS and derived by dividing the 30-degree sign into nine parts, to give an arc of 3 degrees and 20 minutes. This is the equivalent of the Hindu NAVAMSA.

novienic chart   The novienic equivalent chart is one derived in principle from the 'enneal horoscope' of the Greeks, which was the Sanskrit 'navamsa chart'. The novienic chart is used to determine the equivalent of the novienic Moon from the sidereal placing. This novienic Moon is then regarded as being on the Ascendant of the chart, and the houses are disposed according to the Modus Equalis symbolic division of 30-degree arcs. The rules of construction and interpretation were set out by the astrologer Fagan. See also SEXASCOPE.

novienic Moon   See NOVIENIC CHART.

Noviens   Sometimes used of the novienic Moon – see NOVIENIC CHART.

Nubeculae Magellani   Constellation formed by Bayer in 1604 and named in honour of the seafarer who described them 'Magellan's Little Clouds'. The astrological influence is said to be poetical, artistic and imaginative, though it involves much hard work. There are actually two separate 'clouds', the 'Greater' (Nubecula Major) and the 'Lesser' (Nubecula Minor) – the Greater is said to convey isolation and loneliness.

Nubecula Major   See NUBECULAE MAGELLANI.

Nubecula Minor   See NUBECULAE MAGELLANI.

Nubilum   See PRAESAEPE.

number symbolism   See QABBALISTIC ASTROLOGY and TETRACTYS.

nunctator   See NUNTIUS.

Nuntius   One of the medieval esoteric names for Mercury, which pictures the planet in its anthropomorphized role as messenger. It is sometimes called 'nunctator', though in both forms it is found most often in the astrological terminology linked with alchemy. Blavatsky writes of astrological Mercury as 'still more occult and mysterious than Venus', and likens the word Nuntis (a term seemingly derived from the Mystery wisdom) to 'Sun-Wolf' – no doubt a reference to the hermetic Anubis (see NOVEMBER). She is of course touching upon the esoteric tradition which links Mercury in union with the Sun – the St Christopher Mercury, who carries the Sun Child on his shoulders, was originally a dog-headed Latinized Anubis.

Nunu   See PISCES CONSTELLATION.

nuptial number   In effect, the number 60, derived from the properties of the Pythagorean Triangle, a right-angled triangle the shorter sides of which were ascribed the values 3 and 4 the hypotenuse yielding 5. The multiples of these three figures (3 × 4 × 5) equal 60. This mystical number was linked with the so-called SAROS CYCLE, as well as with the harmonics of the spheres.

Oberon Name given to one of the satellites of Uranus.

obeying See COMMANDING.

Oblique Ascension A system of measurement (usually expressed in degrees of arc) of the circle of declination from the first degree of Aries eastwards along the circle of the equatorial plane. If a point to be located by such a co-ordinate is not within the plane of the equator itself, then it must form an angle with its corresponding part of the equator when they rise together: this angle is called the 'ascensional difference'.

oblique sphere Any sphere that is not in the vertical of the Earth poles.

obliquity of ecliptic This is expressed as the angle between the Earth's equatorial plane and the plane of its own orbit – 23.5 degrees.

occidens One of the Latin names for the western angle, meaning 'place of sunset' or 'sunset', derived from an etymology linked with the idea of killing – presumably in reference to the killing of the sunlight on the western horizon at sunset. However, occultists usually read deep significance into the link between the west and death. For various terms, see OCCIDENTAL.

occidental Literally 'westerly' or 'western' (see OCCIDENS) – but distinguish the following derivatives, which are ambiguous : OCCIDENTAL MOON, OCCIDENTAL PLANET and OCCIDENTAL SUN.

occidental Moon The Moon is said to be occidental (whatever its mundane placing) when decreasing in light, and hence to the west of the Sun.

occidental planet A planet may be termed 'occidental of the Sun' when it rises or sets after the Sun – but see OCCIDENTAL.

occidental Sun Specifically the term should be 'western sun' (see OC-CIDENS) but in a specialist (and somewhat confusing) application the term is used of the Sun when it is in any of the two arcs of the 7th, 8th and 9th houses, and the 1st, 2nd and 3rd houses.

occourse A term derived from translation of Ptolemy's *eupantesis*, and used in reference to the celestial changes 'experienced' by the planets, such as aspects, ingresses, occultations, and so on. When an aspect or occulation is involved, the planet bringing about the occursion was called the occursor. This word is now archaic.

occultation When a planet is hidden (the term 'occulta' means 'hidden things') by another body, it is said to be in occultation.

occult house One of the names for the EIGHTH HOUSE.

occursion See OCCOURSE.

Och Name of a Daemon listed in the grimoires as one of the seven supreme Angels of the cabbalists, and said to possess all the attributes of the Sun. He is supposed to have a complete knowledge of medicine and healing, and to be able to change base matter into gold. He is said to be capable of prolonging human life to 600 years – which is a scarcely veiled reference to the ancient NAROS CYCLE, itself linked with the cycles of reincarnation.

Octans Constellation from approximately 20 degrees Cancer to 10 degrees Leo, and from 9 to 22 degrees south, formed by La Caille in 1752 under the original title of Octans Hadleianus in honour of Hadley the inventor of the octant. It is supposed to give a scientific mind, though may also subject the native to psychic disturbances.

octave See CELESTIAL HARMONICS, DIAPASON and PLANETARY TONAL INTERVALS.

octile A modern name for the aspect of SEMI-SQUARE. The term suggests that the aspect is used as a division of the zodiac according to the principles of HARMONICS, being considered as a division of the circle by 8, rather than as a pure aspect.

October The tenth month of the GREGORIAN CALENDAR, the eighth of the JULIAN CALENDAR (hence 'Octo', which means 'eight'), though now with 31 days. The pre-Julian calendar reckoned from March as the first month, hence the origin of the word from the Latin *octo* (eight). Esoterically the month is still linked with the number 8 in that this begins a new septenary cycle (see SEPTEMBER), a deeper descent into matter: in the northern parts of Europe, winter was reckoned from the first full Moon of this month.

octopodos   A term derived from the Greek, referring to a system of HOUSE DIVISION restricted to eight houses (the word *topoi* in Greek meaning 'places' or 'houses' in an astrological sense). As Fagan has shown, the system is unrelated to the conventional zodiacal signs, and is linked with the ancient three-hour 'watches' time system. See also OKTOTOPOS.

Oculus   Fixed star, the pi of Capricornus, set in the right eye of the Goat, and said to be of the nature of Venus conjunct Saturn.

Oculus Leonis   Name given to the 9th of the LUNAR MANSIONS in some medieval lists. It is probably located from the fixed star Alterf in the Lion's mouth, though termed 'eye of the lion' in reference to an earlier image of the asterism. See table 48 under LUNAR MANSIONS, and LUNAR ZODIAC.

Oculus Tauri   See ALDEBARAN.

oikeiosis   See FAMILIARITY.

oktotopoi   The astrologer Manilius termed his system of EIGHT PLACES the *oktotopoi* (eight places).

old Moon   See MOON PERIOD.

old Saturn   See SATURN PERIOD.

Old Style   Term used in reference to the so-called 'old' chronology of the JULIAN CALENDAR which was superseded by the GREGORIAN CALENDAR. It is normal practice for astrologers providing astrological data which might be taken as relating to either system to mark their data OS (or o.s.) for the Old Style, and NS (or n.s.) for the NEW STYLE, to avoid ambiguity.

old Sun   See SUN PERIOD.

omens   See OMINA.

omina   The art of divination by means of omina or celestial omens, is not strictly speaking a part of genuine astrology, though the early forms (widely practised in Babylonian and Egyptian societies) were incorporated into later forms of astrology. According to the historian Pingree, the professional omen-readers in the Mesopotamian royal courts were called the *baru*, and it was through the *baru* that the gods revealed the meanings of the symbolic language of celestial phenomena. The organization of omina material appears to have been fourfold, in terms of phenomena relating to the Moon (Sin), the Sun (Samas), meteorological phenomena (Adad) and Venus (Istar or Ishtar). The system of omina was widely spread in the Middle East, and even into India by the time of the Persian Empire. Pingree says that the main impact of the texts and praxes on Indian astrology was in the field of military astrology (*yatra*) and catarchic astrology (*muhurta*). The influence of certain omina-based ideas may be seen in surviving astrological material derived from 2nd-century Egypt, set in the form of instructions said to have been presented by the priest PETOSIRIS to his king NECHEPSO: several ideas from such sources have supposedly circulated even into modern astrology, as for example, the imposing (if insubstantial) edifice of PRE-NATAL ASTROLOGY which is based on a somewhat obscure principle (see TRUTINE OF HERMES) claimed to have been derived from Petosiris.

one-degree method A method of SYMBOLIC DIRECTION involving the taking of one degree of zodiacal arc as equivalent to one year of life. To be distinguished from DAY FOR A YEAR, which is used in SECONDARY DIRECTIONS. A variation is the so-called RADIX METHOD.

onomantic astrology A form of popular astrology which should not really be called 'astrology' at all, in so far as it is not based on an interpretation of the heavens, but upon an exceedingly simplistic approach to numerology. The practice is based on the conversion of letters of the alphabet (usually the name of the querent, or the names of contestants, and so on) into numerical equivalents, by one or other of the many rules of numerology. From such conversions, simple diagrams (which to some extent resemble astrological charts) are derived from which the onomantist will give response to questions, usually on a dualistic (yes or no) basis. Certain of the onomantic methods, such as the Hellenistic Circle of Petosiris, claim to give response in matters of health, even in matters of life and death. The idea behind this form of onomancy appears to rest on a lost hermetic tradition that in previous times sounds and numbers were cosmically related – several qabbalistic systems of exegesis are based on this axiom.

Ophanim The Hebraic plural for the spiritual 'wheels' of the biblical Ezekiel, which Blavatsky rightly links with the World Spheres: the word is Hebraic for 'Angels of the Stellatum'. Blavatsky identifies these with the spiritual beings now called CHERUBIM (perhaps the Assyrian *Karoubs*, sphinxes) figured in the zodiac as the four Fixed signs – see TETRAMORPHS.

Ophiel   Name (sometimes Oriphiel) given to the fifth of the supreme Angels of the Qabbalists. He was said to possess the attributes of Venus, and to have the curious ability to transmute solar forces (gold) into the Venusian forces (copper). His Seal is given in SIGILS – PLANETARY SPIRITS.

Ophiuchus   Constellation from approximately 27 degrees Scorpio to 27 degrees Sagittarius, straddling the celestial equator in a 40-degree arc. It is sometimes called Serpentarius (serpent-holder), though the serpent itself is usually described as a separate asterism (see DRACO). There are very many variants of the Greek term, which in the Latinized form was very removed from the original, as, for example, Afeichus and Alpheichius. Equally, there were many variants of the original Latin, as for example Serpentiger, Serpentis Lator, and inevitably it was linked with the deified initiate Asculepius, who used healing serpents in his sanctuaries. The history of the Arabian names is the story of the degeneration of the Arabian translation of the Greek into Al Hawwa, which floriated into many terms, including Alangue, Hasalangue, and such as appear in medieval texts. Ptolemy likens its influence to Saturn tempered by Venus, but in literature the constellation has gained unpopularity as the promoter of evil and poison (the snake in the hand?): but see the relevant star RASALHAGUE. In the MODERN ZODIAC derived from Delporte, Ophiuchus has the curious distinction of being made zodiacal – see figure 44 under MODERN ZODIAC.

opposition   An ASPECT maintained by two or more planets so placed that they confront each other across the zodiac, being separated by 180 degrees. It is sometimes called the aspect of separation, and is said to promote tension between the planets concerned, or to intensify their actions and influences. The amount of ORB permitted this aspect is not fully agreed among authorities, but some permit up to 10 degrees.

Orai   One of the gnostic names for a ruler of Venus.

orb   That area of influence (expressed in degrees of arc) within which planets in ASPECT may be said to exert an influence while still forming or separating. Sometimes an aspect is ascribed an orb, at other times the planet is ascribed an orb. There is no fixed or agreed rule by which the orb for each of the aspects may be determined – but some indication of the traditional orbs is given in the relevant entries to the aspects, and table 56 sets out certain of the orbs suggested by modern astrologers, the names relating to the Bibliography. Traditionally allowances are made for orb when any of the planets involved are retrograde. There is much argument as to the orbs which should be allowed fixed stars – Wilson for example will permit a 1st-magnitude star as much as 7.5 degrees, which essentially means that fixed stars become operative in every natal chart. The present author allows only 0.5 degrees orb for a fixed star up to and including 3rd magnitude. The astrologer Libra quite rightly links the specific orbs with the natures of the planets involved. A precise orb is called 'partile', whilst one which is in weak orb is said to be 'platic'. When a planet is moving towards the partile, it is said to be 'in orb of application', and when it is separating from the platic it is said to be 'in orb of separation' – but see also APPLICATION and SEPARATION. The lack of agreement about precise orbs evinced in table 56 may probably be explained by the fact that Ptolemy's view of aspects was coloured by the JAIMINIC rather than by anything geometrical: the traditional Ptolemaic orbs are those given by Devore in table 56, though it is scarcely accurate to describe these planetary orbs as being 'according to Ptolemy', as Devore does. The orb attached to planets (rather than to aspects) may be seen exemplified in the closely studied MORINEAN ORBS, for here we have an excellent practical basis for a theory, which has resulted in tables of orbs appearing in many astrological textbooks long after their calculative basis has been forgotten. However, not all authorities would accept the Morinean orbs, any more than they accept the traditional orbs. Theory is very often different from practice in astrology: for example, Carter (who attempted a systematic portrayal of the aspects) is theoretically quite clear about the orbs he would allow, but the application of orbs in his practical examples of chart interpretation is even wider than those suggested by Hone.

orbit The path apparently followed by a planet in its revolution around the solar centre. In pre-Copernican systems the orbit was explained in terms of *EPICYCLES, since it was believed that supralunar bodies could move only in perfect circles. In post-Copernican systems, which have developed from the COPERNICAN SYSTEM, a theory of elliptical orbit (see ORBITAL REVOLUTION) has developed, and now forms the basis for the modern model of the solar system and for computations: yet see LEMNISCATE.

orbital revolution A term often shortened in astrological texts to 'revolution'; the annual motion of any celestial body around the Sun. A complete orbital revolution is a 'year' of the planet or body concerned, but this is usually translated into Earth years – that is, made into a ratio of the Earth's orbital revolution. The sidereal 'month' orbit of the Moon is 27 days, 7 hours

―――――――――――― TABLE 56 ――――――――――――

	Devore	Wilson	Lilly	Leo	Hone	Cornell	Brau	Libra
MO	12	12	12			12	15	
ME	7	7	7			7		
VE	8	8	7			8		
SU	17	17	15			17	15	
MA	7	7	7			7.5		
JU	12	12	9			12		
SA	9	9	9			10		
UR	5					8	2	
NE	5					8	2	
PL							2	
Conjunction				12	8		8–10	10–12
Opposition				12	8		6–8	10–12
Trine				8	8		4–6	8
Square				8	8		5–6	8
Semi-square				4	2		2	4
Sesquiquadrate				4	2		2	4
Sextile				2	4		4–6	7
Quincunx				2	2		2	2
Parallel				1				1
Quintile					2			
Bi-quintile					2			

and 43 minutes, the geocentric revolutions of the other planets are given in table 57, alongside the sidereal periods derived from Schultz.

Orb of application   See ORB.

Orb of separation   See ORB.

oriens   One of the Latin names (meaning 'rising', in reference to the symbolic and actual rising of the Sun) for the East Angle, the Ascendant, and the cusp of the 1st house. For various related terms, see ORIENTAL.

oriental   Literally 'easterly' (see ORIENS); but distinguish the following derivatives, which are sometimes confusing: ORIENTAL MOON, ORIENTAL PLANET and ORIENTAL SUN.

TABLE 57

Planet	Geocentric	30-degree Arc	Sidereal
ME		14 days	87.9690 days
VE		24½ days	224.7008 days
SU	(365 days, 6 hours, 9 minutes)		
EA			365.256 days
MA	2 years	6 weeks	686.9798 days
JU	12 years	1 year	4,332.588 days
SA	29 years	2½ years	10,759.2 days
UR	84 years	7 years	84 years and 8 days
NE	165 years	13¾ years	164 years and 282 days
PL	248 years	20¾ years	247 years and 257 days

oriental houses See EASTERN HOUSES.

oriental Moon The Moon is said to be 'oriental' (whatever its mundane placing) when it is increasing in light, and hence to the east of the Sun: the term is ambiguous (see ORIENTAL PLANET).

oriental planet A planet may be termed 'oriental of the Sun' when it rises or sets before the Sun – see however ORIENTAL MOON.

oriental Sun Specifically the term should mean 'the easterly Sun' (see ORIENS), or 'the Sun on the Ascendant', but (confusingly) the term is used of the Sun when it is in any of the two arcs of the 10th, 11th and 12th houses, and the 4th, 5th and 6th houses.

Orion Constellation approximately 6 degrees of Gemini to 2 degrees of Cancer, and from 13 south to 24 north of the equator, one of the most impressive of asterisms. The name is from the ancient Greek, which appears to have been derived from an earlier term which was (as Allen says) 'akin to our Warrior' in sound. The giant Orion is said to have died from the sting of Scorpius, consequent to his boast that he could slay any earthly creature. Many variants were used in the earlier texts, Aorion, Arion, Argion and Urion, though to the Arabian astrologers he was Al Jabbar (the giant), which was probably from Ptolemy's term Gigas, and gave rise to several curious Arabian derivatives, such as Algebar (frequent in medieval astrological texts),

Algibber and so on. The interesting explanation for the early Arabian Al Jauzah (wrongly translated as 'giant') is that the word was used of a black sheep with a white spot on its body, and thus made an excellent stellar reference to this brilliant asterism. The names which figure so prominently in medieval astrological and magical texts (Orion being used in talismanic magic to promote warrior strength) are derived from this word: Elgeuze, Jeuze, and so on. In Egypt the constellation was linked with Horus (as, for example, in the DENDERAH ZODIAC), though the asterism appears to have been called Sahu. Related names are the Jewish Gibbor, and the Latin Venator (huntsman) which idea probably links with its reputation in astrology for promoting hunting. The asterism is said to induce a dignified, self-confident and arrogant nature when emphatic in a chart, but its several powerful stars always require special notice: see therefore BETELGEUZE, BELLATRIX, MINTAKA, RIGEL and ALNILAM.

Orion's Belt   Popular name for the Cingula Orionis, distinguished by the stars ALNILAM, Alnitak (the zeta of the constellation) and MINTAKA.

Orion's Foot   Popular name for the fixed star RIGEL, in the foot of ORION.

Oriphiel   See OPHIEL.

Ormazd   Sometimes called Ahura Mazda (Great Being of Light), Ormazd or Ormuzd is the creative power of Zoroastrianism who worked against the darkness and lies of AHRIMAN (more exactly, against the Angrimayu who was later called Ahriman in Western occultism). While the latter is often identified with the lunar forces, Ormazd is linked with the solar forces, and Blavatsky records the esoteric teaching that he is the synthesis of the Elohim, the creative Logoi, which are in turn associated with the realm or sphere of the EXCUSIAI.

Orphic   A term derived from the associations built around the legendary magician Orpheus, who is said to have founded the Greek cult of Orphism, which remained connected with several ancient Mystery centres. When the term is used in (non-scholarly) astrological contexts, it is generally as a synonym for 'esoteric', or as pertaining to the secret doctrines which pervade astrological lore.

Orphic ages   According to an account of the Orphic theogony recorded by Servius, there are four ages of the world. The first is that of Saturn, the second of Jupiter, the third of Neptune, and the fourth that of Pluto. The gods must (for once) be distinguished from the 'planetary beings', since it is clear from the context that they are intended to personify the four elements: SA fire, JU air, NE water, PL earth. These associations have given rise to many misunderstandings in later symbolism, but they are probably linked with the stages of the creation of the Earth outlined in modern esotericism – see, for example, Wachsmuth.

Orphic descent of the soul In his commentary on an esoteric section of Plato's *Republic*, Macrobius quotes what he describes as Orphic initiation knowledge relating to what the descending soul learns in its passage through the planetary spheres, prior to rebirth. A summary of this descent reflects much of the ancient teaching of the influences of the planets themselves. The soul, wrapped in a luminous body, descends into the Sphere of Saturn where it develops the power of reasoning and theorizing (contemplative reason – *intelligentia et ratiocinatio*); in the sphere of Jupiter it develops the power of putting into practice (*vis agendi*, but more specifically in Greek, *to praktikon*). In the Sphere of Mars, it takes on the power of 'ardent vehemence' (though the word translated is actually *Thumikon*, which in esoteric lore is linked with magical powers). In the Sun Sphere, the soul learns 'sensing and imagining' – in a word, what Coleridge would have called 'phantasie', the power of pictorial visualization or representation. In the Sphere of Venus the soul takes on the 'power of desire', which is linked with love. In the next Sphere (Mercury), the soul is given the power of 'giving expression', and of 'interpreting feelings' (those feelings engendered in the previous Sphere). In the final Sphere, that of the Moon, the soul begins to learn about physical bodies – actually about how to make physical bodies move and grow, which is esoterically connected with the etheric forces – see Wachsmuth. In the lower realm, the Sphere of the Earth, where it takes on through incarnation the 'terrene' body, the soul is imprisoned in a dark shroud of physical. As the historian Mead remarks, the 'planetary energies thus imbibed are in no way regarded as being either beneficent or maleficent', and there is no question of good or bad emanation in the spheres: see therefore the contrasting EGYPTIAN DESCENT OF THE SOUL.

orthodox planets In her highly complex system of INTUITIONAL ASTROLOGY, Bailey lists four different classifications of planets which include what she calls the orthodox planets – but see also ESOTERIC PLANETS, NONSACRED PLANETS and SACRED PLANETS. With a single exception, the orthodox planets correspond (at least in their rulerships) to the ancient Ptolemaic model, with Mars ruling Aries, Venus Taurus, and so on. The exception is that Uranus (the only one of the so-called MODERN PLANETS in this classification) is accorded rule over Aquarius, which of course is quite in accordance with the modern astrological view:

AR	ruled by MA	TA	ruled by VA	GE	ruled by ME
CN	ruled by MO	LE	ruled by SU	VG	ruled by ME
LB	ruled by VE	SC	ruled by MA	SG	ruled by JU
CP	ruled by SA	AQ	ruled by UR	PI	ruled by JU

It would appear that this group of orthodox planets are regarded as being operative in the charts of ordinary people – that is, in the charts of those who have not been initiated, or who have not otherwise undertaken to speed up their evolutionary progress. See also UNORTHODOX ASTROLOGICAL RELATIONSHIPS.

ortive difference   A term derived from the Latin *ortus solis* (rising of the Sun) and used to denote the difference involved when directing the Sun at its rising – though the term is also confusingly used of the same procedure in regard to the setting of the Sun. Precise definition of the differences involved is not always clear: Devore suggests that the ortive difference may be an indication of some effort to accommodate the fact of horizontal parallax, but it is equally probable that the difference is to do with the regional variations which were so much a problem for early astrologers. The actual sunrise in a mountainous district is very different from that given in tables designed for the notional horizon derived from consideration of the Celestial Sphere. It is certain (at least to the present writer who has had some practical experience of the matter) that many medieval horoscopes were based not on the calculated sunrise but upon the observational sunrise, and that therefore a difference (which might be calculated) was involved: perhaps this was indeed the ortive difference.

Osiris   Name given to one of the HYPOTHETICAL PLANETS by the astrologer Sutcliffe, as one of the four ultraneptunian bodies. See also ISIS. The remaining hypotheticals of this group were designated only by numbers. The Osiris of Sutcliffe is to be distinguished from its namesake discussed by Thierens in his survey of the esoteric ETHERIC SPHERES, as the ruler of Aries.

Ouranos   The ancient Greek name for the expanse of the heavens. The god URANUS was the husband of Gaia (the Earth), through whom the Titans were fathered. Esoterically, Ouranos and Gaia are the original parents of Man, who belongs partly to the realm of spirit (*ouranos*) and partly to the realm of elemental forces (*gaia*). This is why the esoteric texts give so many descriptive epithets for Ouranos, such as Waters of Space or the Celestial Ocean, for besides relating to boundless space, they relate also to Man. The term Ouranos was sometimes used in the late 18th century as a synonym for URANUS, from which it must be distinguished: see also GEORGIUM SIDUS.

outer planets   See SUPERIOR PLANETS.

Ov   Designation (with the alternative name Ov.O) made by the astrologer Charubel to an extraneptunian planet which he perceived clairvoyantly some years prior to 1897, for which he claimed a revolution of 297 years. He said that it gives 'longevity and magnetic power' – in some respects this hypothetical is similar to the then undiscovered PLUTO (with a revolution period of 248 years), but one wonders if it is merely accidental that Charubel should have found this planet to be in orb of Ascendant (12 degrees of Scorpio) of his own chart? Ov has been identified with MELODIA.

P

Paduan cycle Name given to one of the most ambitious of all medieval astrological frescoes, in the Salone of the Palazzo della Ragione, Padua. Three registers, totalling 333 separate compartments of largely planetary, stellar and cosmic images, extend over 217 metres. On rather dubious grounds, the cycle has been associated with Giotto (14th century), but later fire damage (1426) and heavy cyclone damage (1756) demanded considerable restoration, and at least three major restorations since. The post-1420 painters are unknown (Nicolo Mireti and Stefano de Ferrara are suggested as masters in charge). The cycle is still the most comprehensive surviving medieval attempt to relate zodiacal, constellational and seasonal imagery to Christian theology and symbolism: unfortunately the system is far from clear to modern minds, especially in regard to the symbolism of decans, and the associations drawn between the Apostles and the signs. The sequence is in fact broken to the south by an old ZODIACAL CALENDAR, surrounded by a Christian iconography which points to the ancient link between Christ and the Sun Sphere. Some of the images are no longer understood, but some reflect the iconography traditionally associated with Pietro d'Abano. A useful synopsis of Beltrame incorporates the more scholarly (though incomplete) treatment of Ivanoff and Barzon.

Pagan-Pluto Name given in comparatively recent times to the planet PLUTO. The astrologer Pagan designated as ruler of Scorpio a hypothetical planet in 1911 (prior to the discovery of the 'new' planet Pluto) which she named Pluto − it was a designation which influenced modern astrologers into adopting Pluto as the ruler of this sign. The modern term 'Pluto-Pagan' is used to distinguish the planet from various other Plutonic planets, such as the LOWELL-PLUTO, the WEMYSS-PLUTO, and the THIERENS-PLUTO. Pagan herself suggested a sigil for her Pluto ♇ which whilst graphically more correct than the more usual form ♇ has not been widely adopted.

Pallas Name given to one of the ASTEROIDS, discovered in 1802, and associated by the astrologer Dobyns with the sign Libra, along with an asteroidal co-ruler JUNO.

palmistic zodiac In his published research on ASTROPALMISTRY, Gettings has proposed a schematic coordination between the structure of the hand (chirognomatic and chiromantic) with the zodiac. By means of this model (which has been called the palmistic zodiac) it has been possible to demonstrate convincing relationships between personal charts and hand forms. The schema given is:

AR	Thumb (seat of Will)	LB	Mount of Venus
TA	Mount of Venus	SC	Ring Finger
GE	Line of Life	SG	Line of Head (seat of Mentation)
CN	Mount of Moon	CP	Middle Finger
LE	Line of Heart (seat of Emotions)	AO	Line of Fate
VG	Little Finger	PI	Index Finger

Pan Name given to one of the HYPOTHETICAL PLANETS listed by the astrologer Jayne as a transplutonian.

Pansophia A term used by the historian Haase in his study of the harmonics of Kepler as denoting an attitude of mind arising from the hermetic tradition, which (he says) strives towards presenting a unified harmonic picture of the world, often using fantastical and seemingly obscure materials in the picture-building. It is a holistic and hermetic attitude, which Haase contrasts with the fissiparous attitude of the MATHESIS UNIVERSALIS. In effect it is the late hermetic tradition as recorded in the writings of such occultists and astrologers as Pico della Mirandola, Regiomontanus, Agrippa, Trithemius and, to some extent, Kircher and Fludd – but above all Paracelsus.

pantheons See COSMIC PANTHEONS.

Papilio Name given to the 25th of the LUNAR MANSIONS in certain medieval lists. Probably identified by the asterisms around the urn of the AQUARIUS CONSTELLATION, near to Sadachbia. See, however, table 48 under LUNAR MANSIONS.

parallel An aspect formed between two or more planets with the same degree of declination, either north or south (together or separately) of the celestial equator. Planets so placed are said to be in parallel – however, in modern astrology the term 'in parallel' is used only of those parallel on the same side of the celestial equator: those in opposite declinations are said to be in contraparallel. Traditional astrology teaches that the parallel aspect has much the same force as in CONJUNCTION, though a maximum orb of 1 degree only is permitted. The whole realm of parallel aspects is very much in dispute in modern astrology, and some astrologers regard the parallel as effective only when the planets concerned are otherwise in Angular aspects. See also PARALLEL IN MUNDO.

parallel in mundo   Sometimes called Mundane Parallel, this has nothing to do with the zodiacal PARALLELS related to aspectal theory. A planet is in mundane parallel with another when it is the same distance from one of the four Angles. Thus, a planet on the cusp of the 2nd house may be in parallel (*in mundo*) with the planet on the cusp of the 12th house, since both are equidistant from the cusp of the Ascendant (1st house). The theory and practice of *in mundo* interpretation was developed by Placidus de Tito, and has not gained wide acceptance in modern astrology.

paran   A term derived by the astrologer Hand from the Greek PARANATELLONTA, applied to a potential relationship manifest at a moment of planetary transit, when (and only when) the natal planet is involved, and the planets of transit cross the horizon or meridian at the same time. The term 'paranatellonta', which applied to a temporal sequence (and was sometimes used wrongly of a spatial sequence), has thus been adapted to designate a supposed influence. Whilst the term appears to rest on a confusion of the meaning of the original Greek, it is properly restricted to certain *TOPOCENTRIC ASCENSIONAL TRANSITS. Brau, who gives the alternative term Paranatellon for Paran, defines it as a relationship between two planets which simultaneously cross the same Angle (or different Angles in the same chart). It is reasonable to link this approach to aspects with observational astrology, but it is quite wrong to claim (as Brau does) that Ptolemy wrote about Parans. This confusion might suggest that Brau is misunderstanding Ptolemy's view of the paranatellonta.

paranatellon   A term sometimes used in modern astrology as an alternative for PARAN, though in a way which might suggest a basic confusion with the PARANATELLONTA: for example of misuse, see PARAN.

paranatellonta   A term derived directly from the Greek, meaning approximately 'rising together', or 'rising side by side', and often (wrongly) applied to the sequence of the zodiacal signs or constellations, imagined as rising over the horizon, or (less frequently) over the meridian. The term may, however, have been applied to other phenomena, such as the rising of star groups, with or without planets. Boll, in reference to Ptolemy's use of the word, refers to the paranatellonta in terms of star groups: the word is explained as referring to stars which rise and set at the same time as the degrees of sections of the ecliptic, but to the north or south of these (see also PARAN). In modern astrology, the term is sometimes used to refer to the mere sequence of the zodiacal signs or constellations, irrespective of their rising: for example, Ivanoff refers to the frescoes of the PADUAN CYCLE as a parantellonta. Properly speaking, however, this is a misuse of the term.

parantellonta   A modern version of PARANATELLONTA. But see also PARAN.

parapegma   Originally a Greek-derived Latin term, meaning 'something fixed, or hung for display', and often applied to tables of astrological or astronomical calculations made of brass and fixed to a wall, pillar or floor. In modern astrological use, the term might be applied to the metal or marble slabs containing astrological symbols, descriptive of the meridianal parantellonta, transits, and like phenomena, in ZODIACAL CALENDARS.

Paransara   A name given to the Vedic sage who narrated the Vishnu Purana, but the name appears to have been used for various later astrologers. See, however, JAIMINI.

parilicium   See HYADES.

pars   A Latin term often translated as 'part', derived ultimately from the Greek astrological term *klipoi*. The term is used to cover an almost bewildering range of points derived from the adjusting planetary and nodal points to the Ascendant or other fixed points of reference, and ascribing an interpretative value to the adjusted positions of the revolved planets or nodal points. The PART OF FORTUNE, which is so important in the astrological tradition that it is sometimes merely called the part, is derived from the rotation of the degree held by the natal Sun until this becomes the Ascendant degree: the position of the adjusted Moon marks the point or part – but for typical construction of *Pars Fortunata* see figure 46A under PART OF FORTUNE. There is a corresponding pars for each planet (and, indeed, for several nodal points): in general the planetary pars are derived by revolving the Sun to the Ascendant, and then regarding the new position of the planet concerned as marking the pars. Examples of the general rules for calculating others are set out under individual headings – see therefore BROTHERS AND SISTERS, DISCORD, LOT OF EROS, FATALITY, FATHER, GAVEL POINT, GOODS, HUSBANDRY, INCREASE, INHERITANCE, LIFE, PLAY, POINT OF BONDAGE,

POINT OF FAITH, POINT OF MALE CHILDREN, POINT OF PASSION, POINT OF SERVANTS, POINT OF SICKNESS, POINT OF SPIRIT, POINT OF TRAVEL BY LAND, POINT OF UNDERSTANDING, MARRIAGE POINT, POMEGRANATE and PRIVATE ENEMIES. Many of the points are often (though erroneously) called 'Arabian Pars' simply because the notion entered medieval astrology by way of the Arabian astrologers, who had indeed developed their own idea of pars, but who were also transmitting the original conception from Greek astrology – see LOT OF DAIMON and SEVEN LOTS, for example. The pars are really nodal points, and it is often said that they 'cast no rays' – that is, that they are not subject to influences by aspects other than the conjunction, which may also be made operative by transit or progression.

parsec   A term derived from modern astronomy, denoting a unit expressive of vast stellar distance. One parsec equals 3.26 light years. It corresponds to the distance from which the mean distance of the Earth from the Sun would appear as a parallax of one second of arc.

Pars Fortunae   The Latin term for the PART OF FORTUNE, sometimes *Pars Fortuna*.

part   See PARS. When the English term 'part' is used in an astrological context, and without further qualification, it is almost invariably a reference to the PART OF FORTUNE.

Parthenos   One of the Greek names, meaning 'virgin', used for the zodiacal sign and constellation Virgo. Sometimes the constellation was Parthenos Dios (virgin goddess). See KORE.

partial eclipse   See UMBRAL ECLIPSE.

partile aspect   An exact ASPECT between two or more bodies. In general astrological use the term is applied to aspects which are in the same degree of longitude, but some astrologers insist that the term may be properly used only when the aspect is exact and 'in parallel' – see PARALLEL.

Part of Daimon   See LOT OF DAIMON.

Part of Fortune   The most important of the PARS – a nodal point in the horoscope chart which is the same distance of zodiacal arc from the Ascendant as the Moon is from the Sun (figure 46A). The point marks the place which the Moon would hold at the symbolic sunrise. Now the most important of

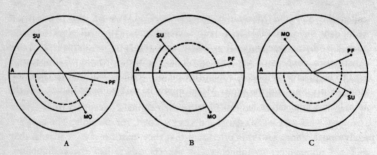

Figure 46: Pars Fortunae, *cast according to the methods of* (A) *Manilius,* (B) *Ptolemy and* (C) *Firmicus.*

the pars, in Hellenistic astrology it vied with the LOT OF DAIMON in importance. The above account of the pars is set out to satisfy the modern conception, but it must be observed that this is merely one of several methods (itself derived from Manilius) used in early astrology for determining the pars. Ptolemy gives a variation on this rule, as may be seen in figure 46B, while many of the ancients distinguished between a pars originated for a day chart from one originated from a night chart – the symbolic sunrise becoming the symbolic moonrise. Figure 46C, derived from Firmicus, is really an extension of the rule of Ptolemy to the night chart.

Part of Jupiter The name for the so-called Arabian PARS, sometimes called the POMEGRANATE.

Part of Love See EROS.

Part of Pluto See GAVEL POINT.

Passer The Latin for 'sparrow', an early name for the asterism later determined by Bayer as VOLANS.

Passiel Name given to two quite distinct spirits, with separate functions. In certain methods of ceremonial magic the name is one of several attached to the Governor of zodiacal Pisces. In the qabbalistic tradition, it is the name given to one of the Angels who has rule over Abaddon.

passive The Sun and Moon are sometimes said to be passive, in that they take their colouring from the signs in which they are placed: Simmonite called the Sun and Moon the 'passive stars'. However, this is clearly a redundant or inapplicable term with such a context, since all planets receive such a zodiacal colouring. It is likely that the term is derived from the Aristotelian concept of the *PRINCIPLES, for the principle of Moisture is said to be passive as indeed is the principle of Dryness. Within this classification the dry Sun and the moist Moon might be said to be 'passive', but the term has no application in modern astrology.

passive stars See PASSIVE.

Pastor et Aries Name given to the 22nd of the LUNAR MANSIONS in some medieval lists. It is probably identified by the multiple star Algedi in the horns of Capricornus – but see Table 48 under LUNAR MANSIONS.

path of rebirth See DESCENT OF THE SOUL.

patterns See FIGURINES and JONES PATTERNS.

Pavo Constellation, the Peacock, situated to the south of constellations Sagittarius and the Southern Crown, and formed by Bayer in 1604. Astrologers say that it gives vanity and love of display.

Peacock See PAVO.

Pegasus Constellation situated to the north of the urn of Aquarius and the eastern fish of the asterism Pisces. This is the Winged Horse or Flying Horse, said in Greek mythological accounts to have been born of the blood of Medusa after her head had been cut off by Perseus: this severed head had dropped into the sea, which may partly explain the suggested etymologies of the word (see SCHEAT). The Greeks called the asterism Hippos, though sometimes Hippos Ieros (divine horse) and there are many related or derived terms, including the Latin Equus, Equus Ales, Alatus (winged), Sonipes (noisy-footed), and so on. It is suggested that the jackal-like animal on the DENDERAH ZODIAC represents at least part of Pegasus. Ptolemy associates its brightest stars with the influence of Mercury conjunct Mars, and the asterism is said to bring ambition, vanity, enthusiasm and bad judgement. Besides SCHEAT, see also MARKAB.

Peg Star One of the names used to denote the fixed star SPICA, when regarded as a zodiacal fiducial.

Pelagus Fixed star, the sigma of constellation Sagittarius, set in the arrow of this image. It is said to be the Nunki of the ancient Euphratean Tablet of the Thirty Stars, associated with the sea, hence the Greek name Pelagus. Ptolemy equates it with an influence similar to that of Mercury conjunct Jupiter: it is claimed that Pelagus gives a truthful and religious mind, and produces benefit with all planets, including even the malefics.

pentacle Specifically, a geometric figure linked with the number five, such as the five-sided pentagon or the PENTAGRAM or the PENTALPHA. However, the term is often used in a general sense for any simple geometric figure constructed for magical or amuletic purposes – thus the six-pointed SEAL OF SOLOMON or the septenary CHALDEAN ORDER is sometimes called a Pentacle – see also the so-called PYTHAGOREAN PENTACLE.

pentades The astrologer Fagan has argued that the so-called Egyptian DECANS, which are supposed to relate to 'ten-day star' measures, mark off only the visible vault of the heavens, and as symbolic projections are divorced from the material world. If this argument was sound, then the entire circle of division would have to be taken into account, making divisions not of 360 degrees, but of 180 degrees, in which case they would be pentades. However, there appears to be little doubt in the later development of the decans which influenced European astrology that they were precisely what their name indicates – 10-degree arcs, or threefold divisions of the signs. Fagan's argument merely calls into question the antiquity of the Egyptian zodiac, and permits certain of the ancient celestial diagrams to be linked with the heliacal rising of SIRIUS.

pentagram The pentagram, sometimes called the pentalpha, is the five-pointed star regarded as a symbol of many esoteric qualities: it is indeed the symbol of the microcosm, as well as the etheric (see MORNING STAR). The pentagram has from very early times been linked esoterically with the planet Venus. In this latter connection, as the work of Schultz demonstrates, when the patterns of conjunctions between Venus and Sun are plotted over synodic periods, a pentagram is traced in the skies around the Earth (figure 47). Sample data (additional to that in figure 47) given for a first order cycle relating to dates of conjunction are: 13.4.53; 15.11.54; 22.7.56; 28.1.58; 1.9.59; 11.4.61. In the 15th century Agrippa (an esotericist masquerading as a magician) published a number of interesting figures relating the planets to the human body and corresponding centres of rulerships. In the relevant figure for the pentagram (figure 48) the planets accorded the points of the pentagram are a sort of exoteric blind, for what is of real importance is the

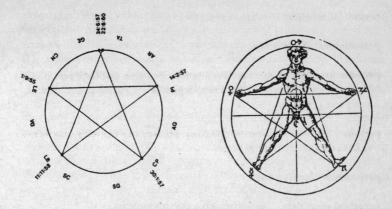

Figure 47 (left): Pentagrammatic form traced by the periodicity of Venus in relation to conjunctions with the Sun (after Schultz).

Figure 48 (right): The human form contained within a pentagram (after Agrippa).

centre, marked with the sigil for the Moon. The esotericists maintained that the pentagram was the symbolic form of the post-mortem state, and it is therefore of deep significance that the centre of this posture (the 'etheric posture', as it is sometimes called in esoteric circles) is placed upon the lunar sigil, for it is through the lunar sphere which the newly departed must first travel after death. The sexual parts, linked with the NEGATIVE MARS of Scorpio, marks the unregenerate man which must be burned away in the purgatorial fires of the lunar sphere. Agrippa makes this clear by implication when he points out that another posture (with the arms raised) puts the centre upon the navel, which represents the solar, life-enhancing position, yet which breaks the pentagram-structure of the lunar-centred figure.

pentalpha   One of the names for the five-pointed PENTAGRAM, so-called because it reproduced in its sigillic form the letter A (the Greek alpha) five times: see figure 48.

Perilous Year   The Point of the Most Perilous Year, one of the so-called Arabian PARS. If the degree of Saturn in a natal chart is rotated to the Ascendant, the position then held by the Lord of the Eighth House of that natal chart marks the point.

periodical lunation   See SIDEREAL LUNATION.

periodicities   See CYCLES.

periodic lunation   Name used for a horoscope figure cast for a period of the Moon, when it returns to the exact degree of its radical place: 27 days, 7 hours and 41 minutes.

periods   See INITIAL CHRONOCRATORS, PLANETARY PERIODS, PLANETARY YEARS and REVOLUTION PERIODS.

perisomata   See HUMOURS.

perpetual noon date   See ADJUSTED CALCULATION DATE.

Persephone   Name given to one of the HYPOTHETICAL PLANETS in the system proposed by the Dutch astrologer Ram, and accorded rule over zodiacal Taurus. This transplutonian is included in an ephemeris, and claimed to relate to 'creativity and spiritual performance'. See also VIRGO CONSTELLATION.

Perseus   Constellation, sometimes called Champion, the killer of Medusa and rescuer of Andromeda. Perseus is from the Greek traced back to the Babylonian Parasiea, though the asterism has been given many names. Some epithets are used in Latin texts, most usually in reference to his victory over Medusa (the ALGOL) – Deferens Caput Algol, Victor Gorgonei Monstri, Gorgonifer, etc. Late and medieval terms, Celeub and Chelub are almost certainly from the Arabic *kullab*, which refers to the weapon he carries. The asterism was also called Cacodaemon, though this probably refers to the head of Medusa (see ALGOL). The astrological association is in accord with the idea of victory – it gives intelligence, strength and an adventurous nature, though with a tendency to deception. In connection with fixed stars, see ALGENIB.

personal degree   See PERSONAL STAR.

personal star   The popular notion that each person is linked with a particular personal star may perhaps be seen as an excessively simplistic version of the idea underlying astrology: see BIRTH STAR. However, Pliny records an interesting variation of the idea, in the view that a star is born with each new human birth, and then dies with the death of that person. In recent years the idea of a 'star' has been extended to the idea of 'degree of the zodiac' (which must surely contain at least one star, if not many), on the basis that the precessional rate links the Sun with one degree over a period of 72 years – see PRECESSION – which has been taken as a notional 'average period of life'. Thus, it is claimed that each person is born with a degree, and dies with that degree.

Petosiris A name or title supposedly of an Egyptian priest who wrote for his King Nechepso an astrological text which greatly influenced the antique notion of popular astrology. A number of semi-astrological hermetic doctrines, replete with magical ideas and cosmogonic formulae, have survived under the general title of *Nechepso and Petosiris*, but the style and ideas are typical of the Alexandrian School, and have been placed by Riess in the 1st century BC. See OMINA.

Phacd One of the seven fixed stars forming the constellation of URSA MAJOR, said by Ptolemy to be of the equivalent influence of Mars.

Phact Fixed star, the alpha of COLUMBA, sometimes called Phaet or Phad, the name of uncertain (Arabian) origin. Lockyer maintains that at least twelve Egyptian temples were orientated towards this star. It is associated with beneficial nature of Mercury and Venus combined.

Phainon The Greek name, meaning 'the shiny one', for the planet Saturn. But see also PHOENON.

Phaleg The third of the seven supreme Angels of the Qabbalists, described as the 'war lord'. His Seal is given in SIGILS – PLANETARY SPIRITS.

Phalguni The Sanskrit name for the constellation Virgo – but see KANYA.

phase-angle A term defined by the modern astrologer Addey relating to the theory of HARMONICS, specifically for astrological use, though derived from general wave-theory. The phase-angle is the angle between the 'peak' of a wave-form and the first degree (zero point) of Aries. In consideration of this system of measurement, he takes the wavelength of each harmonic as being no more than 360 degrees in extent.

phases See PLACES.

phenomenon In ordinary use a phenomenon is something which is perceived or observed, which means that in its astrological sense the term is misused since it denotes such celestial events as ingresses, times of nodes, and so on, which may not be observed. In addition, the term covers such 'observable' events as eclipses. Ephemerides often provide lists of so-called phenomena as supplementary material.

Pherg   Al Pherg, fixed star (double) of 4th magnitude, the eta of Pisces, set in the cord forming the two fishes. Of a beneficial influence.

phlegmatic   One of the four TEMPERAMENTS, arising from an excess of the water element in the psychological make-up of the personality. The phlegmatic temperament is withdrawn, sensitive and unstable, and within the astrological tradition is associated with such keywords as receptive, confused, emotional, fluctuating, lacking in direction, self-protective and slow. The type is usually very methodical, and has a great power when the forces of the temperament are rightly canalized, yet its exterior is often deceptive, unrevealing of the depths within. Faults in the temperament arise from over-sensitivity to physical phenomena, which often produces fear, or an inability to cope, under which circumstances the type is mistrustful and suspicious of radical change.

phlegmatic quadrant   See QUADRANTS.

Phobos   One of the two moons of Mars.

Phoebe   One of the poetic names given to the moon of the Earth, but also an (unpoetic) name of one of the moons of Saturn.

Phoenice   See POLARIS.

Phoenix   This name was not used in an astrological context until the asterism was originated by Bayer in 1604, from stars between ERIDANUS and GRUS. Sometimes called Griffin or (confusingly) Eagle (see EAGLE), this asterism was said to exert an influence conducive to good life, and to bring lasting fame. The term has a rich accretion of symbolism attached to it in both alchemical and occult symbolism, however, and attempts have been made to link the 600-year cycle of the Phoenix with both reincarnation theory and various planetary cycles. See NAROS CYCLE.

Phoenon   According to some authorities, one of the Greek names (meaning 'terrible' or 'cruel') for the planet Saturn. But see also PHAINON.

Phul   The seventh of the seven supreme Angels of the Qabbalists, said to possess all the attributes of the Moon. His Seal is given in SIGILS – PLANETARY SPIRITS.

Phyto   See HYADES.

Pictor   Constellation, formed originally by La Caille in 1752, as Equuleus Pictoris (painter's easel), but now generally known in the shortened version. It is located south of COLUMBA and is supposed to give artistic ability, as well as a reliable nature.

Pisces   The last of the twelve signs of the zodiac, and one of the constellations. It corresponds as a zodiacal sign neither in location nor extent with the asterism of the same name – see PISCES CONSTELLATION. The modern sigil for Pisces ♓ is said by some to be a drawing of the two fishes united by the so-called 'silver cord' which joins together their mouths in the constellation image (figure 49): in this image the two fishes usually face opposite directions to symbolize the altercation between spirit and soul in the human being. Pisces is of the water element and of the Mutable Quality, the influence being pre-eminently emotional, intuitive and insecure. There is a strong tendency to withdrawal and towards artistic expression, and the Piscean type tends to be hypersensitive to the emotional needs of others. The sensitive nature of Pisces is expressed in the many keywords attached to the sign by modern astrologers: sympathetic, imaginative, poetic, suggestible, emotionally malleable, poetic, self-pitying, easy-going, self-indulgent, sentimental, kindly and inconstant.

   In excess, or under pressure, the Piscean nature expresses itself in terms relating to its underlying insecurity and lack of drive, the keywords being: dreamy, impractical, lazy, restless, unstable, chaotic, hypochondriac, fickle, lacking in self-reliance. Pisces is ruled by the modern planet Neptune in most modern astrological systems, though in traditional astrology it was ruled by the beneficent and expansive Jupiter (see RULERSHIPS): the sign marks the Exaltation of Venus and the Detriment of Mercury.

Pisces constellation   Constellation of the Fishes, located approximately 15 degrees Pisces to 26 degrees Aries, the extent of the asterism being set out according to the limits recognized by the MODERN ZODIAC based on Delporte in figure 49. The image of the two fishes, joined by a connecting band, figures on the DENDERAH ZODIAC of Egypt. It was named both Fish and Fishes by the Greeks, though the Romans emphasized the duality with their

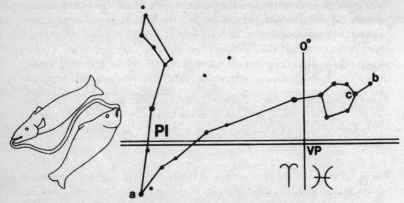

Figure 49: Image for Pisces from the façade of Amiens Cathedral, France, and the constellation of Pisces from the MODERN ZODIAC *of Delporte.*

Piscis Gemellus. To the Babylonian astrologers Nunu was also a pair of fishes, and some even more ancient remains give it as Nuni (fishes) as well as Zib, perhaps meaning 'boundary' and pointing to its marking the 'end' of the zodiacal circle. The Fish which symbolized Christ was from early Christian times linked with Pisces, yet even earlier it had been an esoteric symbol, linked with both the Greek 'fish-goddess' Atargatis, and with the Syrian Dagon, a fish-headed man (said to be the image of the celestial man – the seas and the spiritual realm being regarded as symbolic equivalents). This probably explains why the constellation should have been under the tutelage of Poseidon (and hence the Neptuni Sidus of Manilius). Ptolemy breaks down the influence of the constellation into separate asterisms, with the result that the entire constellation is of variable influence, and may be determined only by reference to the fixed star tradition. The lucida is the so-called 'knot' (the Nodus) in the thread, the Linon of the Greeks, the Nodus Piscium of the Romans. It is the alpha of Pisces, often called Al Rescha from the Arabian meaning 'the cord'. In some esoteric texts, the binding cord is sometimes called the Silver Cord, and is said to be linked with the subtle and invisible connection which holds in union the spirit and soul of incarnate man.

Piscis Name given to the 28th of the LUNAR MANSIONS in some medieval lists. It is probably identified by the fixed star MIRACH. But see also table 48 under LUNAR MANSIONS.

Placidean system

Piscis Australis   Constellation, the Southern Fish, located approximately 15 degrees Aquarius to 5 degrees Pisces and 28 to 36 degrees south of the equator. It has been said that this southern fish was the parent of the PISCES CONSTELLATION, and a curious confirmation of this is found in the legend which has the asterism formed to commemorate the change of bathing Venus into a fish. In late antiquity Venus was linked with Atargatis, and hence in pre-Christian Mystery traditions with the fish-headed Dagon: in Christian Mystery traditions these merged with the Virgin Mary, who gave birth to the Fish who became Christ: hence the name Golden Fish, though Greater Fish is Grecian pre-Christian. Bayer says that the influence of the asterism is the equivalent of Saturn. FOMALHAUT, the fixed star in its mouth, was one of the four ROYAL STARS of the Persians.

Piscis Volans   See VOLANS.

pituita   A late medieval term for the PHLEGMATIC humour.

place   See CHOROGRAPHY.

places   A translation of the Ptolemaic *topoi*, which were arcs of 2.5 degrees, obtained by dividing the zodiacal signs of 30 degrees by 12. As with the system of DECANS, each of the places was ascribed a planetary ruler. These places were also once called phases. Some confusion has arisen from the fact that Ptolemy also used the same term, similarly translated, as the equivalent of 'house'.

places of the moderators   See HYLEGIACAL PLACES and MODERATORS.

Placidean measure of time.   See MEASURE OF TIME.

Placidean system   A widely used method of HOUSE DIVISION, sometimes called the 'Semi-Arc System', though at least three other methods employ the trisection of semi-arcs. It is a time-division, based on the addiction of one third of a semi-arc to the sidereal time of ascension, in a sequence designed to determine the cusps of the 12th, 11th and 10th houses, which of course gives the degrees for the corresponding opposite cusps. The nocturnal semi-arc is treated likewise in terms of the sidereal times of descension to give the remaining cusps. The fact that the Placidean system is the most widely used in modern English astrological circles is largely due to historical accident, rather than to any particular merit in the system itself. From the 19th century most of the available 'Tables of Houses' which were used in ephemerides (especially in the popular 'Raphael's Ephemeris' series) were computed according to this trisection of the semi-arc, and their availability encouraged its wide use by astrologers who did not necessarily understand the principles involved. The system is often said to be named after its (supposed) inventor, the astrologer Placidus di Tito, who lived in the 17th century, but the fact is that this particular trisection method was derived from one (correctly) ascribed to the 8th-century Arabian astrologer ben Djabir (see ALBATEGNIUS SYSTEM). See also HOROPLANETARY DOMIFICATION.

Plaksha   See SECOND ROOT RACE.

planetary Angels   As with the ZODIACAL ANGELS and the ZODIACAL SPIRITS, there is much confusion of nomenclature and role of the planetary Angels. In general terms the whole scheme of angelic rule is dealt with under CELESTIAL HIERARCHIES. The true planetary Angels of medieval astrology are the Governors of the spheres, which are listed under their esoteric name of SECUNDADEIAN BEINGS. However, the list of planetary Angels most widely used in modern astrology is:

| MO | Gabriel | ME | Raphael | VE | Haniel | SU | Michael |
| MA | Camael | JU | Zadkiel | SA | Zaphkiel | | |

planetary chain   A term introduced into esoteric astrology through the theosophy of Blavatsky. It is claimed that each of the physical globes of the cosmic bodies are accompanied by, and indeed nourished by, seven superior globes which remain invisible to ordinary sight. These septenaries are termed a CHAIN. See also GLOBES.

Planetary Chain Logos   In the theosophical cosmoconception, the Planetary Chain Logos is the title given to the individual spiritual entity deputized to take charge of a whole series of Seven Chains or an entire SCHEME OF EVOLUTION.

planetary colours   A number of different associations between the planets and colours has been preserved within the astrological tradition, and it is impossible to untangle the associations in the hope of establishing an order in the chaos. The following are derived from lists as widely divergent as Blavatsky, Cornell, Hall, Libra and Lilly – though it must be admitted that each of the associations would be contested by some modern authorities.

planetary melothesia

MO Silvers, whites, greys, 'green-silver'.
ME Yellow, azure, violet, slate-colours, soft browns, 'spotted mixtures'.
VE Blues, pastel shades, greens, indigo, 'soft pale blues'.
SU Gold, orange, yellow, blue, 'yellow-brown'.
MA Red, scarlet, carmine, magenta, deep orange, 'angry shades (hues?)'.
JU Purple, violet, indigo, sky-blues, 'deep tones of midnight sky'.
SA Black, grey, green, dark brown, sage green, 'dark mottled colours'.

There is much argument among authorities as to the rulerships of the extrasaturnian planets, though the following are often given:

UR Electric colours, light blues, 'a profusion of glaring colours'.
NE Lavender, greys, whites, mauve, 'sea colours'.
PL Black, 'blood reds', 'magmatic colours'.

planetary harmonics   The modern astrologer Walter has done much research into the theory of aspects, especially in connection with the theory of HARMONICS, and has come to the conclusion that there is a relationship between harmonic numbers and the planets, as follows:

ME 2   VE 3   SU 1   MA 6   JU 6   SA 7   UR 8   NE 9   PL 10

planetary heredity   see ANGULARITY.

planetary horas   See HORAS.

planetary hours   In the astrological tradition every hour of the day is said to be ruled by a particular planet. It is the cycle of such planetary hours which has given rise to the name of the DAYS OF THE WEEK. Blavatsky gave an alternative system of planetary hours, dividing the day not into 24 arcs of time, but into four, the sequence beginning the first semi-diurnal arc of Monday with the Moon, then running through the entire week of arcs in the traditional order of the planets. Thus Monday is said to consist of four arcs ruled in sequence by the MO, ME, VE and SU, while Tuesday is ruled by the arcs MA, JU, SA and MO. The names of the days are determined by the opening ruler. The astrologer Leo provides a similar system, allegedly based on the Egyptian time measurement systems of three-arc days, starting Saturday with the ruler SA.

planetary man   See PLANETARY MELOTHESIA.

planetary melothesia   A term derived from Ptolemaic astrology to denote a list (really several lists) of rulerships or lordships held by the planets over various parts of the body. Ptolemy gives a list which has quite evidently influenced the later pictorial imagery of MELOTHESIC MAN:

planetary melothesia

MO rules sense of taste, stomach, belly, womb and left-hand parts.
ME rules speech and thought, the tongue, the bile and buttocks.
VE rules sense of smell, the liver and 'flesh' (corporeality).
SU rules sense of sight, brain, heart, sinews and right-hand parts.
MA rules left ear, kidneys, veins and genitals.
JU rules sense of touch, lungs, arteries and semen.
SA rules right ear, spleen, bladder, phlegm and bones.

Acting through their own natures, the malefics will strike those parts of the body over which they have rule. The astrologer Libra provides a list clearly derived from the zodiacal melothesia, the order of planets following the order of the signs over which they have rule:

MA	rules head and face.
ME	rules lungs and arms.
SU	rules heart and back.
VE	rules veins.
JU	rules thighs and hips.
SA (UR)	rules lower legs and ankles.
JU (NE)	rules feet and tissues.
VE	rules throat and neck.
MO	rules stomach and breasts.
ME	rules bowels and abdomen.
MA	rules generative organs, bladder.
SA	rules knees and bones.

In modern times the planetary melothesia is often called (in connection with MELOTHESIC MAN) the 'microcosmic man' or (alone) 'planetary man'. In modern times the rulerships have been extended to the so-called 'modern planes', as well as to the prevailing image of man in terms of endocrine glands. Collin, for example, gives a melothesic image of man as follows:

MO rules the pancreas of the lymphatic system.
ME rules the thyroid and the pulmonary system.
VE rules the parathyroids, connected with the arterial system.
SU rules the thymus, which he associates with the function of growth.
MA rules the solar plexus and adrenal glands, the cerebro-spinal system.
JU rules the posterior pituitary, and involuntary muscular system.
SA rules the anterior pituitary, cerebral cortex and skeletal system.
UR rules the gonads and genital system, and the function of reproduction.
NE rules the pineal gland, with a function 'as yet unknown'.

———————————— TABLE 58 ————————————

Planet	Moon	Planet	Moon	Planet	Moon
MA	Phobus	JU	Juno	SA	Mimas
MA	Daimos	JU	Europa	SA	Enceladus
		JU	Ganymede	SA	Thetys
		JU	Castillo	SA	Dione
UR	Miranda			SA	Rhea
UR	Ariel			SA	Titan
UR	Umbriel			SA	Hyperion
UR	Titania	NE	Triton	SA	Japetus
UR	Oberon	NE	Nereid	SA	Phoebe

———————————— TABLE 59 ————————————

Planet	Ptolemaic periods	Post-Copernican periods
MO	4 years	25 and 19 years
ME	10 years	10, 20, 27 and 79 years
VE	8 years	8 years
SU	19 years	
MA	15 years	79 years
JU	12 years	83 years
SA	30 years	59 years
UR		84 and 90 years
NE		164 years and 280 days, and 180 years
PL		247.7 years and 360 years

Perhaps the most extensive planetary melothesia, which summarizes much of the ancient tradition, is that given under the separate entries for planets by the medical astrologer Cornell.

planetary moons The names of the satellites (moons) of the planets are given in table 58 above.

planetary orb In modern use the term invariably applies to the ORB ascribed to an individual planet in respect of aspects – as distinct from the orb applied to aspects themselves. In medieval astrology the term was also

TABLE 60

Planets	Years	Days	Synodic	Days	Sidereal	Days	D
ME	7	2,556.75	22	2,549.30	29	2,551.00	−7.45
ME	46	16,801.50	145	16,802.23	191	16,802.08	+0.73
ME	79	28,854.75	249	28,853.48	329	28,941.80	−1.27
VE	8	2,922.00	5	2,919.60	13	2,921.11	−2.40
MA	15	5,478.75	7	5,459.57	8	5,495.84	−19.18
MA	32	11,688.00	15	11,699.07	17	11,678.66	+11.07
MA	47	17,166.75	22	17,158.64	25	17,174.50	−8.11
MA	79	28,854.75	37	28,857.71	42	28,853.16	+2.96
JU	12	4,383.00	11	4,387.73	1	4,332.59	+4.73
JU	71	25,934.75	65	25,927.50	6	25,995.53	−7.25
JU	83	30,315.75	76	30,315.23	7	30,328.12	−0.52
SA	29	10,592.25	28	10,586.60	1	10,759.20	−5.65
SA	59	21,549.75	57	21,551.75	2	21,518.40	+1.54

used to denote the physical body of the planet, the orb of the planet, carried within the ambient of the planetary sphere.

planetary periods The theory of planetary periodicities really falls within the domain of *CYCLES, but the term is often used in reference to a number of periods traditionally assigned to the planets, which are in most cases approximates only to recognized cycles, and called REVOLUTION PERIODS. There are several variant tables in the astrological traditional corpus, the variations seemingly arising from the different ideas of what the approximate periods are supposed to represent. Table 59 sets out the traditional series. Some astrologers appear to use the non-Ptolemaic cycles alongside tables of mean advance (in order to determine an approximate position of a planet) when ephemerides are not available: this, of course, is bad practice, and not the reason why tables of planetary periods were constructed. The so-called PLANETARY YEARS are also similarly misused. In connection with planetary periods, the historian Neugebauer (having referred to the 'astronomical arbitrariness' of their parameters) points out that it is significant that the lunar period is based on Egyptian years only, whilst the Venusian periods are fundamental parameters of Mesopotamian astronomy. He also makes the point that in astrological use the periods may not be regarded only as signifying years – they are also used to represent months or days (though not in modern astrology). This practice corresponds, of course, to the symbolic system of progressions itself, but it is unlikely that the planetary periods were designed for such a purpose. The total of these periods is 129 years, when

—————————— TABLE 61 ——————————

Planet	Hot	Cold	Dry	Wet
MO		5		6
ME		1.5	1	
VE	0.5			4
SU	5.5		2	
MA	2.5		3	
JU	1.5		1	
SA		3.5	3	

counted as months in accordance with the symbolic system. When counted as months, this gives ten years and nine months, which was one of the ancient periodicities. The mean values are the arithmetical mean between the extremes:

SU	19	(19 solar years – 235 lunar months – see METONIC CYCLE)
MO	25	(25 Egyptian years – equals 409 lunar months)
SA	30	(an approximate synodic period)
JU	12	(an approximate synodic period)
MA	15	(an approximate synodic period)
VE	8	(an approximate synodic period of 99 lunar months)
ME	20	(20 Egyptian years, equivalent to 63 synodic periods)

Schultz, who provides so much useful material for the new esoteric approach to astrology, gives the detailed information regarding the synodical and sidereal planetary periods set out in table 60. His interest lies mainly in the relatively small differences (signified under D in table 60) between certain cyclical groups – for example, in table 60 one notes in the seven-year cycle of Mercury, the 22 synodic periods of Mercury and the 29 sidereal periods of Mercury, a difference of only – 7.45 days. See also INITIAL CHRONO-CRATORS and REVOLUTION PERIODS.

planetary phases The phases of the moons are the CRESCENT (shortly before and after lunation) the 'half moon', in which half of the body of the moon is lighted, the 'gibbous', when both halves of the moon are convex, and the 'full moon', which is an opposition with the Sun. See also LUNATION.

While the term 'phases' is most usually applied to the Moon of the Earth, it may also be applied to the inferior planets, and the PLANETARY MOONS.

planetary pictures See COSMIC PICTURES.

planetary qualities The astrologer Morin published a table demonstrating the supposed numerical values of the planets in terms of the QUALITIES described by Ptolemy, as shown in table 61. Each of these qualities is distributed throughout the seven planets in equal numerical equivalents of 10. These Morinean figures of 'weightings' were intended to be adjusted in terms of the planetary Angles, however.

planetary rulership Each planet is said to have RULE over one or more of the signs of the zodiac. The modern system of rulerships has been devised mainly to accommodate the so-called 'modern planets', and is radically different from the traditional schema which was introduced by Ptolemy from antiquity. This Ptolemaic schema was derived from a consideration of the basic duality of each planet (rendered in modern astrology in terms derived from the Ptolemaic system, such as 'positive and negative', 'masculine and feminine', 'diurnal and nocturnal', and so on) which was itself derived from the Egyptian cosmoconception which radically influenced Greek and Roman astrology. The schema set out in table 62 reveals the application of this

---

TABLE 62

Sign (day house)	Planet	Sign (night house)
Cancer	MO SU	Leo
Gemini	ME	Virgo
Taurus	VE	Libra
Aries	MA	Scorpio
Pisces	JU	Sagittarius
Aquarius	SA	Capricorn

---

duality in terms of 'signs' even though the traditional nomenclature used the curious term 'house'. This schema faithfully sets out the traditional system of planetary rulerships over the signs: one sees at a glance that (for example) Jupiter has rule over Pisces and Sagittarius, the former being its day sign (wrongly called 'day house'), the latter being its night sign (wrongly called 'night house'). The duality of the rulerships lingers on in the modern term

NEGATIVE MARS which of course relates to the rule of Mars over the night house of Scorpio. By the time the 'modern planets' had been discovered, and were being incorporated into the system of rulerships recorded by Ptolemy, the ancient esoteric basis for the traditional planetary rule had been largely forgotten. As a result, the fact that the planets (omitting the luminaries, of course, which in any case were accorded only single rulerships) were said to have a positive (diurnal) and negative (nocturnal) side to their natures was something of a puzzle to most astrologers, and the traditional system was readily changed to make way for the new planets, to the detriment of astrological theory. While there is still some disagreement about the rulerships (especially in connection with the rule accorded the ASTEROIDS and certain of the HYPOTHETICAL PLANETS), the majority of modern astrologers appear to subscribe to the system which amends the traditional list in table 62 with the following changes:

> SC ruled by Pluto (originally negative Mars)
> AQ ruled by Uranus (originally positive Saturn)
> PI ruled by Neptune (originally positive Jupiter)

planetary Seals Graphic symbols or sigils denoting planets, and supposed in themselves to exert a corresponding planetary influence. For a sample medieval sequence of planetary Seals, see section ten under CALENDARIA MAGICA. Until this present century, planetary Seals were accorded little serious attention, and were regarded largely as (little understood) survivals from late medieval astrology. However, in our present century, Steiner designed a number of remarkable Seals, one from each planet, the forms arising from deep meditative consciousness (see Kempter). These Steinerian forms are different from the medieval ones, though perhaps the meditative origin and purpose of both groups were much the same.

planetary spheres See SPHERES.

planetary spirit The esotericist Bailey defines the term much in accord with tradition as 'another term for the Logos of a planet'. Such a spirit is on what is esoterically called the 'evolutionary arc', and is more highly developed than the human – but see PLANETARY ANGELS. In spite of the accuracy of this definition, the fact is that the general term 'planetary spirit' is used widely and wildly to cover a myriad of different spiritual beings linked with the planets: see CELESTIAL HIERARCHIES, INTELLIGENCY and SECUNDADEIAN BEINGS, all of whom have at one time or another been called 'planetary spirits'.

planetary strength A general term relating to the assessment of the power of the planets in a given chart in terms of a number of clearly defined factors. The ASTRODYNE, which is rooted in the factors of orb and angularity, is an example of a particular attempt to assess planetary strength, while the work of Volguine is involved with a more complex assessment, which takes into consideration not only orb and angularity, but also sign position, lunar mansions, house position and planetary speed. Each of these modern methods lend themselves readily to computerization.

planetary tonal intervals   In an early phase of the development of the theory of the MUSIC OF THE SPHERES, the 3rd-century AD grammarian Censorinus recorded a number of intervals linked with the planetary DISTANCES which he insisted were derived from Pythagoras. In this system (which relates to the supposed distances not of the planets but of the spheres) the Earth to the Moon was equivalent to one tone; from Moon to Mercury half a tone; from Mercury to Venus half a tone; from Venus to the Sun one and a half tones; from Sun to Mars one tone; from Mars to Jupiter half a tone; from Jupiter to Saturn half a tone and from Saturn to the Stellatum half a tone. For a fundamental change in measurement of intervals, see KEPLERIAN HARMONICS.

planetary year   A large number of (variable) cycles or periods, usually called planetary years have survived from ancient times, and have been radically misunderstood by many modern astrologers. Devore, for example, suggests that these were possibly guides to progression techniques (in the days when ephemerides were not readily available), and clearly does not know what the figures mean. Again, they are explained by Wemyss as relating to the 'probable length of life' of the native, in terms of the natal ruling planet, though it is clear that the figures (see 'Minimum Length' in table 63) are derived from the ancient PLANETARY PERIODS, which have nothing to do with the length of human life. The series of planetary years in table 63 are derived from the Alexandrian Paul (though sometimes attributed to the astrologer Vettius Valens).

Planet of Death   See EIGHTH SPHERE.

planetoids   See ASTEROIDS and LUCIFER.

planets   In modern astronomy, the name planet is applied to all celestial bodies (with the exception of comets and meteors) which revolve around the Sun of our system, and have no light of their own. The ancients, however, believed that the planets did have a light of their own, and called the SUN and

─────────── TABLE 63 ───────────

Planet	Minimum length	Maximum length	Mean age
MO	25	108	66
ME	20	76	48
VE	8	82	45
SU	19	120	69
MA	15	66	40
JU	12	79	45
SA	30	57	43

─────────── TABLE 64 ───────────

Planet	Mean solar distance	Period of revolution	Max. distance from ecliptic	Axial rotation	No. of Moons
MO	238,857	27d. 7h. 43m. 11s.	5 deg. 8 m.	27 d. 7 h. 43 m.	0
ME	35,950,000	87.97d.	7 deg. 00 m. 14 s.	59.4 d.	0
VE	67,180,000	224.70d.	3 deg. 23 m. 39 s.	244 d.	0
MA	141,500,000	686.98d.	1 deg. 51 m.	24 h. 37 m. 22.7 s.	2
JU	483,300,000	11.862y.	1 deg. 18 m. 21 s.	9 h. 55 m. (max.)	12
SA	886,100,000	29.458y.	2 deg. 29 m.	10 h. 39 m. (max.)	10
UR	1,783,000,000	84.015y.	46 m. 23 s.	10 h. 45 m.	5
NE	2,793,000,000	164.788y.	1 deg. 46 m. 28 s.	15 h. 40 m. (?)	2
PL	3,666,000,000	248.4y.	17 deg. 8 m. 34 s.	6.39 d.	?

MOON planets in addition to the five visible bodies MERCURY, VENUS, MARS, JUPITER and SATURN. This traditional seven of the early astrology was in relatively modern times augmented by the so-called MODERN PLANETS, URANUS, NEPTUNE and PLUTO, while some astrologers include the ASTER-OIDS. Within the various specialist and esoteric systems of astrology, there are also the HYPOTHETICAL PLANETS, the SACRED PLANETS, and other planetary bodies, whilst in HELIOCENTRIC ASTROLOGY there is also the Earth, which figures as a planet (though in the case of heliocentric charting the Moon is generally discounted as a separate satellite). Table 64 sets out some useful data. The planets with orbits outside that of the Earth are called the SUPERIOR PLANETS, while those within the orbit of the Earth are called the INFERIOR PLANETS, terms derived from the ancient theory of SPHERES. It is a basic tenet of astrology that each of the planets exercises a distinctive influence upon the Earth and its denizens, and that such influences may be studied, extrapolated and used as the basis for such things as prediction. However, the study of these influences is far from simple, since they cannot be isolated: the planets are always 'set' against some part of the zodiacal belt, and are to be studied only in relation to the influence of this (along with the influence of the fixed stars), which in turn conditions the workings of the planets. Some astrologers indeed view the planets merely as the transmitters and refiners of influences within the stellar spheres. Thus, the Mercury set against the zodiacal area of Aries may in itself be no different from the Mercury set against Pisces, yet the influence of Mercury in these two positions is quite definitely (and demonstrably) different. Other astrologers insist that the planets are not merely 'moving lenses' in this sense, but transmit and transform according to characteristics of their own. If this latter idea is correct, then one may rightly speak of a solar type, or of a Mercurial type in relation to human beings. A vast number of different classifications and nomenclatures relating to the planets have survived from traditional astrology: the names in popular use nowadays are Roman, though the antique order of the planets (each swimming within the spiritual ambient of their corresponding SPHERES) was different from the one in use today, reflecting the order of the spiritual beings we now call the CELESTIAL HIERARCHIES. This order is set out in a woodcut of the late 15th century, used to illustrate a text by Pietro d'Abano (figure 50), for which the (almost modern) sigils give the descending order from the STELLATUM as SA, JU, MA, SU, VE, ME and MO. Among the different nomenclatures are the FEMININE PLANETS, the MASCULINE PLANETS (see SECTA), the BENEFICS and MALEFICS, the GREATER FORTUNE and the LESSER FORTUNE, and so on. For the sigils associated with the planets, see SIGILS – PLANETS. The Greek names for the planets often appear in astrological literature, though there is some disagreement as to their meanings. Firmicus Maternus gives the classical:

MO	Selene	
ME	Stilbon	'the gleaming one'
VE	Phosphoros	'the light bearer'
SU	Helios	
MA	Piroesis	'the fiery one'
JU	Phaethon	'the bright one'
SA	Phainon	'the shiny one'

planisphere In general terms, a plan or map formed by the projection on to a flat surface of a sphere or part of a sphere. In particular the term is applied to the projection of a section of the celestial sphere, as in an astrolabe, or

Figure 50: Theology and astronomy in dispute, with planetary spheres in traditional sequence observed in antiquity (after Abano).

even in a CONSTELLATIONAL MAP. An astrological chart is really a highly schematic planisphere.

platic aspect A weak aspect, in which the two or more aspecting bodies are wide of ORB. The platic lacks the strength of the PARTILE ASPECT.

Platonic age See PLATONIC CYCLE.

Platonic cycle This term is sometimes wrongly identified in popular thinking with the so-called PLATONIC YEAR, which is a period of precession. The Platonic cycle is actually a *magnus annus*, the Great Year, in which it is visualized that the movement of the eight spheres causes them to arrive together in a great union at the same point (the same fiducials) from which they began their period or cycle. In terms of the ancient cosmoconception, once a Platonic cycle was completed a retrograde motion of the spheres set in, and a backward unfolding began, to continue for another vast periodicity until the planetary spheres were once more aligned to their common fiducials. The motion forward was linked in antique times with the AGE OF SATURN, under the direction of an inner harmony. In the retrograde motion, time is visualized as being reversed, so that men and other creatures grow young with the progression (or, rather, retrogression) of time. This account may be easily linked with the esoteric view of the disincarnate soul reaching the last Sphere of Saturn in its post-mortem experience, for with Saturn time ends, and the soul reaches the limits of its outer movement, and begins its descent – see ORPHIC DESCENT OF THE SOUL.

Platonic month A misnomer (in that it has nothing to do with a 'month' or even with the Moon), derived from analogy with the twelve lunar months related to the PLATONIC YEAR. It is used to denote the precession of the Vernal Point through one sign of the zodiac, which takes 2,160 years.

Platonic solids These are the so-called Platonic five regular geometric solids: the tetrahedron, the cube, the octahedron, the dodecahedron and the icosahedron. The importance of these solids to astrology is the repeated attempt made by astrologers and mathematicians to show by graphic demonstrations that these might be inscribed or circumscribed around the planetary spheres. Kepler (in his *Mysterium Cosmographicum*) was the last to attempt such a graphic demonstration, for his later discovery of elliptical orbit so changed both the PTOLEMAIC SYSTEM and the COPERNICAN SYSTEM as to render such demonstration invalid. However, it appears that the underlying idea of geometric harmonies found expression in his later theories of heliocentric planetary harmonies: see KEPLERIAN HARMONICS.

Platonic year The VERNAL POINT retrogrades through the constellations in a period which is approximately 52,920 years (but see PRECESSION): this is the so-called Platonic year, the Annus Platonicus of traditional astrology. This period is divided into twelve equal lengths (the so-called PLATONIC MONTHS), corresponding to the arcs of the zodiac, and the periods during which the Vernal Point retrogrades through the arcs are named after the relevant signs. Unfortunately, there is no generally agreed fiducial (see AYANAMSA), so there is no agreement among astrologers as to when such a month begins and ends; many astrologers insist, however, that the Vernal Point is retrograding through Pisces at the present time, so that mankind may be said to be living under the impress of the Piscean Age. Each of the twelve arcs exude different influences, so that each of the so-called 'months' (periods of 2,160 years in fact) induces fundamentally different attitudes, mores and civilizations. These zodiacal ages or epochs are characterized by more than merely the stamp of the individual sign: the four cardinal points of the horoscope figure are visualized as working in combination to produce what have been called in modern esoteric astrology the THREE PROTOTYPES which periodically repeat themselves, each in four possible variations, thus shaping historical epochs.

Plaustrum One of the several important variant names for the constellation URSA MAJOR.

Play The Point of Play is one of the so-called Arabian *PARS in a chart. If the degree of Mars in a natal chart is taken as the Ascendant, the adjusted position of Venus is the Point.

Pleiades One of the most famous of asterisms, set in the last few degrees of constellation Taurus, the name usually derived from the Greek *plein* (to sail) on the grounds that the heliacal rising of the asterism marked the opening of navigation for the ancient Greeks, while the setting in late autumn marked the close of navigation. This etymology has, however, been questioned. Pleione was the mother of these Seven Sisters: they are sometimes called the Atlantides (their father being Atlas). In the Chinese astronomy they are called the Seven Sisters of Industry. However, their names are almost legion, though emphasis is usually laid on the magical number seven: they are the Seven Virgins, the Virgin Stars, the Seven Atlantic Sisters, the Seven Stars, and so on. Indeed, almost any famous group of seven has been linked with this asterism. The name given to the individual sisters are ALCYONE, Maia, Electra, Merope, Taygeta, Celeno and Sterope. Merope is sometimes called the Invisible Sister because she is the only one who married a mortal, and hid her face in shame. According to Ptolemy, the degree of arc in which the asterism is located gives an influence equivalent to that of Moon conjunct Mars, and its power is meant to be evil, involving disgrace, ruin and violent death.

Pleione See PLEIADES.

Plough One of the several important variant names for the constellation URSA MAJOR.

Pluto Planet discovered in 1930 by the Lowell Observatory, though its existence was suspected from mathematical computations some time prior to this date. When first announced by astronomers it was named Pluto-Lowell or LOWELL-PLUTO, but it had been mentioned before this time under a variety of different names by astrologers – these names included Pluto (see PAGAN-PLUTO). This modern planet Pluto must not be confused with the 'Pluto' described by Thierens as one of the future planets of the ETHERIC SPHERE, where it was to be the ruler of Aries, also named Osiris – see THIERENS-PLUTO. Nor must it be confused with the Pluto described by Wemyss as ruler of Cancer (the so-called WEMYSS-PLUTO). Since the modern Pluto was discovered, it has displaced the traditional NEGATIVE MARS from its rule over the zodiacal Scorpio. Astrologers see it as an eruptive and disturbing influence, involved with the processes of elimination and regeneration. When involutionary, it is connected with the underworld, with crime, and all that is dark and unregenerate in our society: when evolutionary, it is connected with esoteric movements, with all that seeks to redeem. When the planet works through its darker nature, the individual is dominated by obsessive psychic tendencies (which swell from the telluric unconscious realm) – yet when the planet works through its beneficial side the same forces are used to heal, to create and to regenerate. The sigil ♇ is popularly explained as being derived from a combination of the P for Pluto and the L for Lowell. However, in 1911 the astrologer Pagan offered a much more fitting sigil in the form of a reversed Mars ♂. See also MODERN PLANETS.

Pocillator   See ANTINOUS.

point   A modern equivalent for PARS.

Point of Bondage   One of the so-called Arabian *PARS in a chart. If the figure is revolved so that the degree occupied by the dispositor of the Moon is on the Ascendant, then the Moon position in this new arrangement marks the Point.

Point of Faith   One of the so-called Arabian *PARS, obtained by revolving the figure so that the degree occupied by the Moon is on the Ascendant: the position of Mercury relative to this new Ascendant marks the Point.

Point of Female Children   One of the so-called Arabian *PARS in a chart, obtained by revolving the figure so that the degree occupied by the Moon is on the Ascendant: the position of Venus relative to this new Ascendant marks the Point.

Point of Jupiter   See PARS.

Point of Life   A term used by the astrologer Frankland in connection with a measure used in SYMBOLIC DIRECTIONS: it is an arc of $4\frac{2}{7}$ degrees, which he considers as starting at the first degree of Aries, and marking off in terms of such arcs yearly periods. It is a measure linked with the septenary, for it passes through each sign in a period of seven years, forming aspects with the radical positions. The Point (arc or measure, as it should properly be called) is also used in a different way as the SEPTENARY MEASURE.

Point of Male Children   One of the so-called Arabian *PARS obtained in the same way as the POINT OF FEMALE CHILDREN, using Jupiter as the new denotive.

Point of Mars See PARS.

Point of Mercury See PARS.

Point of Moon See PARS.

Point of Passion See SWORD.

Point of Saturn See PARS.

Point of Servants One of the so-called Arabian *PARS in a chart. If the degree of Mercury in a natal chart is taken as the Ascendant, the adjusted position of the Moon becomes the Point of Servants.

Point of Sickness One of the so-called Arabian *PARS in a chart. If the degree of Saturn in a natal chart is taken as the Ascendant, the adjusted position of Mars marks the Point.

Point of Spirit One of the so-called Arabian *PARS in a chart, obtained by revolving the figure so that the degree occupied by the Moon is on the Ascendant: the position of the Sun relative to this new Ascendant marks the Point of Spirit.

Point of Travel by Land One of the so-called Arabian *PARS. If the Lord of the Ninth is taken as the Ascendant of a figure, then the new cusp of the 9th house is said to mark the Point.

Point of Understanding One of the so-called Arabian *PARS in a chart. If the degree of Mercury in a natal chart is taken as the Ascendant, then the adjusted position of Mars in the new chart marks the Point.

polar chart A horoscope chart cast for extreme polar regions. The main problem for horoscopes cast for polar regions is that at certain times, and in certain latitudes, the plane of the ecliptic will coincide or override the horizon plane. If a conventional system of CHART PROJECTION is employed, the Ascendant may under those circumstances appear to descend – which naturally involves the astrologer in symbolic nonsense in the chart. Many attempts to resolve the problem of this 'distortion' of the traditional symbolism have led to a variety of projections being suggested, mainly by redefining the concept of the Medium Coeli.

Polaris Fixed star (double), the alpha of URSA MINOR with many different names: it was the Phoenice of the early Greeks, the Pole Star and even Stella Maris (but see SPICA). For the Anglo-Saxons it was the Scip-steorra (ship star), and for the Latins the Navigatoria, and even Loadstar after the magnetic lodestone. It was for the Arabian astrologers Al Kiblah or Al Rukkabah (the rider), and to some astrologers Cynosura (dog's tail), probably because it was set in the tip of the tail of the lesser bear which was sometimes called a dog. Astrologers say that it is of the nature of Venus conjunct Saturn, and hence a difficult influence, though Robson records that the Arabian astrologers were of the opinion that the contemplation of the star would cure ophthalmia. It is one of the FIFTEEN STARS of medieval astrology, and thus receives a sigil: see SIGILS − FIXED STARS.

polarities The astrologer Leo defines the term 'polarities' from a point of view of solar-lunar combinations, regarding the Sun as a positive, life-giving element and the Moon as a negative, formative element. The sun influences human character, the Moon personality, the blending of these two being the polarity. For another view of polarities, see PRINCIPLES and SECTA.

pole of the Ascendant The geographic latitude of the place for which an Ascendant degree has been determined − usually in preparation for a horoscope: see POLE OF THE HOROSCOPE.

pole of the horoscope An archaic term meaning the latitude of the place for which a figure has been cast.

pole star See POLARIS.

Polis Fixed star (triple), the mu of constellation Sagittarius, set in the upper part of his bow. The Copts gave us the name, attached, however, to a different image of the asterism, for *polis* meant foal, and the word must not be confused with the same Grecian term. It is the Sagittarian impulse which has left its mark on the astrological reading of this star, for it is said to be of the nature of Mars conjunct Jupiter, and to give ambition, martial powers, ability as a horseman and keen perception.

Pollux Fixed star, the beta of constellation Gemini, set in the head of its namesake (the Southern Twin), who was the immortal one, sometimes linked with Hercules (though in fact the son of Jupiter and Leda). Pollux was said to be a keen boxer, which explains one of his early names − Pugil. Among the several names used by the Arabian astrologers were derivatives of

Rasalgeuze. Ptolemy claims that it is of the nature of Mars, and accordingly it is said to have a brave, spirited, audacious though cruel influence: when rising it predicts injuries to the face, or even bad eyesight.

Polyhymnia Name given to one of the ETHEREAL PLANETS perceived clairvoyantly by the astrologer Harris, said to be in orbit beyond his MELODIA: not to be confused with the ASTEROID of the same name. Later astrologers assigned Polyhymnia as the day-house planet of Cancer, the night-house planet of Libra, and said that it was Exalted in Gemini, with a nature equivalent to Moon conjunct Venus, with an admixture of Mercury.

polypolar A term used of a horoscope chart in regard to the distribution of aspects, derived in respect of the so-called *FIGURINES of Meier, though presumably applicable to any figure. A polypolar is a figure which includes at least two oppositions.

Pomegranate In a specialist sense, one of the so-called Arabian *PARS, the Part of Jupiter, sometimes called the Part of Increase. If a chart is revolved so that the degree occupied by the Sun is transferred to the Ascendant, then the position of Jupiter in this new arrangement will mark the beneficial Pomegranate.

ponderable planets These are the planets Uranus, Jupiter and Mars according to Simmonite, who probably forgot about Saturn, for he says that they are 'so called because they move slower than the rest'. One might therefore add to these the two modern planets Neptune and Pluto, both unknown to Simmonite.

Porphyrian system A method of *HOUSE DIVISIONS, the principle of which is the trisection of the ecliptic quadrant (obtained by direct division of the ecliptic). The astrologer Leo rightly describes it as a 'rough and ready' method, yet it appears to have been similar to one of the systems used by the early Greeks, prior to its development by the Neoplatonist Porphyry (4th century AD). However, in respect to antiquity, see also ALCABITIUS SYSTEM and MODUS EQUALIS.

Porrima Fixed star, the gamma of Virginis, set in the left arm of this figure, and sometimes called Caphir.

Poseidon Name given to one of the HYPOTHETICAL PLANETS used in the Uranian system of the HAMBURG SCHOOL. It is said to relate to 'mind, spirit

TABLE 65

Sign	Physical	Mental
AR	Impetus	Enterprise
TA	Inertia	Steadfastness
GE	Expansion/Contraction	Eagerness/Despair
CN	Flexibility	Adaptability
LE	Radiation	Faith
VG	Crystallization	Discrimination
LB	Equilibrium	Balance
SC	Chemical Action	Determination
SG	Spiral Rotation*	Reason
CP	Vibration*	Concentration/Relaxation
AQ	Absorption	Curiosity
PI	Solution	Insight

* Pagan says 'possibly Vibration and Spiral Rotation should be reversed'.

and ideas'. The name of this Greek sea-god was also used by Thierens in connection with his NEPTUNUS (Nereus), which he linked with the principle of absorption. Since the name Poseidon was in Roman times called Neptune, the term is sometimes used as a synonym for the modern planet NEPTUNE.

posited A term almost always used as synonymous with 'situate' or 'placed'.

positive Mars A term which is a survival from the ancient Ptolemaic astrological system, relating to the classification of rulerships. Each of the planets (save for the luminaries) were accorded rule over two signs of the zodiac (see PLANETARY RULERSHIPS), and such pairs were deemed positive or negative, solar or lunar, and so on. Mars was accorded a negative rule over Scorpio and a positive rule over Aries. Thus the term 'positive Mars' really means 'Mars as ruler of Aries'. See, however, NEGATIVE MARS and SECTA.

positive planets See SECTA.

positive signs These are the signs alternating from Aries to Aquarius – also called the diurnal, fortunate and masculine signs. See SECTA.

possessive houses These are the 2nd, 6th and 10th houses, each of which reflects on the mundane level of the influence of the earth triplicity.

Powers A term used of one of the ranks of CELESTIAL HIERARCHIES, sometimes called the Potestates or DYNAMIS. The same term is also used descriptively of a system of astrological KEYWORDS of specific manifestations on the physical and mental plane, of the twelve signs, as listed by Pagan (table 65).

praesaepe A star-cluster located in the head of the constellation Cancer, known as the Beehive or Nubilum (cloudy), the Manger, the Crib, and so on. It is said to be a damaging influence in a chart, bringing disgrace, brutality and even blindness, the nature of the affliction depending upon the planets concerned. See also ASCELLI.

precession The 'precession of the equinoxes', is the term used to denote the retrograde movement of the Vernal Point through the constellations, a phenomenon which is perhaps connected with the mutation of the Earth upon its poles – though the fact is that there is no really convincing explanation of precession. In simple terms one has to visualize the two so-called zodiacs, the tropical and the constellational, one of which 'moves', and one of which is 'fixed' as overlapping. Because of precession, the fixed stars appear to advance (in longitude) one arc of just over 50 seconds each year, which means that the first point of tropical Aries seems to retrograde through the constellations. Since the two zodiacs are of unequal segmentation (the tropical being defined usually in terms of 30-degree arcs, the constellation being defined usually in twelve unequal arcs – see, for example, MODERN ZODIAC) there was never any point in time when the two zodiacs could have been regarded as being coterminous throughout their extent – however, it is clear that there would be some dates (in the past and future) when the two first points would coincide. The Vernal Point – which in terms of the seasons marks the beginning of the year – must therefore be visualized as slipping back (due to this precession) against the constellations. Around this cosmic truth a wealth of astrological symbolism has been woven into a proliferation of ideologies which tends to obscure the fact that no one has adequately dated the time when the two zodiacs coincided at the same fiducial, and that there is little or no agreement as to the extent of the arcs involved in the constellational zodiac. In fact, the Vernal Point slips back against the backdrop of fixed stars in a motion which is theoretically steady, but which (mainly because the ecliptic plane itself is not fixed) is minutely variable. Schultz gives the following details for precession:

in 1 year the rate is 50.25 seconds
in 72 years the rate is 1 degree 0.3 minutes

in 2,160 years the rate is 30 degrees

in 25,920 years (72 × 30 × 12) the rate is 360 degrees

The 2,160 span is called a precessional age or (wrongly) an Astrological Age: the 25,920 span is called a precessional year or Great Year. It must be observed that whilst the figures given by Schultz are accurate, very often the figures given in astrological texts are not so accurate – for example, Brau gives 25,800 for the Great Year. The real problem which arises in connection with the application of precessional information to astrology is that the fiducials concerned are far from agreed: there is no certain showing when a particular cycle may be said to have begun – which is to say that no one knows for sure when the sequences of the so-called PLATONIC YEAR began. The importance of the resultant Platonic year is connected with a derivation from the theory of precession and the important teachings concerned with the natures of the zodiacal ages. Since the Vernal Point retrogrades through one 30-degree arc in 2,160 years it is maintained that the zodiacal sign through which the Vernal Point progresses leaves its own distinctive impress on history – that civilizations of particular characters, with particular natures and purposes succeed each other in epochal series of 2,160 years: see, for example, AGES and THREE PROTOTYPES. It is widely believed that we are at present living in the Age of Pisces (which is to say that the Vernal Point is progressing through the sign Pisces), and that this age is intimately connected with the destiny of Christianity and with Christ (symbolized as a Fish, and the centre of much Piscean imagery). In terms of the theory of precession, therefore, it is clear that the Piscean cycle will eventually come to an end, and civilization (or the world at large) will find itself under the dominion of Aquarius. Unfortunately, because of the difficulty of determining the fiducials, few astrologers agree as to when the Age of Pisces will end and (hence) when the Age of Aquarius will begin (there is even more disagreement among specialists as to precisely what Aquarius, with its ruler Uranus, will import for the world). Some astrological authorities place the beginning of the Piscean epoch in the first centuries prior to Christ – for example, Rudhyar gives 97 BC, Thierens 125 BC, and Massey gives 255 BC. However, others insist on a much later date – the astrologer Leo loosely dates it to 'soon after the dawning of the Christian era', and other esotericists insist that we are only just beyond half-way through the epoch of Pisces, which would mean that it began about AD 1000. It is all a question of fiducials – unless one can actually pinpoint the beginning of a cycle reliably, one cannot say when it will end. It would seem that within very wide latitudes any astrologer is free to propose his own system. Schultz publishes the following list of stellar fiducials for the zero point of the zodiac in relation to the calendar, the dates given being BC:

16,000	Antares	14,000	Zubenelgenubi	12,700	Spica
10,500	Denebola	8,800	Regulus	7,100	Praesaepe
6,000	Castor and Pollux	4,250	Crab Nebula	3,000	Aldebaran
2,300	Pleides	750	Hamal	0	Kaitan

The precision of such tables belies the fact that few astrologers would agree with it, and very many specialists who have laboured over the vexing question of fiducials have established their own lists. See, however, AYANAMSA. Since the effect of precession is that the stars appear to fall back against the Vernal Point, this means that those astrologers who seek to integrate into their charts the effects of fixed stars (translated to Right Ascension) must allow for this precessional rate. Most tables of fixed stars are accordingly dated, so that allowance may be made for precession – for example, the rounded-off degrees given under FIXED STARS are calculated for 1985, the mid-point of the eighties decade, and are therefore tolerably accurate for most purposes of astrology linked with this decade. By simple numerical adjustment it is possible to allow for calculations related to other times. A modern rule of thumb is derived from the rate of precession proposed by Newton, which is approximately 50 seconds of arc per year. However, the stars are far from being 'fixed', and a variety of defined motions – proper and radial motions mainly – do adjust the precessional rate. Even so, in practical terms this has been shown to be sufficient for most ordinary astrological needs. It has been observed that this numerical relationship is the equivalent of 72 years to one degree of precession – approximately the mean length of human life. Hipparchus is usually identified as the first astronomer to note (if not define) the phenomenon of precession (in the 2nd century BC) but it is evident that it was at least partly recognized by Babylonian astrologers (see, for example, SYSTEM A, relating to ayanamsas), and its effects were noted in the temple orientations of Egyptians and Greeks. In any case, long before the time of Hipparchus the two so-called 'zodiacs' had been clearly distinguished – but see GREEK ZODIAC. The rate of precession was not accurately determined until comparatively recent times, so that historically speaking the dates assigned to zodiacal epochs have varied considerably. For example, Albategni gave a precessional rate of 1 degree to every 60 years and four months. Dante was following the Arabian astrologer Alfraganus in assuming an even rate of precession of 1 degree per century, and in anticipating that the twelve precessional ages would be of equal extent. The Arabic astrologer Albumasar suggested a theory of TREPIDATION involving an oscillation around an arc of 7 degrees, in periodicities of 900 years, rather than a direct retrogression through the constellations. Schwabe and Wachsmuth have done much research into the realm of precession, and have distinguished between the zodiac and the constellations in an interesting way: they each make useful and original observations about the bearing which precession has had on historical epochs.

precessional age A period sometimes supposed to be 2,160 years in duration: but see PRECESSION.

precessional year A period of 25,920 years – see PRECESSION.

precessions A term sometimes used as a translation of the Greek *proegeseis* used by Ptolemy in connection with the direct movement of a planet, in contradiction to its retrograde motion.

predictive astrology The branch of astrology which deals with revealing the future trends implicit within a natal chart. This is the astrological realm covered in DIRECTIONS and PROGRESSIONS.

pre-natal chart See PRE-NATAL EPOCH. Not all pre-natal charts claim to be cast for the moment of conception. Some astrologers cast such pre-natals for the so-called quickening (medically and theologically different from the conception) which is regarded generally as taking place about six weeks after conception. Other times are used also, depending upon the individual views of the astrologer as to the nature of gestation: see, for example, Collin.

pre-natal epoch Sometimes simply the epoch, the pre-natal epoch was originally defined as a time. It was claimed to be that moment (said to occur approximately at the beginning of the gestative period) when the degree ascending on the eastern horizon and the longitude of the Moon at that moment were interchangeable with the longitude of the Ascendant degree and the longitude of the Moon (or their respective opposite points) at birth. Some astrologers would cast a chart for such a time, and regard this chart as a conception chart relating to the destiny of the native as would an ordinary birth chart. As formulated by Bailey, the law of the pre-natal epoch has long been supposed to recall or reflect the mythical *TRUTINE OF HERMES, but an investigation into what the general law implies for particular application leads into considerable complexities which belie the simplicity of the definition. In any case, originally the term was applied not to a time but to a chart, cast for this moment of time. Eventually, by extension, the word was applied to the body of interpretative material built around the chart, so that the term 'epochal astrology' eventually emerged, to cover a wide field of astrology concerned fundamentally with the application of astrological doctrines to the supposed conditions of the gestation period, as well as to the relationship which is supposed to exist between the embryonic life and the subsequent natal horoscope. The idea of the epoch was largely popularized by Bailey in a system which does not bear close scrutiny, yet there are now several different forms of pre-natal astrology derived from other views of the periodicities of incarnation other than the precise or postulated moment of conception – see PRE-NATAL CHART. Strictly speaking, however, the pre-natal chart described by Bailey has never pretended to be a horoscope for conception, even though it was later frequently termed a conception chart and was believed by many amateur astrologers to be just that – it was an adjusted chart symbolically linked to the modern notion of conception. The fact that the doctrines of the epoch were popularized in Western astrology mainly by adherents of theosophy has meant that the majority of astrologers, not themselves of such persuasion, have failed to grasp precisely what the epoch is indeed supposed to indicate. It is supposed to mark the 'descent of the Monad to the Astral plane' (in the words of Bailey) – this is by no means 'conception' in any popular sense, but from Bailey's later observations has developed the assumption that the chart for the epoch must therefore show the inherent character of the Ego which is about to reincarnate; in strictly theosophical terms, this is actually a *non-sequitur*. One may therefore reasonably question the validity of the statement of Bailey that 'the epoch has a more intimate relation with the individual than the horoscope of birth'. Considerable research has shown that the work of Bailey was astrologically invalid and formulated on erroneous medical premises. The major objections are set out by Dean who rightly comes to the conclusion that however intriguing the concept of the pre-natal epoch in principle, 'in practice it is simply wrong'. In fact Bailey's significance in the development of pre-natal astrology has been out of proportion to its true worth. It is worth noting that Schwickert later doubled the number of possible epochs (and thereby renders the complicated more complicated) by insisting that all four Angles might be used in the lunar exchange rather than just the horizon polarity. Sepharial was probably the first Western astrologer to use pre-natal astrology or conception charts in modern times (the latter were used in Hellenistic and Roman astrology), but he appears to have used it only as a prop for rectification, and indeed described it as 'the only reliable method of rectification'. However, even here it is questionable whether the method is a system of rectification at all, since the pre-natal epoch itself is furnished in such a way as to be derived from the fiducials of the natal chart, the former being derived by mathematical or geometric laws from the latter.

previous syzygy   See SYZYGY.

Primum Hyadum

Primum Hyadum   Fixed star, the gamma of constellation Taurus, literally 'the first star of the Hyades', set in the forehead of the Bull. It is said to be of the same nature as the parent asterism, afflicting especially the eyes, but generally being a most disturbing influence, tending towards violence and ruin.

primary directing   Term used for the method of astrologically unfolding potential in a radical chart, according to the PRIMARY SYSTEM – to be distinguished from the SECONDARY DIRECTIONS or the SECONDARY PROGRESSIONS. See PRIMARY DIRECTIONS.

primary directions   In the method of computing known as the PRIMARY SYSTEM, used for studying the unfolding of potential within a chart, planets are directed to the precise aspect with other planets – thus, one may direct Mars to the place of Jupiter, to calculate the future time when the conjunction will take place in regard to the specific chart, the arc itself being used as a basis for time conversion (on the ratio of $\frac{1}{90}$ of the arc between sunrise and Midheaven). Such unfolding of potential aspects is the directing of aspects, and the aspects themselves are called DIRECTIONAL. See also PROGRESSED ASPECTS.

primary system   The general theory of astrological prediction from the radical chart is based on the idea that each Earth day is the equivalent to the advance of the Sun by approximately one degree. By analogy, this degree is taken to be equal to the passage of one year in the life of the native: this basic analogy has given rise to a method of prediction called the primary system, which involves studying the unfolding of potentials between bodies and nodal points mainly in terms of aspects, the arcs of distance involved being translated into equivalence of time. In the primary system, the distance of directed aspects is calculated as $\frac{1}{90}$ of the arc between the sunrise degree and the Midheaven point of the same day. Such a method is to be contrasted with the so-called SECONDARY PROGRESSIONS. The latter method is more popular with modern astrologers because the time-conversion factors permit greater accuracy, or at least, greater security. Primary directions are themselves derived from the ancient concept of the PRIMUM MOVENS, and since this means that calculations of time equivalent are made in Right Ascension along the equator (with four minutes equivalent to one year) it is very easy to make an error in conversion, or to continue error derived from wrong data (such as a mistake in the time of birth).

prime essences   One of the names given by Paracelsus to the traditional THREE PRINCIPLES of astrology and alchemy.

principles

Prime Vertical chart   A horoscope chart, or symbolic figure, erected to symbolize a plane through the Prime Vertical. The so-called MUNDOSCOPE is such a chart system.

primordial qualities   Term applied by de Lubicz to the two basic qualities of 'dry and hot' and 'cold and wet', represented in the Egyptian symbolism as two lions, and called Shu and Tefnu. The basic idea was adopted by the Greeks in their more exoterically sophisticated theory of QUALITY, while the Egyptian symbolism survived into Christian art.

primordial seven   In esoteric astrology, the seven SACRED PLANETS.

Primum Movens   In the Ptolemaic system (as in the ARISTOTELIAN SYSTEM), the name given to the Ninth Sphere which was believed to impart movement (literally, 'the first movement') to the spheres. See, however, SPHERE OF THE PRIMUM MOBILE. In later cosmoconceptions, the Primum Movens became the Tenth Sphere – but see also NINTH HEAVEN and STELLATUM.

Princeps   Fixed star, the delta of constellation BOOTIS, set in the club or spearshaft of the giant. The name appears to be fairly modern – Ptolemy views all the stars of this constellaton as being similar to that of Mercury with Saturn, but leaves it unnamed. In the Chinese system the star was merely one of the 'Tseih Kung' (seven princes), and the erudite Allen could find no European name for it.

Princeps Signorum   One of the several titles for the constellation and sign Aries, 'the leader of the signs', sometimes PRINCEPS ZODIACI.

Princeps Zodiaci   One of the many names for the constellation and sign Aries, 'leader of the zodiac'. While it is true that Aries is noted as the 'first' of the zodiacal circle, the term is on a deeper level probably a reference to the ancient idea (recorded by Albumasar in the 9th century), that at the creation of the world all the planets were in conjunction in Aries.

Principalities   See ARCHAI and POWERS.

principal places   The four houses where the two luminaries are said to be most beneficial in a horoscope figure: the 1st, 7th, 9th and 11th houses. See also HYLEGIACAL PLACES.

principles   The early Greek astrologers employed the four Aristotelian principles which ruled matter – terms translated as 'hot', 'moist', 'dry' and 'cold'. These were the pairs of interacting opposites said to underlie all phenomena: the hot and moist united forms and brought increase – the dry and cold separated form and brought destruction. The polarities, or 'principles' (sometimes popularly called the Aristotelian principles, and often confusingly called the qualities – see QUALITY) played a most important part in early astrological doctrines. The planets Jupiter, Venus and Moon were said to be 'hot and moist', while Saturn was said to be 'excessively cold' and Mars 'excessively dry'. The four principles extended into a concept of Ages (for which, see SEVEN AGES), and even into the natures of the four winds, which were linked with the four seasons, the four ANGLES and the cardinal points. The east wind was Apeliotes and dry. The west wind was Zephyrus and moist. The south wind was Notus and hot. The north wind was Boreas and cold. See also PLANETARY QUALITIES and SEVEN VITAL PRINCIPLES.

print-out Usually reference to a computerized print-out, specifically in astrological circles to either such a data sheet giving information relating to a particular horoscope, or to a required tabulation of astrological configuration, or even to a computerized ephemeris. See also COMPUTERIZED CHART.

prison A term now almost obsolete, used as the equivalent for FALL. It originated from the Greek terminology which had planets in *phelakai*, subsequently translated as 'in prison'. See also THRONES.

Private Enemies The Point of Private Enemies is one of the so-called Arabian PARS. If the Lord of the Twelfth House is revolved to the Ascendant of a figure, then the new cusp of the 12th house marks the Point.

processions A term applied by the modern astrologer Addey to the numerical derivations of HARMONICS. A procession of nines is set out under SUBSISTENCES.

Procyon Fixed star (binary), the alpha of CANIS MINOR, set in the body of the dog, the name derived from the Greek *prokuon* (before the dog) in reference to the fact that it was the star which preceded the rising of the 'dog star' SIRIUS. The Latin equivalent name was Antecanis, but the medieval Aschere, Aschemie, Algomeysa and Algomeyla appear in early manuscripts, the first two said to have been derived from the Arabian phrase meaning 'bright star of Syria', so named because the star disappeared from the Arab's view at its setting over that country. A curious duality of influence is recorded: Allen, working from ancient texts not available to Robson, says that it portends wealth, fame and good fortune. Robson, who has influenced a whole generation of astrologers, records Ptolemy's view that it is of the nature of Mercury conjunct Mars, and makes of it a celestial disaster area – violence, 'elevation ending in disaster', and danger from dog-bites.

Proditor See TAURUS CONSTELLATION.

profection Sometimes given as an archaic term for 'progression': but see PROFECTIONS.

profections A term derived from Ptolemaic astrology relating to the rising of the signs in connection with a simple method of progressing the chart by the rate of one sign per year. The hylegiacal degree is advanced 30 degrees for each year, irrespective of latitude: the revolved chart is considered for a given year in the life of the native in terms of this simple system as multiples of 30-degree revolutions for a year, or 2.5 degrees for each month. The astrologer Leo, however, defines profections in terms of regular progressions of the Sun and other significators, allowing to each profection the whole zodiacal circle and one sign over: thus, if the Sun in the first year is in 30 degrees of Aries, the next year it will be in 30 degrees of Taurus.

prognosis A term which, whilst generally synonymous with prediction, is sometimes used of general chart interpretation (see PREDICTIVE ASTROLOGY).

progressed Aspects See SECONDARY PROGRESSIONS.

progressed figure See SECONDARY PROGRESSIONS.

progressed horoscope Sometimes called the progressed figure, or progressed chart, it is a horoscope figure calculated for a specific time after the natal chart, and schematically derived from the same. A progressed horoscope may be calculated and interpreted for any period after the birth of the native, though the progression must always be derived from the radical (the natal) chart by one or other of the several systems of PROGRESSION: for example, in the system used according to the schematic DAY FOR A YEAR, a progressed horoscope is cast for a given number of days after birth equal to the number of years elapsing from birth to the time under scrutiny, and the two charts (radical and progressed) are linked schematically by arcs denoting planetary and nodal movements. Whatever the method used, the progressed horoscope is always a record of the development inherent in the radical chart, and may be interpreted only in relation to the radical: it appears to have no other validity in itself.

progressions In popular use this term is generally regarded as being an equivalent of *DIRECTIONS, and it is true that in both systems of prediction, the positions of the planets and nodal points are moved forward (or backwards) according to certain rules, and both are derived from a natal chart, with a view to studying the potential and 'futurity' promised by that natal chart (often called the 'radical'). However, while progressions are involved with techniques of prediction based on the directing of a radical chart, not all methods of *SECONDARY DIRECTIONS are based exclusively on directing forward in space and time. Some methods of secondary directions are concerned with both the idea of converse directing, and with studying the pre-radical chart. The very etymology of the word 'progressions' (from the Latin *pro-gredior*) suggests the idea of 'moving forward', and astrological purists must regard progressions as only one form of directions. The phrase 'progressed horoscope' is generally favoured over the phrases 'directed horoscope', however.

promittor The promittor is that which promises fulfilment, and in a specialist astrological sense refers to the potential of planets, or to the unfolding of that potential through progressions and directions. Thus, in general terms, the planets Mars and Saturn are rightly called the anaretic promittors, for they promise (among other things) the fulfilment of death – see ANARETIC. All planets promise something or other, and thus in a general sense all planets are really promittors. It is really the astrologer, in his capacity as interpreter of a chart who makes a specific planet into a promittor of a particular thing, event or condition. The term is used in a highly specialist sense of any planet to which another planet (called the 'significator') is directed to form a progressed aspect with the radical positions of the promittor, by which the promise of this letter is fulfilled. The term is used widely in HORARY ASTROLOGY, but rarely in natal horoscopy.

proper face A term derived from Ptolemy, though seemingly wrongly defined by him. A planet is said to be in its proper face when it bears the same *ELEMENTAL ASPECT to either of the luminaries which its own house bears to their houses. This Ptolemaic rule has been added to by later commentators, who insist that in addition to fulfilling the above requirements, the planet must be in its own house, and must also be in the ratio aspect with both the luminaries – all of which is one way of saying that a planet would rarely be in its proper face. This later concept also implies a *DOMAL DIGNITY, which many experienced astrologers regard as chimerical. Some astrologers have simplified the definition of the term (though confusingly using the term 'face' as a synonym – see, however, FACE). Wilson, for example, who seems to be of the opinion (incorrect, as it happens) that he is accurately reporting Ptolemy's view, says that a planet is in its own face (*sic*) when it is at the same distance from the Sun or the Moon as its house is from their house.

proper motion The term 'proper motion' is derived from the old sense in which proper meant 'personal': in simple terms, it is the motion of a planet measurable in the plane at right angles to the line of sight. It is to be distinguished from APPARENT MOTION.

prophecy The foretelling of the future. In relation to astrology this is properly one of the specialist fields connected with the art of PROGRESSIONS and DIRECTIONS, by which the promise of a radical chart is unfolded and revealed in terms of cycles and dates relating to the life of the native. Unfortunately, astrology has been linked with other forns of prophecy – perhaps more accurately termed 'astromancy' – which really plays no part in the philosophical background of the science. For example, a surprising number of people use the horoscope as an 'autoscope', and make predictions according to clairvoyant principles unrelated to genuine astrology.

proportional arcs A term derived from the astrology of Sepharial, which proposes that there is a contra-degree, a nodal point, on the other side of the radical Sun, Moon, Ascendant, and Medium Coeli, equidistant from these positions as the planets on the other side. The proportional arc of a planet at 6 degrees Aries, in relation to (say) the Sun at 18 degrees Taurus, is 42 degrees: this arc extended to the other side of the Sun would give a nodal point (a 'sensitive degree', as Sepharial called it) in 30 degrees Gemini. The sensitive degree is said to be receptive to influences from progressions and transits. But see also CONVERSE DIRECTIONS.

Propus Fixed star (binary), the eta of constellation Gemini, set in front of Castor's left foot. It has many names attached to it, such as Praepes and Tejat (a modern word coined apparently from an Arabian anatomical term *Al Tahayi*). It is said to be of the nature equivalent to Mercury conjunct Venus, and while Allen records that it 'portends lives of eminence to all born under its influence', Robson views it as an evil star, for it causes 'violence, pride, over-confidence and shamelessness'.

prorogator Pingree refers to the prorogator (the Greek *aphetis* described by Dorotheus of Sidon, *c.* AD 75) as a point on the ecliptic, determined by complicated rules, which conditions the length of the native's life. This point was progressed at the rate of one degree of Oblique Ascension per year, towards either the Ascendant or the Descendant: as this point conjuncts a malefic, or is aspected by one, the native's life is threatened or even destroyed. In any event, he dies when the prorogator has reached the horizontal angle to which it is being directed. Pingree points out that this system is a modification of an older method (ascribed to the Babylonians) which fixed the maximum length of life as the number of degrees of Oblique Ascension between Ascendant and Midheaven. In later astrological texts, the prorogator was usually simply the directed APHETA.

Proserpina See VIRGO CONSTELLATION.

Psalterium Georgianum An obsolete constellation, formed in 1781 in honour of George II, apparently in reference to his horoscope figure.

psychic houses Sometimes also misleadingly called Terminal Houses, these are the 4th house, the 8th house and the 12th house – those reflecting on the mundane level the influences of the WATER TRIPLICITY, which is concerned with issues hidden in the depths of the psyche, and is often concerned with the regenerative processes. Devore calls this group the 'Trinity of Psychism'.

psychic zodiac A supposed (and entirely fictitious) zodiac of 13 signs, the interloper being called ARACHNE.

Ptolemaic aspects Sometimes the MAJOR ASPECTS, which were listed by Ptolemy, are called Ptolemaic aspects, but this is misleading as they were neither originated by him nor even well defined by him. The term appears to have become necessary because of the introduction of the quintile aspects, which are (more correctly) called the KEPLERIAN ASPECTS.

Ptolemaic astrology Specifically the name applies to a corpus of astrological traditions gathered together by Claudius Ptolemaeus in the 2nd century of our era (see TETRABIBLOS), and in general to that corpus as it was augmented and refined by Arabic and late medieval astrologers. The astrological conceptions, terminologies and even cosmoconceptions, which go under the name of Ptolemy were not of his making, but merely collected, collated and partly systematized from the earlier Chaldean, Egyptian and Greek traditions. Indeed, it is clear that the system which bears his name is replete with ideas and terms which he himself could not always understand, many of which he defined incompletely. It is often claimed (by those who have not bothered to read Ptolemy, or who are unfamiliar with medieval astrology) that the

Ptolemaic astrology is very little different from the late medieval traditions upon which modern astrology is based. This is simply not true: the system is actually very far from the interpretational level of traditional Western forms. While many of the interpretations and traditions (and especially many of the terminologies) he records have been adopted (for example, the tradition of the major aspects – see PTOLEMAIC ASPECTS) just as many have become obsolete (see, for example, PTOLEMAIC TERMS): just as many ascribed to him by later commentators would have been quite foreign to his form of astrology (see, for example, ZIGIATUS). Only the more important of his terms – or such as a general researcher into astrology is likely to come across – are incorporated into this present work. See also PTOLEMAIC SYSTEM.

Ptolemaic system The most widely adopted model of the solar system (and indeed of the universe) used in medieval astrology (but see also ARISTOTELIAN SYSTEM) until it was gradually deposed by the COPERNICAN SYSTEM, both as a model and as an astrological tool. The system itself is largely set out in Ptolemy's *Almagest* and in his *Hypothesis of the Planets*, but was much adapted and refined by later astrologers and astronomers. It is a geocentric system, with a series of epicycles for each of the known planets (of which there were seven, including the Sun and Moon). The sequence of planets carried in the SPHERES, was in descending order from the Sphere of the Fixed Stars (usually called the STELLATUM): SA, JU, MA, SU, VE, ME, MO. At the centre of the concentric spheres was the Earth, encased in a series of elemental spheres, of which fire was the outer, air the next and water the next. In the later models the interior of the Earth was turned into a reflection of the stellar spheres (inverted), though given up to the various levels of the demons: these, however, played little or no part in astrology. In later models, the positions of Venus and Mercury were transposed. It is sometimes wrongly said that there were eight spheres in the Ptolemaic system, no doubt in reference to the Stellatum of fixed stars, and the seven planets of the ancients – but this is far from the case. The daily motion of each planet within the spheres is a result of the whole system of spheres being carried by the rotation of the Stellar Sphere from east to west, around a static Earth. Ptolemy ascribed to the Aristotelian dictum that the extralunar bodies moved in perfect circles, and in order to account for the appearances which contradicted this theory, he imagined that each planet made an irregular journey: though actually always moving in a perfect circle, they each followed a complicated series of epicycles, the centres for which lay on a DEFERENT. The movement of the superior planets involved an annual revolution of its epicycle, while the inferior planets had epicycles which accounted for their proper periods, the deferent for their annual irregularities. The Sun alone revolved without epicycles, in majestic sweep around the Earth. Ptolemy was familiar with the general idea of PRECESSION, which required in his system that the Stellar Sphere (Stellatum) be carried by a further sphere (the so-called NINTH SPHERE), which gave to the Stellar Sphere a diurnal motion from east to west, while this latter (carrying the planetary spheres) moved slowly in the opposite direction. The elaborate Ptolemaic sytem was rendered more and more complex, and required the additions of more and more spheres, until it actually became unworkable because of its own inherent complexity and was superseded by the more simple heliocentric system of Copernicus. The differences between the complicated epicycles and seemingly simple concentrics was not the only point of difficulty. The theory of precession as proposed within the Ptolemaic model actually involved ascribing the Stellar Sphere with two contradictory movements (see PRIMUM MOVENS). However, complex as the system was, it found favour for many centuries with astrologers because it (or rather the tables derived from it) did offer a reasonable basis for the computation of horoscopes – though these were usually backed by direct observation of the skies, which is far from the case in modern astrology. Indeed, the tables and computations offered by the system, while lacking the precision of modern tables, were reasonably accurate. The cosmoconception of Ptolemy was no more his making than was PTOLEMAIC ASTROLOGY: both were merely collected, partly systematized transmissions from earlier sources – the astrological traditions came mainly from Chaldean and Egyptian sources, the astronomical ideas from Aristotle, Plato, Eudoxus and Apollonius of Perga, to name only the major contributors.

Ptolemaic terms   A title given to a system of degree rulerships called TERMS as reproduced by Ptolemy, in contrast with the related system of termini arcs which he recorded as the CHALDEAN TERMS and the EGYPTIAN TERMS, the former resting upon the government of triplicities, the latter upon the government of houses. Terms are usually explained as being certain degrees within the signs which are supposed to have the power of influencing the nature of a planet posited within them to the nature of the planet which 'rules' the term. However, Ptolemy is far from clear as to what the terms are, or as to how they are to be interpreted: it is unlikely that they are sensitive degrees at all, though this is how they are generally explained by those who quote traditional ideas without attempting to understand what Ptolemy actually said. Placidus (who did try) could not understand what Ptolemy was trying to say, and came up with his own theories instead. The present writer must admit to being more than confused by the whole matter of Ptolemaic terms, or termini, and Wilson admitted that they were for him 'wholly

——————————————— TABLE 66 ———————————————

Sign	Ptolemaic terms					Egyptian terms				
AR	JU6	VE8	ME7	MA5	SA4	JU6	VE6	ME8	MA5	SA5
TA	VE8	ME7	JU7	SA2	MA6	VE8	ME6	JU8	SA5	MA3
GE	ME7	JU6	VE7	MA6	SA4	ME6	JU8	VE5	MA7	SA6
CN	MA6	JU7	ME7	VE7	SA3	MA7	VE6	ME6	JU7	SA4
LE	JU6	ME7	SA6	VE6	MA5	JU6	VE5	SA7	ME6	MA6
VG	ME7	VE6	JU5	SA6	MA6	ME7	VE10	JU4	MA7	SA2
LB	SA6	VE5	ME5	JU8	MA6	SA6	ME8	JU7	VE7	MA2
SC	MA6	VE7	JU8	ME6	SA3	MA7	VE4	ME8	JU5	SA6
SG	JU8	VE6	ME5	SA6	MA5	JU12	VE5	ME4	SA5	MA4
CP	VE6	ME6	JU7	SA6	MA5	ME7	JU7	VE8	SA4	MA4
AQ	SA6	ME6	VE8	JU5	ME5	ME7	VE6	JU7	MA5	SA5
PI	VE8	JU6	ME6	MA5	SA5	VE12	JU4	ME3	MA9	SA2

unintelligible'. The one thing which seems certain is that the terms are not actually sensitive degrees in the ordinary sense of the word, so that the modern commentators are quite wrong in their explanation of what Ptolemy had in mind. In the astrological literature, the notion has developed that a planet 'in its own term' (according to one of the two most widely published tables given above) was particularly powerful. However, this notion does not appear to have any support from the Ptolemaic reporting of the term. In view of this, the terms in table 66 are given more out of antiquarian interest than with any hope to throw light on the subject. For the sake of comparison with the Ptolemaic, the Egyptian system is reproduced in table 66, though this was not the one most favoured by Ptolemy, who found it 'inconsistent': however, the Egyptian system was that favoured by DOROTHEAN ASTROLOGY and was thus adopted by Firmicus Maternus. Ptolemy says that the third system – our so-called 'Ptolemaic' – is not his own, but one adapted from an ancient astrological document of uncertain provenance in his possession. The degrees in this system, as well as the rulerships accorded them, are set out in table 66.

Pugil See POLLUX.

Punarvasu The Sanskrit name for the constellation Cancer – but see KARKATAKAM. The same name is also given to the 5th of the Hindu NAKSHATRAS, and, according to Robson, in this latter context means 'the two good again'.

Puppis   An asterism which was created by the subdivision of the constellation of the ship ARGO into Keel (Carina), Sail (Vela), Mast (Malus), and Stern (Puppis), by the astronomer La Caille. The astrological significance of these separate divisions has never been determined, but the entire Argo is associated with drowning – yet, contrariwise, it is also known to give prosperity in sea voyages and trade.

Purva Ashada   Name given to the 18th of the Hindu NAKSHATRAS (the former unconquered).

Purva Bhadra-Pada   Name given to the 24th of the Hindu NAKSHATRAS (the former beautiful feet).

Purva Phalugni   Name given to the 6th of the Hindu NAKSHATRAS (the former bad one).

Pushya   Name given to the 6th of the Hindu NAKSHATRAS (flower).

Pyroeis   One of the Greek names for Mars, meaning 'fire' and often given variable spellings (and various meanings) in medieval astrological and alchemical texts.

Pythagorean Harmonies   See MUSIC OF THE SPHERES.

Pythagorean pentacle   Name given to a six-pointed star, with an eagle figured at the highest point, below a bull, a lion and the face of man (the image of the four Fixed signs of the zodiac – see, for example, TETRA-MORPHS). The imagery of the Pythagorean pentacle (which has nothing to do with Pythagoras, even though the numbers 4 and 6 were derived in this context from the Pythagorean SOUL-NUMBERS) was widely adopted in proto-Renaissance ecclesiastical sculpture, and appears in medieval astrological, alchemical and occult texts. The esoteric content of the pentacle was directed towards symbolizing the human being – within the esoteric image the eagle represented the 'thinking power' (which might soar); the lion represented the 'emotional depth' (through the zodiacal association with Leo, which rules the heart of man); the bull represented the physical body and will-life (Taurus being an earth sign), whilst the face was linked at once with the 'humanity' of Aquarius, and with the entire human being in which the other three dwell, as thinking, feeling and willing.

Pythagorean triangle   See NUPTIAL NUMBER.

Python See DRACO.

Pyxis Nautica A modern constellation originated in 1752 by La Caille from stars previously forming the ship mast of Argo Navis. Though not particularly distinguished, the group is said to give an interest in nautical and geographical matters.

Q

qabbalistic astrology   The word 'qabbalah' relates to a complex philosophy and cosmoconception which has an oral tradition, transcribed for over a thousand years, rooted in a mystical and esoteric interpretation of scriptures. Some indeed argue qabbalism as the roots of modern occultism. It is usual to distinguish 'qabbalism' from 'cabbalism', for reasons adduced by Fulcanelli – but there are many variant spellings of the name. The influence of astrology on qabbalistic thought – or, more accurately, on Christian qabbalism – has been profound. Equally, the influence of Christian and Jewish qabbalism on esoteric astrology has been extensive, though they have scarcely permeated ordinary genethliacal astrology. So extensive were these fruitful influences, which may be traced through Agrippa, Boehme, Dee, Fludd, Paracelsus and Welling (to mention only those authorities whose terms are touched upon in the present work) that it is entirely beyond the scope of a short article to deal with them. Table 5 under ARCHANGELS OF THE SEPHIROTH sets out the traditional correspondences, but it is significant to observe that Fludd quite rightly (from the Christian qabbalistic point of view) gives the Soul of the Messiah to rule over Malkuth, as Christ is united with the Earth: however, within the ancient Hierarchies, the Archangel of the Earth is called URIEL. Trachtenberg sets out some variations, which are perhaps only of interest to scholars: Myer, dealing mainly with Avicebron, sets out a sephirothic correspondence which ignores the Earth (Malkuth marking the Sphere of the Moon). Figure 51 is the Sephirothic Tree (with the supposed planetary rulers) given by Myer, which shows important differences from that listed under ARCHANGELS OF THE SEPHIROTH. A correspondence between the zodiacal signs and the Hebraic letters (more accurately, their sounds and sound values) has been recorded, in some cases associated with the planetary spirits, as in table 67. It is often claimed in the less scholarly texts that the names of the PLANETARY SPIRITS (see SECUNDADEAIN BEINGS) are qabbalistic in origin, which is far from the truth. Trachtenberg traces the names of these medieval Christian planetary rulers by way of Averroes to the gnostics, who were in turn influenced by the Babylonian planetary spirits, and the Persian Amshaspands. Because of the importance attached to the sound-values of letters and numbers in the qabbalistic tradition, a variety of different correspondences between numbers and zodiacal signs

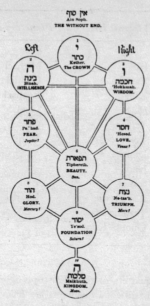

אין סוף
Ain Soph.
THE WITHOUT END.

Left   Right

1
י
כתר
Kether.
The CROWN

3   2
ה   ו
Binah.   Hokhmah.
INTELLIGENCE.   WISDOM.

5   4
פחד   חסד
Pa'had.   'Hesed.
FEAR.   LOVE.
Jupiter?   Venus?

6
תפארת
Tiphereth.
BEAUTY.
Sun.

8   7
הוד   נצח
Hod.   Ne-tza'h.
GLORY.   TRIUMPH.
Mercury?   Mars?

9
יסוד
Ye'sod.
FOUNDATION
Saturn?

10
ה
מלכות
Malkhuth.
KINGDOM.
Moon.

Figure 51: Sephirothic Tree with zodiacal correspondences (after Myer).

and planets have been established. The sequence given by the astrologer
Libra is:

AR 7	TA 6	GE 3 or 12	CN 5	LE 1	VG 1 or 10
LB 8	SC 9	SG 4	CP 3	AQ 2	PI 11

SU 1 or 4   MO 2 or 7   JU 3   ME 5   VE 6   SA 8   MA 9

Quadra Euclidis   See NORMA.

Quadrans Muralis   Constellation, Mural Quadrant, formed by La Lande in
1795 between the right foot of HERCULES and the left foot of BOOTIS. Not
recognized by modern astronomers, but said by astrologers to give rectitude,
justice and idealism.

quadrantine lunation   A term (rarely) applied to the four major aspects
of Sun and Moon – the conjunction, opposition and two squares – which
in the monthly cycle mark the four quarters of the celestial figure – see
QUADRANTS.

——————— TABLE 67 ———————

Sign	Hebraic letter	Planetary spirit
AR	He	Michael
TA	Vau	Gabriel
GE	Zain	Raphael
CN	Heth	Uriel
LE	Teth	Guriel
VG	Yod (Jod)	Nuriel
LB	Lamed	Yeschamiel
SC	Nun	
SG	Samek	Ayil
CP	Ain (Ayn)	Ubariel
AQ	Tzaddi (Tsade)	
PI	Quoph (Koph)	

quadrants The four quarters of the horoscope figure. In reference to the circle of the zodiac, the oriental quadrants are from Aries to Gemini, and from Libra to Sagittarius, inclusive. The occidental quadrants are from Cancer to Virgo, and from Capricorn to Pisces, inclusive. Very many other names have been used to distinguish the quadrants: the first is called vernal, sanguine and infantine; the second is called meridianal, estival, youthful and choleric (Wilson gives 'choloric'); the third is called autumnal, melancholic and mature, while the fourth is called northern, decrepit, wintry and phlegmatic.

quadrated chart See CHART SYSTEMS and QUADRATED CIRCLE.

quadrated circle A term probably originated by Bouche-Leclercq to apply to the early horoscope charts symbolized in a square or otherwise quadrated form, derived from the more usual (and older) system of symbolic projection which was based on a circle: see CHART SYSTEMS. Two quadrated forms have persisted into modern astrology, and each of the divisions within their quadratures define houses in terms of the projected ecliptic. The first is based on what might be called the Decussata (figure 12B under CHART SYSTEMS), the second on the Greek Cross (figure 12C).

quadratures Name given to the DICHOTOMES of the Moon.

quadripartium See TETRABIBLOS.

quadrupedal A term used to denote the five zodiacal signs represented by 'images' which depict four-footed creatures: Aries (the ram), Taurus (the bull), Leo (the lion), Sagittarius (the horse), and Capricorn (sometimes a goat, but more properly a fish-tail – something of a foreigner in this classification). These are also called 'four-footed'.

quadruplicities Each of the twelve signs of the zodiac have been grouped into fours, related in such a way as to reflect the communality of QUALITY: namely, Cardinality, Fixity and Mutability. These three groups of quadruplicities (sets of four) are:

AR, CN, LB and CP   Cardinal quadruplicity.
TA, LE, SC and AQ   Fixed quadruplicity.
GE, VG, SG and PI   Mutable quadruplicity.

qualities See QUALITY.

quality This term has (confusingly) two important technical applications in astrology. The first meaning is related to the idea of 'four qualities', linked with the innate natures of the elements: Fire is said to be of the 'hot and dry' qualities; Earth is of the 'cold and dry' qualities; Water is of the 'cold and humid' qualities; Air is of the 'hot and humid' qualities. In this sense, the four qualities apparently originated by Philistion were a development of Empedocles's theory of ELEMENTS, in which fire was hot, air was cold, water was moist, and earth was dry. In relation to human beings, illness was a sign that a person was 'out of quality', or suffering from some imbalance. These qualities were eventually combined to form the limited number of dual combinations set out above, and it was in this guise that they dominated Roman and medieval astrology, mainly through their link with the theory of bodily humours – see TEMPERAMENTS and PRIMORDIAL QUALITIES. A second meaning is related to the idea of 'three Qualities', by means of which each of the four elements finds expression in the twelve signs of the zodiac. These Qualities are CARDINALITY, FIXITY and MUTABILITY. Thus Aries, Cancer, Libra are of the Cardinal Quality. In this sense, the Qualities are also called the QUADRUPLICITIES and the Modes.

quarterly charts A term used by Bradley to denote the four charts which may be derived from the quarters, or three-monthly periods in the annual SOLAR RETURN.

quartessence The term 'quartessence' is found mainly in the curious astrological terminology of alchemical texts. Philalethes denies the existence of the QUINTESSENCE but describes a quartessence, a 'fourth essence' which is the binder of the THREE PRINCIPLES, and is called by the curious designation 'moist and silent fire'.

quartile An old expression used to denote the SQUARE aspect.

Quaws Al Quaws, the Arabic name for Sagittarius.

Queen of Nebulae See GREAT NEBULA.

querent A specialist term in astrology, theoretically used only to denote the one who asks the question in HORARY ASTROLOGY: sometimes, however, the term is used of any questioner in the various divisions of astrology. In horary astrology the significator of the querent is generally the Ascendant, or its Lord, or planets in the Ascendant sign.

quesited A term usually restricted to HORARY ASTROLOGY, and related to the person or thing enquired about in respect of a horoscopic chart. The term is intended to draw a distinction between the quesited and the QUERENT, the person who puts the question.

quincunx An aspect of 150 degrees, the two aspecting points being separated by five signs of the zodiac.

quindecile An aspect of 24 degrees, derived from dividing the zodiac into 15ths.

quintessence The term means 'fifth element', and is a reference to the invisible power or essence which binds in a unity the otherwise separate FOUR ELEMENTS. The term has had very many synonyms, of which the most frequently used are Aether, Mercury of the Philosophers and the modern ETHERIC, the last of which has been studied in some depth by Wachsmuth: however, see also QUARTESSENCE.

quintile An aspect of 72 degrees, arising from a division of the zodiac into fifths. The astrologer Morin belittled 'the new quintiles' of Kepler (see KEPLERIAN ASPECTS), and has found much support from other traditional astrologers, yet many moderns claim that this particular aspect does convey a quality of single-mindedness and power to those planets forming such an aspect.

quintiles   A general name sometimes given to a range of ASPECTS derived by Kepler to supplement the traditional group – see KEPLERIAN ASPECTS.

R

radical A term with several applications, all etymologically connected with the idea of 'root'. Generally without a qualification or specified context, the term is the equivalent of RADICAL CHART, the name given to the primal horoscope figure. Planets disposed in such a primal figure are usually called radical, so that a prefix radical usually indicates a planetary placing or nodal point within the radical chart: for example, radical Saturn is the placing of Saturn in a given radical chart, and is normally so termed to distinguish it from the progressed Saturn. The most frequently used abbreviation for this example would be rSA and pSA.

radical chart The radical or radix chart is the primal horoscope for a particular birth, the 'root figure' in genethliacal astrology, from which other charts (such as the progressed chart) may be derived, or which form the basis of later exegesis. The radical chart need not be the first chart cast for a birth (or other event) since a later rectification may reveal this to be inaccurate, and in need of adjustment: the finally adjusted recast chart is then the radical.

radical elections A term applied to certain astrologically determined times, calculated relative to a radical chart, and considered in relation to the unfolding tendencies and promises within that chart, expressed through transits or progressions. Usually, radical elections are intended to determine fortunate or otherwise propitious astrological conditions for specific enterprises.

radical position The planetary and cuspal positions in a radical chart (see RADICAL) as distinguished from progressions or directions from that radical chart.

radix In general this is an alternative name for the RADICAL CHART, the radix figure. Sometimes, however, the term may be used of the radix system, a progressional technique (somewhat equivalent to that of the SYMBOLIC DIRECTIONS) which progresses all bodies at a rate of 59 minutes 8 seconds of arc equivalent to a year: see MEASURE OF NAIBOD.

radix cosmogram A radical horoscope figure used in COSMOBIOLOGY.

radix method Term used of a simple method of SYMBOLIC DIRECTIONS, involving the addition of 59 minutes 8 seconds of arc for each year of life: see MEASURE OF NAIBOD.

Rahu The Sanskrit name for the DRAGON'S HEAD (the north node), which is accorded great importance in Hindu astrology. See also KETU, the Sanskrit equivalent of the DRAGON'S TAIL.

Ram school Name given to a Dutch school of astrology, associated with the highly personalized system of T. J. J. Ram, which involves (among other idiosyncratic features) the use of three hypothetical planets (Persephone, Hermes and Demeter) and an entirely new system of planetary rulerships which incorporate these transneptunians.

Rangifer Name of a small and faint constellation between Cassiopeia and Camelopardalis, sometimes called Tarandus, the Reindeer, and Renne, formed by Le Monnier as a memento of his stay in Lapland in 1736. Said by Robson to give a retiring nature and obscure life.

Raphael The Archangel linked with the Sphere of Mercury, and one of the SECUNDADEIAN BEINGS. In the qabbalistic tradition, Raphael is the healer, and is connected with the idea of knowledge gained through experience: he is linked in astrological lore with rule over the element of air.

rapt In its astrological application, this term is derived from the Latin *raptus* (carried away), in reference to the apparent diurnal movement of the heavens, by which it is carried: see, however, RAPT PARALLEL.

rapt parallel This term is applied to two planets of a natal chart which, by direction, form mundane parallels (which is to say that they are equal distances from one of the four Angles): see PARALLEL IN MUNDO. The origin of the word 'rapt' is explained in terms of the 'rapt motion' of the Primum Mobile, which in the Ptolemaic system of spheres was said to carry the planets in motion, and thus formed such parallels. A modern equivalent term would be 'transported parallel'.

Ras Algethi Fixed star (double), the alpha of Hercules, with many variant names. Ebertin says that the influence of the star induces boldness and 'power drive'.

Rasalgeuze   See POLLUX.

Ras Alhagas   See RASALHAGUE.

Rasalhague   Fixed star, the alpha of Ophiuchus, the name from the Arabic Ras al Hawwa (head of the serpent charmer) with many variants, among which Ras Alhagas, Rasalange, Rasalauge and Alangue are prominent. Ptolemy records it as being of the nature of Venus with Saturn, and it is said to bring misfortunes, mental depravities and perverted tastes. Alvidas claims for it a more beneficial influence, however.

Ras Elased Australis   A term used by Ebertin for the fixed star epsilon of constellation Leo, probably due to a misreading of the Arabian term Al Ras al Asad al Janubiyyah, which means 'southern star in the head of the lion'. Ebertin links the influence with higher spiritual gifts for those capable of receiving them, and with 'psychological depressions' and even suicide for those incapable of receiving such gifts.

Rasi   The Hindu term for a birth-chart, usually cast for a constellational zodiac, and using equal house division, the so-called MODUS EQUALIS.

Rastaban   Fixed star (binary), the beta of Draco, set in the head of the asterism, the name from the Arabic Al Ras al Thuban (the dragon's head), but with many variants, including Asuia, Asvia and (presumably from early star maps) Dragon's Eyes. It is said to be of the nature of Mars conjunct Saturn, bringing losses, criminal inclination and (with the Moon) blindness.

rational and universal method   See MORINEAN SYSTEM.

rational method   See REGIOMONTANEAN SYSTEM.

Raven   See CORVUS.

ray   The term, now almost archaic in its original astrological sense, was once used in traditional texts as being synonymous with INFLUENCE. Significantly, the term is still cognate with 'radius', and suggests the idea of a connecting line between the circumference and the centre – an idea of importance only within a cosmoconception involving concentric spheres. The idea of 'influence rays' still persists in the term which has a planet 'falling under the rays' of another planet, when it enters orb of aspect. The same notion persists in the doctrine which maintains that certain nodal points, such as the Ascendant, or fixed stars, or Pars 'cast no rays', which is to say that they operate only through the direct contact of the aspect of conjunction, but do not cast influence rays to form other effective aspects. The term has been widely used in the modern INTUITIONAL ASTROLOGY, though in a specialist sense – see SEVEN RAYS – not to be confused with SEVEN RAY TYPES. The term is also used of the calibrated linear scales within the centre of the COSMO-PSYCHOGRAM.

Ray of Active Intelligence   See THIRD RAY.

Ray of Ceremonial Order   See SEVENTH RAY.

Ray of Concrete Science   See FIFTH RAY.

Ray of Devotion   See SIXTH RAY.

Ray of Harmony through Conflict   See FOURTH RAY.

Ray of Idealism   See SIXTH RAY.

Ray of Jupiter   See SEVEN RAY TYPES.

Ray of Love-Wisdom   See SECOND RAY.

Ray of Mars   See SEVEN RAY TYPES.

Ray of Mercury   See SEVEN RAY TYPES.

Ray of Moon   See SEVEN RAY TYPES.

Ray of Neptune   See SEVEN RAY TYPES.

Ray of Power   See FIRST RAY.

Ray of Saturn   See SEVEN RAY TYPES.

Ray of Uranus   See SEVEN RAY TYPES.

Ray of Venus   See SEVEN RAY TYPES.

Ray of Will   See FIRST RAY.

rays See SEVEN RAYS and SEVEN RAY TYPES.

real zodiac A term used by the astrologer Leo, but unfortunately defined (rather than described) in a most confusing manner. On the one hand he appears to consider it one of the sidereal zodiacs (though of equal division), and on the other hand as the familiar tropical zodiac. He seems to have originated the term in order to distinguish this from the 'supposed zodiac' (as he terms it) of SOLAR BIOLOGY, the so-called LUNAR ZODIAC or 'Earth Zodiac', described by Butler.

received See RECEPTION.

reception This term has two quite different specialist meanings. First, it is used to describe a relationship between planets: when a planet is placed in a sign over which it does not have rule, it is said to be 'in reception' or to be 'received by' the actual ruler of that sign. Secondly, when a planet is being subjected to an aspect from a faster-moving planet, it is said to be 'in reception' of that aspect, or to be 'receiving that aspect'.

Rech See PHOENIX.

rectification A specialist term used to denote the process of adjusting a birth chart (itself believed to be inaccurate) by reference to known events which are dated independently of the horoscope, and hence permit correction of a progressed chart. There are several methods of rectification in modern use: for example, some astrologers use the PRE-NATAL EPOCH as the basis for rectification, whilst others employ a rule-of-thumb method (more of a preliminary to precise rectification) involving adjusting the chart to fit the known characteristics of the native to suitable Angles and intermediate house-positions for planets. Some individuals use radiesthesic methods for determining the Ascendant degree, from which the related chart may be calculated, but this is a method frowned upon by serious astrologers, and, furthermore, a method which must in any case be checked against a rectification if it is to have any practical application.

recurrence cycles A term applied to periods of time in which conjunctions of two planets will recur in (approximately) the same degree of the zodiac. See CYCLES, FIRST ORDER CYCLE, and SECOND ORDER CYCLE.

Red Bird A name given to an arc of seven SIEU or lunar mansions in the Chinese astrological system – sometimes it is wrongly called a 'constellation'.

There are three other such arcs: the AZURE DRAGON, BLACK WARRIOR and WHITE TIGER. See also CORVUS.

re-directed   A term used of the movement of a planet back into direct motion after being RETROGRADE.

reduction   A term sometimes used to denote the conversion of sidereal material to tropical, or tropical material to a sidereal.

refranation   A specialist term used only in horary astrology. Refranation occurs when an aspect does not reach completion, due to the retrogradation of one of the aspecting planets.

Regel   See RIGEL.

regeneration cycle   See CYCLE.

Regia   See REGULUS.

Regiomontanean system   A system of *HOUSE DIVISION, sometimes called the Modus Rationalis (rational method), evolved by the 15th-century astrologer and mathematician Johann Muller, who worked under the pseudonym Regiomontanus. The principle of his system is the trisection of a quadrant of the equator by great circles mutually intersecting at the north and south points of the horizon. The degrees at which these circles cut the ecliptic mark the cusps of the houses, which are regarded as being the centre of the house influence (rather than marking the outer limits of the house). Unfortunately, tables for the Regiomontanean system are not as generally available as those for the PLACIDEAN SYSTEM, though the method is certainly one of the two most rational divisions of the Celestial Sphere for astrological purposes. See also CAMPANEAN SYSTEM.

Regulus   Fixed star (triple), the alpha of constellation Leo, set in the body of the lion, the name meaning 'little king', apparently so called from the belief that it ruled the affairs of the heavens. It was included in the Babylonian starlists under the name Sharru (also meaning 'king'); in India it was Magha (the mighty); in Persia, Miyan (the centre), and so on – a most powerful star, and one of the four ROYAL STARS of the Persians. It is sometimes called in medieval star lore Cor Leonis (heart of the lion), and from this have come many Arabian degenerate terms, such as Kalbelasit, Calb-elez-id, and so on. Sometimes it is called the Heart of the Royal Lion, the Star Royal, and the condensed version of the entire asterism, Domicilium Solis; from the Greek *Basilikos* (king) we have Basilica Stella and Basiliscus, and from the Latin, Rex and Regia. According to Ptolemy it is of the powerful nature of Mars with Jupiter, but it is regarded as being ultimately a violent and destructive star, save when culminating, when it is said to bring high honours. The beneficial planets also bring out the best side of its influence. In contrast to Ptolemy's insistence, the European astrologers associate it with a 'kingly' influence, and (in the words of Allen) 'made it a portent of glory, riches and power to all those born under its influence'.

Reindeer   See RANGIFER.

relationship chart   Term derived by Davison to denote a chart which presents in a single figure material derived from two horoscopes. The figure is cast for the mid-point in time and space between the two relevant birth dates. Sometimes confused with the COMPOSITE CHART.

relative houses   These are the THIRD HOUSE, the SEVENTH HOUSE and the ELEVENTH HOUSE – those reflecting on the mundane sphere the influence of the air triplicity, which calls attention to relationships, partnerships and friends. Devore calls this the 'Trinity of Association'.

Repa   See SPICA.

reproductive trinity   A term used by Butler in his SOLAR BIOLOGY in respect of the zodiacal signs Libra, Scorpio and Sagittarius.

Rete   See ASTROLABE.

Reticulum Rhomboidalis   Constellation, the Rhomboidal Net, formed by Habrecht and adopted by La Caille: it lies north of Hydrus, and is said to give tenacity, selfishness and a 'restricted life'.

retrograde   Specialist astrological term derived from the Latin compound meaning 'to step backwards', and applied to the apparent motion of a planet backwards along the zodiacal belt. In traditional astrological interpretation a planet which was so retrograde carried a somewhat sinister connotation, though the tendency is for modern astrologers to either entirely dismiss the idea that retrogradation should influence the significance of planets, or to greatly amend the sinister reputation attached to the ancient interpretation. When the retrogradation ceases, and direct motion is about to begin, there is a time when the planet appears to stand still: it is then called a stationary planet. The retrograde station is that point of apparent stability at which the planet begins to move into retrogradation: the direct station is that point of apparent stability at which the planet begins to move into direct motion along the zodiacal course.

retrograde application   The *APPLICATION to aspect of a retrograde planet to another retrograde planet.

retrograde arc   The arc of the zodiac in which any planet is retrograde, between its retrograde station and its direct station (see RETROGRADE).

retrograde station   See RETROGRADE.

revati   The 26th of the Hindu NAKSHATRAS.

revolution   In astrological specialist use, the return of any of the planets to their radical places in a chart is termed a 'revolution'. The same term is, however, also applied generally to any orbit – and sometimes (wrongly) to any axial rotation of a planet. See ORBITAL REVOLUTION and REVOLUTION PERIODS. The term is unfortunately also applied to the SOLAR RETURN, which is sometimes wrongly called the 'Solar revolution method': see also SOLAR REVOLUTION FIGURE.

revolution periods   Traditional astrology has provided various scales of revolution periods for the planets, some of them remarkably accurate: see PLANETARY PERIODS. A modern series, set out in days and years, is given in table 68.

--- TABLE 68 ---

Planet	Revolution (days)	Revolution (years)
ME	87.97	0.241
VE	224.70	0.615
MA	686.98	1.881
JU	4,332.59	11.862
SA	10.759.20	29.458
UR	30,685.93	84.015
NE	60,187.64	164.788
PL	90,600.00	248.400

Rhea

Rhea   Name given to one of the moons of Saturn.

Rhomboidal Net   See RETICULUM RHOMBOIDALIS.

Rigel   Fixed star, the beta of Orion, set in the left foot of the figure, the name from the Arabic *Rijl* (foot). It has many variant names, including Regel, Riglon, Rigel Algeuze, Algibbar, Algebar and Elgebar. As with the fixed star Regulus, Ptolemy regarded it more unfavourably than modern astrologers, according it an influence of Jupiter conjunct Saturn, though (as Allen records of the later astrologers) it was 'said that splendour and honours fell to the lot of those who were born under it': certainly this is suggested by the lists of planetary influences set out by Robson, almost all of which give good fortune and lasting honours.

Right Ascension   A system of measurement, designed to determine the stellar system of co-ordinates with earthly longitude. It is expressed (usually) in degrees of declination from the first point of Aries eastwards, within the plane of the celestial equator. In most house systems, the Midheaven is directed by Right Ascension.

Rigil Kentaurus   Fixed star, the alpha of Centaurus, set in the foot of the centaur, the name apparently from the Arabic *Al Rijil a Kentaurus* (centaur's foot). The star, under the name of Serk-t, was highly regarded by the Egyptians, and appears to have been used for temple orientation (at the autumnal equinox) in the 3rd and 4th millenium BC. Due to a catalogue error, the star was named Burrit Bungula – from *beta* and *ungula* (hoof) – by which misnomer it has passed into astrological lore.

Riglon   See RIGEL.

Riksha   The Sanskrit term for an *ASTERISM, though also used to denote a fixed star, as well as the Hindu NAKSHATRAS.

Rishabham   The Sanskrit name for zodiacal Taurus – but see also KRIT-TIKA.

rising sign   A term frequently misused in popular astrology. Properly speaking it should apply only to the sign which marks the Ascendant degree on a particular horoscope. The term is sometimes popularly used as being synonymous with ASCENDANT (especially so in the United States of

America), whereas the rising sign is really a whole sign of 30 degrees, and the Ascendant is actually a single degree (and indeed more properly an intersection of an arc).

rising times In ancient forms of astrology, the rising times were tables derived to show how many degrees of the equator crossed the horizon of a given place, consecutively with the zodiacal signs: such were the basis for chart computations. The rising times varied according to (a) the AYANAMSA used, and (b) the CLIMA, or location for which the information was tabulated – see, for example, SEVEN CLIMATA. Determined by means of spherical trigonometry, the rising times were known to Ptolemy, but previously the Babylonians had used arithmetical approximates. The equation of 1 degree of arc to 4 minutes of time is still used in rule-of-thumb astrology, but is far from exact in latitudes far from the equator.

River of Heaven See VIA LACTEA.

Robur Carolinum Constellation, Charles's Oak, formed by Halley in 1679 in commemoration of the oak in which his patron (Charles II) hid after the battle of Worcester in 1651 – in its original form, this asterism overlaid the constellation of Argo Navis. Astrologers say that it gives an honourable, generous and steady nature.

Rohini The 2nd of the Hindu NAKSHATRAS.

root-earth trigon See SIDEREAL MOON RHYTHMS.

root figure See RADICAL CHART.

root races See SEVEN RACES.

Rosicrucian astrology A term properly restricted to the form of astrology promulgated by the followers of Heindel, whose esoteric content was almost entirely derived from theosophical teachings (after Blavatsky and her systematizers), and from the writings and lectures of Steiner – in both cases usually without acknowledgment. The modern Rosicrucianism must not be confused with the late-medieval stream of Rosicrucianism that flowered in the writings of such occultists as Dee, Fludd and Boehme, who actually influenced esoteric astrology more profoundly than they influenced the so-called 'traditional' astrology.

Rota Ixionis

Rota Ixionis   See CORONA AUSTRALIS.

rotation   In correct use this term is applied to the motion of a body upon its own axis: see also REVOLUTION.

round   See GLOBE PERIOD.

rounding off   Since for most practical purposes the positions of planets in charts are required only in terms of degrees, rather than in terms of subdivisions of degrees, such positions are often rounded off. The two main systems of rounding off result in some cases in different figures being obtained, so that each system employed in a particular chart, and the nature of the conversions, must be noted. The first system (and the most widely used) is really a statement of the nearest whole degree: thus 12 degrees 12 minutes and 12 seconds may be given as 12 degrees, as may be 11 degrees, 43 minutes and 12 seconds. The second system requires that the precise figure be expressed in terms of the actual degree itself: thus, since 12 degrees, 12 minutes, 12 seconds actually marks a point in the 13th degree, then this must be expressed as 13 degrees. This latter system recognizes no *ZERO DEGREE, save perhaps as an intellectual abstraction, for a planet must always be posited in one degree or another. This latter method is of considerable importance in regard to conversion of horoscopic material to degree symbols, and a failure to understand the underlying conditions has frequently led to wrongful conversions. An example of an actual chart will therefore be instructive: the two horoscopes in figure 52 show an example of rounding off the degrees in a chart erected for the birth of Goethe by the astrologer Pearce. It is instructive to note that the corresponding data rounded off in the Notable Nativities of Leo is not accurate, and may be taken as a suitable illustration here:

H: SC 17,   MC: VG 5,   SU: 5 VG,   MO: 12 PI,   ME: 29 LE,
VE: 26.5 VG,   MA: 3.5 CP,   JU: 26 PI,   SA: 15 SC

Whilst MO 12 PI is accurately rounded off, SU 5 VG should actually be SU 6 VG, and the Ascendant should be SC 18. The relevance of this is obvious when one attempts to relate the figure to a degree symbol (or indeed to accurately defined aspects) – the data given by Leo would be misleading, whilst that given in the rounding off diagram in figure 52 would not be misleading.

round table   See GLASTONBURY ZODIAC.

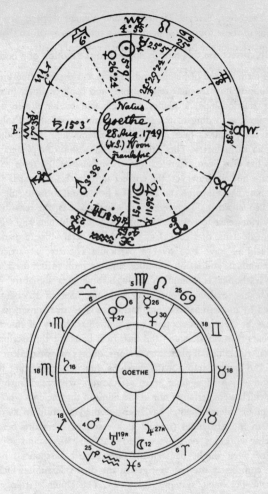

Figure 52: Horoscope for birth of Goethe 28 August 1749 NS, noon, at Frankfurt (after Pearce), with rounded-off version for comparison.

Royal Stars One of the titles given to the four stars of the ancient Persians (beginning of the 3rd millenium BC) which were apparently used as markers, and sometimes called the Watchers of the Heavens, the Four Watchers, and Guardians of the Heavens. The modern names of these stars are ALDEBARAN, ANTARES, FOMALHAUT and REGULUS, but Dupuis and Flammarion named them as Tascheter, Satevis, Hastorang and Venant respectively. The Right Ascensions of these four are approximately six hours apart, so that they mark the four quarters of the heavens, rather than cardinal points.

rule   This term is used of planets and signs in a wide and often unsatisfactory manner to indicate particular relationships and congenialities. In reference to particular relationships, a planet is often said to have rule over certain zodiacal signs, according to the ancient system recorded by Ptolemy, which has in recent centuries been adjusted considerably: in this connection, see PLANETARY RULERSHIP. This particular relationship of a planetary rule over the signs has also been extended into the misuse of the term to give 'rule' over certain houses: however, the correct term in this context is *LORDSHIP. A planet may be said to be the Lord of a house and to have rule over a sign (but see also THRONE). In reference to congenialities, the term 'rule' is used in a very wide sense to cover relationships held by planets and signs through the ancient doctrine of SIGNATURES. In this general sense, a planet or a sign may be said to have rule over a particular country, a particular city, a particular animal, and so on. A more suitable term (often used in the more scholarly textbooks) is 'dominion', though sometimes 'signature' is also used. Virtually every tradition connected with the idea of rulership has been questioned by modern astrologers, and there is no real consensus of opinion which may be quoted in support of a particular theory of rulership. Dean, who appears to misunderstand the historical basis for the ascriptions of the 'traditional' rulerships of planets over signs, quotes many excellent testimonies both for and against traditional rulerships, but it is interesting to note that among those quoted against were astrologers who made use of sidereal zodiacs, which were certainly not in the mind of the ancients who formulated the system of rulerships that has been adopted by modern astrology. The tendency in modern times (mainly as a result of the work being done to classify the workings of the so-called NEW PLANETS, and in some cases because of what might be termed 'bureaucratic aesthetics') has been to eliminate completely the last vestiges of the dual rulerships which lay at the basis of the Ptolemaic system: see SECTA. The Dutch astrologer Ram has suggested a schema which incorporates a number of transplutonian HYPO-THETICAL PLANETS, and which virtually ignores the traditional schema: in this (though with different hypotheticals) he was preceded by Wemyss. The URANIAN ASTROLOGY of Witte has also involved a new schema of hypotheti-cals, but it appears that many of the modern Uranian astrologers no longer regard rulerships as having validity at all. In the midst of so much confusion, the present author would wish to say that a return to the (presently misunderstood) principles evinced by the ancient Ptolemaic school will give excellent results for most purposes of chart interpretation. The fact is undeniable that the majority of modern astrologers who dream up new rulerships and schemas are rarely familiar with the spiritual truths underlying the traditions which they reject.

Ruler of Tension A term adopted by the German astrologer Meier in connection with his aspectal patterns (see FIGURINES). The rules for determining which of the planets in a chart may be termed the Ruler of Tension are complex, seemingly involved with a simplistic numerology, but the interpretation of the effects of the determined ruler is on the whole concomitant with the expectations of traditional astrology: MO, for example, brings unrest and change, whilst UR brings lack of direction or revolution.

Ruler of the Horoscope See LORD OF THE HOROSCOPE.

ruling planet The ruling planet of a particular horoscope is the *LORD OF THE HOROSCOPE, but the ruling planet of a sign or a house (but see RULE for the misuse of this term) is the specific planet assigned to that sign or house in the particular astrological system.

ruminant signs One of the ancient classifications which group the zodiacal signs according to the characteristics of their IMAGE. Aries, Taurus and Capricorn have been identified (through their images) with creatures which chew their cud, and may therefore be called 'ruminant'.

S

Sabathziel   One of several variant names for the spiritual ruler of Saturn.

Sabeans   The Sabeans, or Sabians, are in certain astrological circles regarded as being so deeply connected with early forms of astrology that the name is often used as synonymous with 'astrologers'. No doubt the term SABEAN SYMBOLS was derived from this idea. However, there is no evidence that the ancient Sabeans (who lived in the Yemen well into the first millenium BC) practised astrology in any sense that we would now recognize. It is possible that in common with many pre-Christian tribes they were astrolators, but unlikely that they were astrologers.

Sabian symbols   Name given to a series of *DEGREE SYMBOLS ascribed to each of the 360 degrees of the zodiac, originated by psychic means in the USA by 1925. After some decades of preparatory work by many astrologers, the system was partly developed by the American Rudhyar (1936), and popularized in book form by Jones in 1953. The particular utility of the system lies in the fact that it is attached to a supportive cross-reference of detailed and accurate horoscopes, by which the readings may be checked or exemplified.

Sabik   Fixed star, the eta of Ophiuchus, sometimes called Saik, set in the left knee of the asterism. Said by Ptolemy to be of the nature of Venus with Saturn, it brings perverted morals and 'success in evil deeds'.

Sacrarius   See ARA.

sacred animals   A term carried into English from the Hebraic *Sepher Jesirah* by Blavatsky in reference to the ZODIAC.

sacred planets   The esotericist Bailey distinguishes seven sacred planets (as distinct from her five NON-SACRED PLANETS), which are linked with the SEVEN RAYS of her INTUITIONAL ASTROLOGY:

Sadalsuud

Vulcan is linked with the FIRST RAY.
Jupiter is linked with the SECOND RAY.
Saturn is linked with the THIRD RAY.
Mercury is linked with the FOURTH RAY.
Venus is linked with the FIFTH RAY.
Neptune is linked with the SIXTH RAY.
Uranus is linked with the SEVENTH RAY.

The influences of these so-called 'sacreds' may not be interpreted in the ordinary astrological manner: they are intended to apply to the interpretation of charts linked only with initiates or other 'advanced' persons. Bailey says that 'the sacred planets endeavour to fuse the personality and make it the instrument of the soul and the non-sacred planets influence more specifically the form of nature'. She records that the sacred planets are called the 'seven grades of psychic knowledge' or the 'seven divisions of the field of knowledge'. In another context, she gives several variant names for the informing 'lives' or beings of the seven sacred planets, such as Seven Planetary Logoi, Seven Kumaras, Seven Builders, Seven Manus, Flames, the Primordial Seven, and so on. The sacred planets are not to be confused with the ESOTERIC MOON and ESOTERIC VULCAN.

Sa'd al Ahbiyah　Al Sa'd al Ahbyah, the 23rd of the Arabian MENZILS.

Sa'd al Bula　Al Sa'd al Bula, the 21st of the Arabian MANZILS.

Sa'd al Dhabih　Al Sa'd al Dhabih, the 20th of the Arabian MANZILS.

Sadalmelik　Fixed star, the alpha of constellation Aquarius, set in the right shoulder of the asterism, the name from the Arabic Al Sa'd al Malik (the lucky one of the king), and accordingly sometimes called in the Latin approximation Sidus Faustum Regis, and sometimes simply El Melik. It is equated by Ptolemy with the influence of Mercury conjunct Saturn, and is said to bring complicated lawsuits and persecutions, as well as prominence in occult matters.

Sadalsuud　Fixed star, the beta of constellation Aquarius, set in the left shoulder of the asterism, the name from the Arabic Al Sa'd al Su'ud (the luckiest of the lucky) which accounts for the Latin approximation, Fortuna Fortunarum. It is said to bring troubles and disgrace, and Ptolemy said that it was equivalent to the nature of Mercury conjunct Saturn, though Simmonite likens it to that of Uranus.

Sa'd al Su'ud

Sa'd al Su'ud   Al Sa'd al Su'ud, the 22nd of the Arabian MANZILS.

Sadu cycle   A cycle of 30 years – see EGYPTIAN CALENDAR.

Sagitta. The constellation of the Arrow (from the Latin 'Sagitta') has nothing to do with Sagittarius: it lies to the north of Aquila, and to the south of Cygnus. In some constellation maps it is figured as an arrow grasped in the talons of the Eagle (Aquila).

Sagittarius   The ninth sign of the zodiac, which corresponds neither in location nor meaning with the constellation of the same name – see SAGITTARIUS CONSTELLATION. The sigil for Sagittarius ♐ is said by some to represent the arrow of desire being shot from a vestigial bow, but many occultists maintain that it actually consists of a fourfold cross, with an arrow lifting this symbol of materiality into the upper spiritual realms – an excellent symbol for the spiritualizing and expansive nature of the sign. Sagittarius is of the fire element and of the Mutable quality, the influence being enterprising, open, honest, dignified, optimistic and independent. The nature of Sagittarius as manifest in the microcosm is expressed in the many keywords attached to it by modern astrologers: enthusiastic, philosophic, idealistic, generous, open-minded, loyal, magnanimous, honest, frank, restless – in a word, all the qualities which may be associated with fire working aspirationally. In excess, the Sagittarian tends to be prodigal, sporty, indolent, self-indulgent, conceited, dogmatic, and subject to 'pointless' wanderings. Sagittarius tends to say 'I wish to act with dignity' more insistently than others, and he may do this creatively, and to the benefit of others as a natural part of his being, or he may do it merely to be seen acting in such a way, in a prodigal display of prowess or energy. Sagittarius is ruled by the planet Jupiter.

Sagittarius constellation   Zodiacal constellation, the Archer in many languages, including the ancient Greek, and with many variants in the Latin, along with Arquitenens and Arcitenens the bow holder. Figure 53, derived from Delporte's MODERN ZODIAC, sets out the extent of the constellation. Related terms refer to the half-human, half-animal nature of the image, which lies at the base of the astrological lore relating to the sign: Semivir (half-man), and even Minotaurus have been used, while all too frequently the constellation is confused in literature with CENTAURUS. The Akkadian seems to have been Ban, or more properly Mul-ban (the star of the bow), and the late DENDERAH ZODIAC gives the Archer with the face of a lion. In popular explanations of mythology, the asterism is linked with the centaur Chiron,

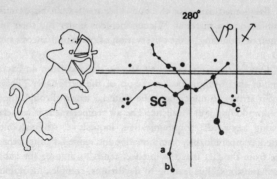

Figure 53: Sagittarius constellation (after Delporte) and the image for Sagittarius from the 13th-century SAN MINIATO ZODIAC.

killed by the dropping of one of the poisoned arrows of Hercules on his foot: however, this may go back to a misreading of Erotosthenes. The image in figure 53 is from the medieval zodiac of SAN MINIATO al Monte, and is that of a centaur. The fixed stars with the asterism which are regarded as being of astrological import are ARKAB, ASCELLA, MANUBRIUM, PELAGUS, POLIS and TEREBELLUM.

Sahu   See ORION.

Saik   See SABIK.

Salamander   The class of soul beings of the fire element (see ELEMENTALS) who correspond to the Trifertes of esoteric lore. They are sometimes called Aetnaei, and the list of correspondences in the CALENDARIA MAGICA names them as Silvani. Paracelsus appears to call them Rolamandri, defined by Waite, however, as 'igneous men, otherwise essences of the race of the Salamander'.

Salibak   See LYRA.

salts of salvation   See TWELVE CELL-SALTS.

Samael   A name, meaning approximately 'venom of God', derived from the Hebraic tradition and applied to the Archangel acting as ruler of the Sphere of Mars. Samael is one of the SECUNDADEIAN BEINGS.

Samas A Babylonian name for the Sun – see OMINA. In his examination of Mesopotamian omina material, Wiener includes under the solar omens the phenomenon of 'doubling' – the observation of two simultaneous suns.

sanguine One of the four TEMPERAMENTS, derived from an excess of the air element in the psychological make-up of the personality. The sanguine temperament strives to understand the material world through the intellect, and is freedom-loving and idealistic. The air temperament is associated with the following keywords: communicative, inquisitive, original, intellectual, quick-witted, companionable, discriminative and refined. Faults in the temperament arise from the fact that the intellect tends to divorce the subject from external compulsions, thus leading to daydreams, complex ideologies, and to an inability to relate emotionally to others: under such circumstances the type may become emotionally arid, diffused and even dishonest in speech. See also HUMOURS.

sanguine quadrant See QUADRANTS.

San Miniato Zodiac The largest medieval (marble) floor zodiac in Europe, set in the nave of the basilican monastic church of San Miniato al Monte, in Florence. The historical importance of the zodiac rests in the fact that it is linked to a highly sophisticated symbolism within the church (and indeed on its façade) relating both to the orientation of the church and to daily sunrise. The arc of Taurus within the zodiac is directed towards the arc of sunrise over Florence, as the schema in figure 54 indicates. While the zodiac is so directed, the church itself breaks all the standards of ecclesiastical orientation, and appears to have been so directed as to permit an extraordinary light effect (a pre-sunset phenomenon) which is also linked with the interior symbolism of the zodiac, but which is manifest perfectly only once each year. The dating of the zodiac (recorded in marble in the nave) is probably intended to point to a major satellitium of planets in Taurus, which occurred on 28th May 1207. For a detailed study of the zodiac and its related symbolism, see Gettings, in the Bibliography.

Saratan As Saratan, the Arabic name for Cancer.

Sarfah Al Sarfah, the 10th of the Arabian MANZILS.

Sariel A name given to one of the Grigori, who (according to Hebraic legend) taught men the nature of the course of the Moon, and its influences, as well as (probably) the nature of the 28 LUNAR MANSIONS.

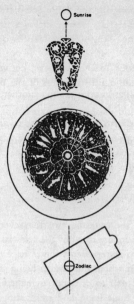

Figure 54: The SAN MINIATO ZODIAC, *schematically related to the orientation of the basilica and the arc of sunrise over Florence.*

Saros cycle   A cycle of eclipses, extended over approximately 19 years, and first described by the Chaldean astrologers. The Moon orbits the Earth in a plane which is inclined by some 5 degrees to that of the Earth's solar orbit: the points where these planes intersect are the nodes (see DRAGON'S HEAD and DRAGON'S TAIL). These two nodes regress along the Celestial Sphere, and the rate of movement is such that it takes 223 lunar months (18 years, 11 days and 8 hours – a figure which must be adjusted for extra or deficient leap years) for the regressing points to return to their original positions. This periodicity is the Saros cycle, during which time there are 14 partial eclipses, 17 annular eclipses, 10 solar eclipses and 29 lunar eclipses, a total of 70 in all. See also METONIC CYCLE. The same name, 'Saros cycle', is also applied by some astrologers to a cycle of 60 days or 60 years ('days' and 'years' and 'ages' being interchangeable in some ancient systems – see for example the 'days' of biblical creation, which are actually 'ages'), or even multiples, such as 60 × 60, or 3,600 days or years. See NUPTIAL NUMBER.

439

Satan

Satan   According to the astrologer Leo, Satan was one of the names of Saturn – but see also DEVIL.

Satan's Head   See ALGOL.

satellite   A name given to a planet or moon (or indeed to any mechanical contrivance) which revolves in orbit around another body.

satellitium   A European astrological term used to denote a group of three or more planets in conjunction (or, loosely speaking, grouped together in one sign, or one house, or even in an arc of less than 30 degrees). The American equivalent term is 'stellium'.

Satevis   See ROYAL STARS.

Saturn   Name given to the outermost of the seven traditional planets, the planetary orb within the sphere ruled by the celestial THRONES. The influence of Saturn in the microcosm represents the restrictive side of his nature, and stands for the manner in which the human being is prepared (or is able) to pay for things received, and to demand payment from others. On a deeper level, it represents the underlying and motivating fears in the life of the native: it governs the materialistic element in life, and in particular the limitations which the native is likely to experience. When Saturn is beneficially emphasized in a chart, it is usually an indication that the subject will be practical, patient, reliable, prudent and honest, if a little austere in his attitude to life. A badly placed Saturn tends towards excessive limitation or restriction, which manifests through a deep-seated fear of life: the subject tends to be emotionally cold or despondent, narrow in outlook, and subject to melancholia. In the melothesic man, Saturn has rule over the knees, as well as over the entire skeletal frame, an expression of the cosmic fact that Saturn lends 'structure' to life's forms. In the earlier astrological tradition, the Sphere of Saturn was said to have rule over time itself – that great 'structurer' of human life. The Saturnine type usually finds expression on the physical plane through activities demanding control and organizational ability: for this reason they tend to make excellent accountants or agents, for they are particularly adapted for dealing in theories of finance, though they are often personally rather parsimonious and perhaps over-careful. On the social plane, Saturn has rule over bureaucracy, which explains why so many Saturnine individuals are found within offices or bureaucratic organizations: this is yet another manifestation of the urge of Saturn to draw demarcations, to establish rigid structures, and to exercise control over life.

Scales

saturninus See LUNATICUS.

Saturn period A term derived from modern esoteric astrology, relating to periods of cosmogenesis. Saturn is said to be the first of the planets – though this is almost a meaningless phrase, since at that early stage of cosmogenesis Saturn was more a state of being than a planet in any sense that we might use the word now. It is described in occult literature as being something like a 'warmth globe', at a time when there was as yet no central sun or planetary system. All potential for the future development of the solar system was contained in this Saturnine warmth globe, a state of being esoterically termed the 'Saturn period'. This period, sometimes also called old Saturn, was overseen by the Hierarchy known as the Lords of the Flame.

Saturn revolution Term derived from modern esoteric astrology relating to cosmogenesis. It is said that the path or evolution in the development of the solar system, with its periods of activity and sleep (prelaya) must recapitulate, however briefly, the entire sequence of evolution up to that point. Since the cosmogenesis of our own system began in Saturn (see SATURN PERIOD), the Saturn revolution is the first of these recapitulations.

Saturn Scheme The Saturn Scheme or Saturnine Chain are terms derived specifically from the theosophical cosmoconception linked with the SCHEME OF EVOLUTION, and are not to be confused with the SATURN PERIOD of the evolutionary sequence.

Saturn-Uranus cycles Saturn conjuncts with Uranus twice in 90.726 years (first order cycle). The period of cycles involving conjunction to opposition is 9.93 years, which has been associated with sunspot activity cycles, while the first order cycle has been linked with fundamental political changes, and even with revolutions. There is a second order cycle of 26 conjunctions giving a periodicity of 1179.440 years.

scala A term used in late medieval astrology and occultism, the late Latin for 'ladder', and closely allied to the CALENDARIA MAGICA systems. The *scalae* give lists of correspondences on a numerological basis, though (unlike the calendaria) use no symbols, signs or sigils in their constructions. The most influential *scalae* were those of Agrippa, which summarize the medieval tradition, yet provide a little more systematization than the normal calendaria.

Scales A popular name for both zodiacal and constellational Libra. For

Scera

North Scale and South Scale, see ZUBENELSCHAMALI and ZUBENELGENUBI, respectively.

Scera   See SIRIUS.

Scheat   Fixed star, the beta of Pegasus, set in the left leg of the winged horse. Alternative names include Al Sa'id, Seat Alpheras, Mankib al Faras (the Arabic for 'horse's shoulder') and the derivative 'Menkib'. It has a most unfortunate reputation, for it is said to cause murder, suicide and drowning, though Ptolemy accords it the equivalent influence of Mercury conjunct Mars. A study of the influence of planetary conjunctions with this star suggests a connection with death through drowning, however. This may even link with an ancient Phoenician etymology – *Pag* (or *Pega*) and *Sus*, the combination referring to the figurehead of a ship – though ancient Egyptian etymologies have also been suggested. See SKAT.

Schedar   See ALPHERATZ.

schema   One of the most frequently used words in medieval astrology for a CHART, the Latin term really applying to the Scheme of the Heavens, as symbolized in the chart.

scheme   In specialist astrological use, a modernization of the common Latin *Schema*, usually intended to mean a 'Scheme of the Heavens', and hence a synonym for CHART.

Scheme of Evolution   According to the theosophical cosmoconception, there are ten separate Schemes of Evolution in our own solar system, these being the so-called Ten Chains (see GLOBE PERIOD), of which only the first seven have been named (for the eighth, however, see ASTEROID SCHEME). These are:

1 The VULCAN SCHEME or Vulcanian Chain.
2 The VENUS SCHEME or Venusian Chain.
3 The EARTH SCHEME or Earth Chain.
4 The JUPITER SCHEME or Jupiterian Chain.
5 The SATURN SCHEME or Saturnine Chain.
6 The URANUS SCHEME or Uranian Chain.
7 The NEPTUNE SCHEME or Neptunian Chain.

It must be emphasized that the Vulcan of this schema (see VULCAN SCHEME) is not the same as the VULCAN hypothetical of other systems, nor is it linked with any of the VULCAN PERIOD traditions of the esoteric lore.

Scorpius

Scorpio The name given to the eighth sign of the zodiac, and sometimes to the zodiacal constellation SCORPIUS. The sigil for Scorpio ♏ is said by some to be a vestigial drawing of the male private parts, over which the sign has rule. However, esoteric astrologers claim that the sigil is really a vestigial drawing of the severed tail-half of a serpent, linking the sigil with the story of the Fall of Man. Scorpio is the only zodiacal sign to have been accorded two quite different images – that of a scorpion and that of an eagle. The former is symbol of the unregenerative Scorpionic urge, while the latter symbolizes the regenerative nature – see EAGLE. Scorpio is of the water element, and of the Fixed Quality, the influence being magnetic, determined, secretive, shrewd, dignified, self-confident, masterful, sensitive and critical. The nature of Scorpio as it manifests in human beings is expressed in the many keywords which have been attached to it by modern astrologers: regenerative, inspirational, tenacious, forceful, magnetic, emotional, penetrating, competitive, extremist, strong in desire – in a word, all those qualities which may be associated with a water type expressing itself powerfully. In excess, the Scorpionic nature may be described in terms which express its underlying cruelty, sense of violence and instinct for crime (esoterically, the desire to have something for nothing): rebellious, degenerative, suspicious, sarcastic, cruel, selfish, violent, indulgent and domineering. The Scorpionic tends to say 'I wish to change' more insistently than the majority, and he may do this creatively, in a healing manner, or destructively, in a criminal manner, depending upon which side of his nature is being developed. In modern astrological systems, Scorpio is usually said to be ruled by the planet PLUTO, though in the traditional form it was accorded the rule of NEGATIVE MARS. It is sometimes said to mark the Exaltation of Uranus, and the Fall of the Moon.

Scorpion's Heart See ANTARES.

Scorpius Zodiacal constellation which in antiquity was the largest of such asterisms, incorporating in its claws (the Greek *Chelae*) what is now our separate asterism of Libra. However, the tradition is confusing, for even in antiquity the constellation was described as double, and Libra is found as a separate figure in Babylonian and Egyptian star maps. The name, properly Scorpius, but also sometimes Scorpio (hence the two genitive Latin forms, *Scorpii* and *Scorpionis*, the latter favoured in medieval astrology) was used in Roman times, the former being conveniently adopted to distinguish nominally between sign and constellation. The extent of the constellation is indicated in figure 55, derived from data furnished by Delporte's MODERN ZODIAC: the image of the scorpion in this figure is from the 13th-century SAN MINIATO

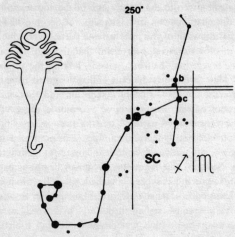

Figure 55: Scorpius, from the MODERN ZODIAC *of Delporte, along with the image of Scorpio from the 13th-century* SAN MINIATO ZODIAC.

ZODIAC. In esoteric astrology, the association between the EAGLE and Scorpio (which figures in the important TETRAMORPH imagery) is extended from the zodiacal sign even into the constellation, for Drummond claims this name from antiquity. The asterism was also identified with the serpent – most especially with the basilisk. The Arabic astrologers called it Al Akrab (the scorpion), however, from which we have many medieval derivatives, such as Alacrab, Alatrap, Hacrab, and so on. The Akkadians called it Girtab (perhaps 'the stinger', though the meaning is in dispute). In popular constellation mythology, the asterism is said to represent the scorpion which killed Orion, but it appears to have been a scorpion long before the Grecian mythology was developed. Ptolemy describes the stars in the front of Scorpius as carrying the influence of Mars, if slightly Saturnine – those in the body are said to be Mars mingled with a little Jupiter, and those in the sting are said to be Mars conjunct Mercury. However, the reputations attached to the fixed stars in the asterism are generally far from pleasant, and appear to emphasize the Martian nature: see ACULEUS, ACUMEN, ANTARES, GRAFFIAS, DSCHUBBA, SHAULA and LESATH.

Scutum Properly, the 'Scutum Sobiescianum' (Sobieski's Shield) is the name given by Hewel (see R. H. Allen) to a small group of stars near the feet of Antinous, between the Caput Serpens and the head of Sagittarius. The asterism is named after the third John Sobieski, King of Poland, in honour of his defensive wars against the Turks. R. H. Allen gives an alternative origin for the constellation, in its other name as The Cross, linking it with an early history of the Franciscan, John Capistrano.

Seal   In a specialist astrological sense, a Seal is one of the graphic symbols, or SIGILS, representing an occult prophylactic force. Just as in a non-specialist sense an official seal might be stamped on a document as a sign of ratification, so a magical Seal was often used for magical purposes to imprint a magical power into some materiality, such as a sheet of paper or a strip of metal, or even a magical gem. Such a Seal might be drawn, printed, incised or moulded. As a magical symbol, it was supposed to be imbued with a secret power, beneficial or maleficent, depending upon how and when it was constructed, and upon the intentions of the magician or astrologer who used it. Ancient Seals, often of deep esoteric significance, have been recorded for planets, planetary spirits (and demons), decans, zodiacal Angels, rulers of the spheres, and so on. It would seem that a knowledge of magical Seals gave the same power over the realm of spirits as did knowledge of the demonic names: this explains why so many of the Seals are so frequently incorporated into amulets and talismans such as the BIRTH STONES, in which the Seal was a sign that a magical VIRTUE had been imprinted into the jewel as an additional power or influence.

Seal of Solomon   Name given to the six-pointed star constructed from two interpenetrating triangles. The symbolism expressed in this figure is extremely rich: on one level it represents the triad of the macrocosm interpenetrating the triad of the microcosm (wherein it becomes a symbolic duad). On another level it is seen as the interaction of the four elements (see SIGILS − ELEMENTS) wherein it is a symbolic quaternary. Within the system of Rosicrucian symbolism, the space within the central hexagon is seen as the repository for the invisible fifth element, the QUINTESSENCE, for which the Seal is often offered as the sigil.

seats   Another name for the THRONES.

secondary directions   Term used of the method of unfolding potential in a radical chart according to the *SECONDARY PROGRESSIONS, to be distinguished from the primary directions of the *PRIMARY SYSTEM.

secondary progressions   A term evolved to distinguish a method of unfolding potential within a radical chart from the more ancient *PRIMARY SYSTEM. The basic assumption underlying the secondary method is that the position of planets (as well as other cosmic and mundane factors) on a certain number of days after birth will find a reflection in the life of the native that equivalent number of years after birth – that, for example, the cosmic qualities of the seventh day after birth will echo precisely in the life of the native during his seventh year. Unlike the primary directions of the primary system, calculations of the unfoldment of potential are based on the movements of the planets along the ecliptic, and thus the time equivalent is approximately that of one day to four minutes (but see DAY FOR A YEAR): conversion in time equivalents is therefore not so liable to error as with the primary system. The secondary progression chart is called the progressed figure (or chart), and the new aspects formed both within this figure, and between this and its radical chart, are called progressed aspects, to be distinguished from the differently derived directional aspects of the primary system. There is much argument about the validity of the system of secondary progressions in predictive work, but much research (and practical use of the method) in recent years has indicated that the traditional basis of the idea (often expressed in the catch-phrase 'Day for a Year') is entirely consistent with astrological symbolism. There is some argument among the astrological practitioners of this difficult branch of astrology about which periodicity should be taken as representing the basic unit of the 'day'. Many modern astrologers appear to favour the TRUE SOLAR DAY as the basis for their computations – however, other astrologers have taken the time measurement wrongly attributed to Ptolemy, and used in primary directions, which equate the day with a transit of one degree of Right Ascension over the Midheaven, which is then regarded as both an arc and time-span equivalent to one year of life. Others have adopted the TABLE OF NAIBOD as a basis for symbolic computation, whilst others use the so-called Equation of Time. A variation is that outlined in the 17th century by Antonio de Bonattis, which involves adding each sequential solar arc to the Midheaven, each arc then being regarded as the equivalent of the years (in sequence). In secondary directions, the progressed movements of the ANGLES are regarded as being of great importance in interpretation: they are visualized as keypoints which discharge the accumulated power of the radical and progressed planets at the time when exact aspects are formed. So important are these that Davison, the leading specialist in this field, suggests that it is possible to predict all the major events of the life of the native by using only directions involving the Angles and the Sun. Davison insists that not only should the progressed Angles and planets be taken into account, but also the CONVERSE DIRECTIONS.

second house The second of the astrological HOUSES, linked with the nature of Taurus, and representing the possessions and feelings of the native, as well as all those factors which lend material support to life. It was often called the House of Money, though it actually relates to far more than financial matters, even though it is indeed a useful index of how the native will earn or otherwise come by money. Esoteric astrology suggests that this house represents the element in life which the native will seek to redeem – certainly it may be used as a measure of how the native will meet his obligations. In regard to the melothesic man, the 2nd house denotes the throat and the larynx, and in some respects the sound of the voice. In MUNDANE ASTROLOGY it rules the nation's wealth, revenues, currency and investments (though not speculations, which belong to the 9th house).

second order cycle See FIRST ORDER CYCLE.

second ray In the esoteric system described by Bailey, the second of the *SEVEN RAYS of her INTUITIONAL ASTROLOGY is called the Ray of Love-Wisdom, and is linked with the will to unify, to cause vision and is said to bring the power to perceive. This second ray is said to arise in Ursa Major, and is transmitted to our solar system by means of the constellations (zodiacal signs?) Gemini, Virgo and Pisces, through their pairs of corresponding ORTHODOX PLANETS and ESOTERIC PLANETS, according to the schema in table 69. The entire second ray is linked with the sacred planet Jupiter: but

————————————— TABLE 69 —————————————

Constellation (sign)	Orthodox planet	Esoteric planet
Gemini	Mercury	Venus
Virgo	Mercury	Moon
Pisces	Jupiter	Pluto

see SACRED PLANETS. In connection with the esoteric ruler of Virgo, above, see also ESOTERIC MOON. Distinguish the second ray from SECOND RAY TYPE.

Second Ray Type The second of the SEVEN RAY TYPES, associated with the planet Venus. Distinguish from SECOND RAY.

Second Root Race In the theosophical cosmoconception, the Second Root Race is the second of the human races, or streams of human evolution, sometimes called the Hyperborean Race, which was said to largely occupy the long-disappeared northern continent named Plaksha.

second station The stationary point (see STATIONS) where a planet becomes direct of motion.

second zodiac See SOLAR BIOLOGY.

sect See SECTA.

secta A Latin term derived from a late Greek specialist term *airesis*, used by Ptolemy of the divisions of signs and planets into the feminine and masculine (nocturnal and diurnal) groups. The Greek term was originally involved with the idea of the 'acquisition of power', and was unfortunately translated by the Latin *conditio* and *secta*, and hence (even more unfortunately) into the English 'condition' and 'sect'. See FEMININE PLANETS, FEMININE SIGNS, MASCULINE PLANETS and MASCULINE SIGNS.

Sectio Tauri See TAURUS CONSTELLATION.

secular horoscope A modern term arising from the proposition that since a chart cast for a human birth is valid as a reflection of the cosmic environment, then that same chart may be cast contemporaneously to apply to simultaneous earthly events. The latter chart has been called the secular horoscope. The proposition, which has attracted some notice from Dean, is based on a misunderstanding of the nature of natal astrology, but is to some extent in accordance with the general tenets of MUNDANE ASTROLOGY, and may even be seen as an astrological equivalent of the Jungian SYNCHRONICITY.

Secundadeian Beings The name (probably originated by the Abbot Trithemius at the end of the 15th century from sources given by Pietro d'Abano) is applied to an important group of spiritual beings assigned rule over a sequence of repeated historical periods of 354 years' duration, sometimes called the Trithemian periods. Each of these beings is said to be of the rank of Archangel, though in respect to their historical charge over periods (rather than over races of people) they are really of the rank of ARCHAI. The names, along with many associations (including predictions relating to certain post-15th century rules) are derived from the Hebraic-gnostic tradition, and are linked with the planetary spheres, rather than with the planets themselves. Trithemius gives a survey of history in terms of the Secundadeian Beings which has been summarized in a somewhat simplified and personalized manner by Stein, as indicated in table 70. Trithemius assigned the rule of

————————— TABLE 70 —————————

Secundadeian	*Planet*	*Stein summary*
Ophiel	Saturn	Light and Darkness. Christ's Time on Earth.
Zachariel	Jupiter	Order and Chaos. Monastic Life.
Samael	Mars	The Crusades.
Michael	Sun	Earth-embracing. A world economy.
Anael	Venus	Persecution of Christians – faith of martyrs.
Raphael	Mercury	Ages of the Grail and Orders of Knighthood.
Gabriel	Moon	Feuds connected with heredity.

solar Michael to our own epoch, which was said to have begun in 1881 (variant interpretations of his figures are recorded). The sequence of ruler-ships follows a special order: SA, VE, JU, ME, MA, MO and SU. The sequence of rule was said to begin at the Creation of the World, which was dated by Trithemius as 5560 BC – though at least five other dates for the Creation are given in pre-Trithemian chronologies.

Sedis   See CASSIOPEIA.

seed trigon   See SIDEREAL MOON RHYTHMS.

see-saw   Term applied to one of the *JONES PATTERNS in which the distribution of planets within the figure form a balance, separated by two 'balancing spaces' of between 60 and 90 degrees. This distribution is said to influence the native into being able to see two sides of a problem, as well as inclining him towards mediation.

Seginus   Fixed star, the gamma of Bootis, set in the left shoulder of the giant, and said by Ptolemy to be of the nature of Mercury with Saturn – a far from beneficial influence.

selenocentric chart   A chart computed for the solar system, presuming the Moon (Greek *Selene*) to be the centre of that system. Such a selenocentric chart was published for the landing of the first man on the Moon, in vol. 11, no. 4 of the *Astrological Journal* (1969).

semester chart   Term used by Bradley for the last demi-solar return period expressed in a SOLAR RETURN, 'effective for the last half of the year'.

semi-arc Half of either the DIURNAL ARC or of the NOCTURNAL ARC, measuring one of the main quadrants of the horoscope figure, from the Ascendant to the Midheaven, from the Midheaven to the Descendant (both semi-diurnal arcs), and from the Descendant to the Imum Coeli, and from the Imum Coeli to the Ascendant (both semi-nocturnal arcs).

semi-arc system An alternative (though rarely used) term for the PLACIDEAN SYSTEM.

semi-decile An aspect of 18 degrees, arising from a division of the zodiac into 20ths. The aspect is sometimes called the Vigintile. See, however, QUIN-TILES.

semi-diurnal arc See SEMI-ARC.

semi-nocturnal arc See SEMI-ARC.

semi-quartile An archaic term for the SEMI-SQUARE aspect.

semi-sextile An aspect of 30 degrees, the two aspecting points being separated by an arc equivalent to one sign of the zodiac. It has been called an aspect of growth, and is generally regarded as beneficial. The amount of ORB permitted this aspect is far from agreed, but some authorities allow as much as 4 degrees.

semi-square An aspect of 45 degrees, the two aspecting points being separated by the equivalent arc of one and a half signs of the zodiac. The amount of ORB permitted this aspect is far from agreed, but some authorities allow as much as 4 degrees. Like the SQUARE, it is generally believed to be a difficult aspect.

Semita lactea See VIA LACTEA.

sennight chart A term used to denote a chart (usually one of a series) which is a record of a weekly progression or transit of a radical chart.

separating See SEPARATION.

separation A term applied to the dissolving of an aspect: the faster moving planet is said to separate from, or to be in separating aspect to, the aspected planet or nodal point. When within ORB of such separation, the planet may be described as being in separating orb. See also APPLICATION.

September The ninth month (30 days) of our GREGORIAN CALENDAR, the name being a throwback to the early Roman (pre-Julian) calendrical system in which it was the seventh month (from the Latin *septem*, seven), the year beginning in March. The esoteric magic of seven is linked with this month still, as a fruition of a cycle in lunar terms, the fruition also of the earth. The Anglo-Saxons called it *gerstmonath* (barley-month) in recognition of the harvesting – a similar idea of 'spiritual harvesting' is to be perceived after the age of 7 × 7 (49) in many human lives.

septenary measure A term applied to an arc of 4⅔ of a degree, used in SYMBOLIC DIRECTIONS. It is derived by dividing a sign by seven, and is also a measure used in the POINT OF LIFE system.

Septentrio See URSA MINOR.

septile An aspect of 51.4 degrees, resulting from the division of the zodiac circle by seven, and used mainly in the theory of HARMONICS. See also BI-SEPTILE and TRI-SEPTILE.

Seraphim A term of much-disputed etymology – though of Hebraic origin, linked both with the idea of 'serpent (ShRPM) and 'burning' (ShRP) – and used to denote the celestial creatures described in Isaiah 6:2. as beings with a human form and three pairs of wings. Blavatsky writes of them as 'the Fiery Serpents of Heaven', and equates them with the four Maharajahs of the Buddhist cosmogeny. Within the occult tradition they are linked with the Ophanim, serpent-beings associated with the four Fixed signs of the zodiac, and thus with the four Evangelists of the Christian tradition (though see also CHERUBIM). The Seraphim are sometimes called Lords of Love, and in some occult systems are named as the rulers of the zodiac.

Serpens The constellation of the Serpent is that asterism figured in the hands of Ophiuchus. DRACO is sometimes called Serpent, and there is often some confusion about its name (of which there are many variants, such as Anguilla, Anguis, Coluber, etc.). In modern times it is usually divided into two asterisms, Caput (head) and Cauda (tail), set on either side of the major figure. It is one of the four serpentine asterisms.

Serpens Lator See OPHIUCHUS.

Serpent See DRACO.

Serpentarius See OPHIUCHUS.

Serpentiger See OPHIUCHUS.

Sertan A name for the fixed star ACUBENS.

serving trinity A term used by Butler in his SOLAR BIOLOGY in respect of the zodiacal signs Capricorn, Aquarius and Pisces.

sesquialtera See DIAPENTE.

sesquioctave See TONUS.

sesquiquadrate An aspect of 135 degrees, the two points being separated by four and a half signs of the zodiac, the angle being compounded of a square and a semi-square. It has been called the aspect of agitation, and is generally regarded as a difficult formation in a chart. The amount of ORB permitted is not agreed, but some authorities allow as much as 3 degrees.

sesquiquintile An aspect of 108 degrees – see QUINTILES.

Set One of the several variant names for the constellation Ursa Major.

seven ages In popular astrology, the term is usually applied to the various doctrines relating to the division of the life of man into septenaries – see CLIMACTERICS. In a specialist sense, the term is sometimes applied to the seven historic ages governed by the *SECUNDADEIAN BEINGS, the so-called Trithemian periods. In its popular sense, the division is in terms of the seven planetary ages, following a sequence from the first Sphere of the Moon, through to the Sphere of Saturn, to each sequence of which is attached a definite number of years (see PLANETARY AGES). Philo Judaeus notes a system provided by Hippocrates which divides man's life unequally, but in multiples of seven, into six ages, with the final seventh age (after 56) as that of the 'Old Man'. A list of the traditional seven is preserved under the name of 'Hermes' from a medieval source, giving the sequence shown in table 71. Saturn is traditionally accorded a period of 30 years, but in this list (and in others) is given rule over any years after 70. Several attempts have been made to link these planetary ages with the list poetically set out by Shakespeare in *As You Like It*, but it is more likely that the poet had in mind one of the several sources from Proclus, which does not give periodicities (perhaps taken for granted) but only the planetary ruler-ships:

Planet	Period (years)	Rules age up to and including
MO	4	4
ME	10	14
VE	8	24
SU	19	43
MA	15	58
JU	12	70
SA	30	

TABLE 71

MO is ruler of 'nutrition and the physical life'.

ME is ruler of 'letters, music and motley'.

VE is ruler of the period when we 'begin to produce seed'.

SU sees 'youth in full vigour and full perfection'.

MA 'aspires after power and superiority over others'.

JU gives himself 'up to prudence, and an active political life'.

SA marks that time when we deem it 'natural to separate ourselves from generations, and transfer ourselves to an incorporeal life'.

The several available lists agree in spirit, if not in actual periodicities, to the above ideals. In more recent times attempts have been made to link such periodicities to the extrasaturnian planets, but much of the ancient wisdom in the earlier lists has thereby suffered. The most frequently quoted of the modern lists is that given by Sepharial, but a well thought out structure which includes the modern planets has been given in diagrammatic form by Collin, who postulates a logarithmic sequence interacting through three fairly distinct periods of gestation (the embryonic form), childhood and maturity. With the modern planets incorporated, in whatever system, the ancient concept of seven ages is of necessity bypassed.

seven-and-a-half-degree aspect This aspect, for which no short name appears to have been originated, became popular in the United States as a consequence of the work done by Nelson on radio disturbances. However, much more astrologically apposite research into the aspect has been done by Wangemann, who shows that there are 48 different multiples of this aspect, which is naturally linked with its parent semi-square and other multiples. Wangemann discovered that a small ASYMMETRIC ORB of 0.7 degrees for the applying, and 0.1 degrees for the separating aspect, was applicable.

Seven Builders

Seven Builders In esoteric astrology, the SACRED PLANETS.

seven climata The so-called seven climata are tabulations of zodiacal rising times computed for certain localities (see CLIMA), valid for Mediterranean areas, and derived ultimately from a method employed by Babylonian astrologers. Several sets of seven climata exist (see SYSTEM A, for example), but the post-Ptolemaic list, derived from ten climata computations, to which later authorities added a necessary interpolation for Byzantium, along with periods of longest daylight, are shown in table 72. This Hellenistic model, apparently

---
TABLE 72

Clima	Zone	Daylight	Notes
1	Meroe	13	
2	Syrene	13.30	
3	Lower Egypt	14	Derived directly from early figures for Alexandria, the old clima 1.
4	Rhodes	14.30	Derived by arithmetical progression from early figures for Babylon, the old clima 2.
5	Hellespont	15	
—	Byzantium	15.15	Interpolated from climata 5 and 6.
6	Mid. Pontus	15.30	Derived by arithmetical progression from early figures for Rome, the old clima 6.
7	Borysthenes	16	

---

based on a division of the habitable quadrant of the Earth into 90 sections made by Hipparchus (who was aware of the effect of latitude on altitude of stars) was adopted by medieval astrologers from the tables converted by the Arabian Alfraganus, who influenced the astrological thought of Dante. The Seven Climata of Alfraganus, the Alfraganian Climata, are the ancient climata lists transformed into degrees of latitude and equivalent distances in miles on the erroneous assumption that 56 miles is the equivalent of one degree. Thus, the influential theories of climata were concerned with latitudes only some 700 miles north of the equator.

seven forms See SEVEN PROPERTIES.

seven globes See GLOBES.

Seven Kumaras In esoteric astrology, the SACRED PLANETS.

seven lots The Alexandrian Paul records a tradition of seven *clipoi*, the Greek term for 'lots', which were the equivalent of the modern PARS. In addition to the prototype of the modern PART OF FORTUNE, these were (translated) AUDACITY, LOT OF DAIMON, LOVE, NECESSITY, NEMESIS and VICTORY.

Seven Manu In esoteric astrology, the SACRED PLANETS.

seven moods In a series of lectures translated under the title *Human and Cosmic Thought*, the occultist Steiner developed a schema which linked what he called the seven philosophic moods with the seven planets:

MO	Occultism	ME	Transcendentalism	VE	Mysticism
SU	Empiricism	MA	Voluntarism	JU	Logicism
SA	Gnosis				

He distinguished the TWELVE PHILOSOPHIC TENDENCIES in terms of the twelve signs or constellations.

seven properties A term derived from the esoteric astrology of the German mystic Boehme, as developed by Freher, and sometimes called the Seven Forms, the Seven Working Properties, the German original term being *wirkende Eigenschafte*. The seven forms are linked with the seven planets, though it is the principles behind the planets which interest Boehme in his esoteric account of the effects produced by their individual working, their polarities, and their balance. The first form is that of Saturn, which is a harsh and hardening principle, a centripetal force, 'indrawing and compacting about its chosen centre': it is Saturn which draws and fixes. The second form is Mercury (Mercurius) which strives for freedom against the restraint imposed by Saturn: it is separative and centrifugal, 'in eternal flight from fixity'. Its role is seen in the thinking mind, in 'divisive analysis'. The third form is Mars, which results from the opposition of the first two, and may be likened to a turning wheel which itself represents a fusion of the centrepetal and the centrifugal impulses: it is the 'turning wheel of desire'. This is the Ternary of Fire, which bursts forth into action into the fourth form which is that of the Sun. Boehme likens this solar bursting to the crack of lightning, which is the explosion of spirit into time. The attainment of desire in this Sun form brings its own joy and harmony, which is the fifth form of Venus. This form is active of harmony-producing, from the resolution of conflict to the 'mirage-paradise of misplaced hopes'. From this develops the sixth form, the fuller self-perception and understanding which is characteristic of Jupiter. The seventh form, that of the Moon, is anti-climactic, yet is involved with 'final production of concrete realization'. A relationship is established between this seventh and the first form, so that the individual actions of the planetary workings may be seen as cyclical. As Muses points out, the teleology of the pairs of antitheses (see TERNARY OF FIRE and TERNARY OF LIGHT) yield an ethically intransitive relationship: for example, the Mercury to Jupiter polarity marks a beneficent and expansive development, a natural growth, but the reverse of Jupiter to Mercury would represent the perversion of Jupiter to the ends of Mercury, so that inherent in each of the seven properties is a duality.

Seven Races In the theosophical cosmoconception, the Seven Races are the seven streams of human evolution, four of which have run their course in the present GLOBE PERIOD, the fifth of which is still in progress (see FIFTH ROOT RACE) and two of which are to be developed in the future. See FIRST ROOT RACE, SECOND ROOT RACE, THIRD ROOT RACE and FOURTH ROOT RACE.

seven rays In the esoteric astrological system of INTUITIONAL ASTROLOGY described by Bailey, a series of seven rays are outlined, each originating in the constellation of Ursa Major (though besides being individually linked with the major seven stars of this asterism, they are also associated with the Pleiades and the fixed star Sirius), and thence transmitted to our solar system by means of three constellations, which are the zodiacal signs, and by a conventional system of planetary rulers paralleled by an unconventional system of rulers, called respectively the 'orthodox' and the 'esoteric'. The entire sequence is set out in diagrammatic form in figure 56 – but see the following entries: FIRST RAY, SECOND RAY, THIRD RAY, FOURTH RAY, FIFTH RAY, SIXTH RAY, SEVENTH RAY. The rays set out in terms of planetary influences by Morrish are discussed under SEVEN RAY TYPES, and must not be confused with the Bailey system. The seven rays especially linked with the theosophical astrology are not to be confused with those of the intuitional astrology. The former are in fact rays emanating from the seven planetary spirits, or Angels. Strictly speaking, these rays directly influence the seven spiritual bodies of man, the first three corresponding to the three as yet undeveloped in ordinary man (Atma, Buddhi and Manas), the last four corresponding to the personality and the so-called 'bridge' which enables a reciprocate action between the higher and lower tertiaries of man's being. These septenaries are associated with eight planets (the Sun overlooking each of the seven rays) and, with an associate colour, are linked with the being of man in the schema of table 73 which is derived by Adam from the Theosophic tradition which was promulgated by Leadbeater, Besant and

TABLE 73

Human Principle		UR	NE	VE	SA	ME	MA	JU	Colour
	Atma	UR							Violet
Ego	Buddhi		NE						Indigo
	Manas			VE					Sky-blue
'Bridge'	Antahkarana				SA				Green
	Mental					ME			Yellow
Personality	Astral						MA		Orange
	Physical							JU	Crimson-rose

Powell. Within this system, the first ray is the Ray of Uranus (linked with the Atma, itself a symbol of the redeemed physical body, and hence with the will). The second ray is the Ray of Neptune (linked with the Buddhi, itself a symbol of the redeemed astral, and hence linked with love). The third ray is the Ray of Venus (linked with the Manas, itself here a symbol of the redeemed mental realm of man, though in other schemas a symbol of the redeemed etheric, and hence linked with developed intuition). The fourth ray is the Ray of Saturn, linked with the so-called 'bridge', that 'ray of light pouring from the Ego on to the personality', in other systems called the Ego itself. The mental realm is governed by the fifth ray, though it is connected with a higher spirituality than is usually connected with the ordinary term 'mental'. The sixth ray is the Ray of Mars, which is concerned with the emotional life, with what is esoterically called the 'astrality' of the personality. The seventh ray is the Ray of Jupiter, concerned with the physical body (in this schema, seemingly not separated from the etheric body, which implies that it should properly be termed the 'living body'). In his treatment of this system of seven rays Adam offers some argument for amending the above classification, which he terms the 'orthodox' classification: he allots the Buddhic principle to Mercury, rather than to Neptune, and the lower mind (the mental realm) to the Moon (which 'we are told is a substitute for Neptune').

Seven Ray Types   In his syncretic ASTRO-PSYCHOLOGY, Morrish sets out what he calls the Seven Ray Types, seemingly derived from a system of occultism which might easily be confused (in nomenclature at least) with the SEVEN RAYS of Bailey or of the theosophical system. Morrish sets out the First Ray as that of Saturn, 'ruler by means of set forms of obedience to

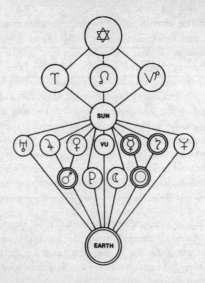

Figure 56: The seven rays of Bailey's INTUITIONAL ASTROLOGY.

established authority'. The Second Ray is that of Venus, the 'philanthropist'. The Third Ray is that of Mercury, the 'adaptable type – the active mental organiser'. The Fourth Ray is that of the Earth, the 'artist' who is 'able to create material forms expressing spiritual values'. The Fifth Ray is that of Jupiter, the 'scientist or philosopher' who is to be distinguished from the Saturn-type scientist, the 'materialist' who works with set forms, with a tendency to suppression. The Fifth Ray type is an idealist 'expressing itself expansively'. The Sixth Ray is that of Mars, the 'devotee-fanatic', the 'object-worshipper', the 'militant religious and partisan'. Morrish notes that while the Venusian type works for the group, the Martian type tries to force others to do so. The Seventh Ray is that of Uranus, the 'occultist', the 'research-worker and electrician' who is at the same time the 'reformer'.

Seven Sages   See PLEIADES.

Seven Sisters   See PLEIADES.

Seven Sleepers   One of the several important names for URSA MAJOR.

Seven Sortes   See HERMETIC SORTES.

seventh house   The seventh of the twelve astrological HOUSES, linked with the sign Libra, the Descendant House, the ancient 'Dysis'. It is associated with the influence of Venus, the ruler of Libra, and representative of the Venusian urge away from the selfhood of Mars, into relationships with others. Such relationships may be harmonious or conflicting, which explains why the house is often called the House of Marriage and Open Conflict. It may be regarded as the lens through which the native sees other people in close relationships, for it has rule over partnerships and friendships – both conditioned by the ruling planet of the sign on the Descendant, and/or by planets in the house itself. At its best, the house marks the attempt to establish a free co-operation with another; at its worst, the attempt to impose ego (properly the domain of the opposite 1st house) on another by means of guile or coercion. It is said also to mark the native's relationship to public weal, and hence with the law, so that it is called by some the House of the Public: it governs also lawsuits and contracts. In *MUNDANE ASTROLOGY it governs the social consciousness of the people, and those who co-operate with, or oppose, the people – it is also the domain of international disputes.

seventh ray   In the esoteric system described by Bailey, the seventh of the SEVEN RAYS is called the Ray of Ceremonial Order, and is linked with the will to express, and with the 'principle of order'. The seventh ray is said to arise in Ursa Major, and is transmitted into our solar system by means of the zodiacal constellations (zodiacal signs?) Aries, Cancer and Capricorn, through their pairs of corresponding ORTHODOX PLANETS and ESOTERIC PLANETS, according to the schema set out in table 74. The entire Seventh Ray is linked with the sacred planet Uranus – see therefore SACRED PLANETS.

TABLE 74

Constellation (sign)	Orthodox planet	Esoteric planet
Aries	Mars	Mercury
Cancer	Moon	Neptune
Capricorn	Saturn	Saturn

Seventh Ray Type   The seventh of the SEVEN RAY TYPES, associated with the planet Uranus. Distinguish from SEVENTH RAY.

Seven Virgins   See PLEIADES.

seven vital principles A term used by Butler in his *Solar Biology*, relating to the planetary powers, in which sevenfold schema he includes the Earth (the Sun being the central source of these vital principles). The septenary is also called the 'seven primate creative principles' and 'creative principles', and finds a solar expression in the seven planets. The argument runs that when the individual conquers or subjugates the lower part of his being, then he attains the rewards expressed by the following principles:

UR	Blessing	SA	Glory	JU	Honour	MA	Strength
VE	Riches	ME	Power	TE	Wisdom		

seven working properties See SEVEN PROPERTIES.

sexascope A term used by Fagan, and applied to a system of astrological interpretation of the NOVIENIC CHART, regarding the novienic moon as the significator of sexual appetite and propensities.

sex degrees A term not to be confused with SEX POINTS, but used in relation to the various systems of DEGREE SYMBOLS as interpreted by Sepharial.

sex points These are the so-called sex degrees used by Bailey in the calculation of the PRE-NATAL EPOCH, said by some to be an extension of the trutine of Hermes. The sex point is to be used (in theory at least) for rectification of a natal chart by means of examination of the pre-natal chart in relation to the known sexuality of a native. The sex point is actually the place of the Moon in regard to a number of arcs determined in the ecliptic in relation to the tropical zodiac by the Hindu NAKSHATRAS (incorporating the asterism ABHIJIT, however) which, for the sake of convenience are regarded as consisting of 12 degrees and 51 minutes. The sex point will theoretically mark an arc in the epochal chart which reveals the actual sexuality of the native, though the laws for constructing and interpreting the chart as a whole are somewhat complex, and perhaps even confused. The term 'sex point' was probably adopted to distinguish it from the term SEX DEGREES, then used in quite a different connection.

sextans Properly speaking, Sextans Uraniae, a constellation formed and named by Hevelius in the 17th century, from twelve stars between Leo and Hydra.

sextile An aspect of 60 degrees – generally regarded as a beneficial influence, and sometimes called the aspect of opportunity. The amount of orb permitted this aspect is far from agreed among authorities, but some allow as much as 4 degrees.

Shamali See THIRD ROOT RACE.

Shamayim One of the names given to the first of the seven Heavens of the Qabbalists, ruled by GABRIEL.

Shamballa A term with two seemingly unrelated meanings. In the theosophical cosmoconception (rooted in Vedic esotericism), Shamballa is said to be a place presently located in an oasis in the Gobi Desert. It is said to be the dwelling of high initiates, such as the Lords of Venus, the Sons of Fire, and those who taught primitive man. Unfortunately, the esoteric aspect of this Shamballa has been exotericized out of all recognition in popular occultism. The same term is used with more caution and precision by the esotericist Bailey, who describes it as 'that major centre of our planetary life', locating it in our solar system: she places it under the 'control' of the fixed star Polaris, as playing a part of the sequence of projection of the *SEVEN RAYS. In another context, she locates Shamballa 'where the Will of God is focused' on our planet.

Shamshiel Name given to one of the Grigori of the Hebraic tradition who taught men 'the signs of the Sun' and (probably) 'the nature of the zodiac'.

Shani The Sanskrit word for the planet Saturn, as used in Hindu astrology.

Shapings The term applied by Jones himself to his so-called JONES PATTERNS.

Sharatain Al Sharatain, the 27th of the Arabian MANZILS.

Sharatan Fixed star, the beta of constellation Aries, the name from the Arabic dual form Al Sharatain which refers to the 'two signs' – this star and the gamma MESARTHIM – both of which appear to have marked the vernal equinox in the days of the astronomer Hipparchus (mid 2nd century BC), who first formulated exoterically the idea of PRECESSION. The star is accorded an unfortunate influence (Mars conjunct Saturn), and brings physical injuries and defeats. It is sometimes called Sheratan.

Shaula Fixed star, the lamda of Scorpius, set in the sting of the asterism. The name has been variously corrupted to Shauka, Alascha and Shomlek. Along

Shaulah

with the upsilon of Scorpius, it marked the MANZIL now called SHAULAH. The star has a very unfortunate reputation.

Shaulah Al Shaulah, the 17th of the Arabian MANZILS.

Shebat See BABYLONIAN CALENDAR.

Shodasvargas A Sanskrit term applied to the Hindu astrological chart presented in terms of 16 divisions, which may have been an elaboration of the Hellenistic *oktotopos* division.

Shukra The Sanskrit word for the planet Venus, as used in Hindu astrology.

sidereal The term sidereal, meaning 'pertaining to the stars', is derived from the genitive form of the Latin *sidus* (star), *sideris*.

sidereal correspondences The historian Powell, having established a SIDEREAL ZODIAC defined in terms of epochal fiducials, provided a useful series of correspondences for the main sidereal zodiacal system. In table 75 the BABYLONIAN ZODIAC is derived from the fixed star fiducial Aldebaran, the (modern) Indian zodiac is derived from the fiducial Spica, for zero degree of Libra, and the MODERN ZODIAC is defined according to the complex IAU system reported by Delporte. The STAR CALENDAR ZODIAC is ultimately derived from the specification given by Vreede.

	Babylonian		Indian		Modern		Star Calendar
	Deg.	Min.	Deg.	Min.	Deg.	Min.	Deg.
AR	0	0	359	03	3	54	4 (5)
TA	30	0	29	03	28	38	29
GE	60	0	59	03	65	21	65
CN	90	0	89	03	93	12	93
LE	120	0	119	03	113	15	114
VG	150	0	149	03	149	04	149
LB	180	0	179	03	193	02	194 (195)
SC	210	0	209	03	216	12	213
SG	240	0	239	03	241	27	243 (244)
CP	270	0	269	03	274	52	275 (274)
AQ	300	0	299	03	302	42	302
PI	330	0	329	03	326	52	328 (327)

TABLE 75

sidereal day   An interval of TIME measured according to the stars – that time taken for the Earth to rotate on its axis once in relation to a stellar fiducial. The sidereal day is the equivalent of 23 hours, 56 minutes and 4.09 seconds of clock time.

sidereal lunation   A period of 27 days, 7 hours, 43 minutes and 11.5 seconds, marking two successive conjunctions of the Moon with the same degree of the zodiac – a periodic lunation.

sidereal month   See LUNAR MONTH.

sidereal moon rhythms   A term seemingly coined in connection with the use of the SIDEREAL ZODIAC in bio-dynamic farming, which has proceeded under the influences of the researches of Thun, supported by Kolisko and Fyfe. The planting of seeds and the preparation of the earth is regulated according to lunar positions in a recently formulated (unequal-arc) constella-tion zodiac, which is not to be confused with either the tropical zodiac or the traditional CONSTELLATIONAL ZODIAC. The sidereal moon rhythm is the lunar relationship to this new zodiac. The results of intensive practical research have enabled astrologers to determine that the form of the plant is linked with the influences of the fixed stars and with the movements of the planets in a most simple way – indeed, it has been shown that the triplicities of the constellations influence radically the development of all seed growth. The earth triplicity is linked with the growth of roots and root crops, and is called the 'root-earth trigon'; the air triplicity is linked with the growth of blossoms and flowers, and is called the 'blossom-light trigon'; the water triplicity is linked with the growth of the leaf, and is called the 'leaf-water trigon', whilst the fire triplicity is linked with the growth of fruit and seeds, and is called the 'fruit-seed warmth trigon'. The extraordinary practical value of these cosmic-agricultural researches is apparently based on considerable esoteric truth: Goethe himself announced a connection between plant growth and the elemental SOUL BEINGS, each group of which is linked with the trigons.

sidereal signs   A term used by Powell in reference to the equal-arc divisions of the CONSTELLATIONAL ZODIAC, linked with the Babylonian system. It is a curious term, perhaps even a contradiction, but is intended to refer to the twelve 'box-type sectors' in the ecliptic circle, which are other than the zodiacal signs of traditional astrology. These sidereal signs appear to be analogous to this ancient constellational system as the zodiacal signs are to the zodiac. See, however, SIDEREAL ZODIAC.

sidereal time Literally 'time by the stars', a term intended to designate a method of computing time based upon the period between specific stars over predetermined points on the Earth. See SIDEREAL DAY and TIME.

sidereal year A period marking two succedent transits of the centre of the Sun over the ecliptic meridian of a fiducial star: a period of 365.256 days. This period is sometimes called an astral year.

sidereal zodiac As the term suggests, this is the zodiac of the stars, of which the CONSTELLATIONAL ZODIAC is just one example. The term is usually applied to one or other of the divisions of the Stellatum into twelve areas in a band along the ecliptic, corresponding in nomenclature (though not in extent or location) to the twelve signs of the zodiac, as in figure 57. There is much disagreement among specialists as to the location and cuspal boundaries of these asterisms, however, and even lack of agreement as to how many such asterisms should be included in the sidereal zodiac. Schmidt is not alone in claiming that 14 constellations cross the ecliptic band, whilst it is evident that the number depends upon (a) the width of the band, and (b) the graphic model of the Stellatum adopted. Messadie has set out calculations proposing as many as 24 asterisms along these lines. A further problem is that there is no agreement as to which stellar fiducials should be used to fix the positions of these variant zodiacal asterisms. A recent attempt to link the sidereal zodiac to the SOLAR APEX has been made by the sidereal astrologer Garth Allen, who has proposed a fiducial called the 'synetic Vernal Point' which lies 90 degrees east of the solar apex. A brief, but well-researched, history of the early zodiacs by Powell shows that several of the early zodiacs, including the prototype of our own modern astronomical zodiac, were sidereal, and linked with sidereal fiducials: see therefore ASTRONOMICAL ZODIAC, BABYLONIAN ZODIAC and GREEK ZODIAC. For a definition of the modern astronomical zodiac, see MODERN ZODIAC. The sidereal zodiac defined and located by Powell is of considerable importance to the development of modern sidereal astrology, and is worth recording here. Powell defines the sidereal fiducial in terms of a particular epoch, restoring the ancient Babylonian fiducial of Aldebaran in 15 degrees Taurus. The formal definition of the zero point is that exactly 45 degrees west of Aldebaran on 1 January 1950. Powell provides a framework of reference for the unequal divisions of this zodiac in terms of the 105 brightest zodiacal stars, computed for the epoch of 1950, and additionally provides a table of correspondences for the main sidereal zodiacs in relation to his formulated sidereal of 1950. This table may be seen schematized under SIDEREAL CORRESPONDENCES. The sequence of unequal asterisms for the Powell sidereal zodiac may be synopsized as follows, in terms of the diagram set out in figure 57.

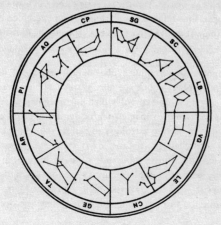

Figure 57: Sidereal zodiac within an equal-arc framework (after Powell).

AR from Mesartim to tau Arietis.
TA from Alcyone to Alhecka.
GE from Propus to Pollux.
CN from Altarf to Acubens.
LE from Alterf to Denebola.
VG from Zavijava to mu Librae.
LB from Zubenelgenubi to Akrab (Graffias).
SC from Dschubba to Shaula.
SG from Alnasl to rho Sagittarii.
CP from sigma Capricorni to Deneb Algedi.
AQ from Sadalsuud to psi Aquarii.
PI from beta Aquarii to Alrescha.

Powell gives all magnitudes, longitudes and latitudes for all these fiducials for the epoch of 1950.

Sidus Faustum Regis   See SADALMELIK.

Sidus Fortunae   Star of Fortune, a name applied to the 24th of the LUNA MANSIONS in some medieval lists. Probably identified by the star Sadalsuud.

Sidus Parvum Lucis Magnae   A term used to denote the 6th of the LUNAR MANSIONS in some medieval lists. Probably located by the fixed star Alhena.

sieu The Chinese term for the equivalent of the LUNA MANSIONS. Unlike the Hindu NAKSHATRAS or the Arabian MANZILS and the LUNAR STATIONS, the Sieu were grouped into huge constellational groups, each consisting of seven sieu: these are the AZURE DRAGON, BLACK WARRIOR, RED BIRD and WHITE TIGER.

sigil A name applied to a graphic cypher or seal or symbol, derived from the Greek term *sigla* (cypher) – but see also SYMBOL. It is a term now used to denote almost any graphic symbol which incorporates a definite meaning, but once it was restricted to magic graphic symbols. From the very earliest times sigils have been used in astrology to denote the whole range of planets, zodiacal signs, constellations, decans, fixed stars, the Celestial Hierarchies, demons, daemons and spirits: thousands of sigils have been recorded from such contexts. By way of example, figure 58 sets out six sigils from a Greco-Byzantine decan list – the three smaller sigils represent Mars, Sun and Venus respectively, while the three encircled sigils represent the decans of Scorpio.

Figure 58: Decan sigils for Scorpio (after Nowotny).

For ease of reference a series of the more important astrological sigils are recorded in the following alphabetical entries: one may, however, gain some idea of the variety involved when one considers that the five historical variants recorded for the planet Saturn (under SIGILS – PLANETS) were selected from over 60 recorded by the present author in a specialist context (see Bibliography). The popular notion that the sigils are derived from vestigial drawings of the images associated with the signs and constellations is quite erroneous, fostered seemingly by Agrippa at the end of the 15th century. Kircher developed upon Agrippa, and established a pseudo-system of correspondences between signs, sigils and associate gods and goddesses, some of which are derived from esoteric sources, some of which are entirely fanciful. On the other hand, it must be admitted that the planetary and zodiacal sigils in use in modern astrology, whilst rarely older than the 14th century in their present forms, were derived from esoteric schools.

sigilla The Latin plural for SIGIL.

sigils – aspects The sigils used to denote the Angular relationships between planets appear to be as old as the aspects themselves – the post-Keplerian aspects being derived contemporaneously:

Conjunction	☌	Opposition	☍	Quincunx	⚻
Semi-sextile	⚺	Semi-square	∟	Sesquiquadrate	⊡ ⊡
Sextile	⚹ ✳	Square	☐	Trine	△

sigils – elements The basic range of sigils for the elements are derived from the Rosicrucian tradition, and are encapsulated in the *SEAL OF SOLOMON, which unites them. Three different systems are given in table 76.

--------- TABLE 76 ---------

Element	Medieval sigil	Rosicrucian sigil	Modern sigil
Air	♯	△ (with bar)	=
Earth	⊖	▽ (with bar)	⊕
Fire	△	△	△
Water	▽	▽	▽
Quintessence		✡	Q

sigils – fixed stars Most of the medieval astrological sigils relating to the stars were limited to the FIFTEEN STARS, and were used mainly in talismanic magic. Agrippa, who passed the sigillic forms into the non-specialist occult stream, made a copyist error with his sigil for Capella, as a result of which the form   has been erroneously accepted for this star. Although there are many variants, the ones below (called 'caracti' in many medieval manuscripts) are the most usual forms. In this list the second variant is that given by Agrippa, the first being from a late 14th-century source:

Aldebaran	⋊	Algol	
Algorab		Alphecca	
Antares		Arcturus	
Capella		Deneb Algedi	
Pleiades		Polaris	
Procyon		Regulus	
Sirius		Spica	
Wega			

sigils – general The following sigils are those most frequently used within the astrological tradition – see, however, the ABBREVIATIONS which are favoured in modern astrology:

Angular	⌐	Ascendant	
Asteroids		Cadent	L
Cardinal	∧	Ceres	
Dragon's Head	♋	Dragon's Tail	☋
Earth		Fixed	⊡
Juno		Mutable	
Pallas		Pars Fortunae	⊕
Succedent	L	Vesta	

sigils – planetary spirits The sigils given in the occult and astrological tradition for the spirits – beneficial celestial beings, or the harmful demonic beings – are about as numerous as the spirits themselves. The following are only those sigilla associated with the important *SECUNDADEIAN BEINGS* mentioned by Trithemius:

Anael		Gabriel	
Michael		Oriphiel	
Raphael		Samuel	
Zachariel			

sigils – planets Many of the planetary sigils used in modern astrology are derived ultimately from Greco-Roman, Byzantine, or even Egyptian prototypes. Table 77 gives only the most frequently used of the Greco-Roman, medieval and modern sources.

TABLE 77

Planet	Greco-Roman	Medieval	Modern
MO	☽	☾ ☽	☽ ☾
ME	☿	☿	☿
VE	♀	♀	♀
SU	☉	⊕	☉ ♂ ♄
MA	♂		
JU		♃	♃
SA	♄		♄

468

The so-called MODERN PLANETS have been accorded many variant sigils of which the following forms are the most usual:

UR ♅ ⚇ ♅   NE ♆ ♆ ♆   PL ♇ ♇ ♇ ♇ ♇

A number of the HYPOTHETICAL PLANETS and FUTURE PLANETS have also been accorded sigils in modern times, as follows:

Aidonius	Apollo	Bacchus
Coelus	Dido	Hercules
Jason	Lilith	Mercurius
Pagan-Pluto	Thierens-Pluto	Wemyss-Pluto

sigils – zodiac The fact is that the signs of the zodiac have often been confused with the sigils for these signs, and even with their IMAGES. Most of the sigils used in modern times date from about the 14th century, though they are often connected by a tortuous graphic etymology with Egyptian (demotic), Greco-Roman and earlier forms: there can be little doubt that the majority of forms in use today were evolved by an esoteric school – possibly that connected with Chartres. Table 78 lists only the most frequently used sigils for specific periods.

| | | TABLE 78 | | |

Sign	Demotic	Greco-Roman	Medieval	Modern
AR				♈
TA				♉
GE				♊
CN				♋
LE				♌
VG				♍
LB				♎
SC				♏
SG				♐
CP				♑
AQ				♒
PI				♓

sign The word 'sign' is perhaps one of the most misused of all astrological terms. Properly applied, it refers to one of the 30-degree arcs of the zodiac, of what Ptolemy called the *zodiacos kuklos*. From early times a distinction was made by thoughtful astrologers between these signs and the constellational asterisms which had the same names as the signs. There was little confusion between the signs of the tropical zodiac and the asterisms of the constellational zodiac. The tropical was always divided into twelve equal arcs, whilst the constellations (not even visualized as a band, as a wide *kuklos* at all) was divided into unequal asterisms, the twelve of the original fourteen which touched upon the zodiacal band proper being called by the same names as those carried by the signs, even though they did not correspond in extent or location, and were only at one remote time united by a common fiducial. A sign is one of the 30-degree arcs of the zodiacal circle: the term which sometimes is used in popular textbooks – constellational sign – is quite meaningless. It has happened of late that popular astrologers have also written and spoken of the signs of the zodiac as though these refer to the IMAGES, or (more frequently, though just as wrongly) to the SIGILS used to symbolize these. A sign is not an image, though an image may at times denote a sign. Properly speaking, the image of a sign is really derived from the ancient constellational pictures – the image of Aries is a ram, for example, that of Taurus is a bull, and so on. The signs have been classified according to ancient principles and given a bewildering variety of different group-names, many of which, whilst virtually archaic, are to be traced back to Ptolemaic astrology. A useful (if archaic) list of such group-names may be abstracted from the 17th-century text of Lilly:

Airy	GE	LB	AQ			
Barren	GE	LE	VG			
Bestial	AR	TA	LE	SG		
Common	GE	VG	SG	PI		
Double-bodied	GE	SG	PI			
Earthly	TA	VG	CP			
Equinoctial	AR	LB				
Feminine	TA	CN	VG	SC	CP	PI
Fiery	AR	LE	SG			
Fixed	TA	LE	SC	AQ		
Fruitful	CN	SC	PI			
Human	GE	VG	AQ			
Masculine	AR	GE	LE	LB	SG	AQ
Moveable	AR	CN	LB	CP		
Northern	AR	TA	GE	CN	LE	VG

Southern	LB	SC	SG	CP	AQ	PI
Tropical	CN	CP				
Watery	CN	SC	PI			

signacular A term often used in medieval astrology as the equivalent of CHARACTERS, sometimes of SIGILS. See also MAGIC SQUARES.

signature In medieval astrology it was taught that every material form bears an outer stamp of the planetary or zodiacal principle which underlies its nature: this is the signature, the signing of nature. The proficient astrologer or occultist would recognize these inner principles from the outer form, and thus obtain knowledge of the real thing – of its quintessential nature – which is represented within the form. The esotericist Boehme developed the idea of 'signature' at great length, and the term has been usefully linked by Muses with the SEVEN PROPERTIES, the seven planetary natures, as a sign of the signature of all things, the outer form being regarded as 'the precise configuration of the differing representation, in relative strengths, of each of the basic properties therein'. In a general sense, the signature is the spiritual imprint of the VIRTUE. Paracelsus anticipates Goethe by visualizing the signature of things as the result of the workings of the Imagination of Nature. 'There is nothing hidden which Nature has not revealed and put plainly forward,' he writes, with the signature in mind. In a later definition Paracelsus equates the 'Doctrine of Signatures' with the entire art of astrology, it being 'the science which teaches one to know the stars, what the heaven of each may be, how the heaven has produced man at his conception, and in the same way constellated him'.

significator In a general sense, each planet is a significator of something or other – for example, the Sun is a significator of the conscious aim of the native, while the Moon is the significator of the personality, and so on. In this sense, the houses are also significators – the 2nd house, for example, is a significator of wealth, among other things. However, the term is often used in a more specialist sense in regard to horary astrology, most frequently in regard to the planets which are significators of the querent: this is usually the strongest planet in the figure, which is in many cases the ruler of the Ascendant. Even within horary astrology each planet remains a significator of some event, quality or person: for example, Venus is the significator of love-affairs, the mistress, the impulse to love, and so on. See also MODERATORS and PROMITTORS.

signs of voice   The astrologer Wilson terms Gemini, Virgo, Libra and Aquarius (as well as the first part of Sagittarius) the signs of voice, because 'if either of them ascend, and Mercury be strong, the native will be a good orator'.

Sihor   See SIRIUS.

Silver Age   See FOUR AGES and TRETA YUGA.

Silver Cord   See PISCES CONSTELLATION.

Silver River   See VIA LACTEA.

Simak   Al Simak, the 12th of the Arabian MANZILS. See also ARCTURUS.

Simha   One of the Sanskrit terms for zodiacal Leo, used in Hindu astrology.

Sin   A general Babylonian term for the Moon.

singleton   A term sometimes used of the planetary 'handle' of the aspectal shaping called a BUCKET in the JONES PATTERNS. The singleton is that planet (or conjunction of planets) which gives drive or 'purpose' to the formation, and hence to the chart as a whole.

sinister aspect   An aspect calculated in the order of the signs, to be distinguished from the reverse computation, which is the DEXTER ASPECT. The change in the meaning of the word 'sinister' (which originally meant merely 'left-handed' or 'left-hand side') has led to certain misunderstandings about this term: there is nothing unwholesome in this aspect.

Sinistra   Name given to the relatively unimportant fixed star set in the left hand of the asterism Ophiuchus, said to give an immoral, mean and slovenly nature.

Sirian cycle   A periodicity observed by the ancients, involving the heliacal rising of the fixed star SIRIUS, in a 162-year periodicity.

Sirius   Fixed star (binary) the alpha of Canis Major, set in the mouth of the dog, one of the most important stars in the heavens so far as the astrological tradition is concerned. It has many names and traditions attached to it. Though now it is called the 'Dog Star', the name is often derived from the Greek *seirios* (sparkling), even though that same term was sometimes used of all the stars (and indeed of the Sun). Some occult etymologists have linked the name with the Egyptian Osiris (which is in any case a Greek version of the Egyptian); it was for Homer the Star of Autumn, to the Romans Canicula (little dog), and they are reported to have sacrificed a dog in their May festivals, during the so-called 'Robigalia', when the Sun reached the orb of Sirius. From these names and associations came a wide variety of derivations, such as Latrator Anubis, Al Shi'ra, Asceher, Aschere, and Scera. In the Egyptian astrology it was symbolized by a dog-like image as hieroglyphic, and was of considerable importance in the middle of the 4th millenium, for its heliacal rising at the summer solstice marked the beginning of both the Egyptian year and the commencement of the Nile flooding. In the so-called zodiac at Denderah the star is figured as a cow in a boat, but it was in earlier forms linked with Isis, both in Denderah and at the important Isis-complex at Philae. It appears to have been linked also with the name of Thoth, and was called Sihor (Nile Star), as well as by such epithets as Fair Star of the Waters, which perhaps explains the reason why Isis was called Stella Maris long before the Virgin Mary (but see STELLA MARIS). As might be imagined from these prestigious names, astrologers link it in general with beneficial influences: it gives honour, renown, wealth and passion. It is linked with guardianship (in accordance with the general association with zodiacal Cancer), and is inevitably associated with the danger of dog-bites. See also in this last connection PROCYON.

Sirrah   A name for the fixed star ALPHERATZ.

Sisterhood   See ETHEREAL PLANETS.

Sivan   See BABYLONIAN CALENDAR.

sixth house   The sixth of the astrological HOUSES, linked with the nature of the sign Virgo, and representing the ideal of service, as well as the kind of relationship the native will establish with those above and below (the 6th house is sometimes said to deal with the chaos left by the Leonine impulse of the preceding 5th house, and to 'love and cherish' the associate Libra of the following seventh house). It is sometimes called the House of Work, and is frequently linked with ideals of hygiene and social care – in the melothesic man it rules the bowels and solar plexus. In *MUNDANE ASTROLOGY it relates to the public health services, to such civil workers as police and voluntary labour. It is said that strikes have their beginnings in this house, but take their 'form' in the 12th house.

TABLE 79

Constellation	Orthodox planet	Esoteric planet
Virgo	Mercury	Moon
Sagittarius	Jupiter	Earth
Pisces	Jupiter	Pluto

sixth ray   In the esoteric system described by Bailey, the sixth of the SEVEN RAYS is called the Ray of Idealism, or the Ray of Devotion, and is linked with the will to create and 'bring into being'. The sixth ray is said to arise in Ursa Major, and is transmitted into our solar system by means of the zodiacal constellations (signs?) Virgo, Sagittarius and Pisces, through their pairs of corresponding ORTHODOX PLANETS and ESOTERIC PLANETS, according to the schema in table 79. The entire sixth ray is linked with the sacred planet Neptune: see therefore SACRED PLANETS. In connection with the rule of esoteric Moon over Virgo, as set out in the above schema, see also ESOTERIC MOON. Distinguish from SIXTH RAY TYPE.

Sixth Ray Type   The sixth of the SEVEN RAY TYPES, associated with the planet Mars. Distinguish from SIXTH RAY.

Sizajasel   In certain systems of ceremonial magic, this is the name given to the Governor of the zodiacal sign Sagittarius.

Skat   Modern name for the fixed star, delta of constellation Aquarius, set in the right leg of the asterism, once confusingly called Scheat (and hence confused with the SCHEAT of Pegasus). The term is probably derived from the Arabic *Al Sak* (the shin bone). Ptolemy says that it is of a nature equivalent to Jupiter with Saturn, which is of a beneficial influence.

slow of course   Said of a planet which is moving more slowly than its mean motion: in horary astrology it is regarded as a debility.

Sol   The Latin name for the Sun. The same name was used by Blavatsky to denote an 'invisible trans-Mercurial planet, nearer to the Sun than Mercury' which was 'one of the most secret and highest planets'. This was probably one of the AROMAL PLANETS, but one said to have 'become invisible at the close of the Third Race' – towards the end of the Lemurian period.

solar apex   A term used to denote the objective of solar motion through space, a motion which has been identified by some astrologers with the target of personality in relation to the zodiacal motion of the Sun as reflected in horoscope charts. The solar apex has been used as fiducial to overcome the anomaly of the slight proper motion of fixed stars, which makes for imprecise definition of the zodiacal fiducial. The sidereal astrologer Garth Allen proposed that the sidereal zodiac should take the solar apex as a fiducial, locating what he called the 'synetic Vernal Point' 90 degrees east of the solar apex. Thus, for the first time in the history of astrology a zodiac was proposed which united the ecliptic divisions with the motion of the Sun in co-ordinates independent of the fixed stars. It was assumed by the siderealists that the solar apex was a point located in the first degree of Capricorn (but see the related figures of Landscheidt in GALACTIC CENTRE), and that this point was not subject to proper motion. However, in spite of the provision of exact degrees, and indeed of a simple ephemeris by Landscheidt, it has been shown that the apex may not be regarded as a point, but as an arc extending to almost 6 degrees, which calls into question the validity of the synetic Vernal Point concept. See, however, SIDEREAL ZODIAC.

solar astrology   A term frequently misapplied as being synonymous with HELIOCENTRIC ASTROLOGY, though more properly it is used to denote that system of astrology which is involved with the casting and interpreting of charts for the moment of sunrise, or for exact local noon time, when the Sun is on the Ascendant degree or on the Medium Coeli, respectively. Usually, such astrology is used only in those cases where precise time of birth is not known, and even then only as a preparation for RECTIFICATION.

solar biology   A term seemingly originated by the astrologer Butler prior to 1887, relating to a form of astrology which, for all its strange concepts (see, for example, EARTH ZODIAC), and for all its being pervaded by biblical and religious references not always apposite to the case, appears to be little other than traditional genethliacal astrology in a new guise. However, Butler says that his astrology is not involved with the prediction of coming events. The positions of the Moon are given in terms of the geocentric system in accordance with the proposition that there is a 'second zodiac' relating to the Earth, while the positions of other planets are listed according to a curious notion which requires the adjusting of the tropical positions by 180 degrees, thus segregating a 'lunar zodiac' (or 'earth zodiac') from the so-called 'solar zodiac', which appears to be the traditional tropical zodiac.

solar body   See LUNAR BODY.

solar cycle   A term most usually applied to a cycle of 28 years, which is not really a solar cycle at all, but a periodicity derived from calendrical systems, in which the days of the week re-occur on the same days of the month as in the preceding period. The genuine solar cycles are the METONIC CYCLE, the PLATONIC CYCLE and the YEAR – but see also TIME.

solar day   The time between two consecutive transits of the Sun over a fixed point on Earth: it is in excess of the SIDEREAL DAY by approximately 1 degree of longitude (roughly 4 minutes of TIME). See also MEAN SOLAR DAY and TRUE SOLAR DAY.

solar distance   The mean solar distance of the planets from the Sun, expressed in thousands of miles, is as follows:

ME	35,950	VE	67,180	TE	92,870	MA	141,500
JU	483,300	SA	886,100	UR	1,783,000	NE	2,793,000
PL	3,666,000						

solar eclipse   When the orb of the Moon interposes itself between the Earth and the light of the Sun, the phenomenon is called a solar eclipse – in effect this is a platic *CONJUNCTION of the luminaries. The ancient astrological rule held that the effect of a solar eclipse lasts as long in years as the eclipse lasts in hours, though this effect is limited to the degree in which the eclipse took place. Eclipses (see also LUNAR ECLIPSE) in a chart are usually regarded as malignant if the degree involved is conjunct the Sun, Moon, Ascendant, Medium Coeli or any of the malefics, either radical or progressed. Experience has shown that an eclipse leaves a sort of 'degree influence' which is only made operational by later transits or progressions which trigger off the line effects. Esoteric astrology teaches that at the time of solar eclipses evil forces are permitted to pour from the Sphere of the Earth into the cosmos, the eclipse acting as a sort of safety-valve.

solar equilibrium   A term of recent origination, defined by Devore as having reference to the SOLAR FIGURE cast for sunrise, and symbolically presented in the equal-house method, with houses of uniform 30 degree arcs.

solar figure   Term used of a particular type of horoscope erected for a nativity for which the time of birth is not known, determining a putative Ascendant by taking it as the Sun's degree. See SOLAR ASTROLOGY.

solar fluid   A term used in an astrological sense by Butler to denote an 'ethereal atmosphere or sea of fluidic element pervasive of, and limited by, our solar system'. The fluid seems to correspond in quality and effect, if not in actual distribution, with the traditional Aether, or QUINTESSENCE, and is almost certainly derived from the occult writings of the American clairvoyant Davis. See, however, SOLAR BIOLOGY.

solar horas   See HORAS.

solar house   A system of dividing the ecliptic circle, which is not to be confused with any of the HOUSE DIVISION methods. The solar houses are equal-arc divisions of 30 degrees, supposedly subdivisions of the two-hour periods of axial rotation of the Earth, and postulating that the Sun is on the Ascendant at the commencement of the division.

solar kalpa   See KALPA.

solar radix method   A system of progressions involving the progressing of the horoscope as a whole by the progressive addition of an arc equivalent to time lapsed, by the conversion of 1 hour into 15 degrees of arc. The measure for a year (that is, the $365\frac{1}{4}$ days of the mean solar year) is equivalent to 59 minutes 8.33 seconds, and the solar month in the system is exactly $\frac{1}{12}$ of that year. Since according to the theory of progressions the diurnal is the equivalent of the annual, the same arc is the measure of the day.

solar return   A solar return is a chart cast for the time at which the Sun returns to its radical position in a preceding chart (usually the natal chart). It may be measured in terms of the tropical or sidereal zodiac, though the astrologer Bradley favours the latter in his own comprehensive survey of the SOLUNAR RETURNS. He says that the solar return chart describes the events to befall the native during the following twelve months. See also LUNAR RETURN. Subdivisions of the solar return described by Bradley are the QUARTERLY CHARTS and the SEMESTER CHART. The solar return is sometimes confusingly called the 'solar revolution' – but see REVOLUTION and SOLAR REVOLUTION FIGURE.

solar revolution   See SOLAR RETURN and SOLAR REVOLUTION FIGURE.

solar revolution figure   A horoscope figure cast for the moment in any year subsequent to the birth of a native when the Sun reaches the precise degree of longitude which it occupied in the radical chart. Such a figure is generally erected for purposes of prediction.

solar semi-circle   Term applied to the zodiacal arc from Leo to Capricorn, inclusive.

solar time   See TIME.

solar year   Another term for the TROPICAL YEAR.

solar zodiac   See REAL ZODIAC, SOLAR BIOLOGY and ZODIAC.

solid signs   Translations of the word *sterea*, used by Ptolemy to denote the four Fixed signs of the zodiac, have given us the unfortunate term 'solid', though the Greek word really meant 'hard' or 'firm'. Ptolemy says that the solid signs are persistent, firm, patient, industrious, self-controlled, hard, inflexible, and unaffected by flattery.

Solomon's seal   See SEAL OF SOLOMON.

solstices   The points in the ecliptic at which the Sun is at its greatest distance to the north or south of the equator are called the solstitial points or solstices (literally, the points where the sun appears to stand still). The summer solstice is marked by the entry of the Sun into Cancer, the winter solstice by the entry of the Sun into Capricorn, round about 21 June and 21 December, respectively.

solstitial Colure   See COLURE.

solstitial point   See ECLIPTIC.

solstitial points   See SOLSTICES.

solstitial signs   According to Ptolemy, the solstitial signs (Cancer and Capricorn) produce souls fitted for dealing with people, fond of political activity and 'attentive to the gods', as well as fitted for the practice of astrology and divination. There is probably some confusion between the entire sign and the specific solstitial degree.

solunar returns   An inclusive term for the two separate techniques involving SOLAR RETURNS and LUNAR RETURNS.

Somerset Giants   See GLASTONBURY ZODIAC.

Sonipes   See PEGASUS.

Sorath Name given by Agrippa (quoting qabbalistic sources) to the Daemon of the Sun, for whom he gives the magical number 666 – but see also NACHIEL.

Sortes The various systems of Sortes (in the Greek *klipoi* – though the term *locus*, plural *loci*, is also sometimes used – but see LOCI) derived from ancient astrology have been studied in modern times by Bouche-Leclercq, who refers to a number of Sortes no longer in use, and especially to two important Sortes recorded by Manilius, the equivalent of the modern PART OF FORTUNE and the interesting (though now defunct) LOT OF DAIMON. Manilius distinguishes between the day horoscope and the night horoscope. For the former he takes the distance between the Sun and Moon, and, following the order of the signs, transfers this arc to the Horoscopos (Ascendant), to locate the Pars Fortunae. In the nocturnal chart, measurement is made widdershins. The modern method does not distinguish between the day and night charts, which in effect give precisely the same point by means of different rules. In connection with the Daimon (sometimes the *klipos daemonos*, translated often in Latin as *sortes genii*, or even *locus daemonis*) Bouche-Leclercq reports also the views of the Alexandrian Paul, who also distinguishes between the day chart and the night chart (apparently following some Egyptian tradition, which may not be fully understood). In the day chart the arc from the Sun to the Moon, measured in the order of the signs, is applied to the Horoscopos against the order of the signs to mark the Daimon. In the night chart the distance from the Sun to the Moon is measured against the order of the signs, the arc being transferred to the Horoscopos along the order of the signs to reveal the Daimon. In these systems the diurnal Daimon is interchangeable with the nocturnal Pars, and vice versa. See, however, TWELVE PLACES. The use of Sortes in Hellenistic astrology was expressed in a refined way in the 1st century AD texts of the astrologer Dorotheus of Sidon, who placed much emphasis on the Fortunae and Daemon (as did most early astrologers). However, as Pingree, in his short history of astrology, points out, the list of Sortes grew to such an extent that by the 9th century the Arabian astrologer Abu Ma'shar was able to enumerate well over a hundred.

Sosol In certain grimoires, the Sosol is named as Governor of Scorpio.

Sothic calendar A term used of one of the calendrical systems used by the ancient Egyptians, involving four-year periods (tetracteres), which began when the fixed star Sirius (Sothis) rose heliacally at Heliopolis. According to Fagan, the rigorous calculations of these periodicities, involving Newcomb's precessional constants and the *arcus visionis* for Sirius at that place, revealed the following series – starting 4228 BC, 2770 BC, 1314 BC and AD 142. Fagan says that this cycle corresponds to the 'Resurrection of the Phoenix' of Egyptian lore: it is a mean cycle of 1,456 years. Fagan further claims that a great many of the celestial diagrams (he is careful not to call them zodiacs or horoscopes) still preserved in ancient Egypt were computed to mark these periodicities.

Sothic cycle A periodicity given as a mean cycle of 1,456 years, and sometimes as 1,460 years. See SOTHIC CALENDAR.

soul beings The four classes of ELEMENTALS, often wrongly called 'elemental spirits', are generally regarded by occultists to be elemental soul beings, since the very quality they lack is that of 'spirit', in that they do not have the ego, or even the ego-potential, which is part of the human constitution. In the Rosicrucian tradition, which has from time to time influenced the development of astrological thought, the four groups are linked with the four elements (hence their generic name). The GNOMES are the earth beings, the SALAMANDERS are the fire beings, the SYLPHS are the air beings, and the UNDINES are the water beings.

soul-numbers A series of numbers linked with the Platonic theory of harmonics and *DESCENT OF THE SOUL which figured in later numerological theories connected with astrology. The numbers are 1, 2, 3, 4, 8, 9 and 27, but the argument as to their mystical natures is beyond the compass of the present treatment. It is perhaps sufficient to set out the schema reproduced by Mead relating to the circular progression and 'conversion of the soul', for this links directly with the original Pythagorean imagery of reincarnation, descent into multiplicity (descent into numbers) which permeates astrological esotericism. The four stages correspond to the point, line, plane and solid:

Point		1		
Line	2		3	
Plane	2		3	(both to the power of 2)
Solid		2	3	(both to the power of 3)

These are called the 'boundary numbers of the soul'. These numbers were adopted in many esoteric schemas in regard to the cycle of descent and re-ascent of the soul (in and out of matter) – for example, they are tacitly assumed in the rationale given by Dee in his MONAS HIEROGLYPHICA.

sound values See QABBALISM.

southern angle The cusp of the 10th house is called the Southern Angle and is sometimes identified with the Medium Coeli: it is associated with the sign Capricorn, and with the Point of Aspiration, at which the native may achieve his 'highest' conscious aims or ambitions.

Southern Claw See ZUBENELGENUBI.

Southern Cross See CRUX.

Southern Crown See CORONA AUSTRALIS.

Southern Fish See PISCIS AUSTRALIS.

Southern Fly See MUSCA AUSTRALIS.

Southern signs These are the zodiacal signs from Libra to Pisces, inclusive, the so-called 'obeying signs', south of the equatorial belt.

Southern Triangle Constellation, the Triangulum Australe, probably formed by Theodor in the 16th century, south of the constellation ARA, between Pavo and Centau. The triangle stars, which define the asterism, are alpha, beta and gamma. It is said to have an influence similar to that of Mercury, and is therefore beneficial.

Southern Twin See POLLUX.

south node The point where a planet crosses the ecliptic into the southern latitude – see NODES.

South Scale See ZUBENELGENUBI.

specialis A term used to denote the 20-year cycle of the CHRONOCRATORS.

speculation In a strictly astrological and occult context Paracelsus defines speculation as an activity arising from the stars 'which are concerned about man'. From the larger context of his writings, we may assume that the term is intended to denote the relationship which exists between what we would now call the 'archetype' (which Paracelsus variously called the Archaeus or the Heavens) and human thinking. Hence, Paracelsus speaks of a speculating man as one having his imagination united with heaven, and 'heaven operates so within him that more is discovered than would seem possible by merely human methods'. The term is of specific interest in that the medieval astrologers regarded astrology itself as one of the 'speculatives', in a sense very different from the modern.

speculum In astrology, a term used for a table attached to a horoscope figure, containing information concerning astronomical data which may be useful in interpreting the radical or progressed figure. The information usually includes such matter as latitudes, declinations, and so on. The word itself is from the Latin for 'mirror' – in some early texts the word is used in a poetic sense to denote the horoscope chart, which 'mirrors' the heavens.

sphaera obliqua See SPHAERA RECTA.

sphaera recta In terms of stereographic projection, the *sphaera recta* relates to zero geographical latitude, with reference to the phenomena of rising and setting: the contrasting *sphaera obliqua* is that projected for other latitudes.

sphaera zodiacus A Latin term used to denote (first) the STELLATUM, and then later (archaically) the ecliptic.

Sphere of Democritus Name applied to an ancient (probably Hellenistic) system of prognostication by means of numbers, only peripherally linked with ONOMANTIC ASTROLOGY.

Sphere of Heaven Sometimes called the Stellatum or Coelum (but see also EIGHTH SPHERE) in the Ptolemaic system, the Sphere in which dwell the beings and orbs of the fixed stars. In the medieval cosmoconception the Sphere was said to be the dwelling of the CHERUBIM, ruled by the spiritual being OPHANIEL, linked with the image of a dove.

Sphere of Jupiter In the Ptolemaic system, the Sphere in which dwell the beings and orb of the planet Jupiter. In the medieval cosmoconception, the Sphere was said to be the dwelling of the DOMINIONS (sometimes the Dominations), ruled by the spiritual being ZADKIEL and the spirit Befor. The Sphere is linked with the image of an eagle.

Sphere of Mars In the Ptolemaic system, the Sphere in which dwell the beings and the orb of the planet Mars. In the medieval cosmoconception, the Sphere was said to be the dwelling of either the POWERS (or Potestates) or the PRINCIPALITIES, depending upon which system of Celestial Hierarchies was adopted – though in this connexion it is worth observing that the Greek term DYNAMIS for these beings appears to have been invariable, and might reasonably be translated as 'Powers'. The Sphere is ruled by the spiritual being SAMAEL, and the spirit PHALEG, and is linked with the image of a horse.

Sphere of Mercury In the Ptolemaic system, the Sphere in which dwell the beings and orb of the planet Mercury. In the medieval cosmoconception, the Sphere was said to be the dwelling of the ARCHANGELS, ruled by the spiritual being RAPHAEL and the spirit OPHIEL, and was linked with the image of a serpent.

Sphere of Saturn In the Ptolemaic system, the Sphere in which dwell the beings of Saturn, and in which the orb of this planet is carried. In the medieval cosmoconception, the sphere was said to be the dwelling of the spiritual beings called the THRONES, under the rule of Zaphkiel, and linked with the spirit ARATRON. The Sphere is sometimes symbolized by the image of a dragon, and was said to mark the limits of time.

Sphere of the Moon In the Ptolemaic system, the Sphere in which dwell the beings and orb of the planet Moon – but see also EIGHTH SPHERE. In the medieval cosmoconception, the Sphere was said to be the dwelling of the ANGELS, under the rule of the Archangel GABRIEL, and the spirit PHUL. The Sphere is linked with the image of an ox.

Sphere of the Primum Mobile Sometimes called the PRIMUM MOVENS (First Mover) in the Ptolemaic system, this is the sphere which was supposed to give motion to all the other spheres which it enclosed. In the medieval cosmoconception, the Sphere was said to be the dwelling of the SERAPHIM, ruled by the spiritual being METATRON, and often linked with the image of a leopard.

Sphere of the Sun In the Ptolemaic system, the Sphere in which dwell the beings and orb of the Sun. In the medieval cosmoconception, the Sphere was said to be the dwelling of either the VIRTUES or the POWERS (but see SPHERE OF MARS), depending upon which system of Hierarchies was adopted. The Greek name EXSUSIAI appears to have been invariable, however. The Sphere is ruled by the spiritual being MICHAEL, and the spirit OCH, and is linked with the image of a lion.

Sphere of Venus In the Ptolemaic system, the Sphere in which dwell the beings and orb of the planet Venus. In the medieval cosmoconception, the Sphere was said to be the dwelling of either the PRINCIPALITIES or the

VIRTUES, depending upon which system of Celestial Hierarchies was adopted. The Greek name for these beings, the ARCHAI, appears to have been invariable, however, and this would therefore suggest 'principalities' as a fair translation. The Sphere is tuled by the spiritual being ANAEL and by the spirit HAGIEL. It is linked with the image of a man.

spheres   The early Greek and Roman cosmoconception postulated that the seven planets (which included the luminaries) were carried around the central Earth in a series of spheres. The body of the planet, often called the 'planetary orb' in later European systems, was carried along by the movement of that sphere —a movement transmitted to it ultimately by the SPHERE OF THE PRIMUM MOBILE, but regulated by spiritual beings. Because seven planets are described in the Ptolemaic system (which is itself an extension of the Aristotelian system) it is generally believed that there were only seven planetary spheres carrying the planetary orbs – however, this is far from the truth. The need to unite the Aristotelian dictum that 'planets move in perfect circles' with the observable movement of these bodies required a most sophisticated system of EPICYCLES and DEFERENT circles, and involved a complex series of spheres within spheres. By the 16th century, shortly before the Ptolemaic model broke down under its own weight of complexity, a model existed which postulated the existence of as many as 79 spheres. Even Aristotle, in his *Metaphysics*, required 55 spheres, though by the time Dante was putting the Arabian astrology into poetic imagery a 'working model' had been constructed with only 27 spheres. Under the impress of the theological theory of the CELESTIAL HIERARCHIES (derived from the pseudo-Areopagite) a system of spiritual beings, with named rulers, were established around the theory of the spheres. One important difference between the modern heliocentric system and the late medieval system inherited by astrology was the sequence of the planets: it was believed that Venus was nearer to the Sun, so that in most of the models of our system prior to the 17th century, the sequence of the spheres was presented in the order set out in figure 59. The numbers listed above the extralunar planets

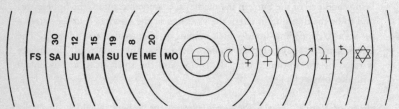

Figure 59: The order of the spheres in the medieval model, along with the planetary periods.

in this figure relate to the traditional PLANETARY PERIODS. The idea that it was the Earth itself which revolved, rather than the vast concentrics of spheres was suggested in Greek times, but was mooted in Europe in the 13th century by the Franciscan monk François de Meyronnes.

spherical descent See DESCENT OF THE SOUL.

spherical regulators See INTELLIGENCY.

Spica Fixed star (binary), the alpha of constellation Virgo, set in the ear of corn held by this female figure. Its symbolic importance is just about equivalent to its astrological importance, for the Christian astrologers linked Virgo with the Virgin Mary, and the Spica (ear of corn) in the arms of the asterism with Jesus, held in the arms of his Virgin mother. This imagery may be traced back to the Egyptian Isis, of course (see, for example, STELLA MARIS), a goddess linked with the constellation and sign Virgo. Variant names include Stachys, Arista, Arista Puella, and in the Arabic tradition, Alk Simak al A'sal, which Allen explains as meaning 'defenceless', in reference to the fact that the star appears to be unattended by others. From this Arabian notion come certain medieval names – such as Asimec, Acimon, Almucedie and so on. One of the names given to the star by the Egyptians was Repa, which means 'Lord', and Lockyer (now so often laughed to scorn by those who know less than he did) shows that the setting of Spica was a fiducial for the temple at Tel al Amarna. A few Grecian temples (including that ancient wonder at Ephesus, later the site for a cult of the Virgin Mary) were also orientated towards Spica. The astrological tradition has almost invariably associated an unfruitfulness with Virgo, and this is extended to her prime star, though it (like the asterism) also brings a noble disposition, an urge to protect the innocent, goodness and (not surprisingly) ecclesiastical preferment. Ptolemy says that it is of the nature of Venus conjunct Mars – a fair, if pagan, statement of the later Christian tradition. The name Spica was also applied to the 14th of the LUNAR MANSIONS in certain medieval lists, which no doubt used the star as fiducial.

Spider See ASTROLABE.

Spirits of Motion See DYNAMIS.

Spirits of Personality A name given by Steiner to the ARCHAI.

Spirits of Revolution of Time See ARCHAI.

Spirits of Wisdom   See DOMINIONS.

Spiritual Hierarchies   See CELESTIAL HIERARCHIES.

spiritual Sun   See ESOTERIC SUN.

splash   Term applied to one of the *JONES PATTERNS, in which the distribution of planets is generally even: this distribution is claimed to give wide interests, and lack of a one-sided nature.

splay   Term applied to one of the *JONES PATTERNS, in which a fairly even distribution of planets is given a focal point by the presence of a satellitium. This distribution is said to make for sturdy individuality.

spring signs   The zodiacal signs Aries, Taurus and Gemini.

square   An aspect of 90 degrees. Since planets (or nodal points) separated by 90 degrees must be in signs which are opposed in their elemental natures, it is clear that such planets must find themselves working inharmoniously; it is for this reason that the square aspect has gained such an ominous reputation in popular astrology: see ELEMENTAL ASPECTS. Serious astrologers, however, insist that the divergent natures of planets under the strain of square aspect may, when properly handled, lead to spiritual advantage and to creativity (creativity itself being a result of the reconciliation of inner tensions, or an attempt to make such reconciliation).

Sravana   Name given to the 21st of the Hindu NAKSHATRAS.

Sravishtha   Name given to the 22nd of the Hindu NAKSHATRAS.

Stachys   See SPICA.

Standard Time meridian   See TIME.

star   See FIXED STARS.

Star Calendar Zodiac   A zodiac of constellations of unequal length, originally specified approximately in its present form by Vreede and Schultz in 1929, and slightly corrected *c.* 1972, but based loosely on indications given by Steiner in 1912 in his own Star Calendar (*Sternkalender*) of that year, which reverted to an astronomical zodiac of unequal divisions. The extent of these asterisms, expressed in sidereal longitude to the nearest degrees, are given in

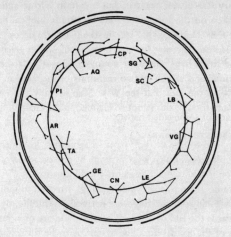

Figure 60: Star Calendar Zodiac (after Steiner and Powell).

the table provided by Powell under SIDEREAL CORRESPONDENCES, and indicated graphically in figure 60. The divisions of the asterisms, which may appear at first glance to be rather haphazard, were apparently derived from a careful comparison of star charts relating to the Ptolemaic listings, yet appear to have been largely justified by the work of Thun and others in connection with bio-dynamic farming – see SIDEREAL MOON RHYTHMS. Many of the charts used by the astrologer Sucher in connection with ASTROSOPHY are based on this zodiac.

star charts   See CONSTELLATION MAPS.

Star of Autumn   See SIRIUS.

Star of Bethlehem   Name given to the so-called 'star' followed by the Wise Men from the East (Matthew 2). Whilst most occultists are aware that the 'star' was actually a unique astral light, there have been many attempts by historians and astrologers to identify the star in ordinary astronomical terms. One explanation is that about 2 BC there was a conjunction between the CHRONOCRATORS – yet such conjunctions are far from infrequent. Other 'explanations' include the idea that the 'star' was a NOVA – but a nova does not adopt 'stations', as is indicated in the biblical description. Diodorus of Tarsus echoes the esoteric tradition when he says quite bluntly that this was not an ordinary star of the heavens, but a certain 'virtue and power' which had merely taken on the form of a star. In strictly biblical terms, the Wise Men say that they had seen 'his star in the East' (Matthew 2:2), and after their visit to Herod they saw the star once again, and it 'went before them, till it came and stood over where the young child was' (Matthew 2:9). This is not the behaviour of a star. The two mentions of the 'star' in verses 2 and 9 have given rise to the idea that the Wise Men actually followed the eastern star of Jerusalem, though this is not so claimed in the Gospel. See MAGIAN.

Star of Mary   See STELLA MARIS.

Star of the Sea   See STELLA MARIS.

star points   Name given to a series of readings attached to individual degrees of the constellational zodiac. They were derived psychically, by the astrologer Sepharial, towards the end of the 19th century, and are obviously linked with the influences of the fixed stars. The system is to be distinguished from the zodiacally based DEGREE SYMBOLS.

Star Royal   See REGULUS.

Starry Cup   A term derived from Macrobius, and usually claimed to relate to the constellation CRATER, though Macrobius himself says that his 'mystic secret' of the 'Starry Cup' is placed between Cancer and Leo, and refers the idea of the cup to the Bacchic intoxication 'first experienced by souls in their descent by the influx of matter into them'. This account is probably linked with the idea of the GATE OF BIRTH, located in Cancer.

station   Cognate with 'stationary', the term is used of those places in the orbit of a planet where it appears to be standing still, prior to direct or retrograde motion. When a planet is in its station, it appears to have no motion, and is said to be stationary: see, however, STATIONS OF THE MOON.

stationary   See STATION.

stations of the Moon   The motion of the Moon is always direct, so that properly speaking it has no STATIONS in the strict astrological meaning of the term, which implies a retrograde motion. However, the first and second *DICHOTOMES of the Moon are sometimes called her first and second stations. Sometimes the points where the Moon makes entry to each of the 28 LUNAR MANSIONS are also called stations or lunar stations: in this sense they are 'passage points' marked by fiducials.

staxeodos koure   See KORE.

Stella Dominatrix   The Latin for 'master star', used of the fixed star ALDE-
BARAN. According to Fagan this name was derived from the fact that the star
was the zodiacal fiducial for the ancients, from which (he argues) 'the oldest
and hence authentic zodiac began'. Certainly Aldebaran was one of the
ROYAL STARS of the Persians, and was called by the Babylonians Ku I-kur,
'the leading star of stars'.

Stella Maris   Latin term usually translated as 'Star of the Sea', and sometimes
applied directly to the Virgin Mary, though more properly to the star which
is associated with the Virgin Mary (most often depicted on her dress or upon
her head or *mapharion*). This star may be traced back to the prototype of
early Marian symbolism, in the Egyptian Isis who was also called 'Star of the
Sea'. Astrologically, the star on the clothing of Mary is derived from the
fixed star SPICA, the lucids of Virgo, which was of course also associated
with Isis and Mary. But see also POLARIS, and in regard to the Isidian
connection, see also SIRIUS.

stellar magnitude   From the very earliest times stars have been classified by
astrologers according to brightness (sometimes called 'apparent magnitude'
in modern times). Ptolemy, for example, records such magnitudes from
ancient traditions. However, magnitude tells us nothing about the actual
luminosity of a star, or about its diameter. In the modern astronomical
system, referred to in most astrological star-lists, the stellar brightness is
expressed by a formula according to which a star differs from another by one
degree of magnitude when it is two and a half times brighter or fainter than
the other. Thus, the star Wega is said to have a magnitude of 0.1 in
comparison with Polaris, which has a magnitude of 2.2. A star of the 6th
magnitude is only just visible to ordinary eyesight. See also ABSOLUTE
STELLAR MAGNITUDE.

stellar music   See MUSIC OF THE SPHERES.

Stellatum   One of the several names given to the Sphere of Fixed Stars in the
cosmoconception derived from the PTOLEMAIC SYSTEM. This was sometimes
called the Ninth Sphere, and the motion of the retrogression of the Vernal
Point, which was viewed as being either continuous or oscillatory (see
TREPIDATION) was said to be transmitted to it by the superior Primum
Mobile. A sphere beyond the Stellatum and this Mobile was considered
theologically necessary, as the 'eternal' realm of God and his ministering

angels, who dwelled in a sphere outside time, and in perfect motionlessness: this was called the Aetherial Sphere, sometimes called the TENTH HEAVEN. It was not until the 13th century that the astrologer Pietro d'Abano suggested that the stars themselves were not limited to a single concentric, but were spread out in space. See also CONSTELLATIONS and FIXED STARS.

stellium See SATELLITIUM.

stereographic chart See THREE-DIMENSIONAL CHART.

sterile signs See BARREN SIGNS.

strong signs The strong signs are traditionally said to be Scorpio and Aquarius, though every sign has its own strengths and weaknesses.

sub-duodenary measure An arc, or measure, used in SYMBOLIC DIRECTIONS, and (probably) originated by Carter. It is derived by dividing the DUODENARY MEASURE of 2.5 degrees by twelve, to obtain an arc of 12 minutes 30 seconds.

sub-ruler A term used to indicate a planet regarded as a secondary or even tertiary influence over a sign, decan or face. Whitman has made the so-called MODERN PLANETS sub-rulers in his own system of ASTRO-KINETICS.

subsistences A term derived from modern HARMONICS (probably) by the astrologer Addey. The subsistences of the number 7 (for example) are the series of harmonics which mark intervals of nine places from the 7th degree – that is, the 16th, 25th, 34th, etc. – all the numbers of which have digits adding up to 7. These subsistences of 7 are derived from a procession of nines. The subsistences of 4 are 13, 22, 31, and so on.

succedent Those houses which follow in succession the four Angles are called succedent: hence, the 2nd, 5th, 8th and the 11th houses.

summer solstice See SOLSTICES.

Sun In astrology, the solar body is termed a 'planet' simply because the traditions and nomenclatures attached to it were established during the time when the majority of astrological systems were geocentric. The sigil for the Sun ⊙ is sometimes said to represent the circle of spirit containing the central seed or nascent soul – though this sigil may not be traced back in such an astrological context beyond the 15th century: see SIGILS – PLANETARY. The

Sun period

Sun has rule over zodiacal Leo, is Exalted in Aries, and has its Fall in Libra. It represents the conscious element in the native, and is usually studied as providing an index of the creative self-expression of the personality. An emphatic Sun in a chart is usually an indication of creative or artistic ability, as of a physical dignity and self-reliance, and often of a generous, affectionate and commanding nature. A badly placed Sun tends towards excess, so that the native is often showy, ostentatious, over-confident, egotistical, selfish and dictatorial. Traditionally, the Sun rules the heart of the melothesic man, and is also connected with the blood, even though it is the opposite sign Aquarius which properly rules the circulation of the blood. The Sun is a useful index of how the vital forces work through the physical body – this because spiritual (as well as physical) energies are derived ultimately from the Sun. The solar type is one who finds such energies easily, and is able to transmit them for the benefit of those around within the immediate environment: he or she is therefore often involved with activities demanding exertion and authority, even with creative enterprises on a large scale, as well as with outlets associated with the FIFTH HOUSE. Solar types seek positions where power, dignity and responsibility may be combined: they make good executives, and whilst generous are not always good accountants or even very careful with money (especially the money of other people). The metal of the Sun is gold, the ruling Archangel is MICHAEL.

Sunbula   As Sunbula, the Arabic name for Virgo.

Sun-forces   A term which originated in the esoteric astrology favoured by Stein, based upon a specific cosmic interpretation of the dualism which pervades all occultism. Stein pictures the fundamental contrast underlying phenomena as the working of the dual Sun-forces and the Moon-forces: the former radiate and liberate, while the latter concentrate and wither. In fact Stein appears to use the terms as modern equivalents for the power of Ormudz (the modern name for which is often given as LUCIFER) and Angrimayu (the modern name for which is often given as Ahriman), derived from the ancient dualism of Zoroastrianism.

Sun period   A term which has nothing to do with the cycle of the Sun, or indeed with any SOLAR PERIOD. As a specialist term it is derived from modern esoteric astrology relating to cosmogenesis. The Sun is said to be the second of the planetary states in the genesis of our world: it was initially a gaseous ball of warm light, from which the present solar system eventually evolved. This early period was overseen by the Hierarchy of CHERUBIM. The Sun period is sometimes called Old Sun.

Sun revolution   Term derived from modern esoteric astrology, relating to cosmogenesis (see SUN PERIOD). It is said that the path of evolution of our present solar system, with its periods of activity and prelaya (sleep), as indeed the growth of every organic being, must recapitulate, however briefly, the entire sequence of evolution. Since the first stage of the cosmogenesis of our present system began in Saturn (see SATURN PERIOD) and reached its second stage in the SUN PERIOD, the term 'Sun revolution' pertains to the second of these recapitulations, which is essentially one of 'warmth-light'.

Sun-sign astrology   A term used to denote a popular form of so-called astrology which is based on the interpretation of character, career and psychological disposition, from the position of the Sun in regard to the tropical zodiac, to the exclusion of all other astrological factors. This system is said to be the second most popular form of astrology in the world – no doubt because of its very simplicity, and for the ease with which it has been adapted for journalistic exploitation. It is interesting to observe that recent judicial interpretation in the English law courts has determined that the popular Sun-sign system may not be properly described as being part of genuine astrological practice.

super chart   A term originated (probably) by Dean to denote a chart which incorporates all known astrological data in one figure. Such a chart would be entirely unreadable by ordinary means. At a very minimum, all the Angles, planes, hypotheticals, nodes, parts, interactions of major and minor aspects, plus antiscions and contrascions, parallels, mid-points, and relevant progressions, would produce a chart which would be a mere plethora of sigils. The hyper chart, also mentioned by Dean, would be even more daunting, for this is conceived as including (in addition to the above) interfaces, perigrees, eclipses, lunations, satellites, fixed stars, etc., plus the imposition of derived figures such as the heliocentric equivalents, pre-natals, harmonics, and so on. The concept of super chart and hyper chart is interesting if only because it indicates how any system of astrology must be selective if it is to be workable.

supercycle   A somewhat misleading modern term applied to a series of (approximately) six sunspot cycles in a periodicity of 132 years.

Superior Midheaven   An unnecessary term, as the MIDHEAVEN is always superior in the sense of location. It appears to have been used to distinguish this Angle from the so-called Inferior Midheaven, which is really a confusing circumlocution for the IMUM COELI.

superior planets   Originally this was a term applied to the three planets beyond the Sun – Mars, Jupiter and Saturn: the term is therefore a relict of the Ptolemaic geocentric system, relating to positions 'higher' than the Sun Sphere in regard to the Earth. See also INFERIOR PLANETS.

supermidpoint A term used to designate the ultimate mid-point, which arises after taking the mid-point of mid-points of mid-points, etc.

Supernatural Ilech Esoteric name for the invisible influence which links the material realm with the stellar realm – but see ILECH.

Surya Sanskrit word for the Sun, as used in Hindu astrology.

Svati Name given to the 13th of the Hindu NAKSHATRAS.

sweet signs Terms used of the AIR SIGNS.

swift A planet may be said to be 'swift of motion' or 'swift in motion' when its speed over a period of 24 hours exceeds its own mean motion.

swift in motion Any planet may be said to be 'swift in motion' when it is SWIFT, but usually the term is reserved for the Moon, when in a motion faster than its mean rate of 13 degrees and 10 minutes. See also INCREASING IN MOTION.

Sword In a specialist sense, one of the so-called Arabian PARS, the Part of Mars, sometimes called the Point of Passion. If a natal chart is revolved so that the degree occupied by the Sun is on the Ascendant, then the position of Mercury in this new arrangement will mark the Sword Point.

sylphs The class of ELEMENTALS linked with air. They correspond to the Nenuphars of esoteric lore, while in the astrological correspondences of the CALENDARIA MAGICA they are often called Aerei.

symbol The term symbol is used very loosely in astrological circles to cover such different concepts as IMAGES, SIGILS, SEALS and even SIGNS. In a strictly astrological context, a symbol is something which stands for, or represents, something else: however, a true symbol is intended to be more than merely a representative, for it should in some way act as a sort of commentary on the nature of that thing. The true nature of symbolism is perhaps most profoundly expressed in the Egyptian 'medu-neters', which the Greeks called 'hieroglyphics', for each symbol used had at least three related meanings (one strictly spiritual, linked with the equivalent of our archetypes, one linked with the potential of the thing symbolized, and a third linked with the hitherto unrealized potential of the thing in itself). In addition to these united triadic meanings, the symbol was intensified by the name ascribed to it, which (in Egyptian symbolism at least) was linked with a spiritual cosmic language which would now be regarded as purely magical. Such an ancient view of symbolism – set out by de Lubicz – is to some extent still preserved in esoteric circles, but has been largely lost to modern thought, so that the union which once existed between the image, sigil, seal and sign (which found a single form in hieroglyphics) has been subject to fissiparous tendencies, so that whilst each word has obviously a common ideological reference, each now refers to a different system of symbolizing – the image through representation, the sigil through graphic symbolism, the seal through magical and numerologically originated symbols, and signs through a (formerly) numinous symbolism. This fragmentation of the original unity of the symbol has given rise to many misunderstandings of ancient astrological symbolisms – for example, among medieval astrologers there was a tendency to infuse Christian symbolism with the old magic of Seals: Arnaldus de Villanova, for example, would link the Christian image of the Lamb of God with the 'lamb' (actually the Ram) of zodiacal Aries in his magical amulets. Again, under the impulse which sought to trace Egyptian origins for the astrological glyphs and sigils, early post-medieval scholars suggested that the ram-horned Ammon was linked with Aries – see, for example, the 'etymological' suggestions of Kircher. However, a more profound knowledge of the medu-neters, and of the union between sound and symbol (which lay at the basis of Egyptian imagery) would have led not to Ammon but to the neter Khnum, who was also ram-horned, and who was specifically linked with the archetype of Aries. The error of linking Aries with Ammon gave rise to a basic misunderstanding of the nature of Aries itself. The term 'symbol', then, is often used in astrological contexts in a rather vague way, to represent almost any of the special forms of the symbol, and a definition of the term must remain a general one. Perhaps the most apposite definition is that provided by the historian Haase, who writes of the symbol as being 'essentially purposive' in that 'it points to some Higher Order for whose characteristics it is a kind of abbreviation'. See ABBREVIATIONS, IMAGE, SEAL, SIGILS, SIGN.

symbolic degrees   See DEGREE SYMBOLS.

symbolic directions   A name given to a method of astrological prediction based on symbolic movements in which the measurement and timing of events is derived from a study of mathematically defined arcs, rather than planetary motions. In primary and secondary progressions, the movements of the planets are regarded as revealing the unfolding of the future from the potential contained within the radical chart, and these are therefore methods based on measurable events within the heavens. Symbolic directions take no such measures as the basis for prediction, however, and rely entirely upon artificial arcs. The astrologer Carter distinguishes several such 'arcs of symbolic direction', five of which are set out in table 80. Three further

--------------------------------- TABLE 80 ---------------------------------

Name of arc	Division of	Division by	Arc
Duodenary measure	Sign	12	2.5 deg.
Naronic measure	Zodiac	600	0.6 deg.
Novenary measure	Sign	9	3 deg. 20 min.
Septenary measure	Sign	7	4⅔ deg.
Sub-duodenary measure	Sign	144	12.5 min.

measures used are FRACTIONAL METHOD, MEASURE OF NAIBOD and the ONE DEGREE METHOD. It will be observed that the SEPTENARY MEASURE above is the same as the POINT OF LIFE used by the astrologer Frankland.

symbolic Moon   An alternative term for the Arabian PART OF FORTUNE.

sympathy   In its specialist astrological sense, this term appears to have two separate applications. The word is used to denote that force, subtle influence, or spiritual empathy, which links all things in union within the CHAIN OF BEING. The same term is used also to describe individual relationships between creatures and things: all sublunary things are visualized as being either 'in sympathy' or 'in antipathy', depending upon the elemental, planetary and zodiacal forces involved. Thus, a dog (under the rulership of Mars) is not in sympathy with a cat (under the rulership of Venus), and so on. This system of division extends even into the heavens, so that planets and elements were represented as being in sympathy or in antipathy (see FRIENDLY PLANET and UNFRIENDLY PLANETS), and it is indeed this simple view of nature which is the tacit rationale behind many of Ptolemy's arguments, while the important law of CORRESPONDENCE is itself derived from the dualistic structure. The twofold meaning of the term has given rise to some confusion: it is true to say that all things are in sympathy, in the first sense that all things are working towards the same end within the Chain of Being, when viewed in the light of the eternal plan of things. Yet in the second sense, it is possible to say that some things are not in sympathy when viewed in the light of transient experience, on the mundane plane. In popular astrological literature, the two distinct uses of the term are often merged and confused.

synastry A term derived from the late Latin *synastria*, a neologism from the Greek meaning approximately 'similarity of stars'. It denotes that branch of astrology which deals with the study of agreements and disagreements reflected in pairs of charts cast for two people (or in charts cast for groups of people). Very often synastry is involved with comparing two charts with a view to establishing compatibility between people, or for studying the causes and likely outcomes of crises between people. It is widely recognized among astrologers that the charts of people who are married, in a specific relation, in love, or otherwise karmically connected, will have certain zodiacal degrees in common. Synastry, in that it is the astrological study of human relationships, may be extended from ordinary radical chart interpretation to the study of progressed charts.

synchronicity A term originated by the psychologist C. G. Jung to designate a psychic parallelism of events which cannot be related causally. The parallelism, or 'connectedness', between such non-causally related events and happenings, he termed 'synchronistic'. Jung quotes astrology as an excellent example of manifest synchronicity – though many astrologers would maintain that astrology is in fact causalistically based, when it is understood in the light of esoteric truths. Jung himself recognized astrology as representing 'the summation of all the psychological knowledge of antiquity', and so it is difficult to understand why he should feel the urge to coin a new term to represent the relationship between the macrocosm and the microcosm. Could it be that the term 'synchronicity' is to be applied to those parallelisms in events which cannot be understood by the intellect? The interesting thing is that the origin of the principle of synchronicity is linked with Jung's (somewhat limited) acquaintance with the Chinese philosophical machine, the I Ching – the Chinese themselves regard the working of the oracle not in terms of an oriental equivalent of synchronicity, but in terms of the operation of spirits: i.e. animistic causality!

synetic Vernal Point See SOLAR APEX.

synodical lunation A term applied to a lunation figure progressed chart, cast for the time when the progressed Moon establishes the same arc from the progressed Sun as that held between the radical luminaries.

synodical returns A term used as an equivalent of LUNAR RETURN, applied to a method of determining likely future events from the monthly return of the Moon to the same elongation (arc) from the progressed Sun as the two bodies held in the radical chart. These 'months' are said by some astrologers to equal a year of life, and the figures are called 'lunar revolutions', presumably to distinguish them from the 'lunar returns', which are concerned with monthly predictions.

synodic cycle A recurrence cycle between two successive conjunctions of the same two planets. The synodic period for Jupiter–Saturn is 19.860 years, for Saturn–Uranus 45.363 years, for Uranus–Neptune 171.403 years, for Uranus–Pluto 127.280 years, and for Neptune–Pluto 49.328 years. See CYCLES.

synodic month A lunation period of 29 days, 12 hours, 44 minutes and 2.8 seconds, the synodic period between lunations.

synodic period See SYNODIC MONTH.

Syntaxis See ALMAGEST.

System A Babylonian astrologers used two different methods for describing the solar motion and its related phenomena, and certain of the implications arising from these two systems continued into tabulations and methods used in Hellenistic astrology, and even into the late medieval forms. They have been called by modern historians 'System A' and 'System B'. In System A, the AYANAMSA is 10 degrees, and in System B it is 8 degrees. These differences are reflected in (among other things) the computations for the earliest systems of the climata (see CLIMA), each of them being linked specifically with localities for Babylon and Alexandria, though later incorporating computations for Rome. The traditional so-called *SEVEN CLIMATA used an ayanamsa of zero degrees, following the method of Ptolemy, who also computed according to spherical trigonometry.

System B See SYSTEM A.

syzygy A term derived from the Greek meaning 'a yoking together', and used most especially of the conjunction between the Sun and Moon. Sometimes the term is extended (with etymological imprecision) to include oppositions between the luminaries. It is a term entirely misused when applied to conjunctions (or oppositions) involving planets other than the luminaries. The term 'previous syzygy' often encountered in predictive work connected with a radical chart refers either to the previous conjunction of the luminaries, or to the opposition, whichever was nearer to the birth time.

T

Table of Houses A set of figures, usually arranged in columnar form, designed to facilitate the calculation (from known sidereal times) of the degrees arising on the Ascendant and culminating on the Midheaven, along with intermediate house cusps, for different latitudes.

table of Naibod A table setting out the MEASURE OF NAIBOD is often published in books dealing with progressions and symbolic directions, yet such a table is scarcely necessary in that it consists basically of the separate additions of the annual arcs, and of the monthly arcs, according to the requirements of the measure. The annual arc is 59 minutes, 8 seconds: the monthly arc is 4 minutes, 56 seconds. The 6-second annual increment is irrationally catered for within the tables, but is of slight importance in predictive work by means of SYMBOLIC DIRECTIONS.

taijak A Sanskrit term applied to the Hindu method of computing aspects according to geometric angles, similar to the traditional method of viewing ASPECTS in Western astrology: see, however, ELEMENTAL ASPECTS.

Takata effect A name given to an effect observed and described by the scientist Takata in relation to the flocculation index of blood serum which varies by sudden increase at sunrise (with an orb of 15 minutes of time on either side of the actual local sunrise). Similar increases have been observed when the Earth moves into line with a group of sunspots: a reduction effect is noted during solar eclipses, maximum reduction being at total eclipse.

talismanic magic From the very earliest times the various astrological SIGILS, SEALS and IMAGES have been engraved or stamped on metal or pottery talismans to make portable charms. A complex procedure was supposed to attend the production of the genuine talisman – not only had it to be made under the correct astrological conditions (under the appropriate planetary hour, and so on), but it had to be manufactured in the metal or gem related to the planetary or stellar forces being evoked in the charm. Wherever possible many of the more potent images, sigils, symbols and names were combined into one talismanic form. The result is that there are literally thousands of medieval manuscripts dealing with the art of talismanic magic which, in one form or another, are linked with traditional methods of astrology. A complex art of talismanic symbolism has been attached to both constellations and fixed stars – in connection with the latter, see FIFTEEN-STARS – post-medieval sigillic forms for these may be studied under SIGILS – FIXED STARS. The talismanic use of seals, sigils and images relating to the DECANS was highly complex, and at times resulted in the most lovely symbols, a few of which have been recorded pictorially by Nowotny: some of the talismanic lore relating to the Egyptian decans appear to have survived even into Renaissance symbolism. The use of zodiacal and planetary sigils and images for talismanic ends is almost in itself a chapter in the history of astrology, but very many of the medieval astrological manuscripts touch in one way or another on the magical power or significance of the basic sigils, as the writings of Agrippa (mainly eclectic) confirm. For specific examples of talismanic bases, see PENTAGRAM, PENTALPHA, SEAL OF SOLOMON and TETRACTYS.

Tammuz   See BABYLONIAN CALENDAR.

Taphthartharath   Name given by Agrippa (quoting ancient qabbalistic sources) to the Daemon of Saturn, for whom he gives the magical number 2080, which is the multiple of the number of Mercury (8) and the linear addition of the Mercurial MAGIC SQUARE (260). See also TIRIEL.

Tarandus   See RANGIFER.

Tarf   Al Tarf, name given to the 7th of the Arabic MANZILS.

Tarot   See ASTROLOGICAL TAROT.

Tascheter   See ROYAL STARS.

Taurids   See METEORS.

Taurus   The name of the second sign of the zodiac, also used to denote one of the zodiacal constellations. As a sign, Taurus corresponds neither in location nor significance with the constellation of the same name: see TAURUS CONSTELLATION. The sigil for Taurus ♉ is said by some to be a drawing of the head and horns of the bull, with which image Taurus is associated. However, in the esoteric tradition it has been suggested that this is a vestigial drawing of the larynx, with the Eustachian tubes, the sigil thus

Taurus constellation

linking the ideas of speech and hearing. This has been used as an explanation as to why the image of the bull is sometimes linked with Christ, for within both Taurus and Christ are contained the idea of the *logos* or 'Word', as well as the idea of blood sacrifice. Taurus is of the earth element, and of the Fixed Quality, the influence being constructive, strong-willed, conservative, slow, practical and sensuous. The nature of Taurus, as it manifests in human beings, is expressed in the many keywords which have been attached to it by modern astrologers: practical, slow, heavy, possessive, conservative, intractable, relentless – in a word, all those qualities which may be associated with an Earth type expressing itself with determination. In excess the Taurean nature may be described in terms which express an underlying inertia, curiously married with a sort of magmatic violence which erupts when the subject is goaded: overbearing, deeply jealous, fixed in attitude, tyrannical, coarse, gluttonous, repetitive, selfish and unexpectedly violent or cruel. Taurus says 'this is mine' to the world more insistently than any of the other types – though he may be gentle or hard in his possessiveness, depending upon which side of his nature is being called into being. Taurus is ruled by the planet Venus, which accounts for the fact that so many Taureans are fair in form, and are so frequently associated with a beautiful or rich voice. The sign marks the Exaltation of the Moon, and (according to modern schemas) the Fall of Uranus. In common with the other three Fixed signs of the zodiac, Taurus is linked with the TETRAMORPH, and specifically with the Evangelist Luke.

Taurus constellation   Constellation linked in name, if not in influence, with the zodiacal Taurus, located approximately between 17 degrees of Taurus and 23 degrees of Gemini, and straddling the equator from 2 degrees south to 34 degrees north: the representative constellation map in figure 61 is derived from data provided by Delporte in his MODERN ZODIAC. In popular language this asterism is often called the Bull, and sometimes (misleadingly) Veneris Sidus (Star of Venus), as it is the asterism associated with the rule of that planet over the sign. In the majority of medieval astrological texts it is linked with the Greco-Cretan myth of Europa, whom Jupiter carried off after he had adopted the form of a bull – however, the link with a bull was established in much earlier Egyptian and Babylonian astrology. The connection with the myth did give rise to several variant names, such as Portitor, Proditor (actually *proditor Europae*), while the fact that the stellar image was often drawn as half a bull gave the term Sectio Tauri. Among the many names furnished by the ancient Egyptians was Hapi, one of the Neters of the Nile. The Akkadian Te Te, given as Bull of Light has been explained as a dual form in reference to the two important asterisms HYADES and PLEIADES within the constellation. The representation of the bull in figure 61 is from the 13th-century SAN MINIATO ZODIAC which is orientated to a Taurean FOUNDATION CHART. In this latter zodiac, as in much astrological imagery, the bull of Taurus is associated with the Mithraic cult, the sign and constellation sometimes being termed the Mithraic Bull. The fixed star within the constellation which is regarded as being of particular importance within the astrological tradition is ALDEBARAN.

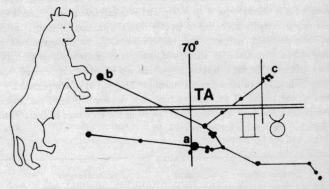

Figure 61: Constellation map of Taurus from the MODERN ZODIAC *(after Delporte) – the image of Taurus is from the 13th-century* SAN MINIATO ZODIAC *(after Gettings).*

Taurus Poniatovii   A constellation formed in 1777 by Poczobut of Wilna, named after the King of Poland, Poniatowski (hence sometimes it is the Polish Bull). It is said to induce obstinacy, changeability, honour and renown. The constellation appears to have several earlier recognized forms – once it was part of the asterism called River Tigris, and once it was set out as a triangular figure.

Taygete   See PLEIADES.

T-cross   Another term used for the so-called COSMIC CROSS. See T-SQUARE.

Tebit   See BABYLONIAN CALENDAR.

Tejat   See PROPUS.

Telescopium Constellation, the Telescope or the Tubus Astronomicus, formed by La Caille between the constellations Ara and Sagittarius in 1752. It is said to give a keen mind and prophetic ability, along with an interest in history and the occult.

Telescopium Herschelli Constellation formed in 1781, in honour of the astronomer Herschel, but now obsolete.

temperament The Grecian medical theory of HUMOURS which so radically influenced the development of early astrological theory was based on a fundamental tetrad which linked bodily fluids with certain mental dispositions, the so-called 'tempers', or temperaments. These temperaments were distinguished eventually as the CHOLERIC, MELANCHOLIC, PHLEGMATIC and SANGUINE, the etymologies for which are in each case derived from a corresponding Greek term for the humour. This tetrad of temperaments was linked to the tetrad of ELEMENTS (through the mediation of the so-called PRINCIPLES), and was accordingly reflected in the structure of the zodiac, as indicated in the schema set out in table 81.

TABLE 81		
Temperament	*Element*	*Zodiacal signs*
Phlegmatic	Water	Cancer, Scorpio, Pisces
Melancholic	Earth	Taurus, Virgo, Capricorn
Choleric	Fire	Aries, Leo, Sagittarius
Sanguine	Air	Gemini, Libra, Aquarius

Templum See ARA.

temporal chart See TOPOCENTRIC SYSTEM.

temporal houses A name applied to the 2nd, 6th and 10th houses. See also POSSESSIVE HOUSES.

ten chains See SCHEME OF EVOLUTION.

Tenth Heaven Name given to the outer one of the SPHERES, which was regarded as being totally motionless. It was sometimes called the Empyrean, though this term (at least after the medieval period) was sometimes used of all the Spheres beyond the Lunar Sphere. In the early diagrammatic and literary accounts of the Spheres, there was no Tenth Sphere, no Tenth Heaven, the Ninth Sphere being the Primum Mobile. Eventually however, perhaps for theological reasons, the PRIMUM MOVENS was separated from the Ninth Heaven (at one time the Stellatum) to become a distinct sphere beyond it. The classification of the basic spheres (in the outer limits, at least) therefore depends upon context, and upon the period under review. The theological need for this outer shell, as a sort of final enclosing of the concentrics of the PTOLEMAIC SYSTEM, found its most perfect exposition in the *Convivio* and *Commedia* of Dante: it was regarded as the abode of God, to which all other spheres 'aspired', their motion being a result of this aspiring love, to wish to participate in this '*ciel della divina pace*' (heaven of divine peace).

Tenth Hierarchy   The realm of man – the lowermost of the Hierarchies (see CELESTIAL HIERARCHIES).

tenth house   The tenth of the astrological HOUSES, its cusp marking the all-important Midheaven in many of the house systems (one exception being that of the MODUS EQUALIS). The 10th house is linked with Capricorn, and has dominion over the status of the native, over the honours of his professional life and career. It relates to the father of the native, and was indeed referred to in early forms of astrology as the House of the Father at a time when the career was so often determined by the father as fiat, or by the tendency to follow a family trade. In the melothesic man, the house is connected with the knees and the skeletal system – with that which lends a structure to the human body. In MUNDANE ASTROLOGY it marks the political controllers, eminent persons, the judge, president and chairmen, as well as communal cultural ideas.

Tenth Sphere   See TENTH HEAVEN.

Terebellum   See TETRAPLEURON.

terminal house   An unfortunate term used of the houses which correspond in the basic horoscope figure to the WATER TRIPLICITY – that is to the 4th, 8th and 12th houses. It would appear that they are so called because they each govern the 'terminations' of certain human activities: the 4th is linked with the end of the physical being of man; the 8th with death and regeneration, and the 12th with the end and purpose of his personal life.

termini   See TERMS.

terms   The planetary terms consist of a series of zodiacal arcs to which are assigned sub-ruling planets which are supposed to influence the nature of these arcs. These subdivisions, unlike those of the DECANS or the FACES, are not of equal arc. They appear now to be used only in horary astrology, yet Ptolemy (who listed three different systems of such *oria*, often translated into Latin as *termini*, and more accurately rendered into English as 'boundaries') seems to have believed that they were valid also for natal astrology. It must be observed, however, that Ptolemy was sceptical about the terms: even so, he lists the arcs and rulerships for the Egyptian terms and for the Chaldean terms, and then reconstructs a further system, derived from an old manuscript which had come into his possession, which might conveniently be called PTOLEMAIC TERMS. Those astrologers who have at times attempted to introduce their own tables of arc termini – such as Wilson, Sepharial and Alan Leo – show little agreement in either arcs or rulerships, and the whole field is now extremely complex. One problem which has arisen is the failure of astrologers to note that Ptolemy was writing about three separate systems, and did not appear to be happy about any one.

ternary of fire   A term derived from Boehme's esoteric astrology and linked with his SEVEN PROPERTIES: it is the ternary of the first three principles (Saturn, Mercury and Mars) which is a ternary harmonized by the Sun, and also balanced against the TERNARY OF LIGHT, as set out in the schema below:

Saturn	Mercury	Mars	SUN	Venus	Jupiter	Moon
	Fire				Light	

The ternary of fire is sometimes called the 'ternary of desire', whilst the ternary of light (VE, JU, MO) is sometimes called the 'ternary of love'.

ternary of light   See TERNARY OF FIRE.

Terrabellum   See TETRAPLEURON.

terrestrial ecliptic   A term coined to distinguish the orbit of the Moon, which marks the terrestrial ecliptic from the celestial ecliptic, which is the ECLIPTIC proper.

terrestrial equator   The great circle of the Earth which is at every point equidistant from the two Poles is called the terrestrial equator.

terrestrial meridian The standard modern terrestrial meridian is the great circle passing through Greenwich and the Poles.

territorial system A method of *HOUSE DIVISION based on the division of the Earth into twelve equal portions by lines passing through the Poles, linked with the ZENITH SYSTEM. Like the latter, the system was proposed by the astrologer Tucker.

tertiary method A term applied to a method of *PROGRESSIONS involved with the projection of lunar returns on to the radical chart. One system is based on adding a mean sidereal day to the birthday for every return of the Moon to its radical place, a tropical month being regarded as 27.3216 days. Another method equates the tropical year with the tropical month.

Teshrit See BABYLONIAN CALENDAR.

testimony The English translation of Ptolemy's specialist use of the Greek *marterion* (proof), a word which is cognate with our 'martyr'. Ptolemy was quite clearly using it as a synonym for ASPECT, or the act of projecting aspectal rays (see ACTINOBOLISM). However, the word has been misunderstood in modern times, and Devore defines the term as a 'partial judgement' or 'synthesis', which, within an astrological context, is quite inaccurate.

Testudo See LYRA.

Tethys Name given to one of the moons of Saturn.

Tetrabiblos The short title given to the four books of astrology ascribed to Ptolemy, consisting of a rationalized compendium of astrological lore which, even by the 2nd century of our era, was already old. The book is often called after its Latin title, the *Quadripartium*, though as Robbins says, it should more properly be called the *Mathematical Treatise in Four Books*, yet many manuscripts use the title (rarely the incipit) *The Prognostics' address to Syrus*. It is one of the most influential of the ancient systematic treatises on astrology, and enjoyed almost the authority of the Bible among the astrological writers of a thousand years or more. Many of the astrological terms and traditions still used in modern astrology have been derived from the *Tetrabiblos*, though a large number are now both misunderstood, misused and archaic. Not all the ideas and traditions ascribed to Ptolemy are found in the *Tetrabiblos*, or even in his other texts, mainly because the so-called Ptolemaic astrology is really an accretion of traditions established around his systematized geocentric cosmoconception. It is a regular occurrence for astrological writers to ascribe ideas to Ptolemy which he would scarcely have recognized, let alone promulgated.

tetractys A name given to a symbolic arrangement of ten points (see diagram below), which, whilst not originally of astrological import, has in one way or another been incorporated into astrological symbolism – especially so in regard to the astrological number-symbolism. Whilst the tetractys is usually traced back to the Pythagorean school, the best exoteric treatment of its symbolism is found in Philo Judaeus, who presents it as merely a matter of numerical sequence, as one of the 'numerical triangles', which have a significance mainly in regard to their corresponding 'numerical squares'. In the tables recorded by Agrippa at the end of the 15th century, 3 is the number of Saturn, 4 is the number of Jupiter, 6 is the number of the Sun, 9 is the number of Saturn and 16 the number of Jupiter. The 10 of the tetractys is not linked in a numerical way with the planets. However, this merely points to a much deeper level of numerical association, which lies at the very roots of astrological numbers. The learned Kircher shows that the tetractys is to be understood in regard to the qabbalistic numerology attached to the tetrad of the word INVH (Jehovah), of which the corresponding numbers are given below:

.	1	H = 10	
. .	2	VH = 15	
. . .	3	HVH = 21	
. . . .	4	IHVH = 26	

The total by means of this association of numbers is 72 – the magical number of the tetractys, and one of the most pure of astrological symbolic numbers. This magical 72 is linked with the pulse rate of the human blood (the microcosmic 72), and with the fact that in 72 years the Sun retrogrades in PRECESSION exactly one degree (the macrocosmic 72): Wachsmuth has dealt with the number within a framework of modern occultism. As Blavatsky says, the tetractys 'has a very mystic and varied significance'. It is sometimes called the 'tetrad' or the 'mystic tetrad'.

tetragon In a specialist sense, a term used as synonymous with SQUARE.

tetragonous In a specialist (an archaic) sense, used as a synonym for 'quartile' or 'relating to the square aspect'.

tetrad   See both TETRACTYS and TETRAMORPH.

tetramorph   A term derived from the Greek, meaning 'of a fourfold form', but in astrology applied to a composite figure which unites in a single form the images of the four Fixed signs of the zodiac, themselves linked with the four elements. The so-called Christian tetramorphs are not really tetramorphs at all, but four separate figures combining imagery related to the four Fixed signs of the zodiac, and associated by Christian apologists with the imagery mentioned in the Bible in Revelations 6:6–8 and Ezekiel 1:5–10, an imagery which is in any case derived from ancient pre-Christian astrological concepts, as Neuss has shown. The schema in table 82 sets out the traditional imagery of the Christian lore which has been associated with the genuine tetramorph:

TABLE 82

Attribute	Evangelist	Element	Zodiacal sign	Archangel	Symbol
Human face	Matthew	Air	Aquarius	Raphael	Dragon
Lion head	Mark	Fire	Leo	Michael	Lion
Bull head	Luke	Earth	Taurus	Uriel	Bull
Eagle head, or wings	John	Water	Scorpio	Gabriel	Eagle

When this table of associations is linked with the tetramorph proper, then it is seen that the wings represent the water element rather than the air element (as might be imagined), since the wings are associated with the Eagle of Scorpio – the symbol of the regenerate Scorpionic nature – see EAGLE. An important change in symbolism was that linked with the adaptation of the image of the winged human water-bearer of Aquarius to symbolize Matthew, for the water urn was transformed into a book (both urn and book being bearers of spiritual energy, the latter linked with the *logos*, or word).

Tetrapleuron   The name given by Ptolemy to an asterism in the form of a quadrangle in the hindquarters of the horse of constellation Sagittarius: Bayer gave the Greek in the meaningless Latin 'Terebellum'. The fixed star Terrabellum, the omega of the asterism, was said by Ptolemy to be of a nature similar to that of Venus with Saturn, the same as the asterism as a whole.

Thawr   Al Thawr, the Arabic name for Taurus.

Theban calendar   A name sometimes given to a system of DEGREE SYMBOLS attributed to the French occultist Christian. The system has not survived (though Sepharial claims that it was used by the occultist Levi). It is possible that the system has been confused with the readings of the DECANS (which have survived) that Christian derived mainly from the occultist Kircher: if this is the case they should properly be called 'decan symbols', and would have had quite a different application from degree symbols. On the other hand, the system may have been confused with his system of KEYS, which Christian derived from Firmicus, Junctinus and Pelusius.

thema   In general use a synonym for CHART – really a 'theme of the Heavens'. Though no early definitions appear to be available, certain figures called 'themae' were actually schematic horoscopes or figures, as for example in the many 'themae mundi' (world charts) which were not so much horoscopes as schematic presentations of the heavens (often cosmically impossible) at the supposed beginnings of the world – in other words, the charts for creation.

Thierens-Pluto   Name developed in comparatively recent times to distinguish the planet Pluto (also named Osiris) described by the esotericist Thierens from the several other Plutonic designations, such as PAGAN-PLUTO, PLUTO and WEMYSS-PLUTO. The Thierens-Pluto related to the future ETHERIC SPHERE.

Third Heaven   Usually this term is used synonymously with the Third Sphere in the concentrics of the Spheres, which in the cosmoconception built around the PTOLEMAIC SYSTEM was the Sphere of Venus, associated in the Christian hierarchy with the ARCHAI. In the days when the sequence of the Spheres was promulgated it was believed that Venus was nearer to the Sun than Mercury.

Third Hierarchy   See CELESTIAL HIERARCHIES.

third house   The third of the astrological HOUSES, linked with the nature of Gemini, and representing the communicative ability of the native, his mental outlook, as well as all blood ties (excluding parents). It is the ruler of short journeys (in medieval astrology, journeys which might be made in less than one day) and was sometimes called the House of Short Journeys. It rules the lungs and arms in the melothesic man, and is often used as an index of the memory capability of the native. In * MUNDANE ASTROLOGY it governs inland traffic, communications systems, public opinion and public relations.

It was sometimes called the House of Brothers and Sisters in connection with the rule over 'blood ties'.

third ray   In the esoteric astrological system described by Bailey, the second of the *SEVEN RAYS is called the Ray of Active Intelligence, and is linked with the will to evolve, and with that which develops sensory perception into knowledge, knowledge into wisdom. This third ray is said to arise in Ursa Major, and is transmitted to our solar system by means of the zodiacal constellations (actually signs) Cancer, Libra and Capricorn, through their corresponding pairs of ORTHODOX PLANETS and ESOTERIC PLANETS according to the schema set out in table 83. The entire third ray is linked with the

──────── TABLE 83 ────────		
Constellation (sign)	*Orthodox planet*	*Esoteric planet*
Cancer	Moon	Neptune
Libra	Venus	Uranus
Capricorn	Saturn	Saturn

sacred Saturn – see, however, SACRED PLANETS. Distinguish from THIRD RAY TYPES.

Third Ray Type   The third of the SEVEN RAY TYPES, associated with the planet Mercury. Distinguish from THIRD RAY.

third root race   In the theosophical cosmoconception, the Third Root Race is the third of the seven human races or streams of human evolution, sometimes called the Lemurian Race, which was said to occupy the almost completely lost continent of LEMURIA (or Shalmali, as it was sometimes called), which covered approximately the present Pacific area. During this Root Race, the separation of humanity into the two sexes took place. Lemuria is said to have perished before the beginning of the Eocene Age, and it was the last sub-race of this epoch which is supposed to have left behind the prototypes of the famous gigantic statues of Easter Island.

Third Sphere   See THIRD HEAVEN.

three crosses   In her account of her INTUITIONAL ASTROLOGY, Bailey says that the three crosses on the hill of Golgotha were biblical symbols of the three astrological crosses – the CARDINAL CROSS, the FIXED CROSS

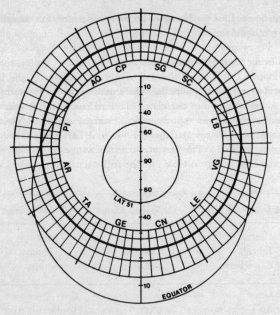

Figure 62: The three-dimensional chart, or stereographic projection (after Blunsdon).

and the MUTABLE CROSS, each of which she linked with a specific Christology.

three-dimensional chart Something of a misnomer, a term applied to a
method of chart projection (figure 62) designed by Blunsdon to reveal
planetary latitudes and ecliptical longitudes in one single projection. The
chart is by no means 'three-dimensional' (save in so far as all chart projections
are symbolic presentations of three-dimensionally conceived data) but stereographic. The symbolic figure may be adapted to fit all the standard methods
of HOUSE DIVISION.

three principles Essentially an alchemical doctrine (developed by Paracelsus
and Boehme) linked with the spiritual triad which, in the history of occultism,
is confronted with the spiritual septenary. The triad consists of Mercury,
Sulphur and Salt, and its relevance to astrology (beyond the mere fact of the
interrelation of alchemy and astrology in many medieval texts) is that the
principle of Mercury relates to the occult and esoteric nature of the planet. It

is Mercury which unites these two polar opposites of Salt and Sulphur: within the microcosm, Mercury represents the realm of emotions. The emotional life of man reconciles the dark will-forces (Sulphur), and the death-producing thought processes (Salt).

three prototypes A term applied by Wachsmuth to denote the three typical combinations in the world-clock, related to the zodiacal points (see PLATONIC YEAR). Prototype 1 visualizes the (so-called) Cardinal points in Mutable signs – for example, with Pisces marking the Vernal Point, and Gemini on the Midheaven (figure 63A). This prototype marks 'historical epochs in which

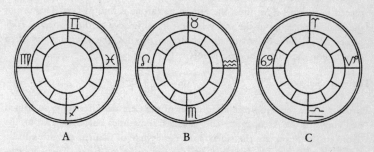

Figure 63: The three prototypes: A: The 'macro-anthropos'; B: The 'cosmo-psychic'; C: The 'geo-sophic' (after Wachsmuth).

important changes take place in man's relation to cosmic processes' along with 'decisive turning points in the evolution of man'. In such periods, man experiences himself as a 'cosmically orientated being', as a 'macro-anthropos'. Mankind lives at the present time within this prototype, with Pisces on the Vernal Point, and is likely to remain in such a position until about AD 2500. Prototype 2 visualizes the Cardinal points in Fixed signs, with (for example) Aquarius on the Vernal Point and Taurus on the Midheaven (figure 63B). During such periods differentiation of soul and body set it: experience is preserved in myths and images. This phase has been described as 'cosmo-psychic'. Prototype 3 visualizes the Cardinal points in the Cardinal signs, with (for example) Capricorn marking the Vernal Point, and Aries on the Midheaven (figure 63C). In such periods, the evolution tends towards the application of experiences to earthly activities: man becomes to some extent independent of the cosmos. This phase has been called the 'geo-sophic', and within its ambient abstract thinking reaches its zenith of activity. The system of three prototypes helps to present a simplified yet cogent view of historical epochs.

throne A term derived from Ptolemaic astrology but widely misunderstood and misapplied, mainly because Ptolemy was far from clear about its definition. Fortunately, it is known that the same term was used in earlier texts than the TETRABIBLOS as the equivalent of *EXALTATION (the corresponding Fall being termed a 'prison', a pictorial polarity of the royal 'throne'). This appears to be far from the original Ptolemaic idea, but it was used in this sense in several medieval sources. A synonym appears to be 'Chariot'. Since post-medieval times, some astrologers have used the term 'throne' as being synonymous with 'rulership', so that a planet in its own throne is a planet placed in its own sign: thus, the Sun in its throne is Sun in Leo. This use of the term is reflected in the many late-medieval images of the planetary gods presented in horse-drawn or dragon-drawn thrones, the related signs often depicted on the wheels of the chariots.

Throne See KURSI.

Thrones The third in descending order of the Celestial Hierarchy, so named by Dionysius the Areopagite. The Thrones, sometimes called the Throni, the Troni or the Seats, are the rulers over the Sphere of Saturn, and are said by occultists to have responsibility for the flow of time itself. Their appearance, unlike that of the celestial beings above them in rank (the SERAPHIM and the CHERUBIM), has received scant attention: they are often symbolized in Christian art by the attribute of a pair of scales (signs of their role in divine justice), and whilst they have six wings (like those of higher rank), they often are distinguished by a yellow tone – perhaps a topaz – on their breasts. The Thrones are sometimes called the Lords of Will in modern esotericism.

Thurayya Al Thurayya, the name given to the 1st of the Arabian MANZILS.

Thurene See HYADES.

Tibetan astrology In Western use, a misnomer for the esoteric astrological system proposed by Bailey, termed by her INTUITIONAL ASTROLOGY.

time A horoscope is usually cast from an accurate birth time, which generally requires some conversion from the given local clock time, to relate it to local sidereal time. That simple-sounding statement requires considerable amplification, however, as in relation to astrological practices, several definitions of time measurement are relevant. Solar time is time measured in relation to the

Sun – the apparent SOLAR DAY is the interval between two successive transits of the Sun – either of Midheaven or Imum Coeli. True local time (TLT) is the apparent solar time (i.e. that recorded by sundials): while not used in modern astrology, TLT is often the basis for many pre-19th-century horoscopes. Mean solar time (which is ordinary block time) is a compensatory measurement, based on the MEAN SOLAR DAY. Local mean time (LMT) is the mean solar time for any given locality – the local variations (not measured by clock time) involve a difference of 4 minutes for each degree of terrestrial longitude, which is one hour for every 15 degrees. Greenwich Mean Time (GMT) is the local mean time at Greenwich, England, which is on the zero line of longitude. 'Universal time' is really 'Earth orbital time', the LMT expressed in 24-hour notation. SIDEREAL TIME is time measured according to the stars (from Latin gen. *sideris*), the axial rotation of the earth being measured by a stellar fiducial. A complete rotation of the earth is the local sidereal day (LSD) – the interval of two successive transits of a fiducial over the meridian of a place. The LSD is said to begin when the zero point of Aries transits the midheaven of a given locality. Greenwich Sidereal Time (GST) is the local sidereal time at Greenwich. The difference between sidereal time (ST) and solar time is important in chart computation. The sidereal day is 23 hours, 56 minutes and 4.09 seconds of mean solar time. This difference is shown in the EPHEMERIS as an increase of (approximately) 3 minutes and 56 seconds of GST for each day, and is accounted for in astrological chart computation. TIME ACCELERATION is a term used in chart calculation to denote the difference between sidereal time and mean solar time which has elapsed since the GST given in the ephemeris. Such conversion is required to convert GST to LST – the time space is fractional – in the region of 10 seconds for each hour of elapsed mean solar time. Ephemeris time (ET) is Universal Time, with the addition of a (gravitational) variation so small as to be insignificant in relation to ordinary chart computation – a list of reductions from 1621 to 1974 is provided in the *Hieratic Ephemeris*, but most modern ephemerides give contemporaneous ET. Some modern ephemerides now list the positions of planets at midnight ET. Standard time (sometimes called ZONE TIME) was introduced at the end of the last century to regularize times throughout the civilized world. Towards this end, every 15 degrees of terrestrial longitude (east or west of Greenwich) was regarded as a Standard Time meridian, but this theoretical structure was not always adhered to, as the TIME ZONES in table 84 below indicate. Daylight-Saving Time (DST), Summer Time and War Time (WT) were in use during summer months in wartime and during energy crises. It is not possible to provide a simple tabulation of the complex effect this has had on determining local time, but in general, when DST was in operation clocks were advanced in some places half an hour, in other places and other times one hour, and (during the Second World War) two hours of advance gave Double Summer Time. The rules for CHART Computation allow for such DST, provided it is known for the year in question. A good easy reference is *Whitaker's Almanac*, though Brau gives a useful table of DST and WT periods in the English-speaking world, with all major variations up to 1980.

time differences   See TIME ZONES.

time zones   For the sake of unity, in the 19th century a system of Standard Time meridians or time zones was established, in which all clocks were set to the same local time zone, which in theory was centred on a meridian which determined a local mean time as standard (see TIME). To cast a horoscope for births outside the British time zone (centred on Greenwich) it is necessary to know the factors conditioning time in the given time zone, to which end a number of tables (as in table 84 below) have been provided. One of the great difficulties experienced by astrologers who regularly cast charts for foreign births is that of establishing the precise Greenwich Mean Time (and hence the sidereal time) from the local time given in accordance with localities and time zones. This difficulty arises because, for all the apparently simple listings given by the popular astrological textbooks, precision is not easily arrived at in view of territorial changes, which take place over comparatively short periods of time. The Preface to the useful list drawn up by Dernay in 1948 points to some of the difficulties in attempting reliable conversion. Dernay has provided an alphabetical list (now somewhat out of date) which has formed the basis for many other abridged lists in popular astrological books: such abridged lists are sometimes misleading, and may result in inaccurate charting. It is therefore perhaps unwise for an astrologer to rely too heavily on such tables (for unfamiliar locations) without checking the chart by means of a reliable method of rectification. The time zones most frequently given in astrological sources, with the correction factors for variation from Standard Time, are given in table 84. During the British Protectorate, Aden Time was + 2 hours 29 minutes and 36 seconds. The USSR has been divided into eleven time zones according to longitude, as set out in table 85. The abbreviations used in connection with the following lists have not been standardized, but the following are frequently used:

A	Atlantic	J	Japanese	So	South
B	Baghdad	M	Mountain	Su	Sumatra
C	Central	N	North	T	Time
E	Eastern	S	Standard	W	Western
Eu	European				

--------- TABLE 84 ---------

Aden Time	(see below)
Atlantic Standard Time	− 4 hours
Azores Standard Time	− 2 hours
Baghdad Time	− 3 hours
Brazil Standard Time	− 3 hours
Central Alaska Time	− 10 hours
Central European Time	+ 1 hour
Central Standard Time	− 6 hours
China Coast Time	+ 8 hours
Eastern European Time	+ 2 hours
Eastern Standard Time	− 5 hours
Guam Standard Time	+ 10 hours
Hawaiian Standard Time	− 10 hours 30 minutes
Indian Standard Time	+ 5 hours 20 minutes
International Date Line East	+ 12 hours
International Date Line West	− 12 hours
Japan Standard Time	+ 9 hours
Java Time	+ 7 hours 30 minutes
Middle Europe Time	+ 1 hour
Mountain Standard Time	− 7 hours
Newfoundland Standard Time	+ 3 hours 30 minutes
New Zealand Time	+ 11 hours 30 minutes
Home Standard Time	− 7 hours
North Sumatra Standard Time	+ 6 hours 30 minutes
Pacific Standard Time	− 8 hours
South Australia Standard Time	+ 9 hours 30 minutes
South Sumatra Time	+ 7 hours
West Africa Time	+ 1 hour
Yukon Standard Time	− 9 hours

Thus, C.Eu.T. is Central European Time. Brau gives an up to date system of Standard Time meridians, with indications of major changes, variations and confusions, up to 1980.

Tiriel Name given by Agrippa (quoting ancient qabbalistic sources) to the INTELLIGENCY of Mercury. He gives the magical number of Tiriel as 260, which is the linear sum of the MAGIC SQUARE of Mercury.

Titan

TABLE 85

Zone	Longitudinal band (in degrees)		Time factor
1	20.30	37.30	− 2 hours
2	37.30	52.30	− 3 hours
3	52.30	67.30	− 4 hours
4	67.30	82.30	− 5 hours
5	82.30	97.30	− 6 hours
6	97.30	112.30	− 7 hours
7	112.30	127.30	− 8 hours
8	127.30	142.30	− 9 hours
9	142.30	157.30	− 10 hours
10	157.30	172.30	− 11 hours
11	172.30	and eastwards	− 12 hours

Titan   Name given to the brightest of Saturn's moons.

Titania   Name given to one of the satellites of Uranus.

tonus   Ratio connected with the planetary system set out in the theory of the MUSIC OF THE SPHERES, and depending upon the supposed planetary distances. It was called also the sesquioctave, the eighth. In the planetary diagrams this ratio is the equivalent of the distance of the Sphere of the Moon and the Sphere of Mars (considered separately), each of which is the ratio of an eighth of the whole series.

topocentric ascensional transits   Mundane transits of Ascension, in connection with (presumably) any of the charts derived from the TOPOCENTRIC SYSTEM, correlating events with transits within an orb of one minute of time. The literature on topocentric astrology is sometimes confusing, but it seems that the topocentric ascensional transits involve the transits of planets, nodal points or even house cusps, when the chart projection has been adapted to the place of event: conjunctions, oppositions, antiscions and contrascions are valid. A development of this particular form of predictive astrology is the so-called PARAN. According to Dean, calculations have shown that allowing an orb of 1 degree, at least 4 parans are present for 91 per cent of the time, which throws some doubt on the utility, if not the validity, of the topocentric ascensional transit theory 'discovered' by Polich and Page.

topocentric astrology   See TOPOCENTRIC SYSTEM.

topocentric house system   A system of *HOUSE DIVISION proposed by Polich and Page for computing cusps for any place on earth – a system which is said to have been derived from experience, and by the projection of known effects on to chart schemas. In the topocentric system the polar axis, which passes through the place (the Greek *topos*) under astrological consideration is regarded as the fixed point of reference. Rotation of the geocentric horizon around this axis produces what is called a 'cone of rotation'. It is this cone which is the basis upon which measurements are made in order to determine the house cusps: hence the houses are determined by the natural rotation of the geocentric horizon around the topocentric axis. The entire system is actually obscure, and leaves many questions unanswered: it is further complicated by the form in which it has been publicized. However, standard reference Tables of Houses are available. See also TOPOCENTRIC SYSTEM.

topocentric system   The term 'topocentric system' should properly be applied to the method of chart origination and interpretation proposed by Polich and Page, but is often used specifically (though wrongly) of the TOPOCENTRIC HOUSE SYSTEM, upon which their methods partly depend. A good synopsis of the entire system has been offered by Love, though the literature dealing with topocentric astrology is curiously obscure. The topocentric astrologer is ultimately furnished with six charts for interpretation and progression – this points to a complexity of approach which is scarcely necessary in any of the traditional systems. Five of the charts used are derived from the tropical radical chart, the ordinary familiar chart of astrology, which is however regarded as relating to the inner life of the native, the aspects acting in an interiorizing manner. From this is derived the so-called topocentric mundane chart, which is held to relate to the outer events arising from the preceding chart. Follows a fixed radical chart, involving interpersonal relationships and 'destiny', to which is related the derived mundane radical fixed chart which is presumably related to events in connection with interpersonal relationships. These last two charts, as their names suggest, are derived from the constellation zodiac. A (more obscure) temporal chart is concerned with suprapersonal relationships, such as births and deaths: from this is derived the mundane temporal chart. The co-ordinate for the chart series is the MEDIUM COELI, which of course presumes an accurate radical chart.

topoi See LOCI.

Torch See ALDEBARAN.

total eclipse See UMBRAL ECLIPSE.

Toucan Constellation to the south of the asterism Phoenix, almost touching Octans, said to have been originated by Bayer in 1604, though sometimes later called Anser Americanus and even American Gans. Robson says that it gives pride with kindness.

Toxotes One of the Greek names, meaning 'archer', for the zodiacal sign and constellation Sagittarius – it is the term used by Ptolemy.

Trabs Name given to the 20th of the LUNAR MANSIONS in certain medieval lists. Probably visualized between delta (Kaus Medius) and sigma (Nunki) of constellation Sagittarius.

transitor The term is sometimes used in a general sense of any planet which is transitting another planet or nodal point, though in a specialist sense it is used of one of the major planets, the aspectal influence of which is slow.

transits Originally this term was used in astrological circles with only its Latin etymology in mind: the word is from the Latin *trans-ire* (to go across) and was applied to the crossing of any planet over any other planet, over any nodal point, or over any place in a horoscope (or, for that matter, in the heavens). The word may still be used in this sense. However, as the specialist techniques of astrological prediction were developed, that branch which was concerned with the study of the actual transits of planets over emphatic points in a radical chart was itself called transits. The word used in this sense applies therefore to one of the subdivisions of the art of *DIRECTIONS. Unlike such predictive techniques as SECONDARY DIRECTIONS, the study of transits is not involved with symbolic directing, using symbolic time co-ordinates. Instead, the actual effects of planets in the stellar spheres – in what the ancients called the ambient – during the period under study is related to the radical chart in a non-symbolic way. Practically this means that if an astrologer wishes to establish by means of transits the influences playing on the life of someone born in, say, 1937 in the 50th life of that person, then he will study the transits over the radical chart of 1937 which occur in the year 1987. In his excellent and influential work on progressions and transits, the modern astrologer Davison says that those who do combine the methods of secondary directions with the methods of transits usually regard the former as arising out of the radical chart, and the latter as having some external force, being representative of the general conditions of the outside world. He maintains, however, that transits are as integral a part of the techniques of progressions as directions of a primary and secondary kind.

translation   In a specialist sense, the transference of influence from a transitting planet which occurs when it separates from aspects and casts its rays to form another aspect with a different planet. This is sometimes called the 'translation of light', though (unless the Sun is involved) no light may really be said to be involved.

Transpluto   Name given to one of the HYPOTHETICAL PLANETS by the astrologer Landscheidt. An ephemeris for the period from 1878 to 1987 has been made available by the Ebertin-Verlag.

tredecile   Another name for the TRESILE.

trepidation   A name used to denote a theory used by Arabian astrologers to account for PRECESSION. Instead of the regression of the Vernal Point being visualized as a steady progression, it was seen as an oscillation (or trepidation) around an arc of 7 degrees. This movement was incorporated into the PTOLEMAIC SYSTEM (AD 950) and was said to be communicated from the Ninth Sphere (in some systems from the Eighth). The theory, derived from Albumasar, gave a precessional rate of 7 degrees in 900 years.

tresile   An ASPECT defined by Simmonite as being derived from a 'quintile and a half', or 3 signs and 18 degrees (108 degrees). See also QUINTILES.

Treta Yuga   Sometimes the Trita Yuga, the Silver Age of the Hindu chronology, a periodicity of 1,296,000 years. See, however, YUGA. The period given includes the twilights, or sandhya and sandhyansa.

Triangulum   Constellation, the Triangle, located south of Andromeda, on the edge of the Milky Way, and said by Allen to be one of the most ancient of the constellations. The connection with the Greek letter delta (itself a triangle) gave the name Deltoton, and an inevitable association with the delta of the Nile, with the names Aegyptus, Nilus and even Nili Donum (Gift of the Nile). Other derivatives are from the Greek *trigonon* used even by

Triangulum Australe

Hipparchus – Trigonum, Trigonus and (confusingly) Trigon. The Arabic Al Muthallath gave rise to a plethora of medieval degenerate forms, including Almutraleh, Mutlat and Mutlatun, etc. Ptolemy said that the stars in the asterism were of a similar influence to Mercury.

Triangulum Australe See SOUTHERN TRIANGLE.

Triangulum Minor Constellation, the Lesser Triangle, formed by Hevelius alongside the major TRIANGULUM. It is now obsolete.

Tribes of Israel Among the several dodecanary lists of zodiacal attributes is that termed the Twelve Tribes or the Twelve Children of Israel, which consists of an attempt (by biblical exegesis) to relate the biblical twelve to the zodiacal twelve. There are many such lists within the astrological (and theological) traditions, but the six listed in table 86 are culled from the

				TABLE 86		
Sign	**Butler**	**Sepharial**	**Thierens**	**Libra**	**Jeremias**	**Durandus**
AR	Gad	Benjamin	Issachar	Gad	Naphtali	Naphtali
TA	Asher	Reuben	Zebulon	Ephraim	Joseph	Issachar
GE	Issachar	Simeon	Joseph	Benjamin	Simeon/Levi	Simeon/Levi
CN	Zebulon	Levi	Benjamin	Issachar	Issachar	Benjamin
LE	Joseph	Judah	Judah	Judah	Judah	Judah
VG	Benjamin	Zebulon	Reuben	Naphtali	Dinah	Dinah
LB	Reuben	Issachar	Gad	Asher	Dan	Asher
SC	Simeon	Dan	Asher	Dan	Benjamin	Dan
SG	Levi	Gad	Naphtali	Manasseh	Gad	Joseph
CP	Judah	Asher	Manasseh	Zebulon	Zebulon	Gad
AQ	Dan	Naphtali	Simeon	Reuben	Reuben	Reuben
PI	Naphtali	Joseph	Levi	Simeon	Asher	Zebulon

sources given under the headings (see Bibliography). This table should at least establish that no single list is really definitive. Only the list of Sepharial follows the order of the children given in the blessing (see Genesis 49), with Taurus as the leader of the twelve, according to the ancient esoteric system (see, for example, ALDEBARAN). The church historian Durandus, who transmits one of the earliest lists, gives the order of the twelve sons named by Jacob and then attached to these a rationale which was supposed to follow the relationship of signs.

Trident In a specialist sense, one of the modern additions to the series of Arabian PARS, the Part of Neptune, sometimes called the Point of Treachery. If a chart is revolved so that the degree occupied by the Sun is transferred to the Ascendant, then the position of Neptune in this new arrangement will mark the Trident within the houses. No doubt the term was originated from the supposed graphic origin of the sigil for Neptune: see SIGILS – PLANÈTS.

trigon A term (now almost obsolete) used to denote the group of three signs linked with the same element: it is therefore the equivalent of triplicity. It is sometimes used (almost archaically) to denote a TRINE aspect. For a peculiarly modern use, see also SIDEREAL MOON RHYTHM.

trigonocrators A term derived from the Greek, meaning 'rulers of trigons', and relating to a doctrine set out by Ptolemy which accords a planetary rule over the groups of TRIPLICITIES, as set out in table 87. Unfortunately, some

—————————————— TABLE 87 ——————————————

Triplicity	Signs			Planet by day	Planet by night
Fire	AR	LE	SG	Sun	Jupiter
Earth	TA	VG	CP	Venus	Moon
Air	GE	LB	AQ	Saturn	Mercury
Water	CN	SC	PI	Venus*	Moon*

* The water triplicity is accorded the single rule of negative Mars, but with the two co-rulers indicated above.

modern astrologers have misunderstood the idea behind the trigonocrators, and have proposed new rulerships which merely accord planetary rule over the individual zodiacal signs rather than over their elemental natures. However, the whole question of trigonocrators is involved with astrological notions which are largely archaic, and have little application to modern practice. It is clear from the TETRABIBLOS that the concept is bound up with the *airesis* theory to which the ancients were partial (see SECTA), but which plays no significant part in modern astrology.

Trigonum See TRIANGULUM.

trimorian A term usually applied to the distance of three signs, and wrongly given as a synonym for the square ASPECT.

trine An ASPECT maintained by two or more planets so placed that they are separated by ⅓ of the zodiac, which is to say by an arc of 120 degrees. This aspect is regarded as being wholly beneficial, and is sometimes called the aspect of good fortune. The amount of ORB permitted is far from agreed among authorities, but some allow as much as 7 degrees.

trinity of association See RELATIVE HOUSES.

trinity of psychism See PSYCHIC HOUSES.

trinity of wealth See POSSESSIVE HOUSES.

Triones One of the several important variant names for URSA MAJOR.

triple Sun A term derived from esoteric astrology, and intended to denote the occult maxim that the visible Sun is only the veil for two other suns, one called by Bailey 'heart of the Sun', and the other 'central spiritual Sun'. See also ESOTERIC SUN.

tri-septile An aspect of 154.3 degrees, arising from the multiple of the SEPTILE by 3, and used in the theory of HARMONICS.

Trita Yuga See TRETA YUGA.

Trithemian periods See SECUNDADEIAN BEINGS.

Triton Name given to the satellite of Neptune.

Troni See THRONES.

tropical radical See TOPOCENTRIC SYSTEM.

tropical signs Since Cancer and Capricorn mark the solstitial points, these are sometimes called the tropical signs.

tropical year A more precise term for the solar year, marking the succedent transits of the centre of the Sun over the Vernal Point, a period slightly shorter than the SIDEREAL YEAR. The tropical year is a period of 365 days, 5 hours, 48 minutes and 4.5 seconds.

true solar day The length of time taken for the Earth to rotate on its axis once in relation to the Sun – a periodicity different from that of the SIDEREAL DAY. Astrologers who require day-equivalent periods for progressions regard the true solar day as the most symbolically appropriate for the Day for a Year system of predictive symbolism. See TIME.

trulli symbols   In the Puglia region of Italy, centred around the town of Alberobello, are a number of conically roofed houses, locally called TRULLI. Many of these have painted upon their roofs a variety of Christian and occult symbols, several of which are astrological: the origin of this practice is unknown, but the use of standard symbols for the modern planets, such as Neptune and Pluto, indicates that whatever the antiquity of the custom, new symbols are added from time to time. A recent survey has been conducted by Goodman.

trutine of Hermes   This is a dictum attributed to the mythical Hermes Trismegistus – 'the place of the Moon at conception was the ascendant of the birth figure or its opposite point'. No mention of this trutine appears in the extensive genuine Trismegistic literature, and indeed the sentiment expressed within it removes it from that type of literature: certain Trismegistian ideas may be confused with this trutine, however, because of the occult connection between the Moon Sphere and the descent into matter (reincarnation) which is explicit in this literature. The idea does not appear to go back beyond late medieval astrological speculation, though it is perhaps Arabian in origin. Though it is often seriously quoted even in modern astrological textbooks – especially in texts dealing with the PRE-NATAL EPOCH – the trutine really belongs to the substratum of mythology of astrology, for it enunciates a principle which is not susceptible to verification by ordinary means. Curiously, the ANIMODER is sometimes also linked with this concept, even though it was originally a method of rectification unrelated to the idea of conception (which in any case, within the framework of the gnostic and Trismegistian literature involved a gradual descent through the Spheres, in which conception was merely a late stage).

T-square   One of the aspectal patterns (see MAJOR CONFIGURATION) in which two planets are in opposition, with a third at their mid-point: the resultant pattern is in the form of a simple T. This formation is also called a T-CROSS – which is perhaps a more accurate designation – though this term is also applied to two pairs of oppositions with axes in square with each other. Since the T-Square pattern combines the restless nature of the Square with the awareness of opposition, it is an indication of a powerful surge of energies: the mid-point planet is usually the significator of the focal point of these energies.

Tual

Tual In certain methods of ceremonial magic, Tual is named as the Governor of the zodiacal sign Taurus.

Tubus Astronomicus See TELESCOPIUM.

Tula The Sanskrit name for zodiacal Libra: but see also CITRA. Sometimes the word is given in European texts as Tulam.

Tulam See TULA.

twelfth house The twelfth of the astrological HOUSES, linked with the nature of Pisces, and associated with the struggle of the individual against the limits imposed upon him by the forms of society: it is also linked with the powers of guilt, which always call for regeneration and rebirth. Devore calls it the House of the Hangover, which corresponds approximately with the traditional term House of One's Own Undoing. The release from guilt often necessitates self-sacrifice, not as a negation of self, but as a catharsis, and such properly belongs to the realm of this house. This would perhaps help explain the old name House of Bondage, which refers to self-imposed bondage, directed from what we would now call the subconscious levels, but which the medieval astrologers would link with the spiritual impetus of Jupiter, the ancient ruler of Pisces. The twelfth is really the house in which secret aspirations are often frustrated, or must be resolved.

twelve cell-salts The twelve cell-salts described and developed for medical purposes by the homoeopathist Schuessler have been linked with the physio-chemical and psychological allocations of the twelve signs of the zodiac. The correspondences drawn by Carey are:

AR	Kali Phos	(Potassium Phosphate)
TA	Nat Sulph	(Sulphate of Sodium)
GE	Kali Mur	(Chloride of Potassium)
CN	Calc Fluor	(Fluoride of Lime)
LE	Mag Phos	(Phosphate of Magnesium)
VG	Kali Sulph	(Sulphate of Potassium)
LB	Nat Phos	(Sodium Phosphate)
SC	Calc Sulph	(Lime Sulphate)
SG	Silica	
CP	Calc Phos	(Lime Phosphate)
AQ	Nat Mur	(Sodium Chloride)
PI	Ferrum Phos	(Iron Phosphate)

Twelve Creative Hierarchies   According to esoteric astrological lore, the Twelve Creative Hierarchies are spiritual beings who participate, or have participated, in the cosmogenesis of our planetary system. Each rank of hierarchy has been linked with one of the twelve signs of the zodiac (see the CELESTIAL HIERARCHIES, who have been linked with the planetary sphẹres). The first two are not named, though linked with Aries and Taurus – the esotericist Heindel, who has most of his learning from Steiner, says that they 'have passed beyond the ken of anyone on Earth'. Gemini is linked with the SERAPHIM, who oversaw the creative phase of the MOON PERIOD. Cancer is linked with the CHERUBIM, who oversaw the creative phase of the SUN PERIOD. Leo is linked with the group of exalted beings known in esotericism as LORDS OF FLAME, who oversaw the creative phase of the SATURN PERIOD, which inaugurated the cosmogenesis of our present system. The remaining seven Hierarchies are called the LORDS OF WISDOM, LORDS OF INDIVIDUALITY, LORDS OF FORM, LORDS OF MIND, the ARCHANGELS, the ANGELS and the VIRGIN SPIRITS. With the exception of the latter (which is the esoteric name for Mankind), these Hierarchies have also been linked with the Spheres.

twelve houses   See HOUSES and HOUSE SYSTEMS. For specific houses, see FIRST HOUSE, which is of the same nature as the all-important ASCENDANT DEGREE, and which relates to selfhood; SECOND HOUSE, which deals with finance and support; THIRD HOUSE, which deals with travel and kindred; FOURTH HOUSE, which deals with parentage and beginnings; FIFTH HOUSE, which deals with creativity and offspring; SIXTH HOUSE, which deals with service to others and health; SEVENTH HOUSE, which deals with marriage and partnerships; EIGHTH HOUSE, which deals with matters pertaining to regeneration and death; NINTH HOUSE, which deals with distant journeys and philosophical considerations; TENTH HOUSE, which deals with reputation and profession; ELEVENTH HOUSE, which deals with social aspirations and friendships, and TWELFTH HOUSE, which deals with mystical elements in the life of the native.

twelve philosophic tendencies   In a series of lectures translated under the title *Human and Cosmic Thought*, the occultist Steiner developed a schema which links what he called the 'twelve different shades of philosophic outlook' with the zodiacal signs (or constellations):

AR	Idealism	TA	Rationalism	GE	Mathematicism
CN	Materialism	LE	Sensualism	VG	Phenomenalism
LB	Realism	SC	Dynamism	SG	Monadism
CP	Spiritualism	AQ	Pneumatism	PI	Psychism

Twelve Places

Steiner distinguished the seven philosophic moods in terms of the seven planets – see therefore SEVEN MOODS.

Twelve Places   The Twelve Places of Manilius consists of a long-archaic system of 'significant degrees' derived from a dodecanary division of the system into 30-degree arcs, commencing at the PART OF FORTUNE. In order of zodiacal arc, these are given as:

1 Fortune and property – including slaves
2 Military affairs and voyages
3 Civil affairs and public business
4 Legal and judicial matters
5 Marriage and social life
6 Acquired wealth
7 Dangers
8 Nobility and reputation
9 Rearing of children
10 Authority
11 Health and sickness
12 Fulfilment of vows

The Twelve Places are sometimes called the Twelve Sortes, but see also SORTES.

Twelve Places of Manilius   See TWELVE PLACES.

twelve signs   See ZODIAC.

Twelve Sortes   See TWELVE PLACES.

Twelve Spiritual Beings   See TWELVE CREATIVE HIERARCHIES.

Twelve Tribes   See TRIBES OF ISRAEL.

two stars   A term sometimes used of the ancient symbol for the Dioscuri, or for the symbolic representation of Gemini, and probably originating in the astrological symbolism of certain ancient Roman coins. Two such symbolic stars are placed alongside constellation Pegasus in the VICENZA CYCLE, and these are known to have been derived from the coinage of Hadrian. In addition to connoting the standard dualistic symbolism proper to Gemini, the two stars (as Ballarin suggests) reflect the ancient duality of the 'day-life' and 'night-death', which is the main theme of the Vincenza constellational fresco.

Tycho's Star A misnomer for a NOVA popularly supposed to have been discovered by the astronomer-astrologer Tycho Brahe in 1572 (see CONSTELLATIONS). Tycho did not discover the nova, since it was noted by many astronomers before him – probably for the first time by Schuler at Wittenberg. It was called New Venus by Cornelius Gemma at least two days before Tycho actually saw it. Some astrologers regarded it as a reappearance of the STAR OF BETHLEHEM, and inevitably predictions were soon broadcast of a Second Coming. In terms of astrology, however, it is unlikely that any other single phenomenon changed the ancient cosmoconception as profoundly as this nova, since until then it was deemed axiomatic that the supralunar spheres were immutable, as Aristotle had proclaimed. The nova exploded the Ptolemaic cosmoconception even more profoundly than did Copernicus.

U

umbral eclipse The Moon is said to be in umbral eclipse when it is in any way touched by the shadow of the Earth (the Latin *umbra* means 'shadow'). The umbral eclipse becomes a total eclipse when the orb of the Moon is completely swathed in the shadow of the Earth. See ECLIPSE.

Umbriel Name given to one of the satellites of Uranus.

unaspected planets In a specialist sense, these are planets which receive no MAJOR ASPECTS. Some astrologers maintain that an unaspected planet is an indication of lack of integration of the quality represented by that planet. Other astrologers ignore the supposed influence altogether. The Australian astrologer Dean made a special study of unaspected planets, however, and suggests that such are usually inducive of a strong planetary influence, which is not always well integrated into the life of the native. If a wide orb is allowed for the major aspects, then unaspected planets are quite rare. Some astrologers maintain that planets unaspected by the major aspects are all the more susceptible to the influences of the MINOR ASPECTS.

under the sunbeams A term used to denote that a planet is within a 17-degree orb of the Sun, which is regarded by some astrologers as a debility.

undines The class of ELEMENTALS linked with water: in the medieval tradition they are often called 'nympha'.

unfortunate A term used of the signs Taurus, Cancer, Virgo, Scorpio, Capricorn and Pisces – the 'negative' signs.

unfriendly planets This is a term derived from the idea of SYMPATHY. The Sun is unfriendly only to Saturn, according to some authorities, yet really the Sun is benign to all planets. The moon is unfriendly to Mars and Saturn. Venus is unfriendly to Saturn. Mars is unfriendly to all planets with the exception of Venus. Jupiter is unfriendly to Mars. Saturn is unfriendly to all save Mercury and Jupiter. See also SYMPATHY.

Unicorn   Name applied (though rarely) to Capricorn. See also MONOCEROS.

unorthodox astrological relationships   In her highly complex system of esoteric *INTUITIONAL ASTROLOGY, the esotericist Bailey presents a system of planetary rulerships over the constellations (which are probably the zodiacal signs), with some planets taken from her five NON-SACRED PLANETS, and others from her series of seven SACRED PLANETS. The sequence of esoteric rulerships contrasts with that of the ORTHODOX PLANETS, as shown in table 88. This system of rulerships must be compared with the series set out in connection with the esoteric SEVEN RAYS.

TABLE 88		
Constellation (sign)	*Esoteric planet*	*Orthodox planet*
Aries	Mercury	Mars
Taurus	Vulcan	Venus
Gemini	Venus	Mercury
Cancer	Neptune	Moon
Leo	Sun	Sun
Virgo	Moon	Mercury
Libra	Uranus	Venus
Scorpio	Mars	Mars
Sagittarius	Earth	Jupiter
Capricorn	Saturn	Saturn
Aquarius	Jupiter	Uranus
Pisces	Pluto	Jupiter

Unuk   An alternative name for the fixed star COR SERPENTIS.

Unukalhai   See COR SERPENTIS.

upagrahas   A term derived from Indian astrology, usually claimed in the West to denote hypothetical planets, though apparently related to the Hellenistic SORTES.

Urania   One of the nine Muses of classical mythology, assigned in Roman times to rule over Astronomy and Astrology, her symbol being the staff pointing to the globe. The symbol is an esoteric allegory, for the staff is straight, and the globe is circular, and they both may be taken as reference to the astrological (Aristotelian) doctrine that bodies in the supra-lunar spheres move in perfect circles, while those in the sublunar spheres (which of course include the Earth) move in straight lines. The symbolism addresses itself to the fact that astrology is concerned with the earthly as much as with the celestial. The word 'Urania' is often applied in modern times to the whole realm of astrology, though when so used it is often in a pejorative sense.

Uranian The term means 'related to Urania', and it was used in this sense at least as late as 1814 by Corfield. The Uranian sciences were thus the various astrologies. In modern times the word has been used as though it meant 'relating to Uranus': see, for example, URANUS SCHEME. See also URANIAN ASTROLOGY.

Uranian astrology Name given to a modern system of astrology derived from the theories propounded by the so-called HAMBURG SCHOOL, under the direction of Witte. The system incorporates the three basic symbolic combinations of traditional astrology, though with important differences. The theory of planetary aspects is not based on consideration of elemental natures, as in the Ptolemaic system, but arises through numerical divisions: a special DEGREE DIAL is used as an aid to recognition and interpretations – see PLANETARY PICTURES and COSMOGRAM. In addition to the standard planets recognized by more conventional astrologers, Uranian astrology postulates eight HYPOTHETICAL PLANETS, all transneptunian (see ADMETOS, APOLLON, CUPIDO, HADES, KRONOS, POSEIDON, VULKANUS and ZEUS). The interpretation of planetary placings is not of the orthodox kind, and there is much reliance upon MID-POINTS, ANTISCIONS and upon a limited number of aspects. The adherents of the system claim that it is empirical, but the edifice is actually so far removed from conventional astrology that it is difficult to evaluate it against normal standards.

Uranian Scheme See URANUS SCHEME.

uranographia A term used to denote the science of delineating or describing the sidereal heavens. See CONSTELLATION MAPS. The modern English equivalent term is 'Uranography'.

uranometria A term used to denote any treatise on the measurement of the relative distances (and magnitudes) of heavenly bodies – though most usually of the fixed stars and the asterisms. See CONSTELLATION MAPS. The modern English equivalent term is 'uranometry'.

uranoscopus An archaic term meaning 'star gazer', sometimes used in modern times pejoratively of an astrologer.

Urion

Uranus The first extrasaturnine planet to be discovered, in orbit between Saturn and the more recently discovered Neptune. It was noted by Herschel in March 1781 – though he did not at first realize that it was a planet. When a year later its true nature was determined, he called it Georgium Sidus (George's Star) after the reigning George III. Continental astrologers called it Herschel, however, as Allen notes, 'in a much varied orthography, strangely erroneous considering the fame of its discoverer'. The same authority notes that the 'star' had been observed 22 times previously by varied scientists, and it appears that the name Ouranos was first used by astrologers in reference to the then hypothetical planet prior to Herschel's discovery. Certainly the eccentric astrologer and friend of William Blake, John Varley, was the first to confirm the destructive and erratic nature of the planet by direct observation of transits. In recent years the planet, under its now-standard name of Uranus, has displaced the traditional Saturn from its rule over zodiacal Aquarius, which perhaps makes sense of the old myths which visualized Ouranos as being even older than the Time God Chronos, one of the ancient names for Saturn. Astrologers claim that it is the planet of change, marking social or personal revolutions and disruptions. When working through its beneficial side, it can be magnetic, original (to the point of genius or madness), reformative and unconventional. Under pressure, however, it lapses into being merely unusual, pointlessly rebellious and given to violence and/or crime. See also NEW PLANETS.

Uranus–Neptune cycles Uranus conjuncts with Neptune to give a first order cycle of 171.403 years, which has been reduced to an average value of 171 years due to the renowned eccentricity of Uranus. The second order cycle involves 21 conjunctions and a span of 3,600 years.

Uranus–Pluto cycles The synodic period for these two planets is 127.280 years, the first order cycle being as a result of two such conjunctions (254.280 years). The second order cycles involves 26 synodics, giving a cycle of 3,432.774 years, with a small deficiency of less than 3 degrees.

Uranus Scheme The Uranus Scheme or Uranian Chain are terms derived specifically from the theosophical cosmoconception linked with the SCHEME OF EVOLUTION, and related to a future development of one of the Ten Chains in the present solar system.

Uriel The Archangel accorded rule over the Earth itself.

Urion See ORION.

Urkab

Urkab See ARKAB.

Ursa Major Circumpolar constellation, the Greater Bear, or Wain, with a plethora of related names, including Arctos, Kallistos (the name of the mortal changed into a bear because of Juno's jealousy), and Lycaonia after the name of Kallisto's father, and (through error) Arcturus, not to be confused with the star of the same name. It was the Phoenician Dub, and the Arabic Al Dubb al Akbar, and a 'bear' in the language of the North American Indians long before America was discovered by Europeans. The magical number 7 of its stars gives us many names, such as Seven Sleepers, Seven Sages, and even the unimaginative Seven Stars, whilst Allen records that they were called Seven Bears in India. In the northern regions it was generally pictured as a cart or wain, and is thus the Wagon, Arthur's Chariot, and even the Romans (touched by their spread to the north) called it Cursus and Plaustrum after forms of chariots: among the Goths it was Himmel Wagen, the 'celestial chariot'. The Charles's Wain is derived from the Saxon word ceorl, an ordinary peasant who eventually, by linguistic adjustment, became either Charlemagne or King Charles, in the same way that Arthur the king was derived from Arcturus, the associate state. From the wagon imagery came also Plough, and many variants, while Triones was popular with the ancients. Allen does a scholar's justice to these names by listing some 150 important variants. In the Denderah planisphere of Egypt the asterism is symbolized as a hindquarter of an animal and linguistically associated with Set, who had rule over all the circumpolar constellations – hence Dog of Set, and Typhon as names. The famous seven stars are, Dubhe, Mirak, Phacd, Megrez, Alioth, Bentenash and Mizar. Ptolemy, who called the entire asterisms Arctos, said that each of these stars was of the equivalent nature to Mars, which would imply that the entire constellation is of the Martian quality.

Ursa Minor Circumpolar constellation, the Little Bear in comparison with the larger URSA MAJOR. It is the Little Dipper against the Big Dipper, but sharing with this companion the name Septentrio, connected with the idea of 'seven' (*septem* in Latin) which eventually gave us the Latin word for the North, so prominent in these asterisms. A popular early name was Cynosura (Dog's Tail) – but see also POLARIS – yet this name could be derived from that of one of the nymphs of Crete who reared Jupiter. Ptolemy tells us that the bright stars in this asterism (which he calls Micros Arctos) are of a similar quality to Saturn, and 'to a lesser degree, to that of Venus', a combination difficult to evaluate.

V

Valens Anthologia   Name given to the anthology of the astrologer Vettius Valens (compiled in the 2nd century AD), one of the earliest collections of horoscopic and astrological material from Hellenistic sources. The historian Sachs has dealt admirably with many of the issues raised by the horoscope figures (as, for example, the question of CLIMATA) which consist of some hundred literary horoscopes, apparently used by Valens as supportive examples of certain astrological doctrines. Sachs comes to the conclusion that Valens was systematically collecting and analysing a large amount of statistical material for births, histories and deaths 'in order to confirm or modify the theoretical structure of astrology'. In a modern context the term 'horoscope' may be misleading, as the material is essentially literary, and many of the individual sets of data include only the horoscopos and the positions of the luminaries. A few more complete sets do contain planetary and nodal positions, however. The basis of the computation of the Horoscopos appears to belong to an astronomical theory already archaic even for Valens.

Vega   See WEGA.

Vela   See PUPPIS.

Venant   See ROYAL STAR.

Venator   See ORION.

Veneris Sidus   See TAURUS CONSTELLATION.

Venter Arietis   Name given to the 2nd LUNAR MANSION in some medieval lists. Probably located by the stars delta and epsilon of constellation Aries.

Venus   The planet in orbit around the Sun at a mean distance of about 67 million miles, with a periodicity of about 244 days. The peculiar brilliance of Venus is due to the reflective power of the atmosphere of clouds in which the body is sheathed: a point which has not gone missed by astrologers, who point to the propensity of Venusians to deal in surfaces, and to be concerned with surface appearances. In astrology Venus represents the spiritual side of the native as manifest in his physical life, and it is an index of the ability of the subject to enjoy beauty and to co-operate with others. It is traditionally termed the 'planet of love', and is indeed a useful indicator of how the subject loves the world and all its rich manifestations: it governs also the higher emotions, and the general refinement of the individual. A Venus which is emphasised in a chart, through an essential dignity, is usually a sign of physical beauty, and of the fact that the native will be in some way creative. Such a well-placed Venus also makes for great popularity with the opposite sex, as well as for a harmonious emotional life. A badly placed Venus tends to restrict in some way the urge towards co-operation with others, and the native may be unpractical, lazy or opportunist. The sigil for Venus ♀ is explained as representing the circle of spirit (which contains all spiritual potential) lifted by the cross of materiality, suggesting that the Venusian nature may be regarded as equivalent to a 'refined or redeemed earthly quality' – it is perhaps no accident that even a modern astronomer like Sidgwick should write of Venus as 'almost the twin sister of the Earth', for as such she has been widely regarded by many esoteric astrologers. Venus has rule over zodiacal Taurus and Libra: in the first sign we find the love of the Earth manifesting as a refining principle which leads to the creative manipulation of materiality, as well as love for sensual experience such as only the physical body may afford. In the second we find the love of the air type, which is for communication and harmony of expression, emphasized and intensified by the rulership. Venus is Exalted in Pisces, and has its Fall in Libra. Traditionally the planet rules the small of the back (but see MELOTHESIC MAN) as well as the entire shape of the physical body.

Venus period   The name given in esoteric astrology to a future condition of our present Earth, beyond the future development of the JUPITER PERIOD, yet prior to the final VULCAN PERIOD, when humanity will attain to objective, self-conscious, creative being.

Venus Scheme   The Venus Scheme or Venusian Chain are terms derived from the theosophical *SCHEME OF EVOLUTION, and claimed to relate to the most advanced of the planetary forms.

vernal equinox   The term means literally 'equal night of spring'. It is marked at the time when the Sun reaches that point at which the plane of the ecliptic intersects the plane of the Earth's equator, which occurs twice in each year – once when the Sun enters Aries, which is the vernal equinox, and once when it enters zodiacal Libra, which is the AUTUMNAL EQUINOX. See also VERNAL POINT.

vernal horoscope   An alternative term for the CHOISNARD CHART, proposed by the astrologer Froger.

Vernal Point   The Vernal Point is the so-called ZERO DEGREE (more properly the zero point) of zodiacal Aries, which marks the VERNAL EQUINOX. The gradual retrogression of this point gives rise to PRECESSION.

vernal quadrant   See QUADRANTS.

vertex   Name given to the GREAT NEBULA situated to the north of the head of ANDROMEDA. Additionally, the vertex is a term originated by the modern astrologer Johndro, arising from a theory (perhaps based on a suggestion made by Collin) that the existence of a magnetic field within and around our Earth postulates that there must be two Ascendants (or more properly, two Ascendant–Descendant axes), one being magnetic, the other electrical. These might, under certain conditions, be visualized as being at right angles to each other. This alternative axis is called the vertex–antivertex axis: the so-called magnetic Ascendant is the traditional Ascendant (that is, the ancient Horoscopos), while the electrical Ascendant is the so-called vertex. Whilst it is widely recognized that Johndro's analogy, upon which this conception is based, is itself invalid, he claims to have shown (using the 'known data method') that the axis is linked with involuntary actions, lack of choice, and fate. As with so many modern astrological speculations, there appears to be a chimerical basis for the idea of the vertex. This curious axis cannot hold any relationship to the Ascendant or to the Ascendant–Descendant axis. The fact is that the Ascendant marks the relationship established between a given locality on the Earth and a point on the zodiacal arc. The vertex–antivertex axis is the supposed expression of a relationship contained wholly within the magnetic field of the Earth: it could in no way (even by forced analogy) be described as an Ascendant, or treated as an equivalent influence, precisely because it is non-zodiacal in origin. See also CO-VERTEX.

vertical circles   Great circles of the Celestial Sphere, passing through the zenith and the nadir, and perpendicular to the horizon, are called the vertical circles.

vertical sphere   One of several CELESTIAL SPHERES, taking as its point of reference a horizon plane as one co-ordinate, and the Midheaven of any given point on Earth as the second. This model is the basis for many of the systems of mundane divisions, and the numerous attempts to reconcile it with the apparent movement of the planets have given rise to the proliferation of different methods of HOUSE DIVISION.

Vesican chart   See CHART SYSTEMS.

Vespa   See MUSCA BOREALIS.

Vesper   The name (with the variant Vesperus) given by Roman and medieval astrologers to the planet Venus when an 'evening star'. See also LUCIFER.

vespertine culmination   Term used by Ptolemy for one of his VISIBLE ASPECTS.

vespertine setting   Term used by Ptolemy for one of his VISIBLE ASPECTS.

vespertine subsolar   Term used by Ptolemy for one of his VISIBLE ASPECTS.

Vesperus   See VESPER.

Vesta   A name given by the esoteric astrologer Thierens to a future development of his Venus (also known as ISIS) and pictured as having a rule over the Taurus of the ETHERIC SPHERES. It is said to be a principle of conservation. The same name Vesta is also used of one of the ASTEROIDS, discovered in 1807, and linked by the astrologer Dobyns with the sign Virgo, along with a co-ruler, the asteroid CERES.

Via Lactea   The Milky Way appears something like a luminous band wrapped around the Stellar Sphere – a luminosity which is caused by the light of the myriads of closely packed stars. In effect, when we look into this luminous band, we are looking into the flattened structure of the galactic system, almost along its wide major axis. Since our own solar system is embedded in this 'flattened pancake' of luminosity, we gain the impression that the Via Lactea itself crosses the ecliptic in (approximately) the last decan of Gemini, and over the last half of Sagittarius. When astrologers speak of the influence of this band, they usually have in mind the wide arcs which intersect the ecliptic. Robson says that this influence is 'sympathetic, human, artistic', and so on, but a survey of the general degree interpretations relating to these arcs (which cannot be separated from the influence of the fixed stars within them) shows that to be not at all beneficial – accidents and blindness being predicted of several of the degrees. The Via Lactea or Galaxy has been given many names, some of them, as Allen remarks, 'arbitrary, descriptive, or fanciful titles'. The white luminosity has linked it with milk, as set out so beautifully in the myth of Jupiter pressing his newborn son Hercules into the breasts of his consort Hera, and the child sucking so fiercely that the milk squirts into the heavens, where it becomes the Milky Way: the milk which falls on earth becomes the lilies. From such stories (of which there are variants) we have Via Lactis, Semita Lactea (milky footpath) and so on. The luminosity marks it out as a river, hence Eridanus, Stream of Ocean and River of Heaven, and the title from the Akkadian, River of the Divine Lady, and the more simple Arabic Al Nahr (the river). In China it was Tien Ho (Celestial River), and the Silver River. The idea that it was a band in the skies gave another set of names – Coeli Cingulum and Orbis Lacteus, both relating to celestial girdles. Yet others saw it as a road, a *via*, and Allen recalls that it was for the Anglo-Saxons the pathway of the giant sons of Waetla, Waetlinga Straet, from which some reasonably draw Watling Street. Langland, in his *Piers Plowman* makes it 'Walsyngham Way to the Virgin Mary', whilst Asgard Brige was used to denote this luminous band as well as the rainbow. Yet, for every term quoted above, a dozen or more variants might be given.

Via Solis   The Way of the Sun – a Latin name for the ECLIPTIC.

Vicakha   The Sanskrit name for constellation Scorpio – but see VRISHIKAM.

Vicenza cycle   In the Palazzo Chiericati, in Vicenza, is the so-called 'Firma-mento Chiericati', a huge ceiling fresco of a well-planned constellational cycle, around an image of Apollo and Diana as personifications of Sun and Moon (this last derived from a 16th-century fresco in Ferrara). The cycle was painted by Domenico Rizzo, *c.* 1560, but was recently restored. Over 80 identifiable constellations are incorporated into 76 frescoes, separated by bands which are themselves painted with related imagery. It has been shown that the 'Firmamento' (in accordance with the influence of Palladio) is largely derived from zodiacal and constellational and polytheistic imagery of ancient Roman numismatics, as well as from the famous Dürer print of 1515. A useful summary of such influences and an extensive bibliography have been given by Ballarin.

Victorian period   A specialist astrological term, referring to a periodicity of approximately 535 years, and called after Victorius of Aquitain (and some-times after Dionysius Exiguus, hence the alternative name Dionysian period). It was a period involved with combining the 19-year METONIC CYCLE with the so-called SOLAR CYCLE of 28 years (the Dominical cycle), to restore the new Moon to the same day of the week of the same month on which the cycle began. After the Gregorian calendrical reformation its use lapsed.

Victory Name given to one of the SEVEN LOTS of Greek astrology, the Lot of Nike, the goddess of Victory. The degree of Nike is the same distance from the Ascendant as Jupiter is from the LOT OF DAEMON.

vigintile The SEMI-DECILE aspect.

Vikasha Name given to the 14th of the Hindu NAKSHATRAS.

Vindemiatrix Fixed star, the epsilon of constellation Virgo, set in the right wing of this celestial being. It has many titles, Vindemiator and Providemiator, meaning 'grape gatherer', and in medieval astrology Alcalst and Almucedie among the related terms from the Arabian *Al Muridin* (those who sent forth) almost certainly a mistranslation of the original Greco-Latin term. Ptolemy says that the star is equal in influence to Mercury with Saturn, and almost all authorities claim it as an unfortunate star, bringing falsity, disgrace and folly in terms of the planetary or nodal points conjuncting it.

Virgin Name given to both zodiacal Virgo and to the Virgo constellation.

Virgin Spirits One of the esoteric names given to present-day humanity, viewed as one of the TWELVE CREATIVE HIERARCHIES.

Virgo The sixth sign of the zodiac. It corresponds neither in location nor significance with the asterism of the same name: see VIRGO CONSTELLATION. The sigil for the sign ♍ is said to have been a contracted drawing of the severed serpent from which the sigils for Scorpio and Virgo were evolved, and the fact is that the Egyptian demotic scripts appear to carry such an association. The link with the head of a serpent postulated in this theory is interesting, for Virgo is indeed a sign deeply committed to the intellectual process. The sigil has also been derived from the merged letters of MV (Maria Virgo) as part of the important Marian symbolism associated with the sign and constellation: this graphic etymology is quite imaginative, however. Virgo is of the earth element, and of the Mutable Quality, the influence being discriminative, quiet, exacting, nervous, shrewd and methodical. The nature of Virgo as it manifests in human beings is expressed in the many keywords which have been attached to it by modern atrologers: dignified, clever, intelligent, nervous (though often good under pressure), graceful, mentally alert, gentle – in a word, all those qualities which may be associated with an earth type expressing itself in the spiritual mental realm. In excess the Virgoan nature may be described in terms which express its underlying critical nature, the keywords being: carping, hyper-critical, 'bookish', shrewish and back-biting. Virgo says 'I want to clarify' more loudly than others tend to, though she may do this to be helpful, or to put others down, depending upon which side of the nature is being called into play. Virgo is traditionally ruled by Mercury, and marks the Fall of Venus.

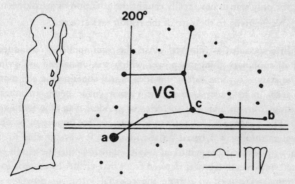

Figure 64: The constellation Virgo from the MODERN ZODIAC *(after Delporte), and the image of Virgo from the* SAN MINIATO ZODIAC.

Virgo constellation Zodiacal constellation, located approximately in the area set out by Delporte in figure 64, straddling the ecliptic over an arc of about 30 degrees. Virtually all the ancient and modern traditions visualize the image of Virgo as a maiden, and a whole string of feminine names have been applied to her, such as Persephone or Proserpina – but in essence she has been regarded as the archetypal woman, the Kore of the Greeks, the Puella of the Romans, and in the Hindu asterisms as Kanya (maiden): from very early Christian times she has been Virgin or Virgin Mary, and in Egyptian times Isis. Some of the more important variant names have revolved around the identification of the lucida in the asterism, the SPICA usually visualized as set in the Virgin's left hand or shoulder, as a child or as a sheaf of corn. The Greek name means 'ear of corn', and this was sometimes adapted by the Christians to the distaff which properly belonged to an heretical gnostic tradition: later Christians visualized the star as an image of the Child in the arms of the Virgin Mary (but see also STELLA MARIS). The link between the star and corn was sufficient to establish the entire asterism as the corn-goddess Demeter, from which we have Arista and Arista Puella, the Maiden of the Wheatfield. But it was mainly in the image of the celestial mother that Virgo was visualized – called even Astraea after the starry daughter of Themis: several of the medieval images of the Virgin Mary depict her dress decorated with stars (even with ears of corn) and such were doubtless derived from the pre-Christian stellar goddesses. Such images were also linked with the asterism as the Egyptian Isis and the Babylonian Ishtar. Ptolemy groups the stars of the constellation, and equates them all with beneficial influences – mainly with Mercury and Venus, yet, besides the lucida, the only star to have really caught the attention of astrologers is ZAVI-JAVA.

virtue In a specialist astrological sense, the term applied to a hidden power within all sublunary things, a power which was visualized as having come from the stars. To some extent this idea is still contained in the modern use of the term in its general sense, the human virtue being an inner quality proceeding from an image of the celestial or ideal man (the Latin *vir* means 'man'). The practice of TALISMANIC MAGIC, so important a branch of medieval astrology, was based on the idea that it was possible to 'inject' a celestial or planetary virtue into an amulet, which would then itself contain a beneficial or protective power derived from that virtue. In esoteric lore many other terms were used to denote the celestial virtue or power – see, for example, ILECH and QUINTESSENCE.

Virtues See VIRTUTES.

virtues and vices In the medieval tradition the seven virtues and vices were linked with the seven planets in a variety of systems, and these frequently appear in medieval astrological symbolism in paintings and sculpture. The correspondences in table 89 are fairly standard, of the kind used by Giotto and his school.

TABLE 89

Planet	Day of the week	Virtue	Vice	Part of man
SU	Sunday	Hope	Indolence	Life-force (spirit)
MO	Monday	Chastity	Envy	Psychic forces (soul)
ME	Tuesday	Wisdom	Gluttony	Intellect (mind)
VE	Wednesday	Love	Lust	Divinity in man
MA	Thursday	Courage	Wrath	Bestial lower element
JU	Friday	Faith	Pride	Higher aspirations
SA	Saturday	Prudence	Covetousness	Physical body

Virtutes Name given to the incorporeal spiritual beings of the series of CELESTIAL HIERARCHIES established by Dionysius the Areopagite. Strictly speaking they should be linked with the EXSUSIAI, as rulers of the Sphere of the Sun – the Golden Legend confirms this by describing the Virtues (as they are sometimes called) with the divine mystery of the Eucharist, for in the esoteric lore Christ is linked with the solar sphere. However, the Virtutes have been wrongly equated with the ARCHAI (for example, by St Gregory), which would give them rule over Venus. The rule of the ARCHAI over Venus, and of the Virtutes over the Sun is beyond dispute in astrological lore, however.

visible aspects A term used by Ptolemy in his *Almagest* to distinguish nine kinds of aspects which might be seen by the astrologer. These (all of which appear to involve aspects with the Sun) play no part in modern astrology, but the aspectal structure is evident from the nine terms used by Ptolemy: they are the matutine subsolar, matutine culmination, matutine setting, meridianal subsolar, meridianal culmination, meridianal setting, vespertine subsolar, vespertine culmination and vespertine setting.

visible horizon See CELESTIAL HORIZON.

vital principles See SEVEN VITAL PRINCIPLES.

vital signs A term applied to the three signs of the fire triplicity, which are supposed to be profligate of their vital forces.

Voel In certain forms of ceremonial magic, the name is given to the Governor of zodiacal Leo.

void of course In modern astrology this old term has been misunderstood. Originally it was derived from horary astrology, and applied to a planet which passes through a zodiacal sign without encountering an aspect to or from another planet. In natal astrology, a planet which formed no aspect before leaving the sign of its radical position was also said to be void of course. Dean, studying the validity of the traditions relating to the tropical and sidereal zodiac controversy, suggests that the 'void of course Moon' might be used as a basis for furnishing useful checks, but makes the mistake of defining void of course in a personal way, and applying a notion derived from horary astrology to realms which are unrelated to this. Dean claims that the Moon is void of course between its last major aspect in one sign and its entry into the next.

Volans Constellation, the Piscis Volans (flying fish), originated by Bayer in 1604, south of Canopus. The asterism appears to have been recognized earlier under the name of Passer (sparrow), however. Robson tells us that it gives a quick mind, and a poetic or artistic ability.

Volasfera symbols Name given to a series of readings ascribed to each of the 360 degrees of the zodiac, according to the system of DEGREE SYMBOLS, in a number of verbal symbols and exegeses by the Italian astrologer Borelli. The system was popularized by the astrologer Sepharial in a translation published at the end of the last century.

Volucris See CYGNUS.

volvelle A name originally applied to medieval parchment dials (found still in certain astrological and calendrical manuscripts) used for computing dates of movable feasts, simple cycles (such as the phases of the Moon), times of sunrise and sunset, tidal changes, and so on. In its simplest form it consisted of one or more movable circles, graduated with figures, which could be swivelled against a similar graduated background, to enable one to make a calibrated reading of co-ordinates. In recent years a complex form of volvelle has been developed to enable the astrologer to compute easily the Ascendant and Midheaven degrees for given places and times. Sophisticated plastic volvelles may be obtained nowadays, but these are rapidly being displaced by computerized charting.

Vrishabham A Sanskrit name for zodiacal Gemini, as used in Hindu astrology.

Vrishikam The Sanskrit name for zodiacal Scorpio – but see also VICAKHA.

Vulcan The term has several applications in astrology. It was a name given by Sutcliffe to one of the intramercurial *HYPOTHETICAL PLANETS. Later astrologers assigned this Sutcliffe Vulcan to the so-called day-house rule over Scorpio, and the night-house rule over Leo, and claimed that it was of an explosive nature. This same Vulcan may be linked with one of the AROMAL PLANETS, among which the astrologer Harris listed his own Vulcan in 1884, though this does not appear to have been intramercurial. The Vulcan described by Bailey in her system of INTUITIONAL ASTROLOGY is actually one of the *SACRED PLANETS, and is connected especially with the *FIRST RAY. The term Vulcan is also used to veil an esoteric astrological doctrine: it is one of the names applied to a planetary state which is supposed to represent the future embodiment of our Earth (see FUTURE PLANETS and VULCAN PERIOD). This latter may be connected with the Vulcan of Thierens, which he sometimes called by the alternative name Mercury, or Mercury-Vulcan. Thierens also called this planet Vulcanus, though it is unrelated to the Vulkanus of the URANIAN ASTROLOGY. None of the hypotheticals listed above appear to correspond with the Vulcan given as an intramercurial planet by the astrologer Weston, round about 1920. Besant mentions the intramercurial Vulcan in her account of Chaldean astrology, and this appears to be different from the Vulcan of the *GLOBES, mentioned in the same text. The astrologer Pagan also mentions a Vulcan, but it is far from clear which she had in mind, though she does link it with the action of Scorpio, a sign to which (even by 1930) she had assigned Pluto as ruler (see PAGAN-PLUTO). The phrase chosen by Pagan to express the Vulcanian element in man is 'the stuff of which he is made', something in man which eludes analysis, and for which one may not give a final account. As she puts it, the 'Smith' (i.e. Vulcan) 'puts everyone into the furnace, and by burning away the dross, gets down to the essence of our [*sic*] being'.

Vulcan period   A term used in esoteric astrology to denote a future (and in some respects final) condition of our present Earth, marking a future evolution of humanity to the highest spiritual consciousness. The term is used only in esoteric astrology, and has nothing to do with any of the hypotheticals listed under VULCAN.

Vulcan Scheme   The Vulcan Scheme or the Vulcanian Chain are terms derived from considerations set out under the theosophical *SCHEME OF EVOLUTION, not to be confused with the hypotheticals listed under VULCAN. It is claimed that this Vulcan Scheme relates to a *CHAIN which will produce entities at a lower level than those of the present EARTH CHAIN.

Vulcanus   The esoteric astrologer Thierens sometimes calls his own hypothetical VULCAN by this name.

Vulkanus   Name given to one of the *HYPOTHETICAL PLANETS used in the URANIAN ASTROLOGY, said to relate to power and strength, and to represent an upper octave of Mars.

Vulpeca   Constellation formed by Hevelius in 1690 as Vulpecula cum Ansere (little fox with the goose) between Sagitta and Cygnus. Inevitably astrologers have linked the asterism with the fox-like characteristics, for it is said to give cunning and a voracious nature.

Vultur Cadens   See FIFTEEN STARS and WEGA.

Vultur Volans   See AQUILA.

wane   When a planet comes to the Descendant, its influence is said to be 'broken', and it is said to wane for the entire semi-arc. But distinguish the waning Moon – see DECREASING LIGHT.

waning Moon   See DECREASING LIGHT.

warmth trigon   See SIDEREAL MOON RHYTHM.

war peak cycle   See CYCLE.

Wasat   Fixed star, the alpha of Gemini, the name for the Arabic Al Wasat (the middle) which some explain as reference to the proximity of the star to the ecliptic. It is a violent star, giving (in the words of Robson) 'destructiveness as a first principle': it is especially connected with chemicals and poisons – the Chinese called it Ta Tsun (great wine jar).

Watchers of the Heavens   See ROYAL STARS.

watchword   An equivalent term for KEYWORD applied in a special sense by the astrologer Pagan. Each of the Ascendant signs are accorded a watchword which is included in an analytic schema and said to be modified (in a particular chart) by the sign occupied by the Moon.

AR Action	TA Stability	GE Variety	CN Sympathy
LE Faith	VG Service	LB Harmony	SC Power
SF Law and Liberty	CP Excelsio	AQ Investigation	PI Unity

Water-bearer   One of the names for zodiacal and constellation Aquarius.

water element   The water element finds expression in the zodiac through the three signs Cancer, Scorpio and Pisces. It is the element most deeply concerned with reflection, sensitivity and fluidity. Water is esoterically an impressionable principle, the important aspect of this tradition revolving around the idea (more fully expressed in alchemical than in astrological texts) that water, to operate in accordance with its true nature, needs to be contained, and thus given form. This is why among the Greeks (at least with Thales) water was the first principle from which form arose, the potentiality of form (in the medieval tradition the first principle was actually the element of fire). This partly explains the esoteric connection drawn between the 'dry' Moon and water, for the Moon is that which gives growth and change to form: it perhaps also explains why the Archangelic ruler of the Moon (Gabriel) should be the announcer of a new form of spiritual existence for the Earth. The esoteric water element is far removed from the fluidic water of the modern chemists, though suitable analogies between the material fluid and the spiritual principle are often found in the popular treatments of the elements.

waters of space See OURANOS.

water signs These are Cancer, Scorpio and Pisces – see WATER TRIPLICITY.

water triplicity This is the group of signs linked with the water element, manifesting different aspects of the PHLEGMATIC temperament: Cancer, Scorpio and Pisces. Water may be seen in three different forms: Cancer is water as a fluid – changeable, tending to reflect the colours of the skies and landscapes around; Scorpio is frozen water, a shield of ice beneath which is a depth unknown; Pisces is vaporous water, almost a dematerialized state of being, often out of touch with (hovering over or disincarnated from) the physical realm. Each of the water signs are extremely emotional and are often described as 'their own worst enemies' or 'victims of circumstances'. Since the water triplicity is concerned with the spiritual element in life, with the dissolving of boundaries, each of the signs reflects an aspect of this in the world. Cancer is connected with the beginnings of life, with the fairyland world of childhood, which dissolves the world in its own mythologies. Scorpio appears to be almost obsessed with death (in reality, with resurrection and redemption), that final dissolving of physical boundaries, with peering beyond the threshold of the material. Pisces is concerned with the idea of the poetical, with that life of spirit which is only tenuously linked with the material realm.

waxing Moon See INCREASING IN LIGHT.

weak degrees See AZIMENE DEGREES.

weak signs The three signs Cancer, Capricorn and Pisces are traditionally

termed weak signs, the basis for the classification being lost. See STRONG
SIGNS.

weekly progressions See WEEKLY SERIES.

weekly series One of the four measures for computing symbolic time, used
in SECONDARY DIRECTIONS, and based on the idea that the true solar day
should be regarded as equivalent to a week. The term is 'Weekly Series of
Secondary Directions', sometimes (confusingly) reduced to 'Weekly Directions'.

Wega Fixed star, the alpha of Lyra, and just as frequently called Vega. The
star is of considerable importance in astrology, mainly because (under the
name Maat) it marked the pole and was also used for Egyptian temple
orientation. The name Maat is often translated as meaning 'vulture star',
which may explain the medieval term Vultur Cadens (falling vulture) in the
list of the FIFTEEN STARS. The Arabian Al Waki itself means 'falling' and
from this we have many derivations. The Greek Lyra was turned into the
late-medieval Allore, Aloshore, and so on, whilst the Roman Fidus became
Fidicula. As a pole-star the Akkadians apparently called it Tir-anna (Life of
Heaven), and the Assyrians Dayan-same (Judge of Heaven). In spite of its
link with the vulture, or grype (a medieval variant), the influence is generally
regarded as beneficial, for it brings refinement and ideality, though a certain
degree of pretentiousness and lasciviousness. The curious thing is that
Ptolemy equates the influence with that of Venus and Mercury, yet the
traditional readings for the conjunctions with the benefics seem to give
unpleasant results.

Western Angle The cusp of the house opposite to the Ascendant is called
the Western Angle: it is associated with the sign Libra, and marks the balance
between day and night.

western houses The western or occidental houses are those between the 5th
and the 9th houses, inclusive. The house is very often called the Western
House.

West Point In astrology, the term West Point is sometimes used as a
synonym for the Descendant, by analogy with EAST POINT.

Wemyss degrees A system of DEGREE SYMBOLS proposed by the Scottish
astrologer Wemyss, the basis of an extraordinary classification of horoscopic

material. In some cases the degree influences listed are spread over a number of degrees, so that in many respects the method should be described as being based on 'arc symbols'. Not all the degrees or arcs are accorded a meaning, and whilst the traditional order of zodiacal signs is adhered to, Wemyss incorporates a number of HYPOTHETICAL PLANETS as rulers in his system, as well as changing one or two of the traditional ascriptions: Leo is given to HERCULES, Virgo to DIDO, Libra to Neptune, Scorpio to Uranus, Sagittarius to JASON and Pisces to the ASTEROIDS.

Wemyss-Pluto   The astrologer Wemyss, writing *c.* 1929, attempted a revolutionary review of planetary rulerships over signs, as indicated in WEMYSS DEGREES. One hypothetical ruler he named Pluto, as ruler of Cancer. It was about this time that the modern planet PLUTO was discovered by Lowell, and there was soon a consensus among astrologers (though see PAGAN-PLUTO) that this new planet should be given rule over Scorpio. Accordingly, there was a need to distinguish the two different 'Plutos', and the modern planet was for a while called the Lowell Pluto, and the hypothetical the Wemyss-Pluto. However, Wemyss claimed a revolution of 1,366 years for his hypothetical, which contrasts with the 248-year revolution of the modern Pluto.

Whale's Jaw   One of the names given in the 19th-century astrological textbooks for the fixed star MENKAR.

White Eagle   Sometimes used by modern astrologers as a term for zodiacal Scorpio. See EAGLE.

White Tiger   One of the so-called constellations of the Chinese Heavens, which is really an arc of seven SIEU, from Goei (the beta of Andromeda) to Shen (the delta of Orionis).

whole signs   A term inexplicably applied to Taurus, Gemini, Leo, Scorpio, Sagittarius and Aquarius – perhaps through some misunderstanding of the classification accorded to the negative-positive polarities arising from the *SECTA. Sometimes the three air signs Gemini, Libra and Aquarius are called whole signs, again with no apparent reason.

Winged Bull   A name (rarely) used in modern times for zodiacal and constellational Taurus, even though the image is not winged, save indirectly through its connection with the TETRAMORPH.

Winged Horse   See PEGASUS.

winter solstice   See SOLSTICES.

wintry quadrant   See QUADRANTS.

Wolf   See LUPUS.

world cycle of mundane events   A designation used by Tucker to refer to his highly personal method of relating world events (mundane events) to celestial phenomena (in his system, aspects of the 'slow moving', extra-jupiterian planets, with the curious absence – in 1936 – of Pluto). Tucker claims to find a cycle of 128 years, relating to the return of Neptune to the same zodiacal sign and degree at the end of every ninth period of 128 years, around which he constructs his own world cycle table of events. As with all similar attempts to establish world cycles against celestial phenomena (see, for example, those of Sucher in regard to the heliocentrically considered cycles of astrosophy), there is an unfortunate simplicity in the approach to historical events. For example, the celestial phenomena (NE opposition UR, UR conjunct SA) is associated with the Crusades, which for some reason Tucker dates as 1148, when this date refers (and then only approximately) to the Second Crusade, which was a disaster. The Crusades were of course conducted over an extensive period, from as early as 1096 until at least 1291.

world cycle   See WORLD CYCLE OF MUNDANE EVENTS.

world spheres   See OPHANIM.

Y

Yatra A Sanskrit term used to denote the Hindu form of military astrology – see OMINA.

Yavanesvara See LORD OF THE GREEKS.

year In general the word is applied in astrology to one of the solar cycles, such as the solar year – but see also ANOMALISTIC YEAR, SIDEREAL YEAR and TIME. Additionally, in esoteric astrology the term is applied to larger periodicities which are usually multiples of the solar-cycle unit: for example, the Divine Year is 311,040 million solar cycles. Again, the PLATONIC YEAR is a similar extension of the solar unit, linked with the periodicities of PRECESSION.

year for a day See MEASURE OF TIME.

yearly chart A term used by Bradley for the SOLAR RETURN.

yearly progressions See YEARLY SERIES.

yearly rhythm A term used by the astrologer Ebertin to denote a system of correspondences linked with the traditional seasons, viewed from a point of view of the zodiacal cycle. In synopsis the sequence runs:

AR The unfolding of energy relating to the time of germination and, within man, the urge to act.

TA The creation of form, through strengthening and invigorating, in the human being constructiveness.

GE The multiplicity and diversity of form and life, expressed in the human being through versatility.

CN Fecundation and fertilization, in the human being receptivity and connubial love.

LE Ripening, harvesting, linked in the human being with self-confidence.

VG The gathering of fruit, linked with diligence and the critical faculty.

LB   Balance and adjustment in the economy of nature, linked on the human plane with the communal sense, and the sense of obligation.

SC   The death process, linked with continuation of life within the seed, as well as psychologically with the endurance principle and overestimation of self.

SG   The season of hibernation, the cultivation of the inner life of the human and (spiritual) expansion.

CP   The season of torpidity, 'invisible' life, which is psychologically concentrated upon the personal sense of self.

AQ   The waiting time and fasting (linked with the Lenten period), on the psychological plane connected with a 'wait and see' attitude, and readiness to help.

PI   The swelling of the seed in the watery earth, which symbolizes on a psychological level the receptivity of the human to many influences, and inner composure.

In connection with the agricultural yearly rhythm, see SIDEREAL LUNAR ZODIAC.

yearly series   One of the four measures for symbolically computing time, used in *SECONDARY DIRECTIONS, and based on the idea that the true solar day should be regarded as the equivalent of a year. In other words, the term is a more precise allocation of the *DAY FOR A YEAR system of progressions. The term is actually 'yearly series of secondary directions', sometimes reduced misleadingly to 'yearly directions'.

years of the gods   See DIVINE YEARS.

Yed Posterior   Fixed star, the epsilon of Ophiuchus, called in Euphratean astronomy Nitach-bat (man of death), a star which Allen notes as one of the several 'singular survivals' in modern astrology, since it is said to be of an evil nature, like the related YED PRIOR.

Yed Prior   Fixed star, the delta of Ophiuchus, the name from the Arabic Yad (hand), the star being set in the hand of the asterism, and distinguished from YED POSTERIOR. Like the latter, it is of an evil nature, associated with an influence of Venus with Saturn, and linked with immorality.

Yliadus   A term used to denote an invisible spirit or agency within the human body (and indeed within the FOUR ELEMENTS): sometimes spelled 'Yliadum' or 'Yleidus', and perhaps linked with the ILIASTER.

youthful quadrant   See QUADRANTS.

Yuga   In the Hindu chronology, which has had some influence on modern esoteric thought, the Yuga is a World Age, a thousandth part of a KALPA. Each Yuga is preceded by a twilight period (sandhya) and terminated by a twilight period (sandhyansa), and the four yugic periods (together called a Mahayuga — which also incorporates the twilights) are named the KRITA YUGA, the TRETA YUGA, the DWAPARA YUGA and the KALI YUGA. The periods are expressed in Divine Years and in the mortal years of ordinary time: in this system the Divine Year is equal to 360 mortal years. There is

TABLE 90		
Yugic period	*Divine years*	*Mortal years*
Krita	4,800	1,728,000
Treta	3,600	1,296,000
Dwapara	2,400	864,000
Kali	1,200	432,000

much obvious confusion among Western occultists regarding the yugic periods. In spite of the reliable figures listed in table 90, one still finds certain writers maintaining that mankind entered the present age, which is that of the Kali Yuga at the death of Krishna (c. 3102 BC) and that this age therefore ended in 1899 (or thereabouts).

Yugic periods   See YUGA.

Z

Zachariel   One of the names of the ruler of Jupiter, especially found in that literature dealing with Trithemian SECUNDADEIAN BEINGS.

Zadkiel   One of the more important names given to the spiritual being ruling the Sphere of Jupiter.

Zaniah   Fixed star, the eta of constellation Virgo, set in the southern wing of the asterism, the name being derived from an earlier Arabian minor asterism (see ZAVIJAVA). It is said by Ptolemy to be of the nature of Mercury conjunct Venus, and brings refinement, order and a lovable nature.

Zanrak   A name used by Ebertin for the fixed star gamma of Eridanus, probably in confusion with the Arabian term Zaurac or Zaurak, which are the standard names for the full Al Na'r al Zaurak (the bright star of the boat). Ebertin associates dire consequences of a prominent Zanrak, such as 'fear of death and suicidal tendencies', though it has not received much notice in traditional astrology. The actual 'case histories' evinced by Ebertin suggest a tendency to isolation, rather than to anything dire.

Zaphkiel   One of the many names given to the spiritual being ruling the Sphere of Saturn. Also known as Zophkiel.

Zariel division   See EQUAL DIVISON METHOD.

Zavijava   Fixed star, the beta of constellation Virgo, set below the head of the asterism, the name from the Arabian term for the 11th manzil (see AWWA), which was 'the barker', the asterism itself being linked with a kennel, of which this was (along with ZANIAH) one of the corner stars. Among the Arab astrologers it was a beneficial star, but for obscure reasons the Europeans have given it an influence equivalent to Mercury conjunct Mars.

zawzahr   The word, in a variety of spellings, is derived from the Arabic term for Gemini, Al Jawza. In recent times, the term has been misused and misunderstood by modern Islamic scholars in reference to a curious survival of Arabic astrological symbolism, in the form of a dragon imagery, which is related to the lunar nodes. These were accorded great importance in Arabic astrology and the dragon imagery entered into the symbolic repertoire of Arabic art, particularly in metalwork. A survey of the so-called zawzahr symbolism, along with an attempt to clarify the terminology relating to this important branch of symbolism, has been published by Gettings in 'A Misunderstood Arabic Astrological Symbol', Dar Al-Athar Al-Islamiyyah, 21 August 1989, pp. 10 ff.

Zazel   Name given by Agrippa (quoting ancient qabbalistic sources) to the Daemon of Saturn, for whom he gives the magical number 45 – see AGIEL.

zenith   The point immediately overhead from any given point on Earth is called the zenith, which is geometrically speaking the Pole of the Horizon. In popular speech, the Sun is sometimes said to reach its zenith, which is merely the highest point in the ecliptic relative to a given place, though it is rarely the zenith proper. In similar popular use, the cusp of the house is sometimes called the zenith.

zenith projection   See NONAGESIMAL.

Zenith Star   See ETTANIN.

zenith system   A method of *HOUSE DIVISION based on the projection of an equally trisected quadrant of the ecliptic by spatial division. The method was proposed (in somewhat obscure terms) by the astrologer Tucker, following the suggestions of his brother (also an astrologer) and seemingly based on a misunderstanding of Ptolemy. Tucker claimed that the system had four main merits: first, it divides the ecliptic by direct application; secondly, the poles of the system coincide with the poles of the ecliptic, and thus coincide with the poles of the horoscope being cast; thirdly, the Midheaven of the system will always be located 90 degrees west of the Ascendant, and fourthly, the permanency of positions thus attained (by the 90-degree angle) will aid purposes of accurate (statistical) research. Unfortunately, the same term had previously been used by the astrologer Leo some 30 years earlier, again for a method of house division – though this was too badly described (in what Leo himself termed a 'somewhat hasty survey') to be of much value. Whilst Leo's method is based on an equal division of the horizon projected on to the ecliptic from the horizon pole, Leo for some reason linked it with the CAMPANEAN SYSTEM, which projects the Prime Vertical division on to the

Zeus

ecliptic from the pole of the Prime Vertical. The astrologer Chandra quite rightly saw Leo's system as one of the two red herrings drawn across the trail of investigations: the other herring was Leo's unusable EAST POINT SYSTEM.

zero ayanamsa   See AYANAMSA and FIDUCIAL.

zero degree   In a practical sense, there is no such thing as a zero degree, even though the term is used widely in certain astrological circles. Almost always, the term really means 'within the degree', or is a reference to the theoretical commencement of a sign. It is not uncommon for astrologers to describe planets as being in the zero degree of a sign, when in fact they mean that the planets are in the 1st degree of that sign. The more astute astrologers therefore speak and write of the location as being the first point of Aries, rather than of the (non-existent) zero point, or zero degree, on the grounds that a circle (such as the ecliptic) may not have a break in it. As a result of the introduction of the term, and arising from the misunderstanding of its application, a misunderstanding has been perpetuated in popular astrological circles which has in turn resulted in various confusions. Frequently (because of this concept of a zero degree) the degree-placing of a planet or nodal point is recorded wrongly. A planet in (say) 10 degrees 12 minutes of Aries, or even in 10 degrees 59 minutes of Aries, is actually in the 11th degree of that sign, though not infrequently an astrologer will describe the first planet as being in 10 degrees, on the principle of 'rounding off' to the nearest degree. A planet in 0 degrees, 0 minutes and 1 second of Aries is, however, still in the 1st degree of that sign: it is not in a zero degree. The problems arising out of the wrong conversion of degrees – especially in regard to correct interpretation of platic aspects or the significance of the DEGREE SYMBOLS – are considerable. See ROUNDING OFF.

Zervana Akarana   Name of an ancient Persian primal god, variously translated as meaning 'without limit of time', or 'duration in a circle', though obviously pointing to an eternal existence beyond the time-barrier of the Sphere of Saturn. Within the realm of astrology this primal Zervana is interesting since images used to depict him combine the symbolism of the so-called TETRAMORPHS into one figure.

Zeugon   Greek term, sometimes Romanized as Zugon, meaning 'beam of a balance', and used to denote both zodiacal and constellational Libra.

Zeus   Name given to one of the *HYPOTHETICAL PLANETS used in the URANIAN ASTROLOGY, said to represent an upper octave of Uranus, and to be related to physical leadership and creativity.

Zib

Zib See PISCES CONSTELLATION.

zigiatus A term now obsolete, yet worth recording if only because it appears in so many early textbooks, is in fact a reflection of the nature of the astrological tradition. According to most accounts, the term appears to be a Latinized form of one of the Greek words meaning 'Libran', and the upshot of its specialist use is that it denotes a person born under Libra who is 'apt to commit suicide', and is therefore zigiatus. There is not a shred of truth in this tradition, nor does Ptolemy himself appear to use the word or even to discuss the idea. It is to be observed that Ptolemy's views on mental disturbances, immorality and demonic possession would find little adherence (and indeed would probably not even be understood) in modern astrology. In general it is his practice to write of crises and of immoralities (which he regards as sickness of the soul) in terms of planetary configurations, and not in terms of the signs, so that it is hardly likely that the idea arose with Ptolemy.

zoas See FOUR ZOAS.

zodiac Properly speaking, the belt centred on the ecliptic, divided into twelve arcs of 30 degrees, called the 'signs of the zodiac', with the following order: Aries, Taurus, Gemini, Cancer, Leo, Virgo, Libra, Scorpio, Sagittarius, Capricorn, Aquarius, Pisces. The term 'zodiac' is shamelessly misunderstood in its strictest meaning and etymology, so that there is now much confusion in the non-specialist mind about a thing which at one time was not confusing at all. The confusion has arisen because of the similarity of the names used to denote the signs of the zodiac and the twelve asterisms which are included among the 14 or so which extend into the zodiacal belt. See, however, CONSTELLATIONS and SIGNS OF THE ZODIAC, and distinguish between TROPICAL ZODIAC and the CONSTELLATIONAL ZODIAC. The zodiac is properly the *zodiacus*, a Latin term derived from the Greek *zodion*, itself etymologically connected with the word *zoon*, which is difficult of translation, but which (as Smith has shown) is certainly linked with the idea of 'life' or with 'living beings'. The popular derivations insist that the Greek zodiac was a circle of animals, which is simply not true: it was a circle of living beings – a concept which of course links with the living beings of Ezekiel, associated with the CHERUBIM as the beings who dwell outside time (the limits of time being marked in the ancient cosmoconception by the Sphere of Saturn). We see, therefore, that when Wilson says that the zodiac takes its name from the animals associated with the signs, he is merely getting the derivation the wrong way round. In any case, it is an elementary mistake to confuse the SIGNS with the IMAGES. The zodiac is technically a belt, traditionally extending 8 degrees on either side of the ecliptic (though some earlier sources give 6 or even 9 degrees) and in modern times the belt would have to be widened considerably to allow for the eccentricities of the so-called NEW PLANETS. This should not, however, disguise the fact that the tropical zodiac was regarded as a division of the ecliptic itself, into twelve equal arcs of 30 degrees. The fact that fixed stars remote in latitude from the ecliptic were (after suitable geometric adjustment to align them to Right Ascension coordinates) regarded as projecting their influences upon this ecliptic, means that in effect the divisions were absolute for all latitudes. The most influential modern attempt to define the sidereal zodiac, which is distinct from the tropical zodiac, is that of Delporte – see MODERN ZODIAC. See, however, SIDEREAL CORRESPONDENCES and SIDEREAL ZODIAC. For a list of other zodiacal systems, see ZODIACS. The names of the twelve signs and the related constellations in early Iranian, Greek, Sanskrit (Hindu) and Babylonian zodiacs, along with the possible translations for the last, are given in table 91.

	Iranian *(Sassanian)*	*Greek* *(Classical)*	*Sanskrit* *(Zodiacal)*	*Babylonian* *(Constellations)*
AR	Varak	Krios	Mesham	Hunga (Hireling)
TA	Tora	Tauros	Vrisha	Gud.an.na
GE	Do-patkar	Didemoi	Mithuna	Mas.tab.ba.gal.gal
CN	Kalakang	Karkinos	Karkata	Al.lul
LE	Ser	Leon	Simha	Ur.gu.la
VG	Khusak	Parthenos	Kanya	Ab.sin (Furrow)
LB	Tarazuk	Zugos	Tula	Zibanitu (Horn?)
SC	Gazdum	Skorpios	Vrischika	Gir.tab
SG	Nemasp	Toxotes	Dhanus	Pa.bil.sag
CP	Vahik	Aigokeros	Makara	Suhur.mas (Goat-fish)
AQ	Dul	Hydroxous	Kumbha	Gu.la (Giant?)
PI	Mahik	Ichthys	Mina	Sim.mah (Swallow?)

TABLE 91 (header)

zodiacal age See PRECESSION.

zodiacal Angels As with the ZODIACAL SPIRITS there is much confusion as to the nomenclature and roles of the zodiacal Angels, and this confusion is at least as old as Agrippa. However, it is the name-list preserved by Agrippa which has been absorbed into modern astrology, to the extent that the names are still important, even if the specific functions of the angels are now confused. This list, derived from his *Duodenarii Scala* (see SCALA) is:

AR	Malchidiel	TA	Asmodel	GE	Ambriel	CN	Muriel
LE	Verchiel	VG	Hamaliel	LB	Zuriel	SC	Barbiel
SG	Adnachiel	CP	Hanael	AQ	Gabiel	PI	Barchiel

The four Archangels, whom Agrippa quite rightly describes as the 'four angels of the cardinal divisions of the Heavens', are sometimes wrongly called 'zodiacal Angels' – these are MICHAEL, RAPHAEL, GABRIEL and URIEL, respectively of the regions of fire, air, water and earth.

zodiacal band   See ECLIPTICAL INCLINATION.

zodiacal calendars   In Italy there are several calendrical devices which employ the sunlight to mark the progression of the days. In almost all cases a small aperture is so arranged as to allow the sunlight to fall upon a line which calibrates the daily zenith transits of the sun. The calibration is generally related to the signs of the zodiac, on the standard 'Sun in sign' method. The largest Italian zodiacal calendar is that designed by Cassendi inside San Petronio, Bologna, but a large and graphically sophisticated external system is found beneath the arches of the Palazzo della Ragione in Bergamo: this not only traces the zenith transits, but also the lemniscate solar motion between the equinoxes and solstices. The small calendar on the south wall of the Duomo in Turin is perhaps more typical, yet this appears to mark with the solar beams only the transits through Taurus (the name Torino is derived from the latin 'Tauro'). There are remains of zodiacal calendars in several other places in Italy – most notably in the Salone at Padua (see PADUAN CYCLE), and it is clear that many such calendars existed prior to the introduction of the mechanical ZODIACAL CLOCKS.

zodiacal chorography   See CHOROGRAPHY.

zodiacal clocks   After the introduction of the mechanical clock to Europe, the medieval system of Sun-driven ZODIACAL CALENDARS gradually gave rise to clockwork calendars, sometimes (confusingly) called Horologia, which also employed zodiacal imagery. Outstanding examples are still found in working condition in Brescia, Venice and Padua, though the latter seems to have imagery influenced by constellational rather than by zodiacal lore. It is difficult to give precise dates for the zodiacal images on such clocks, since frequent overhauls and restorations have usually resulted in several changes in the original images. For example, according to Checchi the zodiacal clock in Palazzo del Capitanio in Padua, which is said to be by Giovanni Dondi, and dated 1344, had been restored at least six times prior to the complete overhaul of 1838.

zodiacal colours   A number of different associations between the signs of the zodiac and colours have been preserved and used within the astrological tradition, though there is more agreement here than in the associations drawn up in connection with the PLANETARY COLOURS. The following may be regarded as representative, being drawn from a number of ancient and modern sources:

AR   Red, brilliant red, orange, and white and red.

TA   Red, red and citrine, white and lemon, yellow, red-orange, green, cream, and brown.

GE   Orange, violet, crystal blue, and 'colours which clash'.

CN   Green, soft greens, white, russet, blue-green, orange-yellow, and 'soft colours'.

LE   Gold, orange, red, yellow, and red and green.

VG   Amber, yellow-green, violet, blue, 'spotted colours', and white flecked with blues. Cornell surprisingly cites black.

LB   Green, yellow, crimson, white, blues, 'dark crimson', and tawny and dark brown.

SC   Murky red, blood-red, brown, green-blue, dark red, and black.

SG   Blue, sky-blue, red, olive, and blue-yellow.

CP   Blue-violet, russet, indigo, greys, and black.

AQ   Indigo, azure, violet, strong blues, and light blues.

PI   White, violet-red, blues, greens, and soft pastel colours, especially light blues and greens.

The Chromatic Chart published by Carey is perhaps the most systematic attempt to range the zodiacal colour associations, resting as it does on a series of complementaries. This series, which has no satisfactory rationalization, and which virtually ignores the traditional associations, is given as:

AR Cerise or magenta	TA Scarlet or cerise-orange	GE Orange
CN Orange-yellow	LE Yellow	VG Yellow-green
LB Green	SC Blue-green	SG Blue
CP Blue-violet	AQ Violet	PI Cerise-violet

zodiacal cycles   We have inherited from medieval astrological lore a number of zodiacal fresco cycles which were painted to serve a variety of different functions. The majority of these cycles are actually compendia of astrological (and sometimes also correlated theological) lore. For example, the largest, the PADUAN CYCLE, incorporates material from zodiacal, planetary, constellational, seasonal and decanate imagery integrated with both esoteric and exoteric Christian theology. In some other cases, the cycles are essentially constellational, as for example in the VICENZA CYCLE, which was greatly influenced by Greek and Roman numismatic imagery. In other cases, the astrological imagery is designed to portray on a vast scale a personal horoscope, as in the Chigi Palace in the Farnesina in Rome, studied (with a few errors) by Gleadow. The so-called zodiacal frescoes in the Palazzo Schiffanoia in Ferrara (now only partly complete, though recently restored) are actually a complex series of images based upon a scholarly reconstruction of medieval decanate systems probably derived ultimately from Egyptian astrology, in turn transformed by Arabian concepts, as suggested by Warburg and Ancona. Distinguish zodiacal cycles from ZODIACAL SCULPTURES.

zodiacal directions   A term applied to a limited number of progressed points and planets, which may be directed in the order of the signs (hence 'direct zodiacal directions'), or against the order of the signs (hence 'converse zodiacal directions'). The term should properly be limited to the directions formed by the Ascendant, Midheaven and the luminaries.

zodiacal houses   An ambiguous and misleading term which the astrologer Leo attempts to define in relation to signs, rather than houses – 'the signs which any planet rules, or in which it has most influence, are said to be its "houses"'. This is sheer nonsense, though it is often repeated by later writers: SIGNS are not HOUSES and never will be. The sign is a subdivision of an orbital revolution, whilst the houses are usually subdivisions of an axial rotation. A planet may be in the sign it rules and in the house it rules (though in this connection, see RULERSHIP and LORD), but never in a zodiacal house.

zodiacal light   An elongated, luminous, triangular figure with its base approximately on the horizon which, as Blavatsky claims 'is entirely unknown to science' and may 'be seen only during the morning and evening twilights'. Blavatsky says that since the nature of this zodiacal light is known only to initiates, the real significance and occult meaning of the zodiac also remains a closed book, save of course to initiates.

zodiacal man   See MELOTHESIC MAN.

zodiacal sculptures   The importance of astrological lore in relation to medieval theology is attested by the vast number of zodiacal sculptures which have survived on both external and internal fabrics of churches, cathedrals and public buildings. This is especially true of France and Italy. Apart from the sculptures and reliefs on the many ZODIACAL CALENDARS and ZODIACAL CYCLES, important astrological reliefs are found on the outer west-work of Amiens, Chartres and Vézelay. The 14th-century zodiac at Chartres is only apparently incomplete, for the 'missing' signs (Gemini and Pisces) are located on the southern door of the west-work, in a medieval esoteric symbolism which has been studied in some depth by Gettings. In the narthex of Vézelay (early 14th century), interspersed with seasonal images and theological symbols, are a series of remarkable zodiacal roundels. Indeed, many French cathedrals incorporate zodiacal cycles or individual sculptures relating to cosmic lore. Italy is particularly rich in zodiacal sculptural cycles, as in Florence, where both the Baptistry and the church of San Miniato preserve 13th-century examples (see SAN MINIATO ZODIAC). In the Baptistry at Parma, above the 12th-century zodiacal sculptures by Antelami, are the remains of a system which was without doubt once complete, though now reduced to only four signs, and vestiges of frescoes. The so-called zodiacal portal at La Sacra di San Michele (Val di Susa) is actually a 12th-century constellational portal, with the twelve zodiacal images and 19 constellational images. It has been claimed by Naylor that the zodiacal sculptures (1497) in the gateway ceiling of Merton College, Oxford, are orientated in relation to a personal horoscope, but as no planets are incorporated (either as symbols or sigils), this is extremely unlikely.

zodiacal spirits　In the esoteric tradition, the true zodiacal spirits are variously the CHERUBIM and the SERAPHIM, depending upon the system of Hierarchies used. However, the term is often in modern times misused for any of the spirits associated with the individual signs of the zodiac (for example Ausiel as zodiacal spirit, or Governor, of zodiacal Aquarius). Such names go back to the time when it was believed that the planetary motions were regulated by spiritual beings. Even by the time of Agrippa, when the last fragments of the esoteric tradition relating to spirits were being put into print, there was much confusion as to spiritual beings, the ZODIACAL ANGELS, the SECUNDADEIAN BEINGS, the PLANETARY ANGELS, the rulers of the DECANS, and even the whole range of CELESTIAL HIERARCHIES being confused in name and operation.

zodiacs　Many different names have been used to describe the various zodiacal systems – for a brief outline of the main historical forms, see ASTRONOMICAL ZODIAC, BABYLONIAN ZODIAC, CONSTELATIONAL ZODIAC, GREEK ZODIAC, MODERN ZODIAC, SIDEREAL ZODIAC, STAR CALENDAR ZODIAC and ZODIAC.

zodiacus vitae   The Latin for 'zodiac of life', sometimes applied to the idea that the various signs of the zodiac symbolize (in their progression from Aries to Pisces) sequential eras or periods which may be related analogously to the life of mankind, individual humans, mythopoetic legends, and so on.

Zona Andromedae   See MIRACH.

zones   See CHOROGRAPHY.

zone times   See TIME ZONES.

zoon   See ZODIAC.

Zophhiel   See ZAPHKIEL.

Zosma   Fixed star (triple), the delta of constellation Leo, set in the lion's back, the name from the Greek for 'girdle'. Another name, Zubra, is cognate with ZUBRAH, which means 'mane', perhaps in reference to a displaced image. It is said to be of an influence equivalent to Saturn and Venus, and is connected with egotism, melancholy and immorality.

Zubenelgenubi   Fixed star (double), the alpha of Libra, set in the southern scale of the balance. The name is from the Arabic phrase meaning 'southern claw', from the Grecian Chela Notios, the reference to the claw deriving from the image of constellation Scorpio, which at one time was pictured as covering the Libran asterism with its own pincers, or claws. There are several derivative names and alternatives. Perhaps because of the ancient link with the claws of Scorpio, the star is said to bring disgrace, sickness and treachery, as well as dangers from poisoning. Ptolemy says that it is of the nature of Mars conjunct Saturn.

Zubenelschamali   Fixed star, the beta of constellation Libra, the name being derived from the Arabic phrase meaning 'northern claw', itself from the Grecian Chela Boreios (see ZUBENELGENUBI). Unlike the latter star, its balance weight, this star is said to bring good fortune, honours and happiness. Ptolemy likens it to the nature of Mercury conjunct Jupiter.

Zubrah   Al Zubrah, the 9th of the Arab MANZILS. See also ZOSMA.

Zugos   One of the Greek names for the sign and constellation Libra, used by Ptolemy. An earlier term, Chelon ton Skorpion (claws of the Scorpion), points to a time when the constellation was regarded as being a part of the Scorpio asterism.

Bibliography

ABANO, PIETRO D', *Lucidator astronomiae*, n.d.
    Tractatulus . . . de aspectibus planetarum versus lunam (trans. of supp. Hippocrates); 1608.

ACOSTA, J., *Histoire naturelle et morale des Indes*, Paris, 1600.

ADAM, C. G. M., *Fresh Sidelights on Astrology*, London, 1916.

ADDEY, J. M., *Astrology Reborn*, Bournemouth, 1973.
    The Discrimination of Birthtypes, Bournemouth, 1974.
    Harmonies in Astrology, 1976.

AGRIPPA VON NETTESHEYM, H. C., *De occulta philosophia*, 1533.

ALBERTI, L. S., *De re aedificatoria*, vol. 9, 1485.

AL BIRUNI, MUHAMAD IBN AHMAD (Abu Al Rashan), *The Book of Instruction in the Elements of the Art of Astrology*, written in Ghaznah, AD 1029, trans. R. R. Wright, London, 1934.

ALFRAGANUS, AHMAD IBN MOHAMMAD IBN KATHIR (Al Farghani), *Elementa Astronomica*, Amsterdam, 1669.

ALLEN, D. C., *The Star-Crossed Renaissance*, Durban, N. Carolina, 1941.

ALLEN, G., see *American Astrology*, Sep. 1957, and extensive references under DEAN (esp. pp. 47–71).

ALLEN, R. H., *Star Names: Their Lore and Meaning*, New York, 1963.

ALVIDAS, H. C., *Science and Key of Life*, 1903, etc.

ANCONA, PAOLO D', *The Schifanoia Months of Ferrara*, trans. L. Krasvik, Milan, 1955.

ANGELUS, J., *Astrolabium planum in tbulis ascendens*, Venice, 1488. See also ENGELS, *The Faces and Degrees of the Zodiac*, reprinted by 'Raphael', *c.* 1902, etc.

ARATUS, *Arati solensis phaenomena, et prognostica . . .*, Paris, 1559. See also MAASS.

ARISTOTLE, *De coelo*, and *De generatione*. In regard to cycles, see HASTINGS.

ARNALDUS, *Arnaldi de Villanova medici acutissimi opera nuperrime revisa . . .*, Lugduni, 1532.

ATHENAGORAS, see GUTHRIE, *Orpheus and Greek Religion*, London, 1952.

BACH, H. I., 'C. G. Jung on "Synchronicity"' (synopsis of 'Synchronizität als ein Prinzip . . .', *Naturerklärung und Psyche*, Jung, C. G., and Pauli, W. P.), 1953.

Bibliography

BACON, ROGER, *et al.*, *The Mirror of Alchimy*, London, 1597.

BAILEY, ALICE, *Esoteric Astrology*, New York and London, 1951.
The Externalization of the Hierarchy, Tunbridge Wells, 1957.

BAILEY E. H., *The Prenatal Epoch*, London, 1916.
Astrology and Birth Control, London, 1929.

BALLARIN, A., *Pinacoteca di Vicenza*, 1982.

BARRETT, B. F., *Heaven Revealed*, London, 1885.

BARRETT, F., *The Magnus, or Celestial Intelligencer*, London, 1801.

BARTSCHIUS (Jakob Bartsch), *Planisphaerium stellatum*, 1624.

BARZON, A., *I Cieli e la loro influenza*, Padua, 1924.

BAYER, JOANNES, *Uranometria omnium asterismorium*, 1603.

BECK, R., 'Interpreting the Ponza Zodiac', *Journal of Mithraic Studies*, vol. 1, no. 1, 1976.

BECKH, H., *Der Kosmische Rhythmus im Markus-Evangelium*, Basel, 1928.

BESANT, A., and LEADBETTER, C. W., *Thought Forms*, London, 1901.
Man, Whence, How and Whither, London, 1913.

BLAKE, W. See DAMON, and GETTINGS.

BLAVATSKY, H. P., *Isis Unveiled*, New York, 1877.
The Secret Doctrine, 1888 (Pasadena edn. 1970).

BLUNDEVILLE, M.B., *His Exercises*. But see GETTINGS, quoted in *Astrology*, no. 1, 1977.

BOBER, H., 'The Zodiacal Miniature of the Très Riches Heures of the Duke of Berry, its Source and Meaning', *Journal of the Warburg and Courtauld Institutes*, XI, 1948.

BOEHME, JACOB, *Three Principles*, n.d. See also GETTINGS, LAW and MUSES.

BOLL, F., BEZOLD, C., and GUNDEL, W., *Sternglaube und Sterndeutung*, Stuttgart, 1966.

BONATUS (Antonius Franciscus de Bonattis), *Universa astrosophia naturalis*, Patarii, 1687.

BONATUS, G., *De astronomia tractatus*, ed. Prueckner, Basle, 1550.

BORELLO, ANTONIO, see 'La Volasfera', trans. 'Sepharia', in 'CHARUBEL', *The Degree of the Zodiac Symbolized*, q.v.

BOSC, E. (J. Marcus de Vege), *La Doctrine ésotérique à travers les âges*, Paris, 1899.
Glossaire raisonné de la divination, Paris, 1910.

BOUCHE-LECLERCQ, A., *L'Astrologie grecque*, Paris, 1899.
Histore de la divination dans l'antiquité, Paris, 1874.

BRADLEY, D. A., *Profession and Birthdate*, Los Angeles, 1950.
Solar and Lunar Returns, Minnesota, 1968.

BRAHE, TYCHO, *Astronomiae instauratae progymnasmata*, Frankfurt, 1610.

BRAU, J. L., *et al.*, *Larousse Encyclopaedia of Astrology* (USA edn), 1980.

Bibliography

BUTLER, E., *Solar Biology, a Scientific Method*, California and London, 1920.

CARELLI, A., *The 360 Degrees of the Zodiac*, Washington DC, 1951.

CAREY, G. W., and PERRY, I. E., *The Zodiac and the Salts of Salvation*, New York, 1971.

CARRUTHERS, P. J., 'The Degrees of the Zodiac', *Journal of Astrological Studies*, no. 1, 1970.

CARTER, C. E. O., *The Principles of Astrology Theoretical and Applied*, London, 1925.

Symbolic Directions in Modern Astrology, London, 1929.

Astrological Aspects, 10th edn, London, 1969.

CENSORINUS, *De die natali*, Venice, 1581.

CHACORNAC, *Tables de maisons 'Chacornac' – pour les latitudes de 0 deg. à 57 deg.*, Paris, 1972.

'CHARUBEL' (pseud.), *The Degrees of the Zodiac Symbolized*, London, 1898.

CHECCHI, M., *et al.*, *Guida ai monumenti e agli oggetti d'arte*, Venice, 1961.

CHOISNARD, P. C. (Paul Flambert), *Langage astral: Traité sommaire d'astrologie scientifique*, Paris, 1930.

La Méthode statistique et le bon sens en astrologie scientifique, Paris, 1930.

CHRISTIAN, P., *The History and Practice of Magic*, critical trans. J. Kirkup and J. Shaw, New York, 1963.

CICERO, see MACROBIUS.

COLEY, H., *Anima astrologiae . . .*, London?, 1676.

Clavis astrologiae elemata . . ., London, 1676.

COLLIN, R., *The Theory of Celestial Influences*, London, 1954.

CONCHES, WILLIAM OF, *Philosophicarum et astronomicarum*, Basle, 1531.

CORFIELD, J., in *Urania*, 1, June 1814.

CORNELL, H. L., *Encyclopaedia of Medical Astrology*, 3rd revised edn, with introduction by L. Lowell, Minnesota and York, 1972.

CROMBIE, A. C., *Augustine to Galileo*, London, 1952.

CUMONT, F. V. M., *Astrology and Religion among the Greeks and Romans*, New York and London, 1912. See also 'Les Noms de planètes, *L'Antiquité classique*, IV, I, n.d.

DAMON, S. F., *A Blake Dictionary*, Brown University, 1965.

DANTE ALIGHIERI, *Divina Commedia*. For specialist astrological background, see MOORE and TOYNBEE.

DAVIS, A. J., *The Principles of Nature, her Divine Relations*, London, 1847.

The Penetralia, Boston, 1858.

The Diakka, and their Earthly Victims, being an Explanation of Much That is False and Repulsive in Spiritualism, New York, 1873.

DAVISON, R. C., quoted in *Astrology*, vol. 36, no. 3, 1962.

The Technique of Prediction, London, 1971.

Bibliography

DEAN, GEOFFREY, *et al.*, *Recent Advances in Natal Astrology*, Subiaco, Western Australia, 1977.

DEE, JOHN, *Monas hieroglyphica,* Antwerp, 1564. (None of the several available translations is reliable.)

DELAPORTA (G. Battista della Porta), *De humana physiognomonia,* Hanover, 1593.

    Coelestis physiognomoniae ... in quibus etiam astrologia refellitur, et inanis et imaginaria demonstratur, Naples, 1603.

DELPORTE, E., *Atlas Céleste*, Cambridge, 1930.

    Délimitation scientifique des constellations, Cambridge, 1930.

DERNAY, E., *Longitudes and Latitudes throughout the World*, New York, 1948.

DEUSEN, E. L. VAN, *Astrogenetics*, London, 1976.

DEVORE, N., *Encyclopaedia of Astrology*, New York, 1947.

DIESCHBOURG, J. For personal communications, see DEAN, pp. 313ff.

DIODORUS OF TARSUS, see 'On Destiny' in Photius, cod. 233, Patr. Grecque, Book CIII, p. 878 (quoted in brief by FULCANELLI).

DIONYSIUS THE AREOPAGITE, *Mystical Theology and the Celestial Hierarchies,* trans. by the editors of the Shrine of Wisdom, Fintry, Brook, Nr Godalming, 1965.

DOANE, D. C., *How to Read Cosmodynes*, San Francisco, 1974.

DOBYNS, Z. P., *The Node Book*, Los Angeles, 1973. For pre-1971 and post-1974 heliocentric advance errors, see DEAN, p. 275.

DRINKWATER, G.N., *Theosophy and the Western Mysteries*, London, 1944.

DURANDUS, WILLIAM, *Rational divinorum officiorum, de ecclesia et eius partibus* (sec. 8), n.d. But see also NISSEN.

EBERTIN, REINHOLD, *Kombination der Gestirneinflüsse* (trans. A. G. Roosedale and L. Kratzsch: *The Combination of Stellar Influences*), Aalen, 1972. See also supplement 1, trans. I. Hodges and C. Harvey, 1961.

    Man in the Universe: An Introduction to Cosmobiology, trans. L. Kratzsch, Aalen, 1973.

    Die Bedeutung der Fixsterne, incorporating material by E. Ebertin and G. Hoffman, *et al.* (trans.: *Fixed Stars and Their Interpretation*), Aalen, 1971.

    Transits, 1972.

EDWARDS, ORMOND, *A New Chronology of the Gospels*, 1978 (trans. from *Chronologie des Lebens Jesu und das Zeitgeheimnis der drei Jahre*, Stuttgart, 1978).

ELFERNIK, M. A., 'La Descente de l'âme d'après Macrobe', *Philosophia Antiqua*, vol. 16, 1968.

ENGELS, JOHANN, *Astrologicall Opticks*, London, 1655.

EPPING, J., *Astronomisches aus Babylon oder das Wissen der Chaldäer über dem gestirnten Himmel ...*, Freiburg im Breisgau, 1889.

Bibliography

ERLEWINE, M., *et al.*, *Interface – Planetary Nodes*, Michigan, 1976.

– and ERLEWINE, M., *Astrophysical Directions*, Michigan 1977.

EUSEBIUS OF ALEXANDRIA, see 'Sermo de astronomis', Patrum Nova Biblio-theca, 1852, and *Peri astronomon quam, praemissa de magis et stella quaes-tione . . .*, ed. T. C. Thilo, Halae, 1834.

EVANS, JOAN, *Magical Jewels of the Middle Ages and the Renaissance*, Oxford, 1922.

FAGAN, CYRIL, *Astrological Origins*, Minnesota, 1971.

Zodiacs Old and New, London, 1951.

FEERHOV, F., *Kursus der Praktischen Astrologie*, Leipzig, 1912.

FIGUIER, G. L., *Keppler, ou l'astrologie et l'astronomie*, Paris, 1889.

FIRMICUS (Firmicus Maternus), *Materni junioris . . . astronomicon libri VIII per Nicolaum Prucknerum astrologum nuper ab innimeris mendis vindicati*, 1533.

Matheseos liber, Venice, 1499.

FLAMSTEED, J., *Atlas Coelestis*, ed. M. F. and J. Hodgson, 1729. See also PEARCE, pp. 17–20.

FLUDD, ROBERT, *Utriusque cosmi minoris et majores technica historia . . .*, 1617–21.

FOERSTER, W. J., *Johann Keppler and die Harmonie der Sphären*, 1862.

FOUCHER, L., *Découvertes archéologiques à Thysaeus en 1961*, Institut d'archéologie, Tunis, notes et documents, vol. V (new series).

FRANKLAND, W., *Astrological Investigations*, London, 1927.

New Meaning in Astrology, London, 1929.

Keys to Symbolic Directing, London, 1930.

FREHER, see MUSES.

FULCANELLI, *Le Mystère des cathédrales: Esoteric Interpretation of the Hermetic Symbols of the Great Work*, trans. M. Sworder, London, 1971.

FYFE, AGNES, *Die Signatur des Mondes im Pflanzenreich*, 1967 (trans.: *Moon and Plant*, Capillary Dynamic Studies, Arlesheim, 1975).

The Signature of the Planet Mercury in Plants, n.d.

'Capillary Dynamic Studies', *British Homoeopathic Journal*, LXII, no. 4, 1973; LXIII, no. 1, 1974; LXIII, no. 2, 1974.

GAFURUS, F., *Practica Musice*, Mediolani, 1496. See also *The Practica Musicae of Franchinus Gafurius*, trans. I. Young, University of Wisconsin Press, 1969.

GARDNER J. E., *Dante's Ten Heavens: A Study of the Paradiso*, London, 1898.

GAUQUELIN, M., *L'Influence des astres: Étude critique et expérimentale*, Paris, 1955.

Cosmic Influences on Human Behaviour, London, 1974.

The Cosmic Clocks: From Astrology to a Modern Science, Chicago, 1967.

GAURICUS, L., *Tractatus astrologicus*, 1542.

Trattato d'astrologia ludiciaria sopra le nativite degli huomini et donne, 1539.

Calendarium ecclesiasticum novum . . ., 1552.

GENTRY, P., *Cosmogonie et astrologie de l'Extrême-Occident dans le voile d'Isis*, Paris, 1922.

Bibliography

GENUIT, H., 'In Memoriam Dr Walter A. Koch', *Astrology*, vol. 45, no. 1, 1971.

GETTINGS, F., *The Hidden Art: A Study of Occult Symbolism in Art*, London, 1978.

The Secrets of San Miniato al Monte: The Christian Mysteries of the Nave Zodiac, (French trans.: N. Marque; Italian trans.: C. M. de Mariassevich; and German trans.: H. Kahnert), Florence, 1982.

Dictionary of Occult, Hermetic and Alchemical Sigils, London and Boston, 1981.

GLEADOW, R., *The Origin of the Zodiac*, London, 1968.

GOAD, JOHN, *Astro-Meteorologica, or Aphorisms and Discoveries of the Bodies Celestial . . .*, London, 1686.

GOODMAN, F., 'Secrets of the Stone Towers', *Unexplained*, vol. 9, 107, 1982.

GRAY, W. G., *The Ladder of Lights (or Qabalah Renovata)*, Toddington, 1971.

GUNDEL, W., 'Dekane und Dekansternbilder', *Studien der Bib. Warburg*, vol. XIX, 1936.

HAASE, R., 'Kepler's Harmonies, between Pansophia and Mathesis Universalis', *Vistas in Astronomy*, vol. 18, 1975.

HADES, A. Y., *Manuel complet d'astrologie scientifique et traditionnelle*, Paris, 1967.

HALL, P., *Astrological Keywords*, London, 1959.

HAND, R. S., 'A New Approach to Transits', *Cosmecology Bulletin*, no. 4, 1976.

'Science and Symbolism in Astrology', *Geocosmic News*, vol. 5, 1, Sept. 1979.

HARRIS, T. L., *A Lyric of the Golden Age*, New York, 1856.

Arcana of Christianity, New York and London, 1867.

Wisdom of the Adepts, Fountain Grove, 1884.

HARRISON, C.J., *Transcendental Universe*, London, 1894.

HARWOOD STEELE, Introduction to *Itinerary of the Somerset Giants*, K. E. Maltwood, British Colombia, n.d.

HASBROUCK, M. M., *The Pursuit of Destiny*, New York, 1976.

HASTINGS, JAMES, *et al.* (eds.), *Encyclopaedia of Religion and Ethics*, Edinburgh and New York, 1971.

HAWKINS, J. R., *Transpluto*, 1976.

HEIDENREICH, A., *The Unknown in the Gospels*, London, 1972.

HEINDEL, M., *The Rosicrucian Cosmo-Conception*, 26th edn, California and London, 1971.

– and HEINDEL, A. F., *The Message of the Stars*, California and London, 1919.

HEIROZ, J., *L'Astrologie selon Morin de Villefranche . . .*, Paris, 1962.

HEVELIUS, J., *Firmamentum sobiescianum, sive uranographia . . .*, Gedani, 1690.

HIERATIC EPHEMERIS, *The Complete Planetary Ephemeris for 1950 to 2000 AD*, The Hieratic Publishing Co., Massachusetts, 1975?.

HITSCHLER, see R. C. Davison, 'Atomic Medicine – the Hitschler Method', *Astrology*, vol. 37, no. 4, 1963.

Bibliography

HONE, M. E., *The Modern Textbook of Astrology*, London, 1951.

HOWE, E., *Urania's Children: The Strange World of the Astrologers*, London, 1967.

HYGINUS, C. J., *Poeticon Astronomicon*, Venice, 1485. See also *The Myths of Hyginus*, trans. M. Grant, 1960.

ISIDORE, *Incipit liber primus ethimologiarum Isidori Hispalensis Episcopi*, Strasbourg, 1470.

Liber ethymologiarum Isidori Hyspalensis Episcopi, Basle, 1489.

IVANOFF, N., 'Il problema iconologico degli affreschi', *Il palazzo della ragione di Padova*, Padua, 1963.

JACOBSON, R. A., 'Using the 90-degree Dial . . .', *Spica*, vol. 9, no. 1, 1969.

JAYNE, C. A., *The Unknown Planets*, New York, 1974.

Horoscope Interpretation Outlined, New York, 1970.

Astrology Now, vol. 1, no. 8, 1975.

JEREMIAS, A., *Das Alte Testament im Lichte des Alten Orients*, trans. C. L. Beaumont: *The Old Testament in the Light of the Ancient East*, London, 1911, Leipzig, 1904.

JOCELYN, J., *Meditations on the Signs of the Zodiac*, London, 1970.

JOHNDRO, L. E., *The Stars: How and Why They Influence*, New York, 1973.

The Earth in the Heavens, New York, 1973.

JONAS, E., see OSTRANDER and SCHROEDER.

JONES, M. E., *Sabian Symbols in Astrology*, New York, 1953.

The Scope of Astrological Prediction, Washington, 1969.

JUNCTINUS, F., *Speculum astrologiae*, Lugduni, 1583.

De divinatione, quae astra diversum, Coloniae, 1580.

JUNG, C. G., see BACH.

KEANE, J. L., *What the Degrees Mean*, Connecticut, 1976.

KEMPTER F., *Rudolf Steiners sieben Zeichen der planetarischen Entwicklung*, Engelberg/Wuerttemberg, 1967.

KEPLER, JOHANNES, *J. Keppleri harmonices mundi*, Lincii, 1619. See also FIGUIER, FOERSTER, HAASE and PAULI.

KIRCHER, A., *Oedipus, aegyptiacus; hoc est, universalis hieroglyphicae veterum . . . instauration*, Rome, 1652.

KLIBANSKY, R., PANOFSKY, E. and SAXL, F., *Saturn and Melancholy*, London, 1964.

KOCH, W. A. and KNAPPICH, W., *Horoskop und Himmelshausen* (T1.2 Regiomontanus und das Häusersystem des Geburtsortes), 1952, etc.

KOCK, W. A. and SCHAECK, E., *Häuser Tabellen des Geburtsortes*, 1971.

KOLISKO, L., *Der Mond und das Pflanzenwachstum* (*The Moon and the Growth of Plants*, trans. Pease and Mirbt), 1936.

The Working of the Stars on Earthly Substance, Stuttgart, 1928.

Gold and the Sun, Stroud, 1947.

Bibliography

Sternewirken in Erdenstoffen, Saturn und Blei, Stroud, 1953.

KONIG, K., 'The Zodiac', in *The Modern Mystic*, Feb.–Mar. 1937.

KRAFFT, K. E., *Traité d'astro-biologie*, Brussels, 1939.

KRUPP, E. C., (ed.), *In Search of Ancient Astronomies*, London, 1979.

KUGLER, F. X., *Die Babylonis Mondrechnung . . .*, Freiburg im Breisgau, 1900.
Sternkunde und Sterndienst in Babel, Münster, 1907.

LACAILLE, L., *A Catalogue of 9766 Stars in the Southern Hemisphere . . .*, London, 1847.

LANDSCHEIDT, T., *Cosmic Cybernetics: The Foundations of a Modern Astrology*, trans. L. Kratzch, Aalen, 1973.

LAW, W., *The Works of Jacob Behmem the Teutonic Theosopher*, 1764.

LEADBEATER, C. W., *The Astral Plane*, London, 1895.
The Hidden Side of Things, Adyar, 1923. See also POWELL.

LEFELDT, H. and WITTE, A., *Rules for Planetary Pictures: The Astrology of Tomorrow*, Hamburg, 1974.

LEINBACH, E. V., *Degrees of the Zodiac*, Virginia, 1973.

LEO, ALAN, see re-issue of the 'Astrology for All' series of 1912: *Casting the Horoscope, How to Judge a Nativity, The Art of Synthesis, The Progressed Horoscope, The Key to Your Own Nativity*, and *Esoteric Astrology*. See also *Alan Leo's Dictionary of Astrology*, ed. Vivian E. Robson, 1929.
'A Thousand and One Notable Nativities', Alan Leo's Astrological Manuals, no. 11, *c.* 1910.

LEWIS, C. S., *The Discarded Image*, London, 1964.

LEWIS, JIM, *Astrocartography*, San Francisco, 1976.

LIBRA, C. AQ. (pseudo.), *Astrology: Its Technics and Ethics*, trans. G. Coba, Amersfoort, 1917.

LILLY, W., *Introduction to Astrology by W. L.*, ed. 'Zadkiel', 1835.
Christian Astrology, Modestly Treated . . ., London, 1659.
– (ed.) *Anima Astrologiae*, London, 1676.

LORENZ, D. M., *Tools of Astrology – Houses*, Eomega Grove Press, 1976.

LOVE, R. G., 'An Outline of the Topocentric System', *Astrology*, vol. 51, no. 2, 1977.

LOVEJOY, A. O., *The Great Chain of Being*, Harvard, 1936.

LUNDY, J. P, *Monumental Christianity, or the Art and Symbolism of the Primitive Church*, London, 1882.

LYNCH, J., *The Coffee Table Book of Astrology*, London, 1962.

MACROBIUS, *Somnium Scipionis ex Ciceronis libro de repubblica exertum*, Venice, 1472. See also ELFERNIK.

MALTWOOD, K., *A Guide to Glastonbury's Temple of the Stars*, London, 1929. See also HARWOOD STEEL.

MANILIUS, I., *M. Manilius astronomicon a Josepho Scaligero*, Venice, 1532.

Bibliography

MASS, ERNEST, *Arati Phaenomena*, Berlin, 1955.

MASSEY, G., *The Natural Genesis*, London, 1883.

MATHEWS, R. H., *Chinese–English Dictionary*, Cambridge, Mass., 1966.

MATTHEWS, E. C., *Fixed Stars and Degrees of the Zodiac Analysed*, Bloomington, Illinois, 1968.

MAYO, J., *The Astrologer's Astronomical Handbook*, London, 1976.

MEAD, G. R. S., *Fragments of Faith Forgotten*, New York, 1960.

MEIHER, H. C., *Spannungsherrscher und Schicksalstypus*, Memmingen, 1939.

'Spannung und Spannungsherrscher', in *Kosmobiologisches Jahrbuch*, 1977.

MESSADIE, *Le Zodiaque a 24 signes*, Paris, 1973.

MEYER, M. R., *A Handbook for the Humanistic Astrologer*, New York, 1974.

The Astrology of Relationships, New York, 1976.

MICHELL, J., *A Little History of Astro-Archaeology*, London, 1977.

MOORE, E., 'The Astronomy of Dante', *Studies in Dante* (Third Series), Oxford, 1968.

MORRISH, L. F.. *Outline of Astro-Psychology*, London, 1952.

MORRISON, A. H., *Astrological Review*, vol. 45, no. 1, 1973.

MUCHÈRY, G., *The Astrological Tarot, Astromancy*, trans. M. Vallior, London, 1928.

Méthode pratique d'astrologie divinatoire, Paris, 1933.

Traité complet de chiromancie déductive et expérimentale, Paris, 1938.

MUNKASEY, M., 'The Houses: The Measurement View – Twenty Different Systems', *Astrology Now*, vol. 1, no. 8, Nov. 1975.

MUSES, C. A., 'Illumination on Jacob Boehme', *The Works of Dionysius Andreas Freher*, New York, 1951.

MYER, I., *Qabbalah: The Philosophical Writings of Solomon ben Yedhuda ibn Gebirol of Avicebron*, London, 1972.

NAYLOR, P. I. H., *Astrology: An Historical Examination*, London, 1967.

NELSON, J. H., *Cosmic Patterns: Their Influence on Man and His Communications*, Washington DC, 1974.

NEUGEBAUER, OTTO, *A History of Ancient Mathematical Astronomy*, London, 1975.

'Demotic Horoscopes', *The Journal of the American Oriental Society*, vol. 63, 1943.

NEUGEBAUER, O., and HOSEN, H. B., *Greek Horoscopes*, Philadelphia, 1959.

'Studies in Byzantine Astronomical Terminology', *Trans. Am. Philos. Soc.*, 50, 2, 1960.

NISSEN, H., *Orientation, Studien zur Geschichte der Religion*, 1906.

NOWOTNY, K. A., *De occulta philosophia: Agrippa ab Nettesheym*, Graz, 1967.

OSTRANDER, S., and SCHROEDER, L., *Astrological Birth Control*, Jonas, M. E. (ed.), New Jersey, 1972.

Bibliography

OUSPENSKY, P. D., *In Search of the Miraculous Fragments of an Unknown Teaching*, London, 1949.

A New Model of the Universe, London, 1931.

OVID, *Metamorphoses*.

PAGAN, I. M., *From Pioneer to Poet*, London, 1911.

Racial Cleavage, or the Seven Ages of Man, London, 1937.

PANOFSKY, E., *Studies in Iconology*, Oxford, 1967.

PARACELSUS (Theophrastus Bombastus von Hohenheim), see WAITE, A. E., *The Hermetical and Alchemical Writings of . . . Paracelsus the Great . . .*, vol. 2: Hermetic Medicine and Hermetic Philosophy, London, 1894.

PAULI, W. P., *The Influence of Archetypal Ideas on the Scientific Theories of Kepler*, 1955.

PAVITT, W. T., and PAVITT, K., *The Book of Talismans*, London, 1914.

PEARCE, A. J., *The Textbook of Astrology*, London, 1911.

PENROSE, F. C., *Temple Orientation*, London, 1851.

PETARIUS, *Dionysii petarii uranologion*, Basle, 1630.

PHILALETHES, see RUTLANDUS and WAITE, A. E. below.

PINGREE, DAVID, see WIEDNER.

PLINY, *Natural History*, II, 13.

POLICH, V., and PAGE, A. P. N., 'The Topocentric System of Houses', *Spica*, vol. 3, no. 3, 1964. See also LOVE.

POWELL, A. E., *The Solar System*, London, 1930.

POWELL, ROBERT, *Mercury Star Journal*, vol. IV, no. 2, 1978.

POWELL, R., and TREADGOLD, P., *The Sidereal Zodiac*, 1978.

PTOLEMY (C. Ptolemaeus), *Ptolemy, Tetrabiblos*, trans. and ed. F. E. Robins, London, 1964.

RAM, T. J. J., quoted in *Astrological Journal*, vol. 15, no. 2, 1973.

RAPHAEL, *Raphael's Astronomical Ephemerides for Given Years*, n.d.

Raphael's Tables of Houses for Northern Latitudes, London, c. 1960.

REGIOMONTANUS (Johannes Müller), see KOCH.

RIESS, E., 'Nechepsonis et petosiridis fragmenta magica', *Philologus*, Supplementband VI, 1891-3.

ROBINS, F. E., see PTOLEMY.

ROBSON, V. E., *The Fixed Stars and Constellations in Astrology*, London, 1969. See also LEO.

A Student's Textbook of Astrology, London, 1922.

ROW, SUBBA, *Collection of Esoteric Writings*, Adya, 1902?.

RUDHYAR, D., *The Astrology of Personality*, New York, 1970.

The Practice of Astrology, London, 1968.

RUPERTI, A., *Cycles of Becoming*, London, 1978.

RUTLANDUS, MARTIN, *Lexicon alchemiae*, Frankfurt, 1583.

Bibliography

SACHS, A., 'Babylonian Horoscopes', *Journal of Cuneiform Studies*, vol. 6, 1952.

SCHNABEL, *Kidenas, Hipparch und die Entdeckung der Praezession*, 1926.

SCHROEDER, L., see OSTRANDER.

SCHULTZ, JOACHIM, *Rhythmen der Sterne*, Dornach, 1963.

SCHWABE, J., *Archetyp und Tierkreis, Grundlinien einer kosmischen Symbolik und Mythologie*, Basel, 1951.

SCHWALLER DE LUBICZ, R. A., *The Temple is Man*, trans. R. and D. Lawlor, Massachusetts, 1977. See also TOONDER.

SENDIVOGIUS, M., *A New Light of Alchymie*, trans. John French, London, 1650.

'SEPHARIAL' (pseud.), *New Dictionary of Astrology*, London, n.d.
   Transits and Planetary Periods, London, 1920.
   The Science of Foreknowledge, London, 1918.

SIBLEY, E., *An Illustration of Astrology*, London, 1798.
   A New . . . Illustration of . . . Astrology, London, 1817.
   Uranoscopia, London, 1780.

SIMMONITE, W. J., *The Complete Arcana of Practical Astrological Philosophy*, London, 1890.

SMITH, E. M., *The Zodia*, London, 1906.

SPIEGELBERG, W., 'Die Ägyptischen Namen und Zeichen der Tierkreisbilder', *Zeitschrift für Ägyptische Sprache und Altertumskunde*, vol. 48, 1911.

STEIN, W. J., *Man and His Place in History*, 1947.

STEINER, RUDOLPH, *The Search of the New Isis – Divine Sophia* (trans. of four lectures given in Dec. 1920).
   Macrocosm and Microcosm (trans. of eleven lectures given in Vienna, March 1910).
   Man in the Light of Occultism (trans. of ten lectures given in Christiania, June 1912).
   Occult Movements in the Nineteenth Century (trans. of ten lectures given in Dornach, Oct. 1915).
   An Esoteric Cosmology (trans. of eighteen lectures given in Paris, 1906, reported by Schure; English edn 1978). See also KEMPTER.

STONE, J. H. VAN, *The Pathway of the Soul*, London, 1912.

SUCHER, W. O., *Cosmic Christianity*, 1970.
   Isis Sophia: An Outline of a New Star Wisdom, 1955 and 1975.
   The Changing Face of Cosmology, 1971.

SUTCLIFFE, G. E., *Two Undiscovered Planets*, n.d.
   The New Astronomy and Cosmic Physiology, London, 1930.

SWEDENBORG, E., *De coelo et ejus miriabilibus, et de inferno*, 1758. See also BARRETT.

SWEENEY, B. M., *Rhythmic Phenomena in Plants*, London and New York, 1969.

Bibliography

TAKATA, M., 'Zur Technik der Glockungszahlreaktion im menschlichen Serum . . .', *Helvetica Medica Acta*, vol. 17, 1950.

TAYLOR, RENE, 'Architecture and Magic', *Essays in the History of Architecture*, presented to Rudolf Wittkower, 1962.

TESTA, P. E., *Il simbolismo dei Giudeo-Cristiani*, Bari, 1962.

THIERENS, A. E., *The Elements of Esoteric Astrology*, London, 1931.

THOM, A., *Megalithic Lunar Observatories*, Oxford, 1973.

THOMAS, K.. *Religion and the Decline of Magic*, London, 1971.

THOMPSON, C. J. S., *The Mystery and Romance of Astrology*, London, 1929.

THORNBURN, J. M., 'Notes of the Heliocentric Co-ordinates of Planetary Positions', *Astrological Journal*, vol. 3, no. 3, 1961.

THUN, MARIA, 'Nine Years of Observation of Cosmic Influence on Annual Plants', *Star and Furrow*, spring 1964.

TOONDER, J. G., and WEST, J. A., *The Case of Astrology*, London, 1973.

TOWNLEY, J., *Astrological Cycles and the Life Crisis Periods*, New York, 1977.

TOYNBEE, P. J., 'Dante's Obligation to Alfraganus in the Vita Nuova and Convivo', *Romania* XXXII, pp. 565ff, n.d.

TRITHEMIUS, J., *De Septem secundadeis. id est intelligentiis . . .*, Nuremberg, 1522.

TUCKER, W. J., *Astrology for Everyone*, Kent, 1960.

The Principles, Theory and Practice of Scientific Prediction, London, 1936.

TUCKERMAN, B., *Planetary, Lunar and Solar Positions – 601 BC to AD 1 – at Five-day and Ten-day Intervals*, Philadelphia, 1962.

VARLEY, J., *A Treatise on Zodiacal Physiognomy*, London, 1828.

VERMASEREN, M. J., see 'Mithraica', II, 1974. Also see BECK.

VILLEFRANCHE, M. DE, *Astrologia Gallica*, Paris, 1661.

VIRGIL, *Georgics*; for 'Ages', see HASTINGS.

VOGH, J., *Arachne Rising: The Search for the Thirteenth Sign of the Zodiac*, London, 1977.

VOLGUINE, A., *Les Significations des encadrements dans l'horoscope*, Paris, 1974.

Astrology of the Mayas and Aztecs, trans. W. J. Tucker, Sidcup, 1969.

The Ruler of the Nativity, ASI Publications Inc., New York, 1973. (Research summarized by Dean, 1973).

VREEDE, E., *Anthroposophie und Astronomie*, Stuttgart, 1954.

WACHSMUTH, G., *The Evolution of Mankind*, trans. from German by Norman Macbeth, Dornach, 1961.

Etheric Formative Forces, Dornach, 1923.

WAERDEN, B. VAN DER, *Science Awakening II – the Birth of Astronomy*, Oxford, 1974.

WAITE, A. E., see PARACELSUS.

WAITE, H. T., *et al.*, *The New Waite's Compendium of Natal Astrology*, London, 1971.

Bibliography

WALTER, H. J., 'Harmonische Aspektfiguren der Quintil . . .', *Kosmobiologisches Jahrbuch*, 1974.

WALTHER, H., 'Towards a New Astrology', *The Modern Mystic*, Dec. 1939.

WANGEMANN, E., 'The Astrological Aspects', *Astrology*, vol. 49, no. 1, 1975.

WARBURG, A., 'Italienische Kunst und internationale Astrologie im Palazzo Schifanoia in Ferrara', *Atto del X Congresso Int. di Storia dell' Arte*, 1922.

WEDEL, T. O., *The Mediaeval Attitude Towards Astrology*, Folcroft, 1920.

WELLING, G. VON, *Opus mago-cabbalisticum et theosophicum . . .*, Homburg vor der Hohe, 1735.

WEMYSS, M., *The Wheel of Life, or Scientific Astrology*, London, 1930, etc.

WESTON, L. H., *The Planet Vulcan*, 1921.

WHITE, G., *The Moon's Nodes*, c. 1950.

WIEDNER, P. P. (ed.), see entry on 'astrology' by David Pingree, *Dictionary of History of Ideas*, New York, 1968.

WILLIAMS, D., *Astro Economics*, Minnesota, 1959.

WILSON, D., *Astro Economics*, Minnesota, 1959.

WILSON, JAMES, *A Complete Dictionary of Astrology, In Which Every Technical and Abstruste Term . . . is . . . Accurately Defined*, London, 1819. See also reprint of Samuel Weiser, New York, 1974.

WIND, E., *Pagan Mysteries of the Renaissance*, Oxford, 1958.

WITTE, A., see LEFELDT.

WITTKOWER, R., *Allegory and the Migration of Symbols*, London, 1977. See also TAYLOR.

WOLFRAM, E., *The Occult Causes of Disease*, London, 1911.

'ZADKIEL' (R. J. Morrison), *An Introduction to Astrology*, London, 1835.

The Grammar of Astrology, London, 1849.